Elementary Statistics
Managing Variability and Error

by Scott Alan Guth

Contents

Contents   iii

Preface   x

I   Introduction   1

1   Basic Notions in Statistical Research   1

1.1   An Introduction to Statistics   1

    1.1.1   Why Study Statistics?   1

    1.1.2   A Few Statistical Terms   1

    1.1.3   Distributional Thinking   3

    1.1.4   Dotplots   3

    1.1.5   Bar Charts   5

    1.1.6   Data Types   6

    1.1.7   The Levels of Measurement   7

    1.1.8   Homework   8

1.2   Where Statistics Fail   11

    1.2.1   Types of Error   11

    1.2.2   Additional Concerns   14

    1.2.3   Homework   18

1.3   The Process of Statistical Analysis   19

    1.3.1   Steps in the Statistical Analysis Process   19

    1.3.2   Analyzing Statistical Studies   21

    1.3.3   Drawing Conclusions from Research   25

    1.3.4   Homework   26

1.4   Random Sampling and Random Assignment   28

    1.4.1   Deciding What to Measure   28

    1.4.2   Gathering Data   28

    1.4.3   Random Selection and Random Samples   29

    1.4.4   Sampling Frame   31

    1.4.5   Types of Statistical Studies   31

    1.4.6   Homework   37

II Descriptive Statistics 41

2 Summarizing Single Variable Data 43
2.1 Frequency Tables and Histograms 44
 2.1.1 Histograms 46
 2.1.2 Relative Frequency Tables and Histograms 50
 2.1.3 Stem and Leaf Plots 53
 2.1.4 Pie Charts 54
 2.1.5 Homework 56
2.2 A First Look at Statistical Inference 59
 2.2.1 Comparing Two Samples with Stacked Dotplots 59
 2.2.2 Homework 63
2.3 Summarizing Data Using Averages 65
 2.3.1 Parameters and Statistics 65
 2.3.2 The Arithmetic Mean 65
 2.3.3 Estimating μ with a Sample Mean 67
 2.3.4 Estimating μ with a Trimmed Mean 68
 2.3.5 Estimating a Sample Mean Using a Frequency Table 68
 2.3.6 The Median 69
 2.3.7 The Midrange 70
 2.3.8 The Mode 71
 2.3.9 The Weighted Mean 71
 2.3.10 The Quadratic Mean 72
 2.3.11 Homework 73
2.4 Summarizing Data Using Measures of Variation 76
 2.4.1 The Range of a Data Set 76
 2.4.2 Deviations from the Mean 77
 2.4.3 Mean Absolute Deviation 79
 2.4.4 Standard Deviation and Variance 80
 2.4.5 Estimating s and s^2 using Frequency Tables 87
 2.4.6 The Empirical Rule 89
 2.4.7 Homework 92
2.5 Measures of Relative Position 96
 2.5.1 The Percentile 96
 2.5.2 Standard Score or z-Score 97
 2.5.3 Homework 99
2.6 Data Exploration: Boxplots and Outliers 102
 2.6.1 The Five Point Summary and Boxplots 102
 2.6.2 Sample Skewness and Pearson's Kurtosis 104
 2.6.3 Outliers 110
 2.6.4 Enhanced Boxplots 112
 2.6.5 Boxplots and Normal Distributions of Values 112
 2.6.6 Comparing Samples with Parallel Boxplots 112
 2.6.7 Homework 116

III Probability 121

3 Basic Rules of Probability 123
- 3.1 Introduction to Probability 124
  - 3.1.1 The Probability of an Event 125
  - 3.1.2 Probability and Two-Way Tables 128
  - 3.1.3 Homework 132
- 3.2 Rules for Computing Probabilities 134
  - 3.2.1 The Addition Rules 134
  - 3.2.2 The Rule of Complements 136
  - 3.2.3 The Multiplication Rules 137
  - 3.2.4 Complements and Compound Events 141
  - 3.2.5 Homework 143

4 Discrete Probability Distributions 147
- 4.1 Discrete Probability Distributions 147
  - 4.1.1 Mean, Variance and Standard Deviation 148
  - 4.1.2 Unusual or Significant Measurements 150
  - 4.1.3 Probability and Area 152
  - 4.1.4 Homework 154
- 4.2 Properties of Expected Value and Variance 156
  - 4.2.1 Combinations of Random Variables 156
  - 4.2.2 The Algebra of Expected Value and Variance 156
  - 4.2.3 Homework 162
- 4.3 Moments, Skewness and Kurtosis 164
  - 4.3.1 Moments 164
  - 4.3.2 Skewness and Kurtosis 166
  - 4.3.3 Homework 171
- 4.4 Binomial Experiments 173
  - 4.4.1 The Binomial Experiment 173
  - 4.4.2 Mean and Standard Deviation 177
  - 4.4.3 Skewness and Kurtosis 180
  - 4.4.4 The History of the Binomial Probability Distribution 180
  - 4.4.5 Homework 182
- 4.5 The Poisson Distribution 184
  - 4.5.1 The Poisson Distribution 184
  - 4.5.2 Skewness and Kurtosis 188
  - 4.5.3 History of the Poisson Distribution 188
  - 4.5.4 Homework 190
- 4.6 Proofs of Theorems 193
  - 4.6.1 Formulas for Expected Value and Variance 193
  - 4.6.2 Proofs of Properties of Expected Value and Variance 194
  - 4.6.3 Sample Variance as an Unbiased Estimator of σ^2 196
  - 4.6.4 Binomial Experiments 197
  - 4.6.5 Deriving the Poisson Distribution 199

4.6.6   Skewness and Kurtosis 201

5   Continuous Probability Distributions   203
  5.1   Continuous Random Variables 203
        5.1.1   The Continuous Uniform Distribution 203
        5.1.2   The Normal Distribution 205
        5.1.3   Skewness and Kurtosis 210
        5.1.4   The History of the Normal Distribution 210
        5.1.5   Homework 213
  5.2   The Standard Normal Distribution 214
        5.2.1   The Empirical Rule and The Normal Distribution 218
        5.2.2   Finding z-Scores When Given Probabilities 219
        5.2.3   Homework 223
  5.3   The Non-Standard Normal Distribution 224
        5.3.1   Finding Probabilities Corresponding to Given Scores . . . 224
        5.3.2   Finding Scores Corresponding to Given Probabilities . . . 225
        5.3.3   Homework 227

IV   Inferential Statistics   229

6   Estimates & Sampling Distributions   233
  6.1   Introduction to Sampling Distributions 233
        6.1.1   Estimates and Their Distributions 233
        6.1.2   Rules for Good Estimators 234
        6.1.3   The General Form of a Confidence Interval 234
        6.1.4   Homework 238
  6.2   Sampling Distributions of Sample Proportions 239
        6.2.1   The Normal Approximation of the Binomial Distribution . . . 239
        6.2.2   The Sampling Distribution of Sample Proportions 246
        6.2.3   The Central Limit Theorem for Sample Proportions . . . 247
        6.2.4   The Confidence Interval for a Population Proportion . . . 250
        6.2.5   Homework 251
  6.3   Intervals for a Population Proportion 254
        6.3.1   Large Sample Intervals for a Population Proportion . . . 255
        6.3.2   Sample Size 257
        6.3.3   Homework 260
  6.4   The Sampling Distribution of Sample Means 262
        6.4.1   The Population of all Sample Means 262
        6.4.2   Mean and Standard Error 263
        6.4.3   The Central Limit Theorem for Sample Means 266
        6.4.4   Homework 272
  6.5   Intervals for a Population Mean (σ Known) 273
        6.5.1   Confidence Intervals for a Population Mean (σ Known) . . . 273
        6.5.2   Sample Size When Estimating a Population Mean 275

        6.5.3   Homework 277

    6.6   Intervals for a Population Mean (σ Unknown) 279

        6.6.1   Confidence Intervals for a Population Mean (σ Unknown) 283

        6.6.2   Sample Size 285

        6.6.3   Homework 286

7   Single Parameter Hypothesis Tests   289

    7.1   Hypothesis Testing – An Overview 291

        7.1.1   Single Parameter Hypothesis Tests 291

        7.1.2   Significant Difference Versus Sampling Error 298

        7.1.3   Errors in Hypothesis Test Conclusions 298

        7.1.4   Power 299

        7.1.5   Choosing α 299

        7.1.6   Real-World Significance vs. Statistical Significance 300

        7.1.7   Sampling Frame and Applicability of Results 300

        7.1.8   Homework 301

    7.2   Large Sample Tests for a Population Proportion 305

        7.2.1   Claims Regarding a Single Population Proportion 305

        7.2.2   Homework 318

    7.3   Tests for a Population Mean (σ Known) 320

        7.3.1   Claims Regarding a Single Population Mean 320

        7.3.2   Homework 331

    7.4   Tests for a Population Mean (σ Unknown) 334

        7.4.1   Claims Regarding a Single Population Mean 334

        7.4.2   Homework 343

    7.5   Tests Comparing Means from Paired Samples 345

        7.5.1   Claims which Compare Means from Paired Samples 345

        7.5.2   Homework 354

8   Tests Comparing Two Parameters   361

    8.1   Sampling Distributions of Differences 361

        8.1.1   The Sampling Distribution of Differences, $\hat{p}_1 - \hat{p}_2$ 362

        8.1.2   The Sampling Distribution of Differences, $\bar{x}_1 - \bar{x}_2$ 364

        8.1.3   Homework 367

    8.2   Large Sample Tests for Two Proportions 368

        8.2.1   Testing the Null Hypothesis $H_0 : p_1 - p_2 = 0$ 368

        8.2.2   Testing the Null Hypothesis $H_0 : p_1 - p_2 = k \neq 0$ 374

        8.2.3   Homework 375

    8.3   Comparing Means from Independent Samples 379

        8.3.1   Comparing Means from Independent Samples 379

        8.3.2   Homework 387

    8.4   Intervals for Differences of Two Parameters 392

        8.4.1   The Difference Between Means with Paired Samples 392

        8.4.2   The Sampling Distribution of Differences, $\hat{p}_1 - \hat{p}_2$ 394

        8.4.3   The Sampling Distribution of Differences, $\bar{x}_1 - \bar{x}_2$ 395

      8.4.4 Homework 399

9 Correlation and Regression   401

   9.1 Linear Correlation and Regression 401

      9.1.1 Relationships Between Quantitative Variables 402

      9.1.2 Finding the Line of Best Fit 407

      9.1.3 Nonlinear Forms 419

      9.1.4 Computational Formulas 420

      9.1.5 More Data Yield Better Predictions 422

      9.1.6 Homework 431

   9.2 Assessing Linear Fit with Residuals 439

      9.2.1 Residual Plots 440

      9.2.2 Sums of Squares as a Measure of Variation. 444

      9.2.3 The Standard Deviation about the Regression Line 445

      9.2.4 A Computational Formula for SSE 448

      9.2.5 Homework 449

   9.3 Inference in Correlation and Regression 453

      9.3.1 Inferences Regarding the Linear Correlation Parameter 454

      9.3.2 Inferences Regarding the Slope Parameter 463

      9.3.3 Why $n-2$ Degrees of Freedom? 464

      9.3.4 Hypothesis Tests Regarding the Slope Parameter 465

      9.3.5 Confidence Intervals for the Slope Parameter 468

      9.3.6 Homework 470

   9.4 The Ryan-Joiner Test for *Non*-Normality 479

      9.4.1 Constructing the Perfectly Normal Sample 479

      9.4.2 Testing for *Non*-Normality 489

      9.4.3 Homework 494

   9.5 Proofs for Correlation and Regression 496

      9.5.1 Correlation and Regression Theorems 496

   9.6 The History of Correlation and Regression 502

10 *Chi-Square* Tests   507

      10.0.1 Karl Pearson 507

   10.1 The Goodness of Fit Test 509

      10.1.1 The Goodness of Fit Test 512

      10.1.2 The Goodness of Fit with One Degree of Freedom 517

      10.1.3 Proof that $\chi^2 = z^2$ for One Degree of Freedom 519

      10.1.4 Homework 520

   10.2 Contingency Tables: Testing For Independence 523

      10.2.1 Contingency Tables 525

      10.2.2 Homework 531

   10.3 A Chi-Square Test for Population Variance 534

11 An Introduction to Analysis of Variance   **537**

   11.1 One Way Analysis of Variance 539

       11.1.1 One-Way Analysis of Variance with Unequal Sample Sizes 546

       11.1.2 Homework 553

12 Distribution Free Tests   **557**

   12.1 The Signs Tests 557

       12.1.1 Determining p-Values 559

       12.1.2 The Signs Test for a Mean Difference 566

       12.1.3 Homework 571

   12.2 The Wilcoxon Signed Rank Test 573

       12.2.1 Determining p-Values 575

       12.2.2 Determining Critical Values 576

       12.2.3 Rank Sums' Mean, Variance, and Standard Deviation 578

       12.2.4 The Signed Rank Test for a Mean Difference 586

       12.2.5 Historical Note 591

       12.2.6 Homework 592

V   Reference Materials   **595**

A Formulas and Tables   **597**

B Solutions to Selected Exercises   **613**

   B.1 Chapter 1 Solutions 613

   B.2 Chapter 2 Solutions 616

   B.3 Chapter 3 Solutions 637

   B.4 Chapter 4 Solutions 639

   B.5 Chapter 5 Solutions 648

   B.6 Chapter 6 Solutions 650

   B.7 Chapter 7 Solutions 653

   B.8 Chapter 8 Solutions 661

   B.9 Chapter 9 Solutions 669

   B.10 Chapter 10 Solutions 683

   B.11 Chapter 11 Solutions 689

   B.12 Chapter 12 Solutions 692

Index   **695**

Preface

To the Reader,

I originally wrote this book because I was frustrated by the price of textbooks. I do not claim to have the resources to compete with the beauty and color of other expensive texts, but I hope that the quality of information in this text is close enough to what others offer to be of some use to you. Please forgive any errors in computation or grammar, and use this book in good health and learning.

My goals in writing this text were as follows. First, I wanted to save you money. Second, I wanted a text that fit well with my lecture style. Next, I wanted to prove as many of the facts covered in the text as possible, and at the algebra level. I have done this for advanced readers and other instructors who may be interested, but all proofs in this text are purely optional. Lastly, I hoped to make the subject as interesting as possible by including as much history as was reasonable.

In the second edition, I have added a great deal of material which involves informal statistical inference. Dotplots are added as well, and the chapter on correlation and regression has been split into two – one descriptive and the other inferential. It is hoped that these enhancements will increase student comprehension, and make the book more interesting.

In the third edition, I have reworked and combined the chapters on correlation and regression.

This book is dedicated to my wife, Alicia, and to my children, Gary and Ryan.

Sincerely,

Scott Guth

Mathematics Department
Mt. San Antonio College
December, 2021

Part I

Introduction

Chapter 1

Basic Notions in Statistical Research

1.1 An Introduction to Statistics

Welcome to Elementary Statistics! In this course we will learn many tools for research, and gain a basic understanding of the techniques of descriptive and inferential statistics. In this first chapter we introduce some of the terms and concepts used in statistics. The ideas that we learn in this chapter will be used throughout this text and in further statistics courses as well.

1.1.1 Why Study Statistics?

Statistics is a collection of tools by which we gather, summarize, and interpret data. Statistics is an important tool used by researchers, and yet it is fraught with many difficulties. When these tools are used improperly, statistical research can yield misleading and even dangerous results. Because of the potential for damage, it is important that researchers are well educated on the proper methods of statistics, and it is of similar importance that readers are aware of the pitfalls into which many statistical studies fall.

Some of us will use the tools of statistics as we conduct research of our own, as undergraduates, graduates, or as professionals. Others of us will simply be readers of research, conducted by members of our own professions or in other areas in which we are interested. The aim of this course is to give a basic understanding of elementary statistics for readers of statistical research, as well as for those who will conduct research themselves.

1.1.2 A Few Statistical Terms

We begin by introducing the most basic terms in the language of statistics. These terms are used throughout this subject, so it is critical that they are understood, and not forgotten.

In statistics, we work with data. The term, *data*, is the plural form of the word *datum*. *Data are* measurements or observations gathered by a researcher. *Statistics* is the discipline of collecting (or *sampling*) data, summarizing data (through graphs, averages, and measures of variation), and making inferences (or *conjectures*) based on collected data.

The primary problem in statistics is that the populations from which we wish to gather information tend to be large. A *population* is the entire collection of common observations of interest. Populations can be extraordinarily large – much too large to study completely. This is the dilemma of statistics – we need information that we cannot possibly get, because populations tend to be too large to measure completely.

Our populations are too large, yet we need information about them. Large companies want to know how many people will buy their product. Politicians want their approval ratings. These measurements are examples of *parameters*. A *parameter* is a *numerical* summary value, computed using data from an entire population. Because populations tend to be extraordinarily large in the context of statistics, we rarely know the value of any parameter.

Our solution to the problem of populations being too large is one which is quite imperfect, and is full of difficulties. This solution will be to gather data, not from the entire population, but from a (relatively) small subset of the population. A *sample* is a subset of a population, and we use measurements from samples to estimate population parameters. In doing so, it is critical that we understand that our estimates will always be somewhat *incorrect* – they will contain error. Because populations tend to be diverse, samples can vary widely. This variation is a source of error in our estimates. Estimates from samples are called *statistics*. A *statistic* is a *numerical* summary value, computed using data from a sample.

Figure 1.1: In statistics, we focus on small subsets, or *samples*, drawn from large populations. (*Photograph by Joh63 of Wikimedia Commons, used by permission*).

Statistics are used to estimate parameters. These estimates are *inferences*. It is highly important to remember that such estimates will always contain error, and that error is the main problem with which we must contend. No matter how well a researcher does her or his job, no matter how we labor to make sure that sample data represent their population well, our sample data will always be just that – sample data, and measurements from sample data always contain error.

Statistical studies use data from samples. Once in a while, attempts are made to study an entire population, but the cost for such a study tends to be extraordinary. When data from an entire population are used in a study, the study is a *census*. In reality, however, a true census of a vast population is quite impossible to achieve, as there are always many individuals who are overlooked as data are gathered. Homeless people are rarely included in a census, along with people with homes in remote areas.

In summary, statistics is the discipline in which we use measurements from samples to make estimates and inferences about population parameters. These estimates can vary widely due to population diversity, and therefore contain error. We, therefore, will learn to manage these errors that arise due to this variability.

Statistics is composed of two major branches. The first branch we will study is *descriptive statistics* where our primary concern will be to summarize large data sets. When we are done with this, we will move on to *inferential statistics*, where we will make inferences or conjectures based on data gathered from samples.

1.1.3   Distributional Thinking

As we have said, population variability is a primary cause of error in statistical estimates. Because of this, it is important that we develop tools for understanding the *distribution* of values in a data set. The distribution of a set of observations is the pattern formed as they are displayed graphically. The understanding of the *shape*, *center* (measured using averages) and *spread* (how widely values are disbursed) of a data set is what is meant by *distributional thinking*.

Visualizing the shape of a population's distribution of values is easiest if we can find some way to display it. There are various ways to graphically portray a distribution of values. Many of these will be discussed in this text, but here we will focus on two: dotplots and bar charts. The graphical display we use depends on the type of data which we are summarizing. For *quantitative data* (numerical quantities) we use dotplots. For *categorical data* (non-quantities), we use *bar or pie charts*.

1.1.4   Dotplots

As we have said, *dotplots* are a tool for visualizing the distribution of a collection of *quantitative data*. For small data sets, dotplots are relatively simple to construct by hand. For larger data sets, computers should be utilized. To draw a dotplot, we round each number in a quantitative data set to a predetermined number of digits. We round the values to the point where those that are *very* close to one another become the same. This effectively groups similar values into categories.

Once values are rounded, we simply plot them as dots above a number line. When values repeat in the rounded data set, we plot their dots in a vertical column. The height of any column of dots represents the *frequency* of occurrences for a corresponding numerical observation.

Example 1.1.1 The following values are randomly selected heights of adult women, measured in centimeters.{148.0, 152.2, 155.2, 152.7, 155.1, 155.5, 156.8, 158.4, 158.3, 156.2, 158.6, 159.9, 164.2, 157.9, 159.8, 160.9, 162.3, 158.0, 159.8, 156.9, 162.4, 160.2, 166.9, 167.2, 167.6, 164.0, 170.4, 175.9, 176.4, 184.1}.

These values are measured to the nearest tenth, and there are many values that are very close together. In order to allow these values to group into categories, we round to the nearest centimeter.

The rounded values are as follows. {148, 152, 155, 152, 155, 155, 156, 158, 158, 156, 158, 159, 164, 157, 159, 160, 162, 158, 159, 156, 162, 160, 166, 167, 167, 164, 170, 175, 176, 184}.

We plot these values on the dotplot in Figure 1.2.

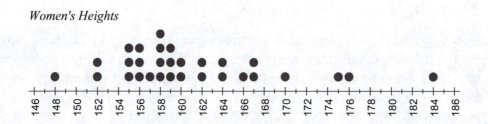

Figure 1.2: A dotplot of women's heights for the data given in Example 1.1.1.

Thinking distributionally, we see a fair amount of variability in these values, ranging from 148 cm to 184 cm. The most frequently observed value is 158 cm, with most other values clustered relatively close to this most frequent value. While there are many ways to measure the *center* of a distribution, we might guess that the center is near 162 cm. The distribution of values has higher frequencies in the center, and lowest frequencies at the extreme left and right. There is one very tall woman in the group, at 184 cm, and no other women are close to her in height. This tallest height might be considered an *outlier* – a value that is extreme and separated by some distance from the other values. Two other tall women have heights which may be outliers as well (at 175 cm and 176 cm). It is interesting to observe that these outliers on the right are not matched by similar outliers on the left. The shape of the distribution as *non-symmetric* and *skewed* to the right, because of the outliers on the right, with no corresponding outliers on the left.

Example 1.1.1 gave us an opportunity to demonstrate *distributional thinking*. Distributional thinking considers of the *center* of a distribution of values, the *shape* of the distribution – whether it is *symmetric*, or *skewed*, and the *spread* - the range over which values are distributed, the locations of highest and lowest frequencies, the locations of possible *outliers*.

1.1.5 Bar Charts

To describe the distribution of non-quantitative data which may be grouped into categories, we use a *bar chart*. A bar chart simply arranges the categories of values along a vertical axis, with rectangular bars plotted above the axis with heights proportional to the number of observations in respective categories.

Example 1.1.2 Below are sample data which show frequencies of births for each day of the week from a random sample of 59,247 babies.

Sunday	Monday	Tuesday	Wednesday	Thursday	Friday	Saturday
7701	7527	8825	8859	9043	9208	8084

The data gathered in this sample are names of days, from *Sunday* through *Saturday*. These are values are non-quantities, and are therefore *categorical* in nature. The bar chart that corresponds to these values is given in Figure 1.3. Category frequencies are plotted above their respective bars.

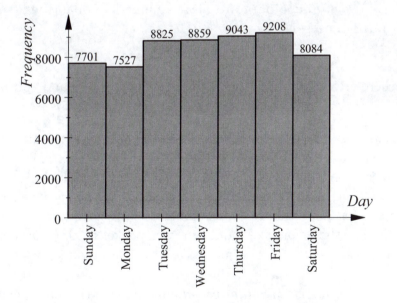

Figure 1.3: The bar chart of data presented in Example 1.1.2.

There are some differences in the frequencies of occurrences for births on various days of the week. The most popular days for births are Tuesday through Friday. The least popular days for births are Saturday through Monday. The question of why this is so is rather interesting. We leave the reader to speculate on the reasons for this mystery. For now, the point to be made is that the bar chart allows us to easily see the categories, days of the week, which contain the highest or lowest frequencies. In real life, the knowledge gained by such information can inform decision making. For example, hospitals ought to consider the possibility that more babies are born on certain days of the week, and on those days they should have more workers available.

Distributional thinking is an important skill in statistical analysis. It helps us make informed inferences regarding large, unknown, populations of observations. With informed understanding, we can make wiser decisions with regard to these populations.

1.1.6   Data Types

We have already referred to two types of data – *quantitative* and *categorical* (or *qualitative*) data. We now clarify these, and provide additional means of classifying data. The classification of data into types is important, as the methods of data analysis depend on data type. We begin by formally defining *categorical* and *quantitative* data.

Definition 1.1.3 *Categorical* or *qualitative* data are non-quantities. Such data may be non-numerical (such as names or colors), or numerical (when the numbers do not represent quantities, i.e. zip codes, phone numbers, or social security numbers).

Definition 1.1.4 *Quantitative* data are always numerical quantities (such as heights, weights, temperatures, ages, etc.).

Additionally, there are two types of quantitative data: *discrete* and *continuous*.

Definition 1.1.5 *Discrete quantitative data* are quantitative numerical values where possible measurements may be ordered consecutively, and gaps exist between consecutive possible measurements where no measurements are possible. Discrete quantitative data are often whole numbers, like numbers of children (you might have 2 or 3 children in your family, but never 2.598 children) or number of correct guesses. Discrete quantities can be fractional, like the amount of money in your pocket (where you might have $1.06 or $1.07 in your pocket, but you could not have $1.06278 in your pocket).

Definition 1.1.6 *Continuous quantitative data* are quantitative numerical values where, within a certain range, any value is possible. The number of centimeters of rain on a given day may be between 0 cm and 100 cm, and there is no value within that range that would be impossible (assuming it was possible to measure values to any level of precision).

1.1.7 The Levels of Measurement

We finish this section by providing an additional method for classifying data into *levels of measurement*. The methods used in descriptive *and* inferential statistics depend on the level of data measurement. Thus, the following levels are referenced from time to time in this text.

Definition 1.1.7 *Nominal data* are characterized by name only. There is no meaningful order to such data, noting that alphabetic ordering is not meaningful, as it gives no insight as to the qualities of the data.

Examples of nominal data include *names*, *countries*, and *colors*.

Definition 1.1.8 *Ordinal data* have a natural and meaningful ordering, but no meaningful differences through subtraction can be computed.

Examples of ordinal data include letter grades, which may be ordered from low to high. This ordering is meaningful as it indicates varying levels of learning. Note that we cannot *subtract* letter grades, so differences are *not* meaningful.

Definition 1.1.9 With *interval data*, there is meaningful order and differences through subtraction, but there is no meaningful zero (where zero means "none") and ratios are not meaningful.

For example, calendar years (as a measure of time) can be meaningfully ordered from low to high, and two calendar years may be subtracted to find the number of years between them. The *zero* year exists, but is not meaningful, as it does not represent *no time*. Equivalently, it does not make sense to say that twice as much time had transpired in the year 20 as in the year 10, even though those values are in a 2 to 1 ratio.

Temperatures, when measured in degrees Celsius or Fahrenheit, are another example of *interval* data. In either temperature scale, 0 degrees does not mean *no heat*. This is emphasized by the fact that 0 degrees corresponds to different heat levels on the two scales. Still, temperatures may be ordered, and it makes sense to subtract temperatures to know their difference. Thus, temperatures satisfy the requirements of *interval* data.

Definition 1.1.10 *Ratio data* are the most refined of all data types. Ratio data have meaningful ordering, differences through subtraction, meaningful zeros and ratios.

Examples of ratio data include heights, ages of people, and weights of birds.

The table below summarizes these levels and their qualities.

Level of Measurement	Meaningful Ordering?	Meaningful Differences?	Meaningful Zero & Ratios?
Nominal	No	No	No
Ordinal	Yes	No	No
Interval	Yes	Yes	No
Ratio	Yes	Yes	Yes

1.1.8   Homework

For problems 1 through 4 below, state whether the collection is a *sample* or a *population*.

1. Seventy-five students from a given college are randomly selected.

2. Grades from all students at a given college are gathered.

3. Two hundred goldfish are gathered to test their response to a new type of fish food.

4. From the entire collection of a new breed of tomato vines, each tomato grown for an entire season is gathered and tested for taste.

For problems 5 through 8 below, state whether the computed value is a *statistic* or a *parameter*.

5. The grade points are averaged for seventy-five college students.

6. The grade points are averaged for all students at a given college.

7. The average age of all registered drivers is computed.

8. The percentage of honors students from 100 randomly selected college students is computed to be 12%.

Answer the following questions.

9. Is it possible to conduct a true census of a large population?

10. Why do we use statistical estimates given that they always contain error?

Are the following data sets quantitative, or categorical?

11. The following are randomly selected favorite primary or secondary colors chosen by randomly selected children.
{yellow, yellow, blue, green, blue, red, yellow, green, green, purple, blue, red, red, green, yellow, purple, purple, purple, red, green, green, orange, orange, orange, orange, purple, green, green, green, purple}

12. The following are foot lengths, in centimeters, of randomly selected adult males.
{24.4, 24.2, 25.4, 25.8, 26.3, 24.7, 25.1, 24.6, 26.1, 26.7, 25.1, 25.8, 26.4, 25.7, 25.6, 26.2, 26.2, 26.8, 26.4, 26.8, 26.3, 26.4, 24.5, 25.6, 25.7, 24.9, 27.0, 27.8, 27.6, 27.6}

13. Below are grade point averages for randomly selected community college students.
{3.2, 2.5, 3.4, 4.0, 2.1, 3.3, 3.2, 3.7, 2.9, 1.4, 2.2, 3.1, 1.7, 2.5, 3.2, 3.9, 3.7, 2.0, 2.6, 3.2, 2.9, 3.1, 2.4, 3.1, 2.4, 2.3, 3.7, 3.9, 2.6, 2.6}

14. Below are letter grades assigned to randomly selected college-level calculus students.
{A, C, B, A, C, B, F, D, F, C, A, C, D, A, D, A, D, D, C, C, D, C, A, A, A, C, C, D, C, C}

Plot the appropriate graphical display, either a dotplot or bar chart, for the data sets referenced below.

15. Plot an appropriate graphical display for the data in Problem 11.

16. Plot an appropriate graphical display for the data in Problem 12.

17. Plot an appropriate graphical display for the data in Problem 13.

18. Plot an appropriate graphical display for the data in Problem 14.

For problems 19 through 22, state whether the data are *quantitative* or *qualitative*.

19. The letter grades of seventy-five college students are collected.

20. The ages of seventy-five college students are collected.

21. Calendar years of students' birthdays are recorded from student applications.

22. The zip codes of all students at a given college are recorded in the school database.

For problems 23 through 26, describe the data as *discrete* or *continuous*.

23. The number of honors students at a given college is counted.

24. Heights of 75 college students are gathered.

25. The amount of change (in coins) in your pocket is counted.

26. The low temperature in degrees Celsius is recorded nightly.

For problems 27 through 34, state whether the level of measurement is *nominal, ordinal, interval, or ratio*.

27. Calendar years are recorded to measure time.

28. Nationalities of students are collected for demographic studies.

29. Weights of babies are recorded at birth to measure development.

30. Letter grades of students are collected to see their distribution.

31. The number of people with whom workers commute to their jobs is gathered as required by the Air Quality Management District.

32. Children learn early to rate as *good, better,* or *best.*

33. The low temperature in degrees Celsius is recorded nightly.

34. The zip codes of all students at a given college are recorded in the school database.

Answer the following questions.

35. What statistical computation would you use to measure the center of a collection of values of the ratio level of measurement?

36. Would it make sense to compute the average zip-code associated to the homes of various people in a sample? Explain why this is so with reference to the level of measurement for zip-codes.

37. What statistical computation(s) could be used to summarize the zip codes associated to the homes of various people in a sample?

1.2 Where Statistics Fail

It could be argued that one of the most important tools that students take from a course in elementary statistics is the ability to read research in a way that is critically minded. An educated reader of research needs to be able to understand the subtleties of such research, so as to not be misled. It is important to note that, even when research is done as perfectly as humanly possible, conclusions are sometimes incorrect. Even in the best of studies, sampling errors will exist and these errors can lead to incorrect decisions. Thus, we read statistical research with a critical mind, making sure that studies are implemented correctly, and noting when they are not. Still, even with proper methods, we remember that conclusions, which are *usually* correct, are sometimes incorrect.

1.2.1 Types of Error

Error is the difference between the true value of a population parameter and an estimate of that parameter (a *statistic* measured from a sample). That sounds simple enough, but note that, because populations tend to be large, a population parameter is generally unknown. Because of this, we cannot know the error in our estimates! This is troublesome, but we will develop a way to estimate a *maximum error* which will, with a certain level of confidence, be an upper limit for the true error.

There are two types of error: *sampling error* and *non-sampling error*.

Sampling Error

Definition 1.2.1 *Sampling error* is error that comes about because *statistics* from samples are being used to estimate *parameters*. In any sample, many population values are absent, and this absence of information causes sampling error.

Sampling error is also due to population variability. Vast variability tends to exist in populations, and samples, which are small subsets of these populations, can never represent all of this variability. We will see that sampling error is directly proportional to population variability. Population variability is the reason that different samples yield different estimates of a single population parameter. All of these differing estimates will contain some error.

While all statistics do contain error, the error can be made smaller by *increasing sample size*. Larger samples give better statistics with smaller sampling error. We will see that the maximum sampling error in two of our most important statistics is inversely proportional to the square root of sample size. Thus, as sample size increases, maximum sampling error decreases.

This does not necessarily mean that it is improper for researchers to study small samples. Indeed, there are many contexts, like medical studies involving patients with rare diseases, where gathering large samples is impossible. In such cases, research should still continue, but readers need to be warned of the small sample sizes and larger errors which correspond. We like large samples, and better quality always comes from large samples. But when we are forced to use small samples it becomes more important to state very clearly the maximum errors in any estimates given.

In summary: *sampling error increases with population variability, and decreases with larger sample size.*

Non-Sampling Error

Definition 1.2.2 *Non-sampling error* is error that arises from sources other than random sampling.

Non-sampling error can be caused, for example, by inappropriate sampling methods, or errors in measurement and computation. Generally speaking, non-sampling error results from *bias* in sampling.

Definition 1.2.3 *Bias* is a tendency for estimates to differ from parameters because of systematic flaws in the design or implementation of a research study.

Bias may be intentional or unintentional. We begin by defining three types of bias.

Selection bias occurs when population members are made less likely to be selected for a sample than other members. Such bias destroys the randomness of the sampling process. This denial of equal access may or may not be intentional. Selection bias can come about simply by researchers not taking care to see that less visible or accessible population members are brought into the sample. This is an easy mistake to make, but avoiding it can be costly in terms of time or money.

Self-selected samples always include selection bias. Self-selected samples consist of members who volunteer to be included, while non-interested individuals choose to not be involved. The problem with self-selecting samples is that, when we only study members who are interested in a research topic, we tend to include individuals who have common traits, neglecting other types of individuals. The result of this is a sample which is representative of individuals who have an interest in the study, but biased against those who do not. Such sample data are unlikely to yield information that represents the population, and therefore inferences made about the population will be unreliable.

Opinion polls often involve self-selecting samples, and also tend to be accompanied by high refusal rates. These facts make it quite difficult to conduct a reliable opinion poll.

Selection bias occurs with *convenience samples* as well. Convenience samples consist primarily of data which were conveniently available to the researcher. Again, such methods usually result in samples which are not representative of their population. Gathering high quality samples is not easy, convenient, or inexpensive. The gathering of quality data can take a great deal of time and effort, and tends to be costly. It is a shame, but in statistics it is very true that *you get what you pay for*, and quality is not cheap.

Non-response bias occurs when sample statistics differ from population parameters because of high refusal rates. We must understand that, while population members generally have the right to refuse involvement in any given study, such refusals can destroy the randomness of the sampling process. For humans, there are personality types that are less inclined toward involvement in statistical research. Because of this, sample data will show bias against such individuals. This is non-response bias – not a personal bias of the researchers – but the end result is the same: certain population members are less likely to be included. The result is a biased sample.

Most samples which involve humans include some non-response bias, but this may not be a problem if the *refusal rate* is small. When refusal rates are close to zero, it is safe to assume that error due to non-response bias will be minimal.

Response or Measurement Bias occurs when statistics are different from population parameters due to flaws in the *measuring* process. This could mean that our measuring instruments are not working correctly or that questions are being asked in a way that influences the response. In one way or another, the method of measuring interferes with the measurements themselves.

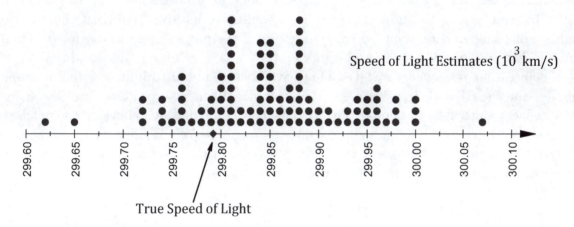

Figure 1.4: One hundred experimental estimates of the true speed of light are represented in the dotplot above. The true speed of light is in their midst, yet the majority of the estimates are greater than the correct value. The tendency for most values to be larger than the parameter they estimate indicates that *bias* is present in the sample. These data contain *measurement bias*.

Response or measurement bias can be very difficult to prevent, because the very act of measuring tends to be an interference, and interference usually changes that which is being observed or measured. There are ways to diminish this problem, but again, it is important that readers of research understand that measurement bias is a common source of non-sampling error. Figure 1.4 depicts a collection of experimentally observed values, of which a majority are larger than the parameter that they estimate. The values tend to be too large because of the machine that was used to measure them. This tendency of most values to be too large indicates that a *measurement bias* is present.

Data gathered from humans can be full of such bias. People love to read opinion polls, whether political or non-political in nature, scientific or not. The difficulty lies in the fact that opinions cannot easily be measured in a truly objective way. It has been shown that the way a question is asked can influence the response. If we ask *loaded questions* (questions that beg a certain response), we introduce measurement bias. The *order of questions* or *responses to multiple choice questions* can have an effect on the way people will answer them. The *gender of an interviewer* has been shown to influence response. In each of these examples, the way in which data are gathered influences the observations. The result is measurement bias.

1.2.2   Additional Concerns

Conflicts of Interest and Self Interest Studies

A conflict of interest occurs when a person or organization in a position of trust has a chance for private gain through some service provided by that position. In statistical research we see that, in numerous cases, researchers often have something to gain by a favorable result of their study. This can be a conflict of interest. We trust them to be honest, but they may be tempted to be dishonest if there is an opportunity for personal gain. When thinking about such *self-interest* studies, be careful to understand that there is a chance that researchers involved may have been tempted to influence the outcome of the study so that their private gain is ensured. When new products are tested for effectiveness and safety, it is important to know how companies who develop the product are involved in the study. If a company does its own research, readers should recognize the conflict of interest, and know that results are sometimes skewed by the company in hopes of making money on their product.

Correlation Does Not Imply Cause!

In Chapter 9 of this text, we will learn methods to determine when two variables *correlate*. This simply means that changes in one variable tend to be accompanied by somewhat predictable changes in another variable. It is important to note that, when variables correlate, correlation does not always imply that the change in the first variable is the *cause* of the change in the second variable. While this may be true at times, we are often unable to determine *when* it is true.

As an example of a relationship which appears to be causal, let x represent the time a student spends studying for a test, and y represent the score that the student receives on the test. We expect that higher values for x would tend to be accompanied by correspondingly higher values of y. Similarly, lower values of x should be accompanied by lower values of y. These variables are probably correlated, and it seems reasonable that the higher value of x may *cause* the higher value for y, but *determining* cause is a complicated matter. The statistical methods for *inferring* (not proving) cause will be discussed later.

As an example when correlation does not imply causation, consider the variable x which represents the number of automobile accidents in which a person has been involved, and the variable y which represents the income level of that same person. It can be shown that there is a positive correlation between these two variables, meaning that people who have experienced more automobile accidents tend to have higher incomes. Now, it would be foolish to recommend to anyone that they should be in more automobile accidents in order to raise their income level, because the accidents are not the cause of higher income. In this case, as in many, many others, there is another unconsidered variable that is causing both of these variables to increase together, and that is the age of the persons in question. Older people are just that – older. They have had more opportunity to be in automobile accidents, and because of their age and experience, tend to make more money. Age is one factor that helps both of these variables to increase. No change in either variable is the cause for a change in the other.

Distorted Representations

There are many ways to distort information, intentionally or not, without actually lying. Improper methods will influence the results of a study in a way that can mislead people. Because of this, it can be unethical to conduct statistical research improperly. Improper methods lead to misinformation, and misinformation can cause great harm. It is the responsibility of the researcher to tell the truth, to not exaggerate results, to use proper statistical methods, and to report the findings of the study as clearly as possible.

There are many ways to misrepresent statistical information. Graphs are often drawn with correct values but incorrect proportions. Consider the bar graphs in Figures 1.5 and 1.6. Both graphs represent the same information, but one is misleading. Which one is it? What is it about the graphs that cause the confusion?

Another way to mislead people is by implying certainty when certainty is not possible. Whenever statistics are reported, it is important for readers to understand that they are estimates of unknown values that include sampling error. When such statistics are reported, their margins of error should be included. Exactness in statistics is not possible, and it is important for readers to understand this. When people speak with great certainty about something that cannot be known, it may be that they are trying to mislead us, and are hiding the fact that their statistics contain large errors.

We would like to think that most often, people who do research try to do it responsibly and desire quality results, but it would be foolish to assume that this is always the case. Sometimes researchers lie, and sometimes lying can be quite profitable. Readers of research must be wary of this, especially when researchers have a great deal to gain personally (money, fame, etc.) by a favorable outcome.

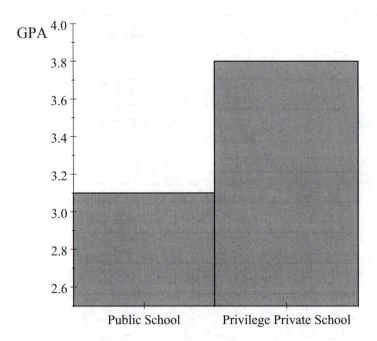

Figure 1.5: The average GPA at a local public high school is 3.1, where at *Privilege Private School* the average is 3.8.

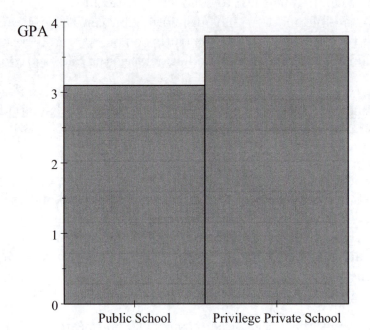

Figure 1.6: The average GPA at a local public high school is 3.1, where at *Privilege Private School* the average is 3.8.

Statistical Ethics

We end this section with a comment on ethics in statistics. We must understand that in many cases, the results of research have consequences out in the real world, and when results are in error, great harm can occur. Many people have been hurt by unsafe medicines, bad engineering, or faith in a cure that turns out to be worthless when other worthy treatments are available. It is important that readers demand that research is done with complete transparency. Every detail of the data gathering process and analysis should be made clear. Readers must hold researchers accountable for doing their work ethically. Much of our time in this subject will be spent discussing the right and wrong ways of doing statistical research, and it is important to take this seriously, because of damage that can be done by incorrect results.

1.2.3 Homework

Give as many problems as you can think of in the following studies.

1. A pharmaceutical company advertises in a newspaper asking users of one of their medicines to log on to a website & fill out a survey to gauge satisfaction and effectiveness of the medicine. Of over a million newspapers sent out, 1532 readers participated in the survey – quite a large sample size!

2. A politician's website has a poll which asks users to give opinions about government policy. One question asks "Do you feel that gun control laws should be strengthened in order to put the law back in the hands of those who are trained to enforce it?" The website receives over 2000 visits per day, and of those, over 1500 people respond to the poll.

For problems 3 and 4, state whether the error is a *sampling error* or *non-sampling error*.

3. Population data can vary widely – this type of error is due to the fact that we are taking relatively small samples from diverse populations.

4. This type of error is due to failings by humans or the equipment they use.

For problems 5 through 7, determine whether the data are influenced by *non-response bias*, *response or measurement bias*, or *selection bias*.

5. A student asks a group of fellow students "Don't you hate this college?"

6. Very few surveys tend to be returned in mail surveys.

7. Honors students tend to gather data that will not be difficult to obtain.

Answer the following questions.

8. It has been shown that children with more toys tend to have a higher level of intelligence. What can be said about the claim that giving children more toys will cause their intelligence to increase?

9. What can be said regarding the claim that cold weather causes people to catch a cold?

10. Is it appropriate for drug manufacturers to validate their products exclusively through their own research? Explain your answer.

11. Is it appropriate and ethical to exaggerate the results of statistical research when used with good intentions or for a good cause?

1.3 The Process of Statistical Analysis

1.3.1 Steps in the Statistical Analysis Process

There is much to be done in the design and implementation of a statistical study, and every study is different. In an attempt to streamline the process of statistical analysis, we provide here an overview of the steps that most statistical studies typically include. Each step in the process is important, and it cannot be overstated that the quality of the final result depends directly on the quality of the work in each step that preceded it. In this sense, a statistical study is like a chain with many links – a chain that is only strong as its weakest link.

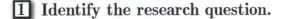

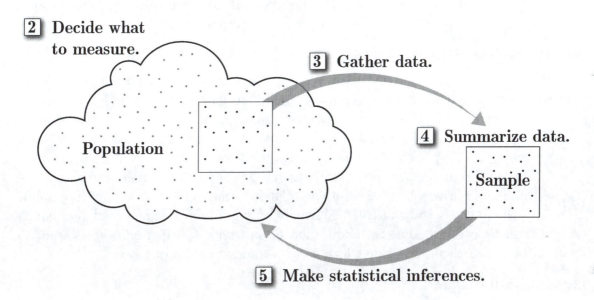

Figure 1.7: The process of statistical analysis.

The steps typical of most statistical studies are: (1) identify a research question to be investigated; (2) decide what to measure and how; (3) formulate a data collection plan & gather data; (4) summarize all data gathered; and (5) make statistical inferences and conclusions. We review each of these steps below.

Step I. **Identify the Research Question**

The context of a statistical study can be simple or complicated, but generally speaking, there is always a question that needs to be answered. In more complicated contexts, researchers need to take the time to understand the context completely before they can properly formulate a research question.

Researchers must ask a question that can be answered by measuring data from a population. The type of question that is asked by researchers determines the type of study they will conduct. If researchers want to ask a question about the nature of some population, then they will conduct an *observational study*. Observational studies allow us to eventually make inferences about a population.

If researchers want to ask a question about the effectiveness of a treatment that is applied to population members, then they will conduct an *experimental study*. Experimental studies randomly divide their samples into groups who receive different levels of a treatment. Experimental studies allow us to eventually make inferences about whether the treatment is the *cause* of an effect.

Thus, researchers must (1) understand the context of their study well, (2) ask a question that can be answered with data, and (3) know whether their question can be answered by an observational or experimental study.

Step II. **Decide What to Measure**

Once the nature of the question to be addressed is understood, researchers must decide what data variables should be gathered that will best answer the question. The variables can represent categorical data or quantitative data. The type of variable will determine the analysis which follows. For example, suppose you wish to measure customer satisfaction of a product. How is customer satisfaction measured? Can we measure it by the number of units sold (a quantitative variable)? Can it be measured by a survey of opinions (which may be categorical)? These are only a few of the many variables which can be used to address the question of customer satisfaction, and it is up to the researchers to decide which will best address their question.

Step III. **Gather Data**

Possibly the most critical step in a statistical study is the data collection process. It is understood that data from the entire population will not be available, so a sample must be chosen which will represent its population well. Researchers must decide the method of data gathering which will give the best representation of the population at hand, yielding an unbiased sample with the smallest error possible in the final results.

Once a plan for gathering data is developed, data are gathered according to the plan. Section 1.4 of this text discusses in detail the process of gathering sample data.

It is often the case that previously gathered data exist which may answer the question at hand. In cases where such data are already available, it is important for researchers understand completely how those data were gathered, what they were used for, and whether they will sufficiently address the question for their study. If such previously gathered are of sufficient quality and appropriateness, they may save the researchers a great deal of time, money, and effort.

Step IV. **Summarize Data**

The data summary process is relatively straight-forward. This involves computing averages, percentages, and other summary values, as well as graphical representations, such as the dotplot or bar chart. Summarizing data is the subject of Chapter 2 of this text.

Step V. **Make Statistical Inferences**

When the chosen sample represents its population well, it can tell us about the population. In statistics, an *inference* is a somewhat reliable *guess* about a population. An inference is used to answer our research question. Sample data give evidence for our inferences, but sometimes such evidence can mislead. Conclusions are made with the understanding that further research may lead to different conclusions.

1.3.2 Analyzing Statistical Studies

Many of us will read statistical studies performed by other researchers in our field of interest. When we read such research, it is important that we read with a critical eye, looking for sources of bias, critiquing their methods. We should consider each of the steps described above, convincing ourselves that the process of statistical analysis was carried out correctly, understanding possible pitfalls, and looking for misinterpretations.

For example, an article released by the National Institute of Environmental Health Sciences on February 28, 2011, detailed a study on the health of cleanup workers in the British Petroleum (BP) oil spill in the Gulf of Mexico during the summer of 2010 (*article is reprinted by permission*).

NIH Launches Largest Oil Spill Health Study
GuLF Study to Follow 55,000 Cleanup Workers and Volunteers for Up to 10 Years

A new study that will look at possible health effects of the Gulf of Mexico's Deepwater Horizon oil spill on 55,000 cleanup workers and volunteers begins today in towns across Louisiana, Mississippi, Alabama, and Florida.

The GuLF STUDY (Gulf Long-term Follow-up Study) is the largest health study of its kind ever conducted among cleanup workers and volunteers, and is one component of a comprehensive federal response to the Deepwater Horizon oil spill. The study is being conducted by the National Institute of Environmental Health Sciences (NIEHS), part of the National Institutes of Health,

Figure 1.8: Contract employees with BP America, Inc. load an oil containment boom onto a work boat at Naval Air Station Pensacola to assist in oil recovery efforts from the Deepwater Horizon oil spill in the Gulf of Mexico. *This image, taken by the US Navy, is in the public domain.*

and is expected to last up to 10 years Many agencies, researchers, outside experts, as well as members of the local community, have provided input into how the study should be designed and implemented.

"Over the last 50 years, there have been 40 known oil spills around the world. Only eight of these spills have been studied for human health effects," said Dale Sandler, Ph.D., chief of the Epidemiology Branch at NIEHS and principal investigator of the GuLF STUDY. "The goal of the GuLF STUDY is to help us learn if oil spills and exposure to crude oil and dispersants affect physical and mental health."

Over time, the GuLF STUDY will generate important data that may help inform policy decisions on health care and health services in the region. Findings may also influence responses to other oil spills in the future.

"We are enrolling workers and volunteers because they were closest to the disaster and had the highest potential for being exposed to oil and dispersants," said Sandler.

The GuLF STUDY will reach out to some of the 100,000 people who took the cleanup worker safety training and to others who were involved in some aspect of the oil spill cleanup. The goal is to enroll 55,000 people in the study. Individuals may be eligible for the study if they: are at least 21 years old; did oil spill cleanup work for at least 1 day; are not directly involved in oil spill cleanup but supported the cleanup effort in some way, or completed oil spill worker training.

Working from lists of people who trained or worked in some aspect of the oil spill response, the GuLF STUDY will contact potential participants by mail, inviting them to take part in the study.

The study was developed to make participation as easy and convenient as possible. In addition, the GuLF STUDY incorporates safeguards to protect the privacy and confidentiality of personal information.

All participants will be asked to complete an initial telephone interview, and provide updated contact information once a year. During the telephone interview, participants will be asked questions about the work they did with the oil spill cleanup, and about their health, lifestyle, and job history. About 20,000 participants will be invited to take part in the second phase of the study, which involves a home visit and follow-up telephone interviews in subsequent years. Small samples of blood, urine, toenail clippings, hair, and house dust will be collected during the home visit, and clinical measurements such as blood pressure, height and weight, urine glucose, and lung function will be taken.

If at any time in the course of the study, the need for mental or medical health care is evident, participants will be given information on available healthcare providers or referred for care. The study leaders have up-to-date information on healthcare providers and a medical referral process in place as part of the study. Materials will be available in English, Spanish, and Vietnamese.

The NIH is funding the GuLF STUDY. A small part of the funds have been provided by BP made to NIH specifically for research on the health of Gulf area communities following the spill, though BP is not involved in the study.

To critique this article, we will address each of the steps outlined for the process of statistical analysis.

Step I. **Identify the Research Question**

The questions that will be addressed by this study are likely to be many, but the article states that "A new study that will look at possible health effects of the Gulf of Mexico's Deepwater Horizon oil spill on 55,000 cleanup workers and volunteers," making it clear that any question to be addressed relates to the long-term health effects upon these workers.

The context of the BP oil spill is one which is quite complicated, with many questions that might be asked. Since this study is one involving medical issues, it is clear that medical professionals should be deeply involved who are aware of the issues of the effects of oil spills on human health. This study is being led by Dale Sandler, chief of the epidemiology branch of the National Institute of Environmental Health Sciences, who seems quite competent and likely to understand the issues completely. The study is an *observational* study, because there are no randomly divided groups who receive different levels of some treatment.

Step II. Decide What to Measure

Much data will need to be gathered from the subjects of this study. Clearly, sample members will be asked if they are experiencing health problems, what they are, and when they started. Objective observations will be made upon "small samples of blood, urine, toenail clippings, hair, and house dust," and measurements "such as blood pressure, height and weight, urine glucose, and lung function" will be taken. All of these will be used to assess the long term health status of cleanup workers. The variables which will be measured will include quantitative variables (blood pressure, height, weight, etc.), and some may be categorical (lung function).

Step III. Gather Data

From the population of around 130,000 cleanup workers, the article states that about 100,000 people would be contacted by mail. It is hoped that about 55,000 people will agree to be involved.

Once the sample members are identified, it is likely that the process will begin with interviews, followed later by medical examinations.

Step IV. Summarize Data

With many variables which may be gathered, there will be many ways to summarize. It is clear that much data to be gathered will be categorical. Many physical ailments encountered will need to be tallied, so that percentages can be computed for each ailment category. Graphical summaries, such as bar charts, could be included as well.

Step V. Make Statistical Inferences

After statistical summaries are computed it will be time to compare proportions of people observed in the sample experiencing a given ailment are significantly different from the proportions of respective ailments in the population in general. When significant differences are observed, it may be time for researchers to make inferences regarding the effects of exposure to so much crude oil. Because this is an observational study, inferences should be made about the population being studied. Such inferences would be statements about the long-term health of those involved in the clean-up effort.

Understanding all of the steps in the statistical analysis process is important for researchers, but also readers of research. Such understanding can help readers of research to know how much faith to place in the results.

1.3.3 Drawing Conclusions from Research

In observational studies, it is possible for statistical inferences to be made when a sample is selected that represents its population well. Inferences can be made regarding the values of unknown population parameters, and the margins of error in such estimates. Cause and effect conclusions should be avoided in observational studies.

The goal of an experimental study is to determine if it is possible for a treatment to be the cause of some effect. When groups containing experimental subjects are created randomly to receive different levels of the treatment, it is possible to infer that a cause and effect relationship exists. This must be done carefully so that there are no other possible explanations for observed effects. The details of experimental studies will be discussed in the next section of this text.

In the following homework section, the student will discuss the statistical analysis process for another research study.

1.3.4   Homework

Read the following article which describes a teen depression study, and answer the question at its end.

Combination Treatment Effective in Depressed Adolescents

Press Release • August 17, 2004

This Article of the National Institute of Mental Health is in the Public Domain

A clinical trial of 439 adolescents with major depression has found a combination of medication and psychotherapy to be the most effective treatment. Funded by the NIH's National Institute of Mental Health (NIMH), the study compared cognitive-behavioral therapy (CBT) with Prozac, an SSRI antidepressant whose generic name is *fluoxetine*, currently the only antidepressant approved by the Food and Drug Administration for use in children and adolescents. John March, M.D., Duke University, and colleagues, report on findings of the multi-site trial in the August 18, 2004, Journal of the American Medical Association (JAMA).

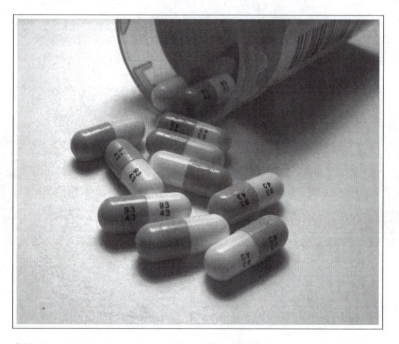

Figure 1.9: The SSRI antidepressant known as Prozac. *Photo by Tom Varco, licensed under the Creative Commons Attribution-Share Alike 3.0 Unported license.*

The results of the first 12 weeks of the Treatment for Adolescents with Depression Study (TADS), conducted at 13 sites nationwide, show that 71 percent responded to the combination of fluoxetine and CBT. The other three treatment groups, of participants between the ages of 12 and 17, also showed improvement, with a 60.6 percent response to fluoxetine-only treatment, and 43.2 percent response from those receiving only CBT. The response rate was

34.8 percent for a group that received a placebo. The difference in response rates for the latter two treatment groups was not statistically significant.

The $17 million study is the first large, federally funded study using an anti-depressant medication to treat adolescents suffering with moderate to severe depression. TADS was conducted between the spring of the year 2000 and the summer of 2003.

Clinically significant suicidal thinking, which was present in 29 percent of the volunteers at the beginning of the study, improved significantly in all four treatment groups, with those receiving medication and therapy showing the greatest reduction.

Please discuss how this study appears to address each of the steps (I through V described in this section) in the process of statistical analysis. Make note as to whether any of these steps are not clearly addressed in the article. Additionally, address whether this study is *observational* or *experimental*.

Step I. **Identify the Research Question**

Name a question which researchers might have asked that is answered by data in this study. Was the study observational or experimental?

Step II. **Decide What to Measure**

What data were used to answer this question? Are the data quantitative or categorical?

Step III. **Gather Data**

Is there information regarding how the data were gathered? Was the sample divided into groups? How were the data gathered from each group?

Step IV. **Summarize Data**

What summary values were used to describe the data sets? What graphical displays could be used to represent the data from each group?

Step V. **Make Statistical Inferences**

Should an inference be made about a *population* or a *treatment*? What inference is suggested by the data?

1.4 Random Sampling and Random Assignment

1.4.1 Deciding What to Measure

Once a research question has been chosen for a statistical study, it is time to decide what data can be measured that will answer the question best. Data come in two types: categorical (or qualitative) data, summarized by proportions or percentages, and quantitative data, summarized by averages and measures of spread. We need to decide whether categorical data with proportions or quantitative data with averages serve our needs best, noting that everything that follows (sample sizes, methods of analysis, interpretations, etc.) will depend on the type of data that we gather.

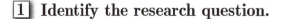

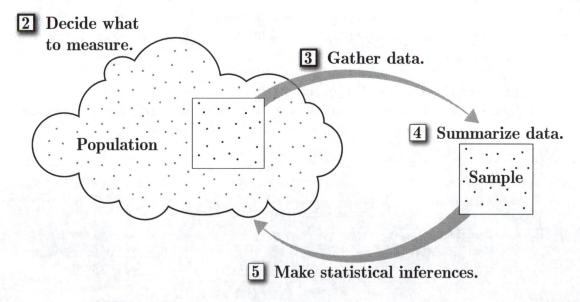

Figure 1.10: The process of statistical analysis, steps 2 & 3.

1.4.2 Gathering Data

Once we have decided what data will be measured, it is time to devise a plan for gathering it. The process of gathering statistical data is *sampling*. When devising a plan for sampling, it is important to remember the goal of the sampling process: *to create a sample which represents its population well*. Representative samples yield estimates with minimal error, and provide reliable information about their populations.

In general, the best samples are samples where observations are gathered randomly. The methods of inference in this text assume that samples consist of random observations. Below, we define the terms *random selection*, and *simple random sample*.

1.4.3   Random Selection and Random Samples

A population member is *randomly selected* when that member has the same chance of being selected as any other population member. At the heart of statistics is random selection. All of our methods assume that population members are selected randomly. A *random sample* is one where every member of the population has the same chance of being chosen for inclusion in the sample. Obviously this is a lofty goal, and can be nearly impossible to achieve, but the theories of statistics which we will be learning assume that our samples are random.

Simple Random Sample

A *simple random sample* of size n is one which is chosen in a way such that every possible sample of that size has the same chance of being selected. Simple random samples are assumed in all statistical inference.

It is important to note that this definition implies that in a simple random sample, every member of a population has the same chance of being selected as a member of the sample, so this selection process ensures *random selection*. Note, however, that simply giving every member of a population an equal chance of being selected is not enough to ensure that a sample is *simple random*. This is illustrated by the following example.

According to the Central Intelligence Agency, the world is approximately 49.8% female. A researcher wants to create a sample carefully to ensure that the proportions of female to male in the sample match the proportions of female to male in the Earth's overall populations. By doing this, the researcher helps to ensure that females and males all have the same chance of being selected in the sample. So, she decides to use a sample with 1000 people, 498 of which are females, randomly selected, and 502 of which are males, randomly selected. The question is: Can such a method of sampling give a *simple random sample?*

While this method seems like a good way of ensuring that the gender proportions in the sample accurately represent the gender proportions in the general population, in reality this removes the simple randomness from the sample! Whenever a researcher attempts to engineer the proportions of groups within a sample, the simple randomness of the sample is destroyed. To illustrate, simply note that no sample of 1000 people with any other gender breakdown (than the 498 women and 502 men) could be selected, so a bias is being shown against those other samples. A sample with 500 women and 500 men could never be selected, and so a bias is being exercised, not against the individuals, but against this particular type of sample. This engineering of the makeup of a sample means that the sample will not be *simple random*. Still, even with this engineering, it is still possible for the individuals to have been randomly selected.

Methods of Sampling

Our goal in sampling is a *simple random sample*. As mentioned previously, a simple random sample of size n is one which is chosen in a way such that every possible sample of that size has the same chance of being selected. Simple random samples are assumed in all statistical inference. This definition implies two things. First, that all individuals have the same chance of being selected for the sample, and second, that no engineering be done to influence the nature of the sample. This is assumed in all of the methods of inference starting in Chapter 6 and assures a sample that is free of *selection bias.*

One way to achieve a simple random sample is to create a list containing all population members. Each member of the list is then assigned a number. Next, a computer may be used to randomly select a subset of these numbers, and the sample is generated by including the population members corresponding to those random numbers. That sounds easy enough, but the difficulty lies in obtaining such a list. If the population consists of people, remember that there is no list containing absolutely all of the members of the Earth's population, nor even any individual country's population. There are many people in the world who are quite inaccessible, and this inaccessibility is what makes random selection so difficult.

Stratified Sampling

While random selection is assumed in most methods of statistical inference, we will find that many researchers gather samples using techniques designed to help ensure randomness, but do not give a truly simple random sample. One such method that can give good results is called *stratified sampling.* Stratified sampling can be easier to implement than simple random sampling, and can sometimes give more reliable statistics. With stratified sampling, the population is divided into groups, called *strata,* and then a simple random sample is taken from each group (or stratum). To help preserve randomness, it is a good idea to select a number of members from each group which is proportional to the size of that group in the population.

As an example of stratified sampling, suppose a population contains 12,000 men, and 11,900 women. A proportional stratified sample might contain 120 men and 119 women, randomly selected. While such a sample cannot be a *simple random sample* (its composition is not random), it still has a good chance of representing the population well in terms of gender diversity.

Cluster Sampling

While stratified sampling can give good results, sometimes the number of groups to choose from can be so vast, that it is impractical to sample from each group. When this is the case, *cluster sampling* can be used, where we again divide the population into groups, but now we randomly select some, but not all, of the groups to sample.

When the United States gathers census information from households in non-incorporated areas, cluster samples are used so that workers are not required to visit every street in these areas. This helps lower the cost, but obviously contradicts the idea of a census where all population members are supposed to be included.

Systematic Sampling

With *systematic sampling*, the population is lined up in some way, usually on a list. Researchers pick every n^{th} name on this list. This can only result in a random sample if the list is first put in random order.

1.4.4   Sampling Frame

In sampling, the goal should always be to gather a sample that is representative of its population. Unfortunately, this goal can be extraordinarily difficult to achieve. Samples usually contain some bias, no matter how careful we are. Whenever samples are gathered, it is important to recognize this. When we cannot argue that a sample represents its population well, we should consider whether there is a subset of the population that the sample *does* represent. Such a population subset is referred to as the *sampling frame* of the study.

If we gather data by calling phone numbers randomly, our sample cannot be assumed to represent all population members. Still, it may be possible to argue that the sample represents the sampling frame of individuals who have phones. If we are conducting a study about college students, but can only gather random data from students of a single college, then that college's student population is the sampling frame. In both cases, results of the study may not give accurate information about the larger population, but should have a good chance of accurately representing the sampling frame. When this is the case, statistical inferences should be directed toward the sampling frame, and *not* toward the larger population.

1.4.5   Types of Statistical Studies

Now we focus on a the different types of statistical studies, how they are implemented, and problems that can arise in each.

Observational Studies

In an observational study, the researcher's goal is to use statistics to describe a population as it is, without influencing it. Observations are measured from sample members in a way that will influence the measurements as little as possible. Every effort should be made to avoid response or measurement bias, as described previously. Depending on the type of study, avoiding such influence in measuring can be difficult, because the very act of measuring can be disruptive. Even in scientific experiments, the act of measuring can cause measurement bias. Thermometers can change the temperature of that which is being measured. Speedometers create a small amount of *drag* or resistance in the vehicles they measure. When there is a better way to take measurements from sample members without influencing them, those methods should be used.

There are three types of observational studies: *retrospective studies* (where data from the past are used), *cross-sectional studies* (where data are gathered in a single moment in time), and *prospective* or *longitudinal studies* (where researchers gather data from sample members over a period of time into the future).

Cause and effect relationships cannot be determined in observational studies, as tempting as such conclusions may be. Observational studies are rather naive in that way – they can help us understand what may be true about a population, but they do not provide insight as to *why* something may be true. The reason for this lack of insight is *confounding*, which we will define shortly.

Experimental Studies

An *experimental study* is very different from an observational study. In an experiment, the researcher's goal is to cause some change in sample members by introducing a *treatment*. A treatment is some act that is meant to bring about a change or response in measurements taken from sample members. When observations are measured after the treatment is applied, the observation is called a *response*, and the variable representing all possible responses is the *response variable*. The variable representing all levels (i.e. *dosage* levels) of the treatment is called the *explanatory variable*. The idea behind this is that the values observed by the response variable are, hopefully, *explained* by the explanatory variable. When experimental studies are done correctly, it is possible to infer a cause and effect relationship between the treatment and response.

Confounding   One problem that can make an experiment difficult to interpret is *confounding*. Confounding occurs when an observed effect is due, possibly, to causes other than the treatment. Possible causes, other than the treatment, for such an effect are *confounding variables*. Suppose, for example, that researchers are testing a new flu treatment. Patients receive the treatment, and then are observed by researcher to measure any effects. It is to be assumed that, with or without the treatment, patients may get better. If the treatment is slow to bring about improvement, it is difficult to know whether improvements are due to the treatment, or simply due to the fact that patients tend to get better over time, even without a treatment. In such a case, it is unclear if the treatment is the cause for the improvement, and we are *confounded*.

In 2001, the Associated Press printed a story about a group of Danish researchers who found fault with the many studies done over the previous decade which concluded that a bit of wine every day could be beneficial in lowering risks for certain types of heart disease, cancer, and other health problem. The Danish researchers found that in most of those studies, that people who drink wine tend to be more educated, earn more money, eat healthier food, and have better health care. In short, we are confounded by each of these variables, because it is unclear that the wine is responsible for causing these positive health benefits. There are many confounding variables, that tend to accompany the explanatory variable of wine drinking, that also may contribute to the positive health effects that have been observed.

The lesson to be learned is that we must be cautious about interpreting the results of such research. Inferences about treatments may be true, and they may not. As we have said, statistics can be used to determine correlation between two variables, but that does not prove *cause*. There is a correlation between wine drinking and some positive health effects, but it may not be that wine is the primary cause.

The Placebo Effect The *placebo effect* is an example of a confounding variable in medical research. Exactly what is the placebo effect is can be a difficult question to answer. A *placebo* is an inert, or ineffective, treatment, such as a sugar pill. It has been often observed that sometimes, after taking a placebo, a patient appears to get better. This perceived improvement is the placebo effect. The difficulty in understanding the placebo effect is due to the fact that we are confounded regarding its cause. When we read research about patients getting better after receiving a placebo, we are not sure whether the improvement is real or only perceived. Even if an improvement is real, is the placebo truly the cause?

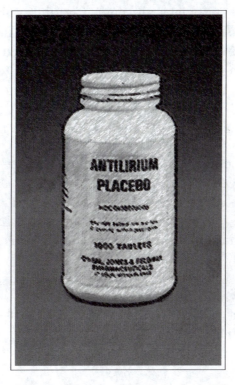

Figure 1.11: The placebo *Antilirium*. (*Original work in the public domain by the U.S. Federal Government*).

An article written in 2001 and published by the Associated Press, again gave news from Danish researchers who were contradicting beliefs widely held by the scientific community. The researchers claimed that there was no *true* placebo effect. The article states the following:

> *One of the most strongly held beliefs in medicine — that dummy pills or other sham treatments greatly help many patients — has been called into question by Danish researchers who found little or no placebo effect in dozens of studies.*

The article goes on to say the following:

> *Many past studies and textbooks suggest that about one-third of patients given placebos in medical experiments get better, presumably because they believe they are getting an effective treatment. But the new research casts doubt on this long-held belief. "The high levels of placebo effect which have been repeatedly reported in many articles, in our mind are the result of flawed research methodology," said Dr. Asbjorn Hrobjartsson, a professor of medical philosophy and research methodology at University of Copenhagen who ran the study with colleagues at the Nordic Cochran Center there.*

Is the *placebo effect* a *real* phenomenon? For years and years people have believed that the mind has the power to cure many ailments. Certainly this is true for some types of diseases. Many medical problems are stress-related, and the mind has the power to learn to avoid stress, and exercising this power may represent a cure. Additionally, it is well established that depression can increase mortality rates for terminal patients. It seems clear that the mind has some part to play in the cure of some illnesses. If a placebo can trigger a mental process which brings about an improvement in patient health, then the placebo effect is truly a confounding variable which must be managed.

It is true that the placebo effect is difficult to define, because we often do not know the reason why people get better after taking a placebo. We do not know whether the placebo causes a true cure because of the power of the mind to heal, or simply a perception of improvement. If the latter is true, then researchers should make every effort to correct such misperceptions by taking objective measurements to determine whether true medical improvements have actually occurred.

Regardless of what the placebo effect is, we still have the problem of confounding when patients receive a treatment – we do not know whether patients' improvements are due to the treatment or the placebo effect. We will gain control over this confounding situation by using *control groups*.

Control and Treatment Groups   To control confounding caused by the placebo effect, researchers divide samples into groups, or *experimental blocks*. Some experimental blocks are called *control groups*. Control group members do not receive the treatment that is being studied, but may possibly receive a placebo. A control groups is used as a basis for comparison to other experimental blocks, *treatment groups*, whose members receive experimental treatment. When treatments at various levels are to be tested, then multiple treatment groups are employed.

In dividing patients into control and treatment groups, it is important that patients do not know the type of group in which they have been placed. This is done so that the magnitude of the placebo effect can be measured. *Blinding* is the process of concealing from members the type of experimental block in which they are placed. When neither experimental subjects nor those who interact with them know the type of group in which subjects are placed, the experiment employs *double blinding*. It is usually important (when working with humans) that those who interact with subjects do not know to which group they belong. Patients receiving a placebo should not know this, as such knowledge will spoil the effect of having a control group.

Cause and Effect Conclusions in Experiments When dividing a sample into experimental blocks, we control confounding variables by creating blocks of *relative homogeneity*, or *sameness*. When experimental blocks are similar in nature, confounding variables are held constant from block to block. For example, if gender is a possible confounding variable, then the experimental blocks should be created with similar proportions of females and males. If age is a confounding variable, then experimental blocks should include members of similar ages.

An experiment is *controlled* when all experimental blocks are as similar to one another as possible, and the only true variable between the groups is the type of treatment (or non-treatment) that they receive. If the groups are similar, then we can conclude that differences are likely to be *caused* by the treatment being tested.

In any given experiment, the list of possible confounding variables can be quite long, so researchers must decide which variables are most important, and make sure that they are held constant from group to group. The groups can never be perfectly identical, of course, with respect to all variables, but efforts must be made to ensure that the most important confounding variables are controlled in this way.

Example 1.4.1 We would like to design a controlled experiment where the health effects of wine are measured. By "*controlled*" we mean that experimental blocks should be created so as to minimize the effects of any confounding variables.

We have discussed the fact that wine drinkers tend to make more money and be more highly educated than non-wine drinkers. Because of this, dividing people into groups based on whether they drink will introduce confounding variables. We can attempt to control these variables to make sure that the groups are similar in terms of education, income, age, gender, and any other variables which may confound.

With all of this effort, this experiment is still flawed because members are effectively choosing their group based on whether they drink wine or not. This process of self assignment into groups can be responsible for the introduction of additional confounding variables. Because of this, it is uncertain that a cause and effect relationship can be established between wine drinking and its possible effects upon health.

The previous example illustrates an important point – when members effectively *choose* their group based on some habit, like drinking wine, confounding variables may be introduced, and cause and effect conclusions are harder to make.

When we wish to determine a cause and effect relationship between some treatment and its supposed outcome, a good way to create experimental blocks is through *random assignment* into groups. This method is called *randomized block design.*

When experimental blocks are created through random assignment, it is reasonable to attribute any differences between the blocks as being caused by the treatment. In such cases, cause and effect relationships may be inferred. When groups are not formed by random assignment, care must be taken to ensure that blocks are similar with regard to possible confounding variables. The less these confounding variables are controlled, the more the study becomes *observational* (and less *experimental*) where cause and effect conclusions are not appropriate.

Definition 1.4.2 In *randomized block design*, confounding variables are held constant by creating experimental blocks through random assignment. When a significant difference in an experimental effect is observed in an experimental block, we may infer that its treatment is the cause.

Finally, it is important to understand the goal of an experiment, and how it differs from the goal of an observational study. An observational study's goal is to describe a population, and thus its random sample must be representative of the population. The goal of an experiment is not primarily to describe a *population*, but the *effects* of a treatment. Because of this, random sampling in experiments is less important than *random assignment into experimental blocks* which are similar to one another.

Replication   Even when research is performed to the best of human standards, sometimes data will lead researchers to make incorrect conclusions. Readers are advised to remember this, and when studies are published which indicate some important breakthrough, it should be understood that the case is not settled until the study is repeated by a group of independent researchers who are able to replicate the results from the first study. When results can be *replicated* by independent researchers, they are much more credible.

1.4.6   Homework

For problems 1 through 4, determine whether the study is *observational* or *experimental*. For experiments, identify the *explanatory* and response *variables*.

1. A random sample of voters is divided into two blocks through random assignment. A company shows a T.V. commercial to one group to see how their views regarding a product are influenced. Their views are compared to the second group who have not seen the commercial.

2. A company measures T.V. ratings by monitoring family viewing habits.

3. People are polled to measure support for a political issue.

4. A random group of voters are polled to measure how a politician's speech impacted their views on a political issue. Their views are compared to another random group who have not heard the speech.

Answer the following questions.

5. Please explain the key difference between these types of studies: retrospective; cross-sectional; prospective.

6. For experimental studies, how do we control the effect of the confounding variable known as the *placebo effect*?

7. Why is it important for medical experiments to be *double-blind*?

For problems 8 through 15, state whether the method of sampling is *systematic*, *convenience*, *stratified* or *cluster*. State also whether the sample is representative of its population, and the *sampling frame* for the study.

8. A research organization randomly collects data from local residents to study how Americans feel about their communities.

9. A statistician gathers data from 100 randomly selected students at each of 25 randomly selected colleges to make inferences about all U.S. college students.

10. Names corresponding to every 1000th social security number are selected.

11. Students from every college in the United States are randomly selected to make inferences about all U.S. college students.

12. When gathering data from geographic areas which are outside of city limits, census workers systematically visit homes on a collection of randomly selected streets.

13. At a given college, 5 random students from every section of every course offered are questioned to measure the college's effectiveness in achieving student learning outcomes.

14. Honors students gather data by randomly polling students in the classes that they attend.

15. Students from a given college are sampled by selecting names corresponding to every 200th student ID.

Answer the following questions.

16. Explain the importance of *replication* in successful research experiments.

17. A researcher wants to study people from the United States, so she picks 100 people from each state randomly. Could this result in a random sample?

18. Is it humanly possible to gather a random sample of 1000 people who live on this planet? Why?

For the following problems, determine whether the conclusion is appropriate. Explain your answers.

19. The average IQ for a random sample of Internet users is higher than the average IQ of a random sample of non-Internet users. Because of this, we can conclude that the Internet is effective in raising average user IQ.

20. A random sample of preschool children is divided into two groups randomly. One group receives quality preschool care, while the other group receives no preschool care at all. Years later it is observed that a significantly higher percentage of the children who had quality preschool eventually went to college. We conclude that quality preschool causes a larger proportion of children to eventually go to college.

21. A sample of 100 randomly selected cars are randomly divided into two groups. The first group receives high octane gasoline, while the second receives low octane gasoline. The mean miles per gallon achieved by the first group is significantly higher than the mean miles per gallon for the second group. It is concluded that the high octane gasoline is the cause of the higher gas mileage.

22. Two college professors give the same final exam. The first professor has a significantly higher average score than the second. The college administration concluded that the first professor's quality of instruction is the cause for her higher scores.

Answer the following questions.

23. It has been widely claimed that drinking grape juice every day can improve a person's health and lower risk of many diseases. List possible confounding variables.

24. It is well documented that depression rates are lower among the wealthy in comparison to the poor. Does this imply that wealth reduces the risk of depression? List possible confounding variables.

25. People who drink diet soft drinks don't often lose weight. In fact, they usually gain weight. These findings come from eight years of data collected by Sharon P. Fowler and colleagues at the University of Texas Health Science Center, San Antonio. Fowler reported the data at the 2007 annual meeting of the American Diabetes Association in San Diego. "What didn't surprise us was that total soft drink use was linked to overweight and obesity," Fowler told WebMD. "What was surprising was when we looked at people only drinking diet soft drinks, their risk of obesity was even higher." Given that diet drinks contain no calories, please provide possible confounding variables which may explain this increase in weight.

26. Referring again to the confounding variables given in the previous problem, design an experiment which would control these confounding variables, and determine the effect of drinking diet soft drinks.

27. Explain why double blinding is preferable to blinding in quality medical research.

28. Suppose that a study is conducted to determine the effectiveness of a given diet in helping patients lose weight. Could the placebo effect be a factor in confounding the results of this study? How could a control be implemented to measure and control this effect?

Part II

Descriptive Statistics

Chapter 2

Summarizing Single Variable Data

Summarizing Data

Having discussed the first three steps in the process of statistical analysis, we move on to the next step: *summarizing data*. The statistics of summarizing data is typically called *descriptive statistics*. The goal of descriptive statistics is to summarize large data sets in a way that is easy to grasp.

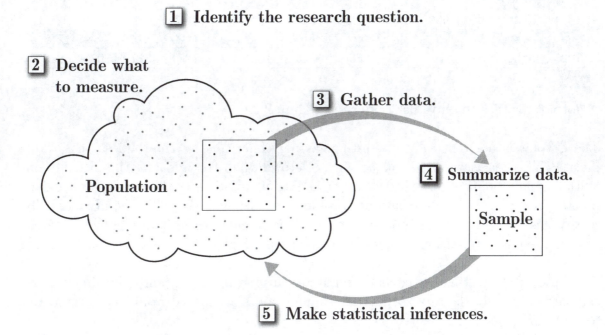

Figure 2.1: The process of statistical analysis, step 4.

In this chapter, we summarize single variable data, which are data sets which consist of one observation per sample member. Our main descriptors of a data set will provide descriptions of the *shape* of a data set – how data are distributed amongst the range of possible measurements, a measure of *center* for a data set, and a measure of variation in a data set (which will be directly related to sampling error). In addition, we will seek to find ways to describe the position of a given value, relative to the rest of the data set. We begin this process with frequency tables, which are an important step toward accomplishing all of these goals.

2.1 Frequency Tables and Histograms

Frequency tables and *histograms* are used to help describe the distribution of data values within the range of possible measurements. As with dotplots, they are used to help us visualize the *shape* of a distribution of values.

Example 2.1.1 As an example, consider the values below, which represent contributions collected for 52 weeks for a certain charitable organization. These data were collected by Daniel Vega, an honors student at Mt. San Antonio College.

Weekly Contribution Totals												
1287	987	899	906	1157	932	951	993	1226	994	860	1247	1769
1054	1053	908	900	1287	1224	939	919	661	947	1541	960	1112
1080	817	912	953	1087	1114	991	843	980	1521	846	866	917
1025	895	1103	1195	53	806	1140	914	1099	1271	887	2075	849

A dotplot for these data values is given in Figure 2.2.

The dotplot is not especially interesting, as only one value in the data set has a duplicate. There is, however, a fair amount of crowding in the midst of the distribution. With so many points crowded together, it is difficult to quantify the density of the data points, giving us the need for another tool for visualizing the distribution of values. To rectify this, we construct a frequency table and *histogram* of these data points which does not require exact repetition of points to demonstrate high data density regions.

The process of creating a frequency table and histogram is simplified by first sorting the data values. In the table below, the sample values are sorted from left to right, top to bottom.

Donation Amounts

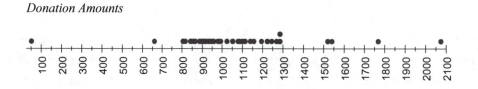
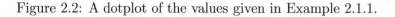

Figure 2.2: A dotplot of the values given in Example 2.1.1.

Sorted Weekly Contribution Totals												
53	661	806	817	843	846	849	860	866	887	895	899	900
906	908	912	914	917	919	932	939	947	951	953	960	980
987	991	993	994	1025	1053	1054	1080	1087	1099	1103	1112	1114
1140	1157	1195	1224	1226	1247	1271	1287	1287	1521	1541	1769	2075

The minimum data value is 53, and the maximum value is 2075.

Next, we compute the range. The *range* is the distance from the minimum to the maximum values, so $Range = MaxValue - MinValue$. In this case, $Range = 2075 - 53 = 2022$. Next we would like to take this range and divide it into roughly 5 to 10 (or more – depending on the size of the sample) non-overlapping intervals, called *classes*. Each class has *class width* computed by the formula:

$$Class\ Width = \frac{Range}{\#\,of\ Classes}$$

(always round up to a slightly larger number), and we round up, not too much, to a convenient value. In this example we will use 7 classes, so

$$Class\ Width = \frac{2022}{7} \approx 288.86,$$

which we will round *up* to 290. It is important that we always make the class width a little larger than the value calculated by the formula above, so that our classes cover the entire range. The class width should be rounded to the same number of places after the decimal as the data values. Our table will include 7 classes, each with width 290. The minimum value in the data set is 53, so our first class will include an interval beginning at 50. For each data value, we will place a tally mark in the corresponding class. Once the values are tallied, we sum the tallies to find the frequency of values for each class.

Class		Tally	Frequency
50	– 339	I	1
340	– 629		0
630	– 919	IIII IIII IIII III	18
920	– 1209	IIII IIII IIII IIII III	23
1210	– 1499	IIII I	6
1500	– 1789	III	3
1790	– 2079	I	1

Immediately, we see that our first goal in describing this data set is being met – we have a visual description of how the data values are distributed amongst the possible classes that we defined. The shape is one which is quite important in statistics – it is roughly what is called *bell-shaped* (or *normal*), meaning that the highest frequencies are in the middle classes, and that the lower frequencies are in the extreme classes (the lowest and highest). As a result of this process, we see glimpses of answers to our other goals as well. Could you guess the *center* of the distribution, or give a measurement for the variation in the data set? As a measure of center, we might guess a value somewhere in the middle class, from 920 to 1209. This is very imprecise, but it is a start. To measure variation, we might remember that we have already computed the *range* of this data set, which was 2022.

Some facts about the above frequency table need to be mentioned. The *lower class limits* for each class are the values 50, 340, 630, 920, 1210, 1500, and 1790. The distance between consecutive lower class limits is the class width. The *upper class limits* for each class are the values 339, 629, 919, 1209, 1499, 1789, and 2079. The distance between consecutive upper class limits is, again, the class width. In this example, the data values are whole numbers, so the upper limit is one less than the lower limit of the next class. The *class boundaries* are the average of the upper limit of any class and the lower limit of the next consecutive class. Once again, the distance between them is the class width, and note that we calculate a class boundary before the first class, as well as one after the last class. The class boundaries are: 49.5, 339.5 629.5, 919.5, 1209.5, 1499.5, 1789.5, and 2079.5. The *class midpoints* of any given class are the average of the lower and upper limits for that class. So the midpoint of the first class is $(50 + 339)/2 = 194.5$, and the distance between the midpoints is equal to the class width. The class midpoints are: 194.5, 484.5, 774.5, 1064.5, 1354.5, 1644.5, and 1934.5.

When making a frequency table, please keep the following rules in mind:

- The classes should not overlap, so that no value could fall into more than one class;

- The classes should all have the same width;

- The number of decimal places for the class limits should match the number of decimal places for the sample data;

- If data are recorded to the nearest tenth, then the upper limit of one class should be 0.1 less than the lower limit for the next class. If the data are recorded to the nearest one hundredth, then the distance between these limits would be 0.01, and so on;

- Choose between 5 and 20 classes, (or more if the sample size is very large);

- Include all classes in the table, even empty classes.

2.1.1   Histograms

Our most important tool for learning the shape of a distribution of data values amongst classes is the histogram. A histogram is a bar graph which includes a horizontal axis, with the class boundaries marked as x values. A vertical axis will also be included to represent frequencies for each class. Rectangular "bars" will be drawn from class boundary to class boundary, with heights corresponding to the frequency of each class. The bars should touch one another. Notice that the class with a corresponding frequency of zero is present with no bar showing.

The histogram for the weekly contributions in Example 2.1.1 is plotted in Figure 2.3.

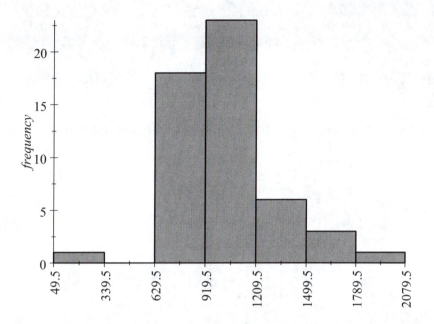

Figure 2.3: The frequency histogram for Example 2.1.1.

Example 2.1.2 The values in Table 2.1 are sorted from left to right, then top to bottom. They represent experimental measurements of the speed of light in air, in thousands of kilometers per second. These values are the result of the classic study conducted by Albert Michelson on the speed of light in air in 1879, in order to find a precise value for the speed of light. This classic study was conducted by directing a beam of light from the top of Mt. Wilson to a mirror on the top of Mt. Baldy (both mountains are in Southern California) which reflected the light back to Mt. Wilson where an apparatus was used to measure the time taken for the light to travel this distance. It is important to understand that, while the speed of light is a single value, there are many possible estimates of that value, but that each estimate would contain some error. The hope is that the *average* of these estimates would be quite close to the true speed of light in air.

The dotplot of these values is given in Figure 2.4.

From the table of values, we see that the minimum data value is 299.62, and the maximum value is 300.07.

Again, we begin by computing the *range*:

$$range = MaxValue - MinValue = 300.07 - 299.62 = 0.45.$$

Speed of Light Estimates (1000 km/s)									
299.62	299.65	299.72	299.72	299.72	299.74	299.74	299.74	299.75	299.76
299.76	299.76	299.76	299.76	299.77	299.78	299.78	299.79	299.79	299.79
299.80	299.80	299.80	299.80	299.80	299.81	299.81	299.81	299.81	299.81
299.81	299.81	299.81	299.81	299.81	299.82	299.82	299.83	299.83	299.84
299.84	299.84	299.84	299.84	299.84	299.84	299.84	299.85	299.85	299.85
299.85	299.85	299.85	299.85	299.85	299.86	299.86	299.86	299.87	299.87
299.87	299.87	299.88	299.88	299.88	299.88	299.88	299.88	299.88	299.88
299.88	299.88	299.89	299.89	299.89	299.90	299.90	299.91	299.91	299.92
299.93	299.93	299.94	299.94	299.94	299.95	299.95	299.95	299.96	299.96
299.96	299.96	299.97	299.98	299.98	299.98	300.00	300.00	300.00	300.07

Table 2.1: Speed of light estimates from a classic experiment by Albert Michelson.

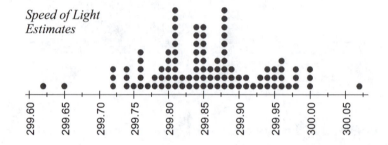

Figure 2.4: The dotplot of the speed of light estimates given in Example 2.1.2.

This time, we will use 16 classes, so the class width is:

$$Class\ Width = \frac{Range}{\#\ of\ Classes} = \frac{0.45}{16} \approx 0.028$$

which we *round up* to *Class Width* = 0.03. Our table will include 16 classes, each with width .03. The minimum value in the data set is 299.62, and that will be our first lower class limit.

Class	Tally	Frequency
299.62 – 299.64	I	1
299.65 – 299.67	I	1
299.68 – 299.70		0
299.71 – 299.73	III	3
299.74 – 299.76	IIII IIII	9
299.77 – 299.79	IIII I	6
299.80 – 299.82	IIII IIII IIII II	17
299.83 – 299.85	IIII IIII IIII III	18
299.86 – 299.88	IIII IIII IIII II	17
299.89 – 299.91	IIII II	7
299.92 – 299.94	IIII I	6
299.95 – 299.97	IIII III	8
299.98 – 300.00	IIII I	6
300.01 – 300.03		0
300.04 – 300.06		0
300.07 – 300.09	I	1

The class boundaries are 299.615, 299.645, 299.675, 299.705, 299.735, 299.765, 299.795, 299.825, 299.855, 299.885, 299.915, 299.945, 299.975, 300.005, 300.035, 300.065, and 300.095.

Using the class boundaries as tic marks on the horizontal scale, we plot the histogram in Figure 2.5. Note again the rough "bell shape" which is so often seen in the distribution of statistical data.

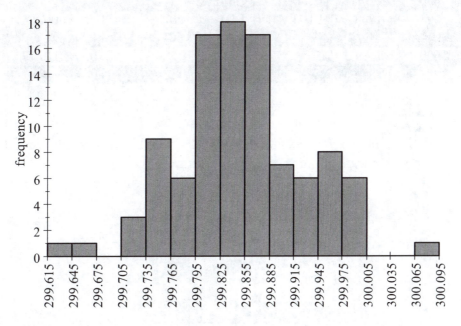

Figure 2.5: The histogram of data values summarized in Example 2.1.2.

There are other shapes of distributions that we will mention from time to time. Some important shapes include *skewed left* (Figure 2.6), *skewed right* (Figure 2.7), *uniform* (Figure 2.8), *bell-shaped with heavy tails* (Figure 2.11), *bell-shaped with light tails* (Figure 2.9), and *bell-shaped with moderate tails* also known as *normal* (Figure 2.10).

Another aspect of the distribution of a data set is the *mode*. The mode of a distribution corresponds to a peak on the distribution's graph. When a distribution has one peak, it is *unimodal*, when there are two peaks as demonstrated in Figure 2.12, it is *bimodal*. If there are more than two peaks, we say that the distribution is *multimodal*.

2.1.2   Relative Frequency Tables and Histograms

In Chapter 3, we will be estimating probabilities using *relative frequencies*. A relative frequency for a given class is equal to frequency, f, divided by sample size, n. Thus,

$$Relative\ Frequency = \frac{f}{n}.$$

If we return to the weekly contributions from Example 2.1.1, we see that the sum of the frequencies is $n = 52$, so when we compute relative frequencies, we divide each frequency by 52.

Class			Tally	Frequency	Relative Frequency
50	–	339	I	1	$1/52 = 0.0192$
340	–	629		0	$0/52 = 0.0000$
630	–	919	IIII IIII IIII IIII	18	$18/52 = 0.3462$
920	–	1209	IIII IIII IIII IIII III	23	$23/52 = 0.4423$
1210	–	1499	IIII I	6	$6/52 = 0.1154$
1500	–	1789	III	3	$3/52 = 0.0577$
1790	–	2079	I	1	$1/52 = 0.0192$
Totals:				$n = 52$	$52/52 = 1.0000$

When we plot the *relative frequency histogram*, the heights of the bars are the relative frequencies, but the rest of the histogram is as before.

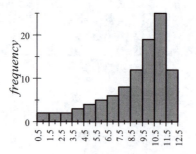

Figure 2.6: A distribution which is *skewed to the left*, or *negatively* skewed.

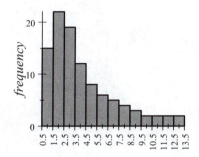

Figure 2.7: A distribution which is *skewed to the right*, or *positively* skewed.

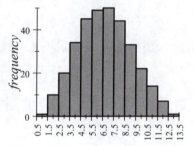

Figure 2.8: A uniform distribution is nearly flat on top.

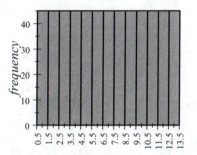

Figure 2.9: This data set is roughly symmetric, has a wide peak and short (or light) tails (indicating few outliers). We will call it *platykurtic*.

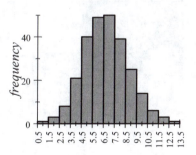

Figure 2.10: This data set is roughly symmetric, has a less wide peak and moderately long tails. It is roughly normal and we will call it *mesokurtic*.

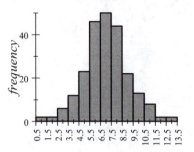

Figure 2.11: This data set is roughly symmetric, has a narrow peak and long (or heavy) tails (indicating outliers). We will call it *leptokurtic*.

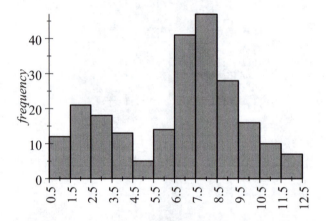

Figure 2.12: A *bimodal* distribution.

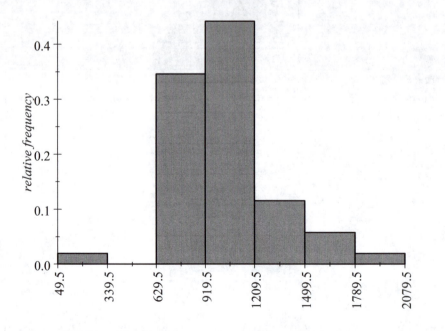

Figure 2.13: A *relative frequency* histogram.

2.1.3 Stem and Leaf Plots

One drawback of histograms or frequency tables is that they hide the data that they represent. While readers have an idea of how the values are distributed, there can still be misunderstandings about the way the values are distributed due to the choices made for the limits on the classes. Sometimes slight modifications to the class definitions can have surprising effects on the histogram. To illustrate, the *speed of light* histogram is reproduced in Figure 2.14, with 13 classes of width 0.035, while another *speed of light* histogram, representing the same data set, with 13 classes, but with width of 0.034, is shown in Figure 2.15. It would be easy to assume that these represent different sets of data, but they do not.

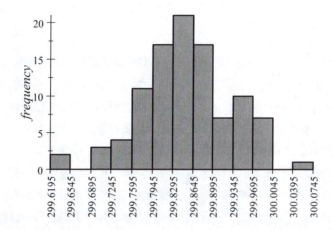

Figure 2.14: The speed of light data, with 13 classes and class width equal to 0.035.

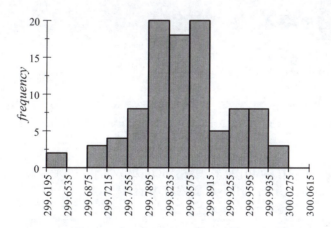

Figure 2.15: The speed of light data, with 13 classes and class width equal to 0.034.

This is one of the problems in any summary – hiding details hides the subtleties in the data. One might wonder if there is a way to display the shape of a distribution of data, while not hiding the data. The best answer to this question is the *stem and leaf* plot.

Example 2.1.1 Continued: The stem and leaf plot below represents the same *weekly contributions* from Example 2.1.1, with class width equal to 100. (Generally speaking, a stem and leaf plot usually has class widths equal to some power of 10). In the plot below, the stems are 100's place, and the leaves are 1's. Look at the stem, 8. With leaves 06, 17, and 43, these represent the values 806, 817, and 843 – all values from the original data set. All of the *weekly contributions* data are represented in the table, so nothing is hidden, plus we have the advantage of being able to see how the values are distributed. If a reader has any questions about the specifics in the data, there is no trouble in looking at the details on the plot to answer those questions. Of course, the drawback of the stem and leaf plot is that it is not as pretty as the histogram. That is just one of the many trade-offs in statistics – the more detailed a description is, the harder it is to look at.

Stem (100s Place)	Leaves (1s Place)
0	53
1	
2	
3	
4	
5	
6	61
7	
8	06, 17, 43, 46, 49, 60, 66, 87, 95, 99
9	00, 06, 08, 12, 14, 17, 19, 32, 39, 47, 51, 53, 60, 80, 87, 91, 93, 94
10	25, 53, 54, 80, 87, 99
11	03, 12, 14, 40, 57, 95
12	24, 26, 47, 71, 87, 87
13	
14	
15	21, 41
16	
17	69
18	
19	
20	75

2.1.4 Pie Charts

While *pie charts* will not be used much in this text, they are useful for describing proportions which a collection of categorical observations make up in a sample.

Example 2.1.3 Below are sample data which shows proportions of births for each day of the week from a random sample of 59,247 births.

Day:	Sunday	Monday	Tuesday	Wednesday	Thursday	Friday	Saturday	Total
Births:	7701	7527	8825	8859	9043	9208	8084	59247
Proportion:	13%	13%	15%	15%	15%	16%	14%	100%

These values are summarized with a pie chart in Figure 2.16. The central angle of each slice of the pie is equal to the corresponding relative frequency, times $360°$. It is interesting to note that the lowest proportions of births occurred on the weekend. In Section 10.1,

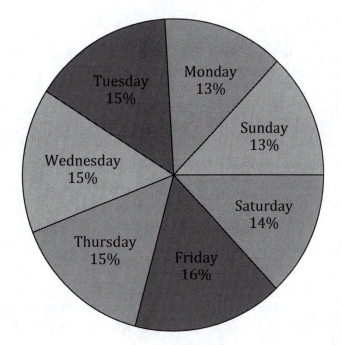

Figure 2.16: The pie chart corresponding to the data presented in Example 2.1.3.

we will present a method which we will use to determine if the lower proportions observed over the weekends is significant enough to be considered more than random fluctuations in these sample data (due to sampling error). (Thanks to Jeffrey S. Simonoff, Professor of Statistics, Leonard N. Stern School of Business, New York University, for providing these values on the StatLib Archive, and permitting their use here. These values come from his book *Analyzing Categorical Data*, published by Springer-Verlag, July 2003).

2.1.5   Homework

For the following problems, give the class width, and class limits for a frequency table with the specified number of classes. Use the given minimum and maximum values.

1. Minimum = 3.9, maximum = 12.7, number of classes = 9.

2. Minimum = 0.051, maximum = 0.934, number of classes = 8.

3. Minimum = 1024, maximum = 3305, number of classes = 6.

4. Minimum = 21.7, maximum = 216.4, number of classes = 7.

For the following data sets, tabulate a frequency table, including frequencies and relative frequencies, and plot a frequency histogram, and a relative frequency histogram. For each example, classify the distribution as, roughly, one of the following: bell-shaped (normal), skewed left, skewed right, uniform, or multimodal. In each case, give a guess for the center, or *average*, of the data set.

5. **Biology**: The journal *Environmental Concentration and Toxicology* published the article "Trace Metals in Sea Scallops" (vol. 19, pp. 326 - 1334), and gave the following cadmium amounts (in mg) in sea scallops observed at a number of different stations in North Atlantic waters.

   {5.1, 14.4, 14.7, 10.8, 6.5, 5.7, 7.7, 14.1, 9.5, 3.7, 8.9, 7.9, 7.9, 4.5, 10.1, 5.0, 9.6, 5.5, 5.1, 11.4, 8.0, 12.1, 7.5, 8.5, 13.1, 6.4, 18.0, 27.0, 18.9, 10.8, 13.1, 8.4, 16.9, 2.7, 9.6, 4.5, 12.4, 5.5, 12.7, 17.1}

6. **Engineering**: The data below consist of lifetimes (in thousands of cycles) of 101 rectangular strips of aluminum subjected to repeated alternating stress at 21,000 psi, 18 cycles per second. These data were taken from an article in the Journal of the American Statistical Association, (1958, p. 159). Use 8 classes. These values are built into the statistics software *Statcato*, which is available for free download at http://www.statcato.org. Alternately, these values are stored in an *Excel* spreadsheet, and may be downloaded at http://math.mtsac.edu/statistics/data.

   {370, 706, 716, 746, 785, 797, 844, 855, 858, 886, 886, 930, 960, 988, 990, 1000, 1010, 1016, 1018, 1020, 1055, 1085, 1102, 1102, 1108, 1115, 1120, 1134, 1140, 1199, 1200, 1200, 1203, 1222, 1235, 1238, 1252, 1258, 1262, 1269, 1270, 1290, 1293, 1300, 1310, 1313, 1315, 1330, 1355, 1390, 1416, 1419, 1420, 1420, 1450, 1452, 1475, 1478, 1481, 1485, 1502, 1505, 1513, 1522, 1522, 1530, 1540, 1560, 1567, 1578, 1594, 1602, 1604, 1608, 1630, 1642, 1674, 1730, 1750, 1750, 1763, 1768, 1781, 1782, 1792, 1820, 1868, 1881, 1890, 1893, 1895, 1910, 1923, 1940, 1945, 2013, 2023, 2100, 2215, 2268, 2440}.

7. **Meteorology**: The following are rainfall totals (in inches) for clouds chemically seeded to increase rainfall amounts. (Data are taken from Miller, A.J., Shaw, D.E., Veitch, L.G. & Smith, E.J. (1979). "Analyzing the results of a cloud-seeding experiment in Tasmania", Communications in Statistics - Theory & Methods, vol.A8(10), pp. 1017-1047). Use 8 classes. These values are built into the statistics software *Statcato*, which is available for free download at http://www.statcato.org. Alternately, these values are stored in an *Excel* spreadsheet, and may be downloaded at http://math.mtsac.edu/statistics/data.

{0.05, 0.07, 0.13, 0.24, 0.26, 0.36, 0.37, 0.38, 0.40, 0.41, 0.42, 0.47, 0.58, 0.59, 0.61, 0.71, 0.76, 0.79, 0.80, 0.80, 0.81, 0.84, 0.86, 0.88, 0.90, 0.90, 0.91, 1.00, 1.08, 1.09, 1.11, 1.16, 1.23, 1.25, 1.32, 1.43, 1.46, 1.48, 1.50, 1.62, 1.63, 1.67, 1.83, 1.87, 2.04, 2.06, 2.08, 2.16, 2.21, 2.22, 2.25, 2.26, 2.35, 2.36, 2.39, 2.48, 2.56, 2.79, 2.87, 2.97, 3.17, 3.25, 3.31, 3.56, 4.16, 4.24, 4.29, 4.61, 5.48, 6.00}

8. **Politics**: In the 2000 U.S. presidential elections, the inaccuracy of voting technology became the subject of worldwide interest. In Florida, the two front-runner candidates, George W. Bush and Al Gore, were so close in the tally that votes had to be recounted by hand to verify the totals. In the process of recounting, it was found that the most unreliable voting machine was the *Votomatic*, with the most under-counted (votes placed but not counted) and over-counted (votes counted that were never placed) votes of any other technology. The under-counted votes for 15 different Florida counties are listed below. Use 5 classes.

{4946, 2070, 5090, 466, 5431, 1044, 1975, 2410, 10570, 634, 10134, 1763, 4240, 1846, 596}.

9. **Politics**: The over-counted votes for the same *Votomatic* machines are listed below. Again, use 5 classes.

{7826, 1134, 21855, 520, 3640, 790, 2531, 890, 17833, 1039, 19218, 2124, 4258, 994, 170}.

10. **Economics**: The following are yearly income levels for twenty randomly selected Alaska residents for the year 1999 (selected randomly from the 2000 U.S. Census): Use 6 classes.

{9000, 10600, 60000, 80000, 5000, 2600, 14800, 4500, 45000, 23000, 21300, 50000, 17000, 49000, 85000, 38000, 64000, 25000, 23000, 55000}.

11. **College Demographics**: The following values are hours worked per week for randomly selected students at Mt. San Antonio College in Walnut, California. Use 5 classes.

{15, 20, 40, 16, 20, 25, 19, 40, 40, 20, 20, 40, 0, 40, 35, 30, 35, 25, 29, 0, 0, 40, 16, 0, 40, 30, 25, 16, 35, 0}.

12. **College Demographics**: The following are grade point averages for the same students whose work hours are listed above. Use 6 classes.

{2.7, 1.7, 2.0, 3.7, 3.1, 4.0, 3.4, 3.7, 2.5, 2.8, 3.0, 3.6, 3.3, 3.2, 3.1, 3.2, 2.6, 2.8, 2.0, 4.0, 2.7, 3.9, 3.1, 2.5, 3.7, 2.9, 3.5, 3.7, 2.2, 3.8}

Data suggest that the average *eye gain ratio* of people with schizophrenia is significantly different from the average for people without this disorder. An *eye gain ratio* is the ratio of an observer's angular eye velocity to the angular velocity of an object that the eye is following.

13. **Psychology:** For this problem, load the schizophrenic patient eye gain ratio observations, with included control group, in *Statcato* (available at *www.statcato.org*), or open the corresponding spreadsheet at *http://math.mtsac.edu/statistics/data*. Construct a frequency table and histogram for the *schizophrenic patients*, using 10 classes. Give a description of the approximate shape of the distribution, along with a guess of the average value.

14. **Psychology:** For this problem, load the schizophrenic patient eye gain ratio observations, with included control group, in *Statcato* (available at *www.statcato.org*), or open the corresponding spreadsheet at *http://math.mtsac.edu/statistics/data*. Construct a frequency table and histogram for the *control group*, using 10 classes. Give a description of the approximate shape of the distribution, along with a guess of the average value.

2.2 A First Look at Statistical Inference

An important goal of an elementary statistics course is the development of tools with which we make compare data from multiple samples and make inferences about their populations. In this section, we make our first attempt at such comparisons. In doing so, we begin to learn the structure of the process of statistical inference.

Whenever we wish to compare two samples drawn from, for example, treatment and control groups, our first and easiest analysis can be made by comparing their corresponding dotplots. Parallel dotplots of multiple samples drawn above number lines of the same scale allow us to make comparisons of two or more distributions of values. With such comparisons, we are sometimes able to make informal inferences with regard to the populations from which the respective samples are drawn.

2.2.1 Comparing Two Samples with Stacked Dotplots

When using dotplots to make comparisons of multiple distributions, the following questions should be addressed.

Procedure 2.2.1 Comparing distributions of values using stacked dotplots with a common coordinate system.

Step I. **Hypothesis**

Hypothesis generation may be accomplished at any point in the comparison process. At some point, the observer begins to make assumptions regarding the relationship between the various samples being studied. Such assumptions may compare the data sets' *centers* (*averages*) or their *spread* or *variability*.

Step II. **Summary**

The primary summary statistics which should be compared, for now, are the minimum values, the maximum values, and an informal evaluation of the distributions' centers. Comparing these values, respectively, in context can often lead us to making inferences about the samples.

Step III. **Shift**

Compare the entire collection of values from each sample. Are the distributions shifted left or right of one another as they are plotted above a common axis? What does this mean in the context of the data?

Step IV. **Overlap**

From the graphs, consider how much of the sample distributions overlap one another. Are all of the values in one sample larger than all of the values in the other sample, or do they overlap? How does this influence any inferences made regarding the centers of the distributions of values?

Step V. **Spread**

In examining the amount of spread or variability, ask which distribution of values is wider? How does the variability of each sample effect the certainty that you have in your inference? In answering this question, remember that the more variability in a given sample, the more error we tend to see in corresponding statistics.

Step VI. **Sample Size**

To counter the uncertainty that is brought about by increasing variability or spread in a data set, it is important to remember that larger data sets give added certainty the statistics which are measured from it. The more data points in a given dotplot, the more sure we can be about our observations.

Step VII. **Meaning in Context**

After making visual comparisons regarding a collection of samples using their dotplots, it is important to remember that all comparisons should be interpreted in the context of the populations that they represent. What is the real-world meaning behind the numbers and inferences made?

Step VIII. **Extreme Cases**

In many datasets, *outliers* often exist as values that are extremely large or small relative to the rest of the data set. What do those extreme values mean in context? Can they be considered reliable? Given that outliers can have a great deal of influence on measures of center and spread, how influential have outliers been in any conclusions or inferences made?

Step IX. **Inferences**

With all of the considerations made above, we may choose to carefully make informal inferences regarding the centers of the data sets, or their spread. In making inferences, it is important to remember that these inferences are made from sample data, which provide only a partial picture of what actually may be true. In light of this, it is important to make such inferences using qualifying statements such as "the data appear to suggest...".

We next provide an example which will allow us to demonstrate the consideration of the steps outlined above.

Example 2.2.2 Below are heights (in cm) for randomly selected adult females and for randomly selected adult males.

Females: {148, 152, 155, 153, 155, 156, 157, 158, 158, 156, 159, 160, 164, 158, 160, 161, 162, 158, 160, 157, 162, 160, 167, 167, 168, 164, 170, 176, 176, 184}

Males: {168, 164, 160, 165, 167, 171, 171, 169, 169, 172, 168, 173, 170, 176, 174, 173, 178, 179, 177, 176, 179, 175, 176, 182, 178, 179, 176, 178, 177, 189}

In Figure 2.17, dotplots are provided for the female and male heights, with a common coordinate system. We would like to use the process outlined above to compare these samples, and possibly make an inference which compares average male and female heights.

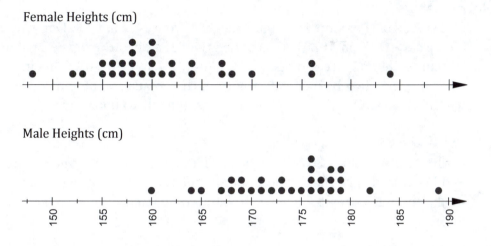

Figure 2.17: The dotplots above share a common coordinate system, representing the heights provided in Example 2.2.2.

Step I. **Hypothesis**

A simple hypothesis which compares the centers of the given data sets seems rather obvious from the scatterplots above. We hypothesize that the average height for female adults is smaller than the average height of male adults.

Step II. **Summary**

An informal evaluation of the centers of these distributions would place the center of the female distribution at around 160 cm, while the male distribution seems to be centered at around 175 cm.

Step III. **Shift**

Both distributions of values are somewhat similar, but contrasts can be made. The distribution of all female heights seems to be shifted to the left of the distribution of male heights. The distributions are otherwise somewhat similar, being moderately bell-shaped with slight skewing in each case. In context, this helps us to see that the population of female heights appears shorter overall in comparison to the population of male heights which seems taller overall.

Step IV. **Overlap**

While the female heights are shifted to the left of the male heights, clearly there is a fair amount of overlap. There is a female who is 184 cm tall, and she is taller than all males, except one. There is a male who is 160 cm, and is shorter than many females observed. While there is a shift to the left for females, it is not as simple as saying that "females are shorter than males," when there are many exceptions to this rule.

Step V. Spread

The spread of the sample of female heights (from 148 cm to 184 cm with a range of 184 cm − 148 cm = 36 cm) is wider than the spread of male heights (from 160 cm to 189 cm with a range of 189 cm − 160 cm = 29 cm). This implies that there may possibly be more variability in the population of female heights, but with such a small collection of values it is difficult to know this with certainty.

Step VI. Sample Size

Both samples include 30 observations. These are relatively small sample sizes, and because of this, the evidence for any inferences that we will make is rather weak, and our estimates will likely be accompanied by rather large sampling error.

Step VII. Meaning in Context

The primary comparison we have made involves average heights. It seems reasonable that the center of the female distribution is smaller than the center of the male distribution, and this is a notion that most of us will be comfortable with. The idea makes sense in context, and the data seem to support it.

Step VIII. Extreme Cases

There is a tallest female height (184 cm) which is somewhat separated from the others by a little less than 10 cm. We have no reason to believe that there is anything unreliable about this observation. If this value was removed from the data set, the center of the data set would be moved to the left, so it is somewhat influential. This would not, however, alter our inference that the average height for females is smaller than that for males. The outlier does influence our inference, but in this case, it does not alter it. There is also a tallest male in the group, but the separation of this value from the rest of the male heights seems less extraordinary, and is unlikely to have much influence on any decisions we might make.

Step IX. Inferences

With all of the considerations made above, there is no clear reason that our inference regarding the average heights of adult females in comparison to adult males. The data do appear to suggest that the average height of females is indeed smaller than the average height for males.

The procedure above for comparing distributions is an important beginning for us in the discipline of statistical inference. The homework for this section provides a collection of problems with which you will have the opportunity to refine your understanding of this critical skill.

2.2.2 Homework

Use Procedure 2.2.1 to compare the samples using stacked dotplots of the samples provided in the problems below.

1. The following are average tip rates for a random selection of female and male food servers. Compare the samples.

   Females: {14%, 10%, 14%, 12%, 15%, 13%, 13%, 13%, 14%, 14%, 13%, 12%, 17%, 12%, 12%, 14%, 15%, 12%, 11%, 14%, 13%, 13%, 8%, 7%}

   Males: {10%, 11%, 12%, 13%, 11%, 11%, 12%, 14%, 15%, 10%}

2. The following are grade point averages for smoking and non-smoking community college students. Compare the samples.

   Smokers: {2.0, 2.0, 2.0, 2.0, 2.1, 2.2, 2.4, 2.4, 2.4, 2.4, 2.5, 2.5, 2.5, 2.5, 2.5, 2.5, 2.6, 2.6, 2.6, 2.6, 2.6, 2.6, 2.6, 2.7, 2.7, 2.7, 2.7, 2.7, 2.8, 2.8, 2.8, 2.8, 2.8, 2.8, 2.8, 2.8, 2.9, 2.9, 2.9, 3.0, 3.0, 3.0, 3.0, 3.1, 3.1, 3.1, 3.2, 3.2, 3.2, 3.2, 3.2, 3.3, 3.3, 3.3, 3.4, 3.4, 3.4, 3.4, 3.5, 3.5, 3.5, 3.5, 3.6, 3.6, 3.6, 3.6, 3.6, 3.7, 3.7, 3.7, 3.7, 3.8, 3.8, 3.8, 3.8, 3.8, 3.9, 4.0, 4.0}

   Non-smokers: {2.3, 2.6, 2.6, 2.7, 2.7, 2.7, 2.8, 2.8, 2.8, 2.8, 2.8, 2.9, 2.9, 2.9, 3.0, 3.0, 3.0, 3.0, 3.0, 3.0, 3.0, 3.0, 3.1, 3.1, 3.2, 3.2, 3.2, 3.3, 3.3, 3.3, 3.3, 3.3, 3.3, 3.4, 3.4, 3.4, 3.4, 3.4, 3.5, 3.5, 3.5, 3.5, 3.5, 3.6, 3.6, 3.6, 3.6, 3.7, 3.7, 3.7, 3.7, 3.7, 3.7, 3.7, 3.7, 3.8, 3.8, 3.8, 3.8, 3.8, 3.8, 3.8, 3.8, 3.9, 3.9, 3.9, 3.9, 3.9, 3.9, 3.9, 3.9, 3.9, 3.9, 3.9, 4.0, 4.0, 4.0, 4.0, 4.0, 4.0, 4.0}

3. Below are foot lengths (in cm) for male and female adults. Compare the samples.

   Male feet: {24.4, 24.3, 25.4, 25.8, 26.3, 24.7, 25.1, 24.6, 26.1, 26.8, 25.1, 25.8, 26.4, 25.7, 25.6, 26.2, 26.2, 26.8, 26.4, 26.8, 26.3, 26.4, 24.5, 25.6, 25.7, 24.9, 27.0, 27.8, 27.6, 27.6}

   Female feet: {18.2, 20.4, 21.5, 20.4, 21.1, 19.1, 20.6, 20.4, 21.7, 21.6, 23.0, 20.9, 20.0, 21.8, 21.0, 22.3, 22.7, 22.8, 21.8, 23.7, 21.3, 22.7, 22.1, 22.6, 22.0, 21.6, 22.8, 22.1, 24.2, 23.7}

4. Below are lipoprotein levels for subjects who eat corn flakes every morning for breakfast, and also for subjects who eat oat bran (Pagano & Gauvreau, 1993, p. 252). Compare the samples.

   Corn Flakes: {4.6, 6.4, 5.4, 4.5, 4.0, 3.8, 5.0, 4.3, 3.8, 4.6, 5.4, 3.9, 2.3, 4.2}

   Oat Bran: {3.8, 5.6, 5.9, 4.8, 3.7, 3.0, 4.4, 3.7, 3.5, 3.8, 5.3, 3.7, 1.8, 4.1}

5. When college class locations are inconvenient for students, does student success decline? The following samples include student success rates for two different populations teachers – those whose classrooms are in a convenient location, and those whose classrooms are in an inconvenient location. Compare the samples.

Convenient: {37%, 51%, 47%, 61%, 63%, 59%, 32%, 70%, 46%, 44%, 68%, 68%, 56%, 89%, 76%, 32%, 57%, 61%, 79%}

Inconvenient: {42%, 66%, 46%, 57%, 73%, 63%, 39%, 66%, 64%, 55%, 61%, 34%, 36%, 44%, 25%, 44%, 30%, 44%, 69%}

6. Are grocery prices higher in more affluent neighborhoods? Thirty-one randomly selected products are purchased in two grocery stores – first in an affluent neighborhood, then again with the same products in a non-affluent neighborhood. Both stores are members of the same grocery store franchise. Compare the samples.

Affluent: {$3.00, $2.80, $3.30, $3.00, $1.80, $1.40, $4.70, $3.80, $1.80, $2.10, $1.10, $1.10, $2.80, $1.90, $0.90, $0.80, $1.40, $2.30, $1.80, $1.90, $0.20, $2.40, $1.30, $3.00, $2.80, $1.30, $1.70, $0.90, $0.70, $1.70, $1.80}

Non-Affluent: {$3.00, $2.80, $3.30, $3.10, $2.00, $1.10, $4.40, $3.30, $1.60, $1.80, $0.90, $0.90, $2.60, $1.30, $0.90, $0.80, $1.40, $2.10, $1.60, $1.90, $0.30, $2.00, $1.00, $3.30, $3.00, $1.40, $1.60, $0.90, $0.70, $1.70, $1.60}

7. Data suggest that the average *eye gain ratio* of people with schizophrenia is significantly different from the average for people without this disorder. An *eye gain ratio* is the ratio of an observer's angular eye velocity to the angular velocity of an object that the eye is following.

Compare the gain ratios of the eyes of schizophrenic people to the control group provided in Statcato (available at www.statcato.org), or using the Excel spreadsheets at http://math.mtsac.edu/statistics/excel.

2.3 Summarizing Data Using Averages

Our next goal of descriptive statistics is to provide some measurement of the location of the *center* of the distribution of values in a data set. Exactly what is meant by the "center" is open to interpretation, and we will examine a few important attempts to answer this question.

Before we begin, it is necessary to present some notational conventions which we will follow throughout this book.

Sample size will be denoted by n, while population size will be N. We will use x to denote each value, in turn, in a data set. In addition to these values, there is the summation operator used in mathematics, Σ, which denotes of the sum of whatever expression follows it. So Σx represents the sum of all values, x, in our data set.

2.3.1 Parameters and Statistics

It is probably a good idea to recall two definitions from Chapter 1, which will become more important in this section. A *parameter* is a numerical measurement computed using data from an entire population. Because populations tend to be extraordinarily large in the context of statistics, we rarely know the value of any parameter. A *statistic* is a numerical measurement computed using data from a sample. Statistics are used to estimate parameters.

Whenever possible, we like to use statistics which are *unbiased estimators* of their corresponding population parameters. An estimator is unbiased when, though always in error, the average of all possible estimates is equal to the parameter that they are estimating.

Additionally, in realizing that estimates vary, we like our estimators to vary as little as possible. In the following section of this text, we will introduce our most important measure of variation, and our goal will be for this measure to be as small as possible for any statistic that we use.

In this section, we begin discussing specific statistics for measuring the center of a data set's distribution, and their corresponding parameters. The first and most important measurement of a distribution's center is the *arithmetic mean*.

2.3.2 The Arithmetic Mean

The idea for the *arithmetic mean* comes from the center of mass formulas in physics. The arithmetic mean of a data set is the center of mass of the data set. This means that if a data set is represented by a dotplot where the dots have physical mass, then the plot could be balanced at a point on the horizontal axis. The balancing point would be at the *center of mass*, which we call the *arithmetic mean*.

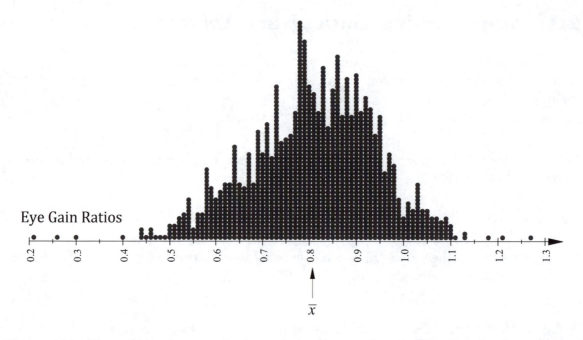

Figure 2.18: The arithmetic mean is the *center of mass* of a distribution of values of equal weight.

On the dotplot in Figure 2.18, the arithmetic mean/center of mass is represented by the small arrow pointing to the horizontal axis. If the dotplot was a physical object, it would balance perfectly at this point.

The *arithmetic mean* will be referred to as, more simply, the *mean*.

The parameter of interest is the *population mean*. The symbol for this will be μ, a Greek lower case letter, pronounced "*mu*." Because populations tend to be so large, there is really no possibility of computing μ. Still, there are formulas for μ. The simplest formula for μ applies to a finite population of size N, and is given below. We will often refer to the population mean as $E(x)$, the *expected value* of the variable x. The symbols $E(x)$ and μ are interchangeable, and we will use both from time to time.

Definition 2.3.1 The population mean, also known as the *expected value* of x, from a finite population of size N is given by the formula below.

$$\mu = E\left(x\right) = \frac{\sum x}{N}$$

Remember that the symbol, $\sum$, is the Greek uppercase letter *sigma*, meaning "*the sum of all ...*". Thus, $\sum x$ represents the sum of all values, x, in our data set.

There are other ways of computing μ, especially when populations are effectively "infinite" (where it is impossible to enumerate all possible values in a population), and some of these we will cover in Chapter 4, while other such methods require more advanced math than we assume for this course, so they will not be discussed.

2.3.3   Estimating μ with a Sample Mean

Since we are unlikely to ever compute μ due to the size of populations, we are forced to estimate it using a statistic. Since statistics always contain sampling error, it is unlikely that our estimate will be exactly the value of μ, so we will give it a different symbol. When we estimate a population mean using an arithmetic mean from a sample of size n, the mean is called a *sample mean*, and represented by the symbol $\bar{x}$ (pronounced "*x-bar*"). The formula for this sample mean is $\bar{x} = \frac{\sum x}{n}$.

Definition 2.3.2 The *sample mean* is given by

$$\bar{x} = \frac{\sum x}{n}.$$

The sample mean, $\bar{x}$, is an *unbiased* estimator of the population mean, μ.

As mentioned, it is our goal to use statistics that are unbiased and that vary as little as possible. The sample mean is an unbiased estimator of μ, and when a sample is roughly normal, or bell-shaped, it varies less than other estimators of μ. In that sense, we will consider the sample mean to be the *best point estimate* of the population mean.

Example 2.3.3 The following values represent maximum elevations (in meters) of a sample of islands in the Galapagos Archipelago: {109, 114, 46, 119, 93, 168, 112, 198, 1494, 49, 227, 76, 1707, 343, 25, 777, 458, 367, 716, 906, 864, 259, 640, 186, 253}

In this sample, there are $n = 25$ elevations, and the sample mean is computed below:

$$\begin{aligned}
\bar{x} &= \frac{\sum x}{n} \\
&= \frac{109+114+46+119+93+168+112+198+1494+49+227+76+1707+343+25+777+458+367+716+906+864+259+640+186+253}{25} \\
&= \frac{10306}{25} \\
&\approx 412.24
\end{aligned}$$

Now that we are actually computing something, it is time to mention a few issues regarding the rounding of answers. First of all, it is important to remember that rounding an answer is the same as introducing error into the answer. The more we round an answer, the larger the error will be. We must be careful not to round too much, so we introduce the following rounding rule.

Procedure 2.3.4 Rounding Rule for a Sample Mean
When rounding a mean, carry at least one place more after the decimal than present in the given data set. Try to not round any values until all calculations are complete.

Of course, the last sentence of the above statement may be impossible, but still we want to avoid rounding intermediate values as much as possible, because each rounding that we do makes the final answer more and more incorrect. This building up of errors in rounding is called *error propagation*.

The strengths of the arithmetic mean are that the formula is easy and well recognized, and that the average takes every value in the data set into account. The main drawback of the mean is that, in taking every datum into account, it is sensitive to extreme values, also known as *outliers*. Given that outliers are often erroneous, this sensitivity can be a problem. Outliers can be quite influential over the various computations which are used in statistics – do we really want these unstable values to have such influence over our statistics? One attempt to solve this problem of outliers is the *trimmed mean*.

2.3.4   Estimating μ with a Trimmed Mean

When the shape of a sample's distribution has *heavy tails* (with a large proportion of values distributed far from the mean), there is a better estimator of μ, called the *trimmed mean*. In such a case, the trimmed mean will generally vary less than $\bar{x}$, and it is still an unbiased estimator of μ. A trimmed mean is computed by first sorting the sample from low to high, then deleting a specified number of values from the beginning of the list, and then deleting the *same number* of values from the end of the list. After this, the arithmetic mean is computed from the remaining values.

Example 2.3.3 Continued: The Galapagos Islands elevations are sorted as follows: {25, 46, 49, 76, 93, 109, 112, 114, 119, 168, 186, 198, 227, 253, 259, 343, 367, 458, 640, 716, 777, 864, 906, 1494, 1707}. If we wish to trim 10% of the smallest values (that is 2.5 which we round to 3 values) and 10% of the largest values (another 3 values), the data set becomes: {76, 93, 109, 112, 114, 119, 168, 186, 198, 227, 253, 259, 343, 367, 458, 640, 716, 777, 864}, and our mean now will only include $n = 19$ values. The trimmed mean is therefore:

$$\frac{\sum x}{n} = \frac{6079}{19} = 319.947.$$

The advantage of the trimmed mean is that it is not so easily influenced as the sample mean by extreme, and possibly erroneous, values. Again, the trimmed mean is a better estimate of μ only when the sample has heavy tails (indicated by a collection of highly extreme values). When samples appear bell-shaped or normal, the best estimate is the sample mean, $\bar{x}$.

2.3.5   Estimating a Sample Mean Using a Frequency Table

When data are summarized by a frequency table, but are not available for one reason or another, it is still possible to estimate the sample mean of the unknown values. We do this by using class midpoints from each class to represent values in those classes. The variable, x, will represent the midpoint of each class. It is important to note that each midpoint must be counted a number of times, according to the frequency, f, of that midpoint's respective class. So, we should add each x to itself f times, or we may equivalently multiply each x by its respective frequency f, then add these values together to get $\sum f \cdot x$. Next, we divide by the sample size. Because each x occurs f times, the sample size is the total of the frequencies, so $n = \sum f$.

Definition 2.3.5 To estimate a sample mean using a frequency table, use the formula $\bar{x} \approx \frac{\sum f \cdot x}{n}$, where $n = \sum f$.

In estimating a mean in this fashion, it is useful to add another column to the frequency table, labeled $f \cdot x$, and containing this product for every row in the table. To estimate the sample mean, we will first need to total the f column and the $f \cdot x$ column.

Example 2.3.6 If we return to the weekly contributions from Example 2.1.1, we have the frequency table below, with another column added for the products $f \cdot x$.

Class		$f = frequency$	$x = midpoint$	$f \cdot x$
50	– 339	1	194.5	194.5
340	– 629	0	484.5	0.0
630	– 919	18	774.5	13941.0
920	– 1209	23	1064.5	24483.5
1210	– 1499	6	1354.5	8127.0
1500	– 1789	3	1644.5	4933.5
1790	– 2079	1	1934.5	1934.5
Totals:		52	—	53614.0

With the totals above we have $\sum f \cdot x = 53614$ and $n = \sum f = 52$, and we may estimate the sample mean.

$$\bar{x} \approx \frac{\sum f \cdot x}{n} = \frac{53614}{52} = 1031.04$$

In rounding, note again that in averaging the midpoints of each class (which include one place after the decimal), we have rounded our average to two places after the decimal – one place more than the class midpoints.

2.3.6 The Median

The *median* of a sample (represented by the symbol, $\tilde{x}$), loosely speaking, is the value in the middle of the sample, after the sample is sorted. That sounds easy, but in reality, the problem can be more complicated when there is not a single value in the middle of a list. Whenever the list has an even number of values, there are *two* middle values, and in this case, the median will be the mean of these values.

To compute the median of a data set, the following procedure should help.

Procedure 2.3.7 Computing the Median, $\tilde{x}$, of a Data Set of Size n

  1. Sort the data set from low to high.

  2. If n is odd, the median is the value in position $(n+1)/2$ in the data set.

  3. If n is even, the median is the mean of the values in positions $n/2$ and $n/2 + 1$.

Example 2.3.3 Continued: Returning to the original $n = 25$ *sorted* elevations of the Galapagos Islands, the median is in position $(n+1)/2 = (25+1)/2 = 26/2 = 13$ in the data set. The 13th position is highlighted in the list below:

$$\{25, 46, 49, 76, 93, 109, 112, 114, 119, 168, 186, 198, \mathbf{227}, 253, 259, 343, 367, 458, 640,$$
$$716, 777, 864, 906, 1494, 1707\}.$$

The value in the 13th position of this data set is 227, so the median is $\tilde{x} = 227$.

Example 2.3.8 As another example, consider the sample of miles per gallon of measurements for a 2005 Toyota Prius Hybrid vehicle, for ten round-trip commutes from Walnut, California to Lake Arrowhead, California: $\{46.2, 44.9, 47.0, 46.3, 45.6, 46.0, 44.6, 45.9, 46.3, 45.6\}$. Again, to compute the median, we first sort the values. The result is: $\{44.6, 44.9, 45.6, 45.6, 45.9, 46.0, 46.2, 46.3, 46.3, 47.0\}$. Since $n = 10$, an even number, the median is the mean of the values in positions $n/2 = 10/2 = 5$, and $n/2 + 1 = 10/2 + 1 = 5 + 1 = 6$. The values in positions 5 and 6 are 45.9 and 46.0. The median is the mean of these values. So, the median is

$$\tilde{x} = \frac{45.9 + 46.0}{2} = 45.95.$$

The importance of the median is that it is not easily influenced by extreme values in a data set, making it a highly stable statistic, and we will use it periodically as we proceed. The difficulty with the median is that it has no formula which can be used to study it analytically. There are formulas for median *position*, but not for the median itself. Procedures exist for making statistical inferences regarding medians, but these are left for later.

2.3.7   The Midrange

The *midrange* of a data set is the mean of the minimum and maximum values in a data set, thus,

$$midrange = \frac{maximum + minimum}{2}.$$

The midrange is simple and quick, but since it only considers two values in a data set, it is highly sensitive to extreme values (it uses the two most extreme values in the set), and is not generally used in statistics.

2.3.8   The Mode

The *mode* of a data set is the value that occurs most frequently. When two or more values occur with an equal highest frequency, the data set has two or more modes and is bimodal or multimodal. If no value repeats, the set has no mode. Referring to the Toyota Prius mileage values in Example 2.3.8 above, the value 45.6 occurs twice, and 46.3 also appears twice. Thus there are two modes, and the sample is bimodal.

2.3.9   The Weighted Mean

The *weighted mean* is the formula for center of mass again. To develop this, we need only adapt the formula for a mean from a frequency table, $\bar{x} \approx \frac{\sum fx}{n}$, where $n = \sum f$. Simply put, the frequencies are weights in this formula – we are counting each value x a total of f times by multiplying x times f. In doing this, values of x with higher frequencies are counted with more *weight*, so in some sense, they are more important. If we call the frequencies weights, and denote them by w, the formula becomes *weighted mean* $= \frac{\sum w \cdot x}{\sum w}$. The denominator was $n = \sum f$, but this is replaced by $\sum w$.

Definition 2.3.9 The *weighted mean* is computed using the following formula:

$$Weighted\ Mean = \frac{\sum w \cdot x}{\sum w}.$$

Here, the weights, w, are values assigned to give more influence to more important values of x.

Example 2.3.10 Example 2.2.4: As an example of this, suppose that an instructor computes an average grade according to the following weights: 3 midterms at 20% each, homework at 15%, and a final exam at 25%. Next, suppose a student has 75%, 89%, and 92% on her midterms, 98% on her homework, and 86% on her final exam. What is her overall weighted mean?

We will use 20, 20, 20 as weights for the 20% midterms, 15 as the weight for the 15% homework, and 25 for the 25% final. The student's percentage scores are converted to decimal form in the table below.

x	w	$w \cdot x$
0.75	20	15.0
0.89	20	17.8
0.92	20	18.4
0.98	15	14.7
0.86	25	21.5
Totals :	100	87.4

We may now compute her weighted mean.

$$Weighted\ Mean = \frac{\sum w \cdot x}{\sum w}$$
$$= \frac{87.4}{100}$$
$$= 0.874$$
$$= 87.4\%$$

2.3.10   The Quadratic Mean

The *quadratic mean* of a collection of *positive* values is computed by first computing the mean of the squares of the values, then taking the square root of this result.

$$Quadratic\ Mean = \sqrt{\frac{\sum x^2}{n}}$$

This gives a slightly different result from the arithmetic mean, and is used mainly when averaging a collection of deviations from the mean, which we will discuss in the next section of this text.

In this section, we have introduced the summation operator, $\sum$, and thus we must introduce its properties, as they will be encountered throughout this text.

Proposition 2.3.11 The properties of the summation are given below. In these properties, x and y are assumed to vary from term to term, but k is a constant.

1. The sum of all values of x plus (or minus) y is equal to the sum of all values of x plus (or minus) the sum of all values of y.

$$\sum (x \pm y) = \sum x \pm \sum y$$

2. The sum of a constant, k, times all values of the variable, x, is k times the sum of all values of the variable, x.

$$\sum kx = k \sum x$$

3. When the constant, k, is added to itself n times, the sum is nk.

$$\sum_{i=1}^{n} k = nk$$

2.3.11 Homework

Compute the mode, midrange, median, and mean for each of the data sets below.

1. Data set: { 2, 5, 5, 11, 15 }

2. Data set: { 4, 6, 11, 25, 26, 26 }

3. Data set: {68.94, 97.56, 92.56, 64.75, 88.67, 80.76, 62.06, 69.50, 84.73, 78.31, 89.35, 96.18, 80.36, 95.70, 84.80, 82.37, 91.11}

4. Data set: { 66, 116, 121, 81, 114, 81, 69, 85, 95, 93, 119, 122, 99, 125, 95, 92 }

5. Biology: The journal *Environmental Concentration and Toxicology* published the article "Trace Metals in Sea Scallops" (vol. 19, pp. 326 - 1334), and gave the following cadmium amounts (in mg) in sea scallops observed at a number of different stations in North Atlantic waters. The values are given below.

{5.1, 14.4, 14.7, 10.8, 6.5, 5.7, 7.7, 14.1, 9.5, 3.7, 8.9, 7.9, 7.9, 4.5, 10.1, 5.0, 9.6, 5.5, 5.1, 11.4, 8.0, 12.1, 7.5, 8.5, 13.1, 6.4, 18.0, 27.0, 18.9, 10.8, 13.1, 8.4, 16.9, 2.7, 9.6, 4.5, 12.4, 5.5, 12.7, 17.1}

6. Engineering: The data below consist of lifetimes (in thousands of cycles) of 101 rectangular strips of aluminum subjected to repeated alternating stress at 21,000 psi, 18 cycles per second. This data were taken from an article in the Journal of the American Statistical Association, (1958, p. 159). These values are built into the statistics software *Statcato*, which is available for free download at http://www.statcato.org. Alternately, these values are stored in an *Excel* spreadsheet, and may be downloaded at http://math.mtsac.edu/statistics/data.

{370, 706, 716, 746, 785, 797, 844, 855, 858, 886, 886, 930, 960, 988, 990, 1000, 1010, 1016, 1018, 1020, 1055, 1085, 1102, 1102, 1108, 1115, 1120, 1134, 1140, 1199, 1200, 1200, 1203, 1222, 1235, 1238, 1252, 1258, 1262, 1269, 1270, 1290, 1293, 1300, 1310, 1313, 1315, 1330, 1355, 1390, 1416, 1419, 1420, 1420, 1450, 1452, 1475, 1478, 1481, 1485, 1502, 1505, 1513, 1522, 1522, 1530, 1540, 1560, 1567, 1578, 1594, 1602, 1604, 1608, 1630, 1642, 1674, 1730, 1750, 1750, 1763, 1768, 1781, 1782, 1792, 1820, 1868, 1881, 1890, 1893, 1895, 1910, 1923, 1940, 1945, 2013, 2023, 2100, 2215, 2268, 2440}

7. Meteorology: The following are rainfall totals (in inches) for clouds chemically seeded to increase rainfall amounts. (Data are taken from Miller, A.J., Shaw, D.E., Veitch, L.G. & Smith, E.J. (1979). "Analyzing the results of a cloud-seeding experiment in Tasmania", Communications in Statistics - Theory & Methods, vol.A8(10), pp. 1017-1047). These values are built into the statistics software *Statcato*, which is available for free download at:

http://www.statcato.org.

Alternately, these values are stored in an *Excel* spreadsheet, and may be downloaded at http://math.mtsac.edu/statistics/data.

{0.05, 0.07, 0.13, 0.24, 0.26, 0.36, 0.37, 0.38, 0.40, 0.41, 0.42, 0.47, 0.58, 0.59, 0.61, 0.71, 0.76, 0.79, 0.80, 0.80, 0.81, 0.84, 0.86, 0.88, 0.90, 0.90, 0.91, 1.00, 1.08, 1.09, 1.11, 1.16, 1.23, 1.25, 1.32, 1.43, 1.46, 1.48, 1.50, 1.62, 1.63, 1.67, 1.83, 1.87, 2.04, 2.06, 2.08, 2.16, 2.21, 2.22, 2.25, 2.26, 2.35, 2.36, 2.39, 2.48, 2.56, 2.79, 2.87, 2.97, 3.17, 3.25, 3.31, 3.56, 4.16, 4.24, 4.29, 4.61, 5.48, 6.00}

8. **Politics**: In the 2000 U.S. presidential elections, the inaccuracy of voting technology became the subject of worldwide interest. In Florida, the two front-runner candidates, George W. Bush and Al Gore, were so close in the tally that votes had to be recounted by hand to verify the totals. In the process of recounting, it was found that the most unreliable voting machine was the *Votomatic*, with the most under-counted (votes placed but not counted) and over-counted (votes counted that were never placed) votes of any other technology. The under-counted votes for 15 different Florida counties are listed below.

{4946, 2070, 5090, 466, 5431, 1044, 1975, 2410, 10570, 634, 10134, 1763, 4240, 1846, 596}.

9. **Politics**: The over-counted votes for the same *Votomatic* machines are listed below.

{7826, 1134, 21855, 520, 3640, 790, 2531, 890, 17833, 1039, 19218, 2124, 4258, 994, 170}.

10. **Economics**: The following are yearly income levels for twenty randomly selected Alaska residents for the year 1999 (selected randomly from the 2000 U.S. Census):

{9000, 10600, 60000, 80000, 5000, 2600, 14800, 4500, 45000, 23000, 21300, 50000, 17000, 49000, 85000, 38000, 64000, 25000, 23000, 55000}.

11. **College Demographics**: The following values are hours worked per week for randomly selected students at Mt. San Antonio College in Walnut, California.

{15, 20, 40, 16, 20, 25, 19, 40, 40, 20, 20, 40, 0, 40, 35, 30, 35, 25, 29, 0, 0, 40, 16, 0, 40, 30, 25, 16, 35, 0}.

12. **College Demographics**: The following are grade point averages for the same students whose work hours are listed above.

{ 2.7, 1.7, 2.0, 3.7, 3.1, 4.0, 3.4, 3.7, 2.5, 2.8, 3.0, 3.6, 3.3, 3.2, 3.1, 3.2, 2.6, 2.8, 2.0, 4.0, 2.7, 3.9, 3.1, 2.5, 3.7, 2.9, 3.5, 3.7, 2.2, 3.8 }

Data suggest that the average *eye gain ratio* of people with schizophrenia is significantly different from the average for people without this disorder. An *eye gain ratio* is the ratio of an observer's angular eye velocity to the angular velocity of an object that the eye is following.

13. **Psychology:** For this problem, load the schizophrenic patient eye gain ratio observations, with included control group, in *Statcato* (available at *www.statcato.org*), or open the corresponding spreadsheet at *http://math.mtsac.edu/statistics/data.* Compute the measures of center for the schizophrenic patients.

14. **Psychology:** For this problem, load the schizophrenic patient eye gain ratio observations, with included control group, in *Statcato* (available at *www.statcato.org*), or open the corresponding spreadsheet at *http://math.mtsac.edu/statistics/data.* Compute the measures of center for the control group.

Estimate the mean of the values summarized in the frequency tables below.

15. Ages of students in a randomly selected evening class are summarized below.

Class	Frequency
15 – 19	4
20 – 24	8
25 – 29	16
30 – 34	14
35 – 39	8
40 – 44	6
45 – 49	1

16. Temperatures for randomly selected days in a desert city are summarized below.

Class	Frequency
12.0 – 16.9	2
17.0 – 21.9	0
22.0 – 26.9	7
27.0 – 31.9	8
32.0 – 36.9	4
37.0 – 41.9	3
42.0 – 46.9	1

2.4   Summarizing Data Using Measures of Variation

One of the biggest troubles in statistics is the fact that, when estimating parameters with statistics, sampling error is always present. Because of this, a considerable amount of theory in this text will be developed in an attempt to quantify this error. In this attempt, one of the things we will learn is that sampling error is directly proportional to population variability. Thus, our first step toward managing sampling error is to find a way of measuring this variability in a data set.

2.4.1   The Range of a Data Set

Our first measure of variability in a data set is the *range*, which we have already defined. By way of reminder, the range of a collection of values is simply the width of the distribution of these values. To compute the range, we simply subtract the minimum value in a data set from the maximum value.

Definition 2.4.1 The *range* of a data set is computed according to the formula:

$$Range = Maximum\ Value - Minimum\ Value$$

Unfortunately, the range suffers the same problem as the midrange of a data set: it uses only two of the values, in fact the most extreme values, and thus is easily influenced by outliers.

Another measure of variability, which again uses only two values, is the *interquartile range*. This measure of variability is much more stable than the range because it is not influenced by extreme values. Instead of using extreme values, it uses the *lower* and *upper quartiles* in the data set.

Definition 2.4.2 The *lower quartile*, Q_1 (or *first quartile*), of a data set is the median of the lower half of a sorted data set. The *upper quartile*, Q_3 (or *third quartile*), of a data set is the median of the upper half of a sorted data set. The lower and upper halves of a sorted data set are values in positions that are either below or above $\frac{n+1}{2}$. The value in the median position is not included in either half when n is odd.

Definition 2.4.3 The *interquartile range, or IQR*, is given by

$$interquartile\ range = upper\ quartile - lower\ quartile$$

or, using the notation defined above,

$$IQR = Q_3 - Q_1.$$

For reference, the *middle* or *second quartile* is the median of the full data set.

Example 2.4.4 Returning to the Galapagos Island elevations, with median $\tilde{x} = 227$, we note again that $n = 25$, is an odd number, so we will exclude the value in the median position, $\frac{n+1}{2} = \frac{25+1}{2} = 13$, from the lower and upper halves as we determine the lower and upper quartiles. The original data set follows, with the median highlighted: {25, 46, 49, 76, 93, 109, 112, 114, 119, 168, 186, 198, 227, 253, 259, 343, 367, 458, 640, 716, 777, 864, 906, 1494, 1707}.

The lower half of the sorted data values is: {25, 46, 49, 76, 93, 109, 112, 114, 119, 168, 186, 198}. There are 12 values, so the median of these values is the mean of the values in positions $\frac{12}{2} = 6$, and $\frac{12}{2} + 1 = 7$. These values are 109 and 112, and so the lower quartile is the mean of these,

$$\frac{109 + 112}{2} = 110.5.$$

The upper half of the sorted data values is: {253, 259, 343, 367, 458, 640, 716, 777, 864, 906, 1494, 1707}. Returning again to positions 6 and 7 gives us the values 640 and 716, so the upper quartile is

$$\frac{640 + 716}{2} = 678.$$

Finally we may compute the interquartile range.

$$interquartile\ range = upper\ quartile - lower\ quartile = 678 - 110.5 = 567.5.$$

2.4.2   Deviations from the Mean

We wish to consider a measure of variation that uses every value in a data set. There are a few measures which do this, and each of them uses the idea of *deviations from the mean*, which measure exactly how much a given value differs from the mean.

Definition 2.4.5 For the sample{ x_1, x_2, $\cdots$, x_n }, with sample mean $\bar{x}$, the n deviations from the mean are: $\{x_1 - \bar{x}, x_2 - \bar{x}, \cdots, x_n - \bar{x}\}$.

The remainder of our measures of variation in a data set will all involve some average of these deviations. There is a problem, however, if we attempt to average the deviations directly, and this will impact much of the theory which we are to develop. This problem is demonstrated in the following example.

Example 2.4.4 Continued: Returning to the Galapagos Islands, the mean maximum elevation is $\bar{x} = 412.24$. Table 2.2 shows each value, x, along with its respective deviation from the mean. The difficulty becomes apparent when we see that the total of the deviations is zero, $\sum (x - \bar{x}) = 0$.

x	$x - \bar{x}$
109	−303.24
114	−298.24
46	−366.24
119	−293.24
93	−319.24
168	−244.24
112	−300.24
198	−214.24
1494	1081.76
49	−363.24
227	−185.24
76	−336.24
1707	1294.76
343	−69.24
25	−387.24
777	364.76
458	45.76
367	−45.24
716	303.76
906	493.76
864	451.76
259	−153.24
640	227.76
186	−226.24
253	−159.24
Total:	0.00

Table 2.2: Deviations from the mean of the Galapagos Island elevations.

This is always true, as is easily shown below.

$$\sum (x - \bar{x}) = \sum x - \sum \bar{x}$$
$$= \sum x - n \cdot \bar{x}$$
$$= \sum x - n \cdot \frac{\sum x}{n}$$
$$= \sum x - \sum x$$
$$= 0$$

This sample has 25 values, and thus 25 deviations from the mean. Noting that the last deviation is -159.24, consider the first 24 deviations. The sum of these must be equal to 159.24, because otherwise the overall sum could never be zero. The way we will view this is as follows: the first 24 deviations are free to be any value at all, but the last deviation, the 25th, must be whatever value is necessary to make the sum of the deviations equal to zero. Thus, we will say that the 25 deviations have 24 *degrees of freedom*. When a sample is of size n, there are n deviations from the mean, but only $n - 1$ of these deviations contain independent information about variation from the mean. We therefore say that the n deviations exhibit $n - 1$ *degrees of freedom*.

Definition 2.4.6 From a sample of size n, containing n deviations from the mean, $x - \bar{x}$, there exist $n - 1$ degrees of freedom.

The fact that the sum of all deviations, $\sum (x - \bar{x})$, is always zero implies that the mean of these deviations will always be zero, and so the mean deviation could never be used as a measure of variation in a data set.

2.4.3 Mean Absolute Deviation

The natural fix for this problem is to consider a deviation from the mean as a distance from the mean, and as distance is always positive, we should take the absolute values of the deviations before averaging. So the *absolute* deviation from a sample mean, $\bar{x}$, is $|x - \bar{x}|$, while the absolute deviation from a population mean, μ, is $|x - \mu|$.

To quantify the variation in our data set, we would like to average these absolute deviations. The question is, which average shall we use? We will consider two possible averages: the arithmetic mean, and the quadratic mean.

First, we will average these absolute deviations using the arithmetic mean. The result is

$$\frac{\sum |x - \bar{x}|}{n},$$

the sample *mean absolute deviation*, while the population mean absolute deviation is

$$\frac{\sum |x - \mu|}{N},$$

where N is the size of a finite population.

Example 2.4.4 Continued: Returning to the Galapagos Islands, with absolute deviations in Table 2.3, the sum of the absolute deviations is 8528.16, so the sample mean absolute deviation is

$$\frac{\sum |x - \bar{x}|}{n} = \frac{8528.16}{25}$$
$$= 341.1.$$

Note again how this value is rounded, we have carried one place more after the decimal than was present in the data values themselves. This is the same rule used for averages.The mean absolute deviation is an average deviation that includes every value in a data set. That is good, but we will not be using this measure of variation because of a reason that may seem strange. The difficulty with this statistic is due to the fact that absolute values are present in the formula. Absolute value is a non-algebraic operation that makes any kind of analysis difficult. So we abandon the mean absolute deviation for an average deviation which uses only algebraic operations. (The algebraic operations are addition, subtraction, multiplication, division, raising to integer powers, and applying radicals).

2.4.4   Standard Deviation and Variance

Instead of using the arithmetic mean to average the absolute deviations, the standard deviation uses the *quadratic mean* as defined previously. The quadratic mean averages the squares of the deviations from the mean, still accomplishing the goal of making the values to be averaged positive, but without using absolute values. The squared deviations for a finite population of size N are $(x - \mu)^2$, and the squared deviations for a sample of size n are $(x - \bar{x})^2$. The quadratic mean averages the squared deviations, and then takes the square root of the average. Thus, for a finite population, the population quadratic mean of deviations, called the *population standard deviation*, represented by the Greek lowercase letter σ (*sigma*), is given by

$$\sigma = \sqrt{\frac{\sum (x - \mu)^2}{N}}.$$

Because of its excellent properties, we next define the *population variance* to be the *square* of the standard deviation. As with the population mean, the population variance has two symbols: σ^2 and $Var(x)$, and they are interchangeable. We will discuss the important properties of population variance in Chapter 4. For now, the variance of a finite population is: $Var(x) = \sigma^2 = \sum (x - \mu)^2 / N$.

Next, we would like to define a *sample standard deviation* and *sample variance* as estimates of their respective parameters above. For sample variance, we will use the symbol, s^2, while sample standard deviation will be represented by s. The strange thing about sample variance is that we use a slightly different formula from what might be expected.

x	$|x - \bar{x}|$
109	303.24
114	298.24
46	366.24
119	293.24
93	319.24
168	244.24
112	300.24
198	214.24
1494	1081.76
49	363.24
227	185.24
76	336.24
1707	1294.76
343	69.24
25	387.24
777	364.76
458	45.76
367	45.24
716	303.76
906	493.76
864	451.76
259	153.24
640	227.76
186	226.24
253	159.24
Total:	8528.16

Table 2.3: The absolute deviations for Example 2.4.4.

One might expect sample variance to be $\sum (x - \bar{x})^2 / n$. We will not use this definition however, because it turns out that this estimate of σ^2 is *biased*. This estimate tends to be too small because the sample x values are closer to $\bar{x}$ (since $\bar{x}$ is computed from sample x values), on average, than they are to μ (which is *not* computed from the *sample x* values). That is, $(x - \bar{x})^2$ tends to be smaller than $(x - \mu)^2$, when x is part of a sample from which $\bar{x}$ is measured, and this smallness is the source of bias. But σ^2 is a measure of variation from μ, and s^2 is supposed to estimate σ^2. It will be shown that this bias is removed by dividing by $n - 1$ instead of n. This is proven in Theorem 4.6.4, in Chapter 4 (page 196), after necessary theory is developed. It should be additionally noted that the bias is removed in this average not by dividing by n, the number of deviations, but by $n - 1$, the number of *independent* deviations, the degrees of freedom.

Definition 2.4.7 *Population variance*, σ^2, and *sample variance*, s^2, are given by:

$$Var(x) = \sigma^2 = \frac{\sum (x - \mu)^2}{N}$$

and

$$s^2 = \frac{\sum (x - \bar{x})^2}{n - 1} = \frac{S_{xx}}{n - 1},$$

where N is the size of a finite population, and n is sample size. The sample variance, s^2, is an *unbiased estimator* of population variance, σ^2.

Note the use of the symbol $S_{xx} = \sum (x - \bar{x})^2$, in the numerator of the sample variance formula above. This notation will be used from time to time throughout this text. Finally, we formally state the definitions of population and sample standard deviations below.

Definition 2.4.8 The *population standard deviation*, σ, and *sample standard deviation*, s, are given by:

$$\sigma = \sqrt{\sigma^2}$$
$$= \sqrt{\frac{\sum (x - \mu)^2}{N}},$$

and

$$s = \sqrt{s^2}$$
$$= \sqrt{\frac{\sum (x - \bar{x})^2}{n - 1}}$$
$$= \sqrt{\frac{S_{xx}}{n - 1}},$$

where N is the size of a finite population, and n is sample size. Note that s is unfortunately a *biased* estimator of σ – the taking of square roots is the cause of this.

x	$(x - \bar{x})^2$
109	91954.4976
114	88947.0976
46	134131.7376
119	85989.6976
93	101914.1776
168	59653.1776
112	90144.0576
198	45898.7776
1494	1170204.6980
49	131943.2976
227	34313.8576
76	113057.3376
1707	1676403.4580
343	4794.1776
25	149954.8176
777	133049.8576
458	2093.9776
367	2046.6576
716	92270.1376
906	243798.9376
864	204087.0976
259	23482.4976
640	51874.6176
186	51184.5376
253	25357.3776
Total:	4808550.5600

Table 2.4: The sum of squared deviations from Example 2.4.4.

Example 2.4.4 Continued: Returning once again to the Galapagos Island elevations, we see the squared deviations in Table 2.4, and their sum, $\sum (x - \bar{x})^2 = 4808550.56$.

The sample variance is

$$
\begin{aligned}
s^2 &= \frac{\sum (x - \bar{x})^2}{n - 1} \\
&= \frac{4808550.56}{25 - 1} \\
&= 200356.2733,
\end{aligned}
$$

not rounded yet as it is to be used to compute standard deviation,

$$
\begin{aligned}
s &= \sqrt{s^2} \\
&= \sqrt{200356.2733} \\
&= 447.6.
\end{aligned}
$$

The standard deviation is rounded to one place more after the decimal than was present in the original sample.

Procedure 2.4.9 Rounding Rule for Sample Standard Deviation and Variance
When rounding a standard deviation or variance, carry at least one place more after the decimal than present in the given data set. Try to not round any values until all calculations are complete.

The standard deviation formula above appears simple, but can be a bit tedious due to the fact that the squared deviations $(x - \bar{x})^2$ tend to be fractional, and in summing them to compute S_{xx}, there may be much propagation of rounding error. To help diminish this error, and to ease the difficulty of working with many fractional values, a *computational* formula for S_{xx} is developed below. This is called a *computational* formula because it requires fewer computations than the original formula.

Theorem 2.4.10 A computational formula for S_{xx} is as follows.

$$
S_{xx} = \sum x^2 - \frac{1}{n} \left(\sum x \right)^2
$$

Proof. Our proof is an algebraic manipulation of the definition of $S_{xx} = \sum \left(x - \bar{x} \right)^2$.

$$
\begin{aligned}
S_{xx} &= \sum \left(x - \bar{x} \right)^2 \\
&= \sum \left(x^2 - 2x\bar{x} + \bar{x}^2 \right) \\
&= \sum x^2 - 2\bar{x} \sum x + \sum \bar{x}^2 \\
&= \sum x^2 - 2\bar{x} \sum x + n \cdot \bar{x}^2 \\
&= \sum x^2 - 2\frac{\sum x}{n} \sum x + n \left(\frac{\sum x}{n} \right)^2 \\
&= \sum x^2 - \frac{2}{n} \left(\sum x \right)^2 + \frac{n}{n^2} \left(\sum x \right)^2 \\
&= \sum x^2 - \frac{2}{n} \left(\sum x \right)^2 + \frac{1}{n} \left(\sum x \right)^2 \\
&= \sum x^2 + \left(-\frac{2}{n} + \frac{1}{n} \right) \left(\sum x \right)^2 \\
&= \sum x^2 - \frac{1}{n} \left(\sum x \right)^2
\end{aligned}
$$

■

With the *computational* formula for S_{xx}, our formulas for sample standard deviation and variance are updated to the following *computational* formulas:

Theorem 2.4.11 When computing the sample standard deviation or variance, the following computational formulas are recommended.

For the sample variance, the formula is:

$$
\begin{aligned}
s^2 &= \frac{S_{xx}}{n-1} \\
&= \frac{\sum x^2 - \frac{1}{n} \left(\sum x \right)^2}{n-1}.
\end{aligned}
$$

For the sample standard deviation, the formula is:

$$
\begin{aligned}
s &= \sqrt{s^2} \\
&= \sqrt{\frac{\sum x^2 - \frac{1}{n} \left(\sum x \right)^2}{n-1}}.
\end{aligned}
$$

Example 2.4.4 Continued: Using our computational formulas, we return once more to the Galapagos Islands, whose elevations x, and squared elevations, x^2, are computed in Table 2.5.

Note that the values are much neater than the values in the previous table for $(x - \bar{x})^2$. Using $n = 25$, $\sum x = 10306$, and $\sum x^2 = 9057096$ we use the computational formulas to,

x	x^2
109	11881
114	12996
46	2116
119	14161
93	8649
168	28224
112	12544
198	39204
1494	2232036
49	2401
227	51529
76	5776
1707	2913849
343	117649
25	625
777	603729
458	209764
367	134689
716	512656
906	820836
864	746496
259	67081
640	409600
186	34596
253	64009
10306	9057096

Table 2.5: The sum of x and x^2 values from Example 2.4.4.

once again, compute variance and standard deviation. First the variance is computed.

$$
\begin{aligned}
s^2 &= \frac{S_{xx}}{n-1} \\
&= \frac{\sum x^2 - \frac{1}{n} \cdot \left(\sum x\right)^2}{n-1} \\
&= \frac{9057096 - \frac{1}{25} \cdot 10306^2}{24} \\
&= \frac{4808550.56}{24} \\
&= 200356.2733
\end{aligned}
$$

We next compute the standard deviation.

$$
\begin{aligned}
s &= \sqrt{s^2} \\
&= \sqrt{200356.2733} \\
&\approx 447.6
\end{aligned}
$$

2.4.5 Estimating s and s^2 using Frequency Tables

There will be times when we wish to compute s, but will have no sample data – we will only have data summarized by a frequency table. When this is the case, we will estimate each value, x, using the class midpoints for each class, and will associate to each x, the corresponding frequency for its class, f. We will then estimate the sample standard deviation using the updated formula below.

Procedure 2.4.12 When estimating the sample variance and standard deviation from a frequency table, the following formula is used.

$$
\begin{aligned}
s^2 &= \frac{S_{xx}}{n-1} \\
&\approx \frac{\sum f \cdot x^2 - \frac{1}{n} \cdot \left(\sum f \cdot x\right)^2}{n-1}
\end{aligned}
$$

Here $n = \sum f$, and x represents each class midpoint. Likewise,

$$
\begin{aligned}
s &= \sqrt{s^2} \\
&\approx \sqrt{\frac{\sum f \cdot x^2 - \frac{1}{n} \cdot \left(\sum f \cdot x\right)^2}{n-1}}
\end{aligned}
$$

is our estimate for sample standard deviation.

We may also develop a new formula for *population* variance using frequencies, merely for theoretical reasons:

$$
\sigma^2 = \frac{\sum f \cdot x^2 - \frac{1}{N} \cdot \left(\sum f \cdot x\right)^2}{N}.
$$

Here each x in the population occurs f times, and again we have $N = \sum f$. We will use this formula for a few necessary developments in Chapter 4.

Example 2.4.13 Returning to the frequency table for 52 weekly contributions to a charitable organization, we have added two new columns to the table, fx, and fx^2, where x represents the class midpoints of each class.

Class		f	x	$f \cdot x$	$f \cdot x^2$
50 – 339		1	194.5	194.5	37830.25
340 – 629		0	484.5	0.0	0.00
630 – 919		18	774.5	13941.0	10797304.50
920 – 1209		23	1064.5	24483.5	26062685.75
1210 – 1499		6	1354.5	8127.0	11008021.50
1500 – 1789		3	1644.5	4933.5	8113140.75
1790 – 2079		1	1934.5	1934.5	3742290.25
Totals		52	–	53614.0	59761273.00

The columns are totaled, and we estimate the variance below.

$$s^2 \approx \frac{\sum f \cdot x^2 - \frac{1}{n} \cdot \left(\sum f \cdot x \right)^2}{n-1}$$

$$= \frac{59761273 - \frac{1}{52} \cdot (53614)^2}{52-1}$$

$$= 87905.4299.$$

Thus, the standard deviation is estimated to be $s = \sqrt{s^2} \approx \sqrt{87905.4299} = 296.49$. The exact value for s, using the original sample data from Example 2.1.1, was $s = 283.06$. Note the considerable difference between the correct s and the estimated s from the frequency table. Clearly, whenever possible, and if precision is important, we should use the sample data, and not a frequency table, to compute our estimates.

Sometimes, it is nice to have a quick estimate of the sample standard deviation. For this, we introduce the *range rule of thumb*.

Proposition 2.4.14 The Range Rule of Thumb
The sample standard deviation of a sample which does not have extreme outliers may be roughly estimated using the sample's range divided by four.

$$s \approx \frac{Range}{4} = \frac{Maximum - Minimum}{4}$$

For the Galapagos Island elevations, with $s = 447.4$, this could have been estimated using the range rule of thumb as $s \approx \frac{1707-25}{4} = \frac{1682}{4} = 420.5$. This is not a bad estimate, but the rule does not always work so well. Obviously, the range rule of thumb uses only the two most extreme values in a data set, and therefore is very sensitive to outliers and errors. Because of this, the range rule of thumb can give highly inaccurate estimates of the sample standard deviation. Still, it *is* simple and quick!

2.4.6 The Empirical Rule

We conclude this section by discussing two rules which describe how data are distributed relative to the data set's standard deviation. The first rule, the *empirical rule*, applies only to bell-shaped, or normal, data sets.

Proposition 2.4.15 The Empirical Rule: For a normally distributed data set, the following has proven to be true, based on practical experience:

1. Approximately 68% of the values are within one standard deviation of the mean.

2. Approximately 95% of the values are within two standard deviations of the mean.

3. Approximately 99.7% of the values are within three standard deviations of the mean.

The Empirical rule can be depicted graphically using the smooth curve in Figure 2.19.

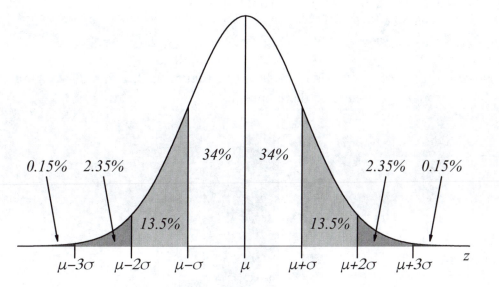

Figure 2.19: A graphical depiction of the *Empirical Rule*.

In this picture, the proportion of values that lie within one population standard deviation, σ, from the population mean, μ, (between $\mu - \sigma$ and $\mu + \sigma$) is equal to

$$34\% + 34\% = 68\%.$$

Likewise, the proportion of values that lie within two standard deviations from the mean (between $\mu - 2\sigma$ and $\mu + 2\sigma$) is equal to

$$13.5\% + 34\% + 34\% + 13.5\% = 95\%.$$

Example 2.4.16 If we apply the empirical rule to one of our data sets, we may see how well this rule, which is based on practical experience, fits. When we apply the empirical rule to sample data, the three intervals use a sample mean $\bar{x}$, and sample standard deviation, s, and become $\bar{x} - s$ to $\bar{x} + s$ for one standard deviation about the mean, $\bar{x} - 2s$ to $\bar{x} + 2s$ for two standard deviations about the mean, and $\bar{x} - 3s$ to $\bar{x} + 3s$ for three standard deviations about the mean.

Below are listed the $n = 100$ estimates of the speed of light, in thousands of kilometers per second, sorted left to right, top to bottom.

				Speed of Light Estimates					
299.62	299.65	299.72	299.72	299.72	299.74	299.74	299.74	299.75	299.76
299.76	299.76	299.76	299.76	299.77	299.78	299.78	299.79	299.79	299.79
299.80	299.80	299.80	299.80	299.80	299.81	299.81	299.81	299.81	299.81
299.81	299.81	299.81	299.81	299.81	299.82	299.82	299.83	299.83	299.84
299.84	299.84	299.84	299.84	299.84	299.84	299.84	299.85	299.85	299.85
299.85	299.85	299.85	299.85	299.85	299.86	299.86	299.86	299.87	299.87
299.87	299.87	299.88	299.88	299.88	299.88	299.88	299.88	299.88	299.88
299.88	299.88	299.89	299.89	299.89	299.90	299.90	299.91	299.91	299.92
299.93	299.93	299.94	299.94	299.94	299.95	299.95	299.95	299.96	299.96
299.96	299.96	299.97	299.98	299.98	299.98	300.00	300.00	300.00	300.07

The histogram for these values is reproduced in Figure 2.20, and we see again that the values have a roughly bell-shaped distribution.

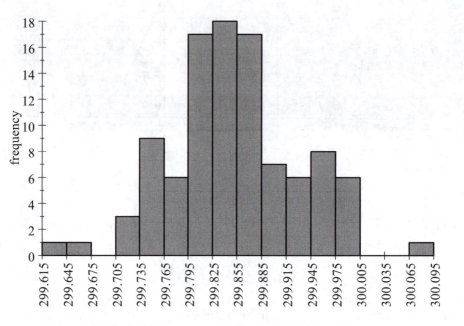

Figure 2.20: A histogram of the speed of light estimates.

The mean of the values is $\bar{x} = 299.8524$, and the standard deviation is $s = 0.079011$. The interval, *one standard deviation from the mean*, becomes $\bar{x} - s = 299.7734$ to $\bar{x} + s =$

299.9314. In the table, these values correspond to the white cells, and there are exactly 67 values in this range, or 67% of the sample.

Compare this to the predicted 68% by the empirical rule, and be impressed! Let us do another application by considering the *two standard deviation from the mean* interval, which is $\bar{x} - 2s = 299.6944$ to $\bar{x} + 2s = 300.0104$. The values in this range are the light grey shaded cells, together with those in the white cells. Altogether we have 97 values, which are 97% of the sample, and we compare this to the 95% which is predicted by the empirical rule. This time the prediction is close to what we observed, but not quite as close as the last prediction. The three standard deviation from the mean interval is $\bar{x} - 3s = 299.6154$ to $\bar{x} + 3s = 300.0894$, bringing in the cells which are shaded darker grey, and now all 100 values, 100% of the sample is included, and we compare this to the 99.7% which was predicted by the empirical rule. Again, the prediction is quite close to what we observe.

Proposition 2.4.17 First Rule for Unusualness or Significance
Because the empirical rule states that, with a normal population, around 95% of data values lie within two standard deviations from the mean, we conclude that measurements outside of this interval would only occur about 5% of the time. Since this is rather unlikely to happen (having a 5% probability), we will call it *unusual* when it does. As we proceed, we will use the word *significant* more and more frequently, with increasing sophistication, but for now, *statistically significant* simply means *unusual*.

Proposition 2.4.18 A measurement, x, is *unusual*, or *significant* if it is around two or more standard deviations from the mean. For normal populations, this happens with only 5% (approximately) of all observations, so any event with a probability less than around 5% will also be considered *unusual* or *significant*.

Definition 2.4.19 The *usual range* of values will be any value from $\bar{x} - 2 \cdot s$ to $\bar{x} + 2 \cdot s$.

Reiterating a result from the speed of light data, the usual range of values would be from $\bar{x} - 2 \cdot s = 299.6944$ to $\bar{x} + 2 \cdot s = 300.0104$. Any measurement outside of this range (there were only three out of 100) would be considered unusual or significant.

Theorem 2.4.20 Chebyshev's Theorem: Not all data sets are normal, and in such cases the empirical rule does not apply. For non-normal populations, a weaker rule exists, but its weakness makes it less useful to us. This rule is called *Chebyshev's theorem*, and it is as follows. In any data set, the proportion of values that lie within k standard deviations from the mean is *at least* $1 - 1/k^2$, where $k \geqslant 1$.

The proportion of values that lie within at least $k = 2$ standard deviations from the mean is at least $1 - 1/2^2 = 1 - 1/4 = 3/4 = .75 = 75\%$. That is, at least 75% of the values in *any* data set lie within two standard deviations of the mean.

2.4.7 Homework

Compute the range, the lower and upper quartiles (Q_1 and Q_3) (use the medians computed in the previous section), the inter-quartile range, the usual range, standard deviation, and variance for each data set below.

1. Data set: { 2, 5, 5, 11, 15 }

2. Data set: { 4, 6, 11, 25, 26, 26 }

3. Data set: { 68.94, 97.56, 92.56, 64.75, 88.67, 80.76, 62.06, 69.50, 84.73, 78.31, 89.35, 96.18, 80.36, 95.70, 84.80, 82.37, 91.11 }

4. Data set: { 66, 116, 121, 81, 114, 81, 69, 85, 95, 93, 119, 122, 99, 125, 95, 92 }

5. Biology: The journal *Environmental Concentration and Toxicology* published the article "Trace Metals in Sea Scallops" (vol. 19, pp. 326 - 1334), and gave the following cadmium amounts (in mg) in sea scallops observed at a number of different stations in North Atlantic waters. The values are given below.

{5.1, 14.4, 14.7, 10.8, 6.5, 5.7, 7.7, 14.1, 9.5, 3.7, 8.9, 7.9, 7.9, 4.5, 10.1, 5.0, 9.6, 5.5, 5.1, 11.4, 8.0, 12.1, 7.5, 8.5, 13.1, 6.4, 18.0, 27.0, 18.9, 10.8, 13.1, 8.4, 16.9, 2.7, 9.6, 4.5, 12.4, 5.5, 12.7, 17.1}

6. Engineering: The data below consist of lifetimes (in thousands of cycles) of 101 rectangular strips of aluminum subjected to repeated alternating stress at 21,000 psi, 18 cycles per second. This data were taken from an article in the Journal of the American Statistical Association, (1958, p. 159). These values are built into the statistics software *Statcato*, which is available for free download at http://www.statcato.org. Alternately, these values are stored in an *Excel* spreadsheet, and may be downloaded at http://math.mtsac.edu/statistics/data.

{370, 706, 716, 746, 785, 797, 844, 855, 858, 886, 886, 930, 960, 988, 990, 1000, 1010, 1016, 1018, 1020, 1055, 1085, 1102, 1102, 1108, 1115, 1120, 1134, 1140, 1199, 1200, 1200, 1203, 1222, 1235, 1238, 1252, 1258, 1262, 1269, 1270, 1290, 1293, 1300, 1310, 1313, 1315, 1330, 1355, 1390, 1416, 1419, 1420, 1420, 1450, 1452, 1475, 1478, 1481, 1485, 1502, 1505, 1513, 1522, 1522, 1530, 1540, 1560, 1567, 1578, 1594, 1602, 1604, 1608, 1630, 1642, 1674, 1730, 1750, 1750, 1763, 1768, 1781, 1782, 1792, 1820, 1868, 1881, 1890, 1893, 1895, 1910, 1923, 1940, 1945, 2013, 2023, 2100, 2215, 2268, 2440}

7. Meteorology: The following are rainfall totals (in inches) for clouds chemically seeded to increase rainfall amounts. (Data are taken from Miller, A.J., Shaw, D.E., Veitch, L.G. & Smith, E.J. (1979). "Analyzing the results of a cloud-seeding experiment in Tasmania", Communications in Statistics - Theory & Methods, vol.A8(10), pp. 1017-1047). These values are built into the statistics software *Statcato*, which is available for free download at http://www.statcato.org. Alternately, these values are stored in an *Excel* spreadsheet, and may be downloaded at:

http://math.mtsac.edu/statistics/data.

{0.05, 0.07, 0.13, 0.24, 0.26, 0.36, 0.37, 0.38, 0.40, 0.41, 0.42, 0.47, 0.58, 0.59, 0.61, 0.71, 0.76, 0.79, 0.80, 0.80, 0.81, 0.84, 0.86, 0.88, 0.90, 0.90, 0.91, 1.00, 1.08, 1.09, 1.11, 1.16, 1.23, 1.25, 1.32, 1.43, 1.46, 1.48, 1.50, 1.62, 1.63, 1.67, 1.83, 1.87, 2.04, 2.06, 2.08, 2.16, 2.21, 2.22, 2.25, 2.26, 2.35, 2.36, 2.39, 2.48, 2.56, 2.79, 2.87, 2.97, 3.17, 3.25, 3.31, 3.56, 4.16, 4.24, 4.29, 4.61, 5.48, 6.00}

8. **Politics**: In the 2000 U.S. presidential elections, the inaccuracy of voting technology became the subject of worldwide interest. In Florida, the two front-runner candidates, George W. Bush and Al Gore, were so close in the tally that votes had to be recounted by hand to verify the totals. In the process of recounting, it was found that the most unreliable voting machine was the *Votomatic*, with the most under-counted (votes placed but not counted) and over-counted (votes counted that were never placed) votes of any other technology. The under-counted votes for 15 different Florida counties are listed below.

{4946, 2070, 5090, 466, 5431, 1044, 1975, 2410, 10570, 634, 10134, 1763, 4240, 1846, 596}.

9. **Politics**: The over-counted votes for the same *Votomatic* machines are listed below.

{7826, 1134, 21855, 520, 3640, 790, 2531, 890, 17833, 1039, 19218, 2124, 4258, 994, 170}.

10. **Economics**: The following are yearly income levels for twenty randomly selected Alaska residents for the year 1999 (selected randomly from the 2000 U.S. Census):

{9000, 10600, 60000, 80000, 5000, 2600, 14800, 4500, 45000, 23000, 21300, 50000, 17000, 49000, 85000, 38000, 64000, 25000, 23000, 55000}.

11. **College Demographics**: The following values are hours worked per week for randomly selected students at Mt. San Antonio College in Walnut, California.

{15, 20, 40, 16, 20, 25, 19, 40, 40, 20, 20, 40, 0, 40, 35, 30, 35, 25, 29, 0, 0, 40, 16, 0, 40, 30, 25, 16, 35, 0}.

12. **College Demographics**: The following are grade point averages for the same students whose work hours are listed above.

{ 2.7, 1.7, 2.0, 3.7, 3.1, 4.0, 3.4, 3.7, 2.5, 2.8, 3.0, 3.6, 3.3, 3.2, 3.1, 3.2, 2.6, 2.8, 2.0, 4.0, 2.7, 3.9, 3.1, 2.5, 3.7, 2.9, 3.5, 3.7, 2.2, 3.8 }

Data suggest that the average *eye gain ratio* of people with schizophrenia is significantly different from the average for people without this disorder. An *eye gain ratio* is the ratio of an observer's angular eye velocity to the angular velocity of an object that the eye is following.

13. **Psychology:** For this problem, load the schizophrenic patient eye gain ratio observations, with included control group, in *Statcato* (available at *www.statcato.org*), or open the corresponding spreadsheet at *http://math.mtsac.edu/statistics/data*. Compute the measures of variation for the schizophrenic patients. Compute the usual range, and note the proportion of values within this range. Does this proportion agree with the empirical rule?

14. **Psychology:** For this problem, load the schizophrenic patient eye gain ratio observations, with included control group, in *Statcato* (available at *www.statcato.org*), or open the corresponding spreadsheet at *http://math.mtsac.edu/statistics/data*. Compute the measures of variation for the control group. Compute the usual range, and note the proportion of values within this range. Does this proportion agree with the empirical rule?

Estimate the sample standard deviation and variance of the values summarized in the frequency tables below.

15. Ages of students in a randomly selected evening class are summarized below.

Class	Frequency
15 – 19	4
20 – 24	8
25 – 29	16
30 – 34	14
35 – 39	8
40 – 44	6
45 – 49	1

16. Temperatures for randomly selected days in a desert city are summarized below.

Class	Frequency
12.0 – 16.9	2
17.0 – 21.9	0
22.0 – 26.9	7
27.0 – 31.9	8
32.0 – 36.9	4
37.0 – 41.9	3
42.0 – 46.9	1

The data are from an article on peanut butter from *Consumer Reports* (Sept. 1990), which reported the following scores for various brands: {22, 30, 30, 36, 39, 40, 40, 41, 44, 45, 50, 50, 53, 56, 56, 56, 62, 65, 68}. The next collection of problems all refer to this data set.

17. Compute the mean absolute deviation of these scores.

18. Compute the sample standard deviation of these scores.

19. Estimate the sample standard deviation of these scores using the range rule of thumb.

20. From the problems above, which gives a better estimate of the standard deviation – the mean deviation or range rule of thumb?

21. For the sample above, how many degrees of freedom do the deviations from the mean include?

22. Compute the *usual range* based on the values above. Are there any values in this sample that are unusual? What proportion of this sample is unusual?

23. Does the proportion of unusual values agree with the empirical rule? Use this agreement/disagreement to argue whether you feel that the sample has a normal or bell-shaped distribution.

2.5   Measures of Relative Position

We have already begun to discuss, using the so-called *usual range*, a method by which we may classify values as unusual or significant. While useful, the usual range method is rather crude, because it only allows for an absolute *yes* (when a value lies outside the usual range), or an absolute *no* (when a value lies within of the usual range). Unfortunately, in statistics, conclusions about significance are often unclear, and absolute yes or no answers can be misleading, especially when a value is on or near the boundary of the usual range. We need a measure to determine exactly how unusual a value is. This will take a few chapters to refine, but for now we begin by introducing *measures of relative position*.

Our first and simplest measure of relative position is the *percentile*.

2.5.1   The Percentile

For any value, r, between 1 and 99, the rth *percentile* is a value, P_r, such that r percent of values in a data set fall less than or equal to that value.

Percentiles are like dividers that separate data values into groups. With 99 percentile *dividers*, the population is divided into 100 groups of equal proportion. To determine the percentile to which a given value, x, belongs, we use the following definition.

Definition 2.5.1 The percentile for a given value, x, is the percentage of values in a data set that are less than x.

$$Percentile \ for \ x = \frac{values \ less \ than \ x}{n} \cdot 100,$$

and we round this to a whole number.

Example 2.5.2 If we return again to the Galapagos Islands, and are interested to know the percentile for $x = 343$, we simply note that there are 15 values, out of 25, that are less than 343 in the sample. These are printed in italics in the following list: {*25, 46, 49, 76, 93, 109, 112, 114, 119, 168, 186, 198, 227, 253, 259,* 343, 367, 458, 640, 716, 777, 864, 906, 1494, 1707}.

The percentile for 343 is

$$\frac{15}{25} \cdot 100 = 60.$$

That is, 343 belongs to the 60th percentile.

It is worth noting that we have seen three percentiles previously, under different names. The *lower quartile* and *upper quartile* (from the previous section) are the 25th and 75th percentiles, respectively, while the median, or the *middle quartile*, is the 50th percentile. Returning to the diagram that accompanied the empirical rule for bell-shaped distributions of data (Figure 2.21), we consider our usual range for a population to be values within two standard deviations from the mean. The lower limit for this range is $\mu - 2 \cdot \sigma$,

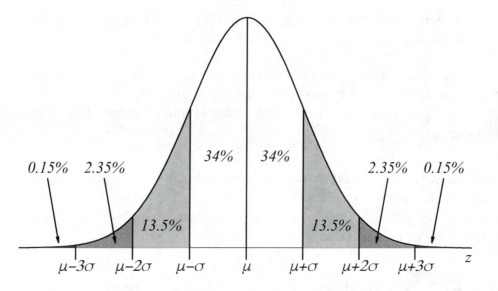

Figure 2.21: A graphical depiction of the *Empirical Rule*.

and the percentage of values that are below this limit is $0.15\% + 2.35\% = 2.5\%$. So 2.5% of the values lie more than 2 standard deviations below average. There is no 2.5^{th} percentile, but we can still say that any value below this would be outside of the usual range, so this includes percentiles less than or equal to the 2nd percentile. For the upper limit for the usual range, we see that

$$13.5\% + 34\% + 34\% + 13.5\% + 2.35\% + 0.15\% = 97.5\%.$$

Again, there is no 97.5^{th} percentile, but still, any value in the 98th percentile or higher would be outside of the usual range.

2.5.2 Standard Score or z-Score

Next, we turn to a more direct method of determining unusualness. Instead of determining whether a value lies within the usual range of two standard deviations from the mean, we will determine the number of standard deviations that a given value lies from the mean. This will be called the *z-score* or *standard* score of that value.

To compute the z-score of a value, x, we first measure the distance from x to the mean, μ. This distance will be $x - \mu$. We need to determine how many standard deviations this distance represents. This is determined by dividing this distance by the standard deviation, σ.

Definition 2.5.3 The z-score of a value x, using population parameters is:

$$z = \frac{x - \mu}{\sigma}.$$

The z-score of a value, x, using sample statistics is

$$z = \frac{x - \bar{x}}{s}.$$

It should be noted that if $x > \mu$, the z-score of x will be positive, but if $x < \mu$, the z-score will be negative. Also, if $x = \mu$, the z-score will be zero.

We have already said that a value is unusual if it lies more than two standard deviations from the mean. Now that we have a tool for measuring this number of standard deviations for any particular value, x, we can clarify our method for determining unusualness or significance.

Proposition 2.5.4 Updated Rule for Unusualness or Significance
When the z-score of x is *greater than* 2 or less than -2, we will consider the value to be unusually or significantly *large* or *small*, respectively. We will consider values which *nearly* satisfy these criteria as *moderately unusual or significant*.

Example 2.5.2 Continued: Returning to the Galapagos Island elevations, suppose we wish to know the z-score for $x = 1494$. For this, we compute

$$z = \frac{x - \bar{x}}{s}$$
$$= \frac{1494 - 412.24}{447.6}$$
$$= 2.42,$$

meaning that an elevation of 1494 is 2.42 standard deviations above average, just outside of the usual range, so this value is unusually high.

Next, suppose we wish to compute the z-score of $x = 114$. We note that, because this value is below the mean, the z-score will be negative. We compute

$$z = \frac{x - \bar{x}}{s}$$
$$= \frac{114 - 412.24}{447.6}$$
$$= -0.67.$$

This means that 114 is 0.67 standard deviations *below* average, and while the value is low, it is not unusually low.

While we are busy updating rules, we might consider updating the empirical rule to use the language of z-scores.

Proposition 2.5.5 Updated Empirical Rule for z-Scores
If x is randomly measured from a normal (bell-shaped) population, and z is the z-score of x, we may reasonably expect the following. There is a 68% probability that $-1 < z < 1$. Also, there is a 95% probability that $-2 < z < 2$. Finally, there is a 99.7% probability that $-3 < z < 3$.

2.5.3   Homework

For each data set and given value, x, compute the percentile and z-score for x, and in each case, use the z-score to determine if the value is unusual.

1. Data set: { 2, 5, 5, 11, 15 }; value: $x = 11$.

2. Data set: { 4, 6, 11, 25, 26, 26 }; value: $x = 26$.

3. Data set: { 68.94, 97.56, 92.56, 64.75, 88.67, 80.76, 62.06, 69.50, 84.73, 78.31, 89.35, 96.18, 80.36, 95.70, 84.80, 82.37, 91.11 }; value: $x = 96.18$.

4. Data set: { 66, 116, 121, 81, 114, 81, 69, 85, 95, 93, 119, 122, 99, 125, 95, 92 }; value: $x = 69$.

5. Biology: The journal *Environmental Concentration and Toxicology* published the article "Trace Metals in Sea Scallops" (vol. 19, pp. 326 - 1334), and gave the following cadmium amounts (in mg) in sea scallops observed at a number of different stations in North Atlantic waters. The values are:

{ 5.1, 14.4, 14.7, 10.8, 6.5, 5.7, 7.7, 14.1, 9.5, 3.7, 8.9, 7.9, 7.9, 4.5, 10.1, 5.0, 9.6, 5.5, 5.1, 11.4, 8.0, 12.1, 7.5, 8.5, 13.1, 6.4, 18.0, 27.0, 18.9, 10.8, 13.1, 8.4, 16.9, 2.7, 9.6, 4.5, 12.4, 5.5, 12.7, 17.1}; value: $x = 21.0$.

6. Engineering: The data below consist of lifetimes (in thousands of cycles) of 101 rectangular strips of aluminum subjected to repeated alternating stress at 21,000 psi, 18 cycles per second. This data were taken from an article in the Journal of the American Statistical Association, (1958, p. 159). These values are built into the statistics software *Statcato*, which is available for free download at http://www.statcato.org. Alternately, these values are stored in an *Excel* spreadsheet, and may be downloaded at http://math.mtsac.edu/statistics/data.

{370, 706, 716, 746, 785, 797, 844, 855, 858, 886, 886, 930, 960, 988, 990, 1000, 1010, 1016, 1018, 1020, 1055, 1085, 1102, 1102, 1108, 1115, 1120, 1134, 1140, 1199, 1200, 1200, 1203, 1222, 1235, 1238, 1252, 1258, 1262, 1269, 1270, 1290, 1293, 1300, 1310, 1313, 1315, 1330, 1355, 1390, 1416, 1419, 1420, 1420, 1450, 1452, 1475, 1478, 1481, 1485, 1502, 1505, 1513, 1522, 1522, 1530, 1540, 1560, 1567, 1578, 1594, 1602, 1604, 1608, 1630, 1642, 1674, 1730, 1750, 1750, 1763, 1768, 1781, 1782, 1792, 1820, 1868, 1881, 1890, 1893, 1895, 1910, 1923, 1940, 1945, 2013, 2023, 2100, 2215, 2268, 2440}; value: $x = 2215$.

7. Meteorology: The following are rainfall totals (in inches) for clouds chemically seeded to increase rainfall amounts. (Data are taken from Miller, A.J., Shaw, D.E., Veitch, L.G. & Smith, E.J. (1979). "Analyzing the results of a cloud-seeding experiment in Tasmania", Communications in Statistics - Theory & Methods, vol.A8(10), pp. 1017-1047). These values are built into the statistics software *Statcato*, which is available for free download at:

http://www.statcato.org.

Alternately, these values are stored in an *Excel* spreadsheet, and may be downloaded at http://math.mtsac.edu/statistics/data.

{ 0.05, 0.07, 0.13, 0.24, 0.26, 0.36, 0.37, 0.38, 0.40, 0.41, 0.42, 0.47, 0.58, 0.59, 0.61, 0.71, 0.76, 0.79, 0.80, 0.80, 0.81, 0.84, 0.86, 0.88, 0.90, 0.90, 0.91, 1.00, 1.08, 1.09, 1.11, 1.16, 1.23, 1.25, 1.32, 1.43, 1.46, 1.48, 1.50, 1.62, 1.63, 1.67, 1.83, 1.87, 2.04, 2.06, 2.08, 2.16, 2.21, 2.22, 2.25, 2.26, 2.35, 2.36, 2.39, 2.48, 2.56, 2.79, 2.87, 2.97, 3.17, 3.25, 3.31, 3.56, 4.16, 4.24, 4.29, 4.61, 5.48, 6.00 }; value: $x = 4.61$.

8. **Politics**: In the 2000 U.S. presidential elections, the inaccuracy of voting technology became the subject of worldwide interest. In Florida, the two front-runner candidates, George W. Bush and Al Gore, were so close in the tally that votes had to be recounted by hand to verify the totals. In the process of recounting, it was found that the most unreliable voting machine was the *Votomatic*, with the most under-counted (votes placed but not counted) and over-counted (votes counted that were never placed) votes of any other technology. The under-counted votes for 15 different Florida counties are listed below.

{ 4946, 2070, 5090, 466, 5431, 1044, 1975, 2410, 10570, 634, 10134, 1763, 4240, 1846, 596 }; value: $x = 10,134$.

9. **Politics**: The over-counted votes for the same *Votomatic* machines are listed below.

{ 7826, 1134, 21855, 520, 3640, 790, 2531, 890, 17833, 1039, 19218, 2124, 4258, 994, 170 }; value: $x = 520$.

10. **Economics**: The following are yearly income levels for twenty randomly selected Alaska residents for the year 1999 (selected randomly from the 2000 U.S. Census):

{ 9000, 10600, 60000, 80000, 5000, 2600, 14800, 4500, 45000, 23000, 21300, 50000, 17000, 49000, 85000, 38000, 64000, 25000, 23000, 55000 }; value: $x = 85,000$.

11. **College Demographics**: The following values are hours worked per week for randomly selected students at Mt. San Antonio College in Walnut, California.

{ 15, 20, 40, 16, 20, 25, 19, 40, 40, 20, 20, 40, 0, 40, 35, 30, 35, 25, 29, 0, 0, 40, 16, 0, 40, 30, 25, 16, 35, 0 }; value: $x = 40$.

12. **College Demographics**: The following are grade point averages for the same students whose work hours are listed above.

{ 2.7, 1.7, 2.0, 3.7, 3.1, 4.0, 3.4, 3.7, 2.5, 2.8, 3.0, 3.6, 3.3, 3.2, 3.1, 3.2, 2.6, 2.8, 2.0, 4.0, 2.7, 3.9, 3.1, 2.5, 3.7, 2.9, 3.5, 3.7, 2.2, 3.8 }; value: $x = 4.0$.

Data suggest that the average *eye gain ratio* of people with schizophrenia is significantly different from the average for people without this disorder. An *eye gain ratio* is the ratio of an observer's angular eye velocity to the angular velocity of an object that the eye is following.

13. **Psychology:** For this problem, load the schizophrenic patient eye gain ratio observations, with included control group, in *Statcato* (available at *www.statcato.org*), or open the corresponding spreadsheet at *http://math.mtsac.edu/statistics/data*. For the schizophrenic patients, is the value: $x = 0.520$ unusual?

14. **Psychology:** For this problem, load the schizophrenic patient eye gain ratio observations, with included control group, in *Statcato* (available at *www.statcato.org*), or open the corresponding spreadsheet at *http://math.mtsac.edu/statistics/data*. For the control group, is the value: $x = 0.520$ unusual?

2.6   Data Exploration: Boxplots and Outliers

As we proceed into more advanced topics pertaining to data analysis, the nature of a population's distribution will become more and more important. We shall wish to know how values are distributed, and we will be particularly interested in whether or not our values come from a normal or bell-shaped population. We can use histograms to determine the shape of a distribution, but histograms suffer from the weakness of being affected by arbitrary choices regarding class limits and width. These arbitrary choices can influence the shape of the histogram somewhat, and in some cases such changes in shape can give differing opinions about the nature of the distribution of values.

2.6.1   The Five Point Summary and Boxplots

In this section we introduce a new useful tool for analyzing the distribution of a data set – the boxplot. The boxplot is not influenced by such arbitrariness, and still has the advantage of providing us with a picture of the nature of a sample's distribution.

Before we can define the boxplot, we must begin by defining the *five point summary*.

Definition 2.6.1 The *five point summary* of a data set consists of the minimum value, the lower quartile (Q_1), the median, the upper quartile (Q_3), and the maximum value.

Example 2.6.2 For the data set {**1.52**, 2.21, 2.68, **3.08**, 3.4, 3.71, 4.00, **4.29**, 4.58, 4.87, 5.18, **5.51**, 5.91, 6.38, **7.07**}, the five point summary is: *minimum* = 1.52, *lower quartile* = 3.08, *median* = 4.29, *upper quartile* = 5.51, *maximum* = 7.07.

The five point summary can be quite useful in determining the nature of a distribution of values. The problem, however, with the five point summary is that the relation between them is hard to see without some sort of graphical depiction. To solve this problem, we introduce the *boxplot*.

Definition 2.6.3 The boxplot corresponding to a five point summary consists of:

1. A horizontal line, called the *whisker*, drawn above a number line from the minimum data point to the maximum data point.

2. The whisker passes through the center of a rectangle whose vertical sides extend from the lower quartile to the upper quartile.

3. A vertical line drawn within the rectangle at the median.

The boxplot for Example 2.6.2 is plotted in Figure 2.22.

To understand the way that a boxplot can be used to give insight into the distribution of data values, it helps to plot the boxplot under a histogram representing the same data set. The boxplot and histogram for the data in Example 2.6.2 are plotted in Figure 2.23.

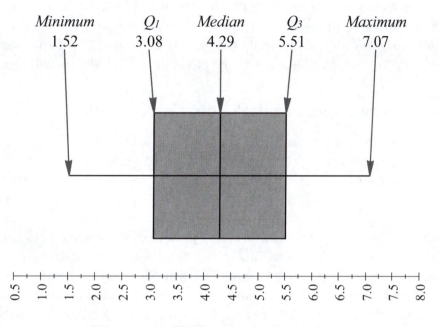
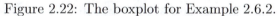

Figure 2.22: The boxplot for Example 2.6.2.

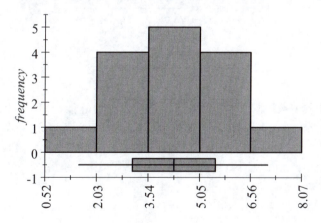

Figure 2.23: The histogram and boxplot corresponding to the values from Example 2.6.2.

Note that the distribution is roughly bell-shaped and symmetric. The box plot is symmetric as well. When a data is truly normal or bell-shaped, the box length, the *inter-quartile range*, will be somewhat smaller than the range, but outliers can have a strong influence on this picture – as always we must be careful when our data set contains outliers as they can lead us to make incorrect conclusions. When there are no outliers, a *normal* data set usually exhibits an inter-quartile range that is about 25% of the overall range. In the example above, the inter-quartile range is about 40% of the overall range.

2.6.2 Sample Skewness and Pearson's Kurtosis

There are two measurements which help us see objectively how values are distributed without actually needing to see a histogram or boxplot for a sample. The *sample skewness* and *kurtosis* are measures of symmetry and peakedness, respectively. A nearly symmetric sample will have skewness nearly equal to zero. If a sample is skewed right or left, the skewness will be positive or negative respectively. *Pearson's sample kurtosis* measures how sharply defined the mode of the sample is. A perfectly bell-shaped, or normal, sample will have a kurtosis equal to three, and this is the value which we consider standard. Data sets with lower, wider peaks and lighter tails (allowing few outliers) will have a kurtosis less than 3, and data sets with higher, narrower peaks and heavier tails (allowing more outliers) will have a kurtosis greater than 3. In all cases, the kurtosis is greater than 1. It should be noted that some texts subtract three from the formula for Pearson's kurtosis below (giving *Fisher's Kurtosis*), so that the standard value is 0 (creating a *zero centered kurtosis*), but we will not adopt this convention. We will apply these formulas in Section 4.3 with examples and homework problems, but for now we simply give the formulas for future reference.

Definition 2.6.4 The sample skewness and Pearson's kurtosis are given by the formulas below.

$$skewness = \frac{\sqrt{n} \cdot \sum (x - \bar{x})^3}{\left(\sum (x - \bar{x})^2\right)^{3/2}},$$

$$kurtosis = \frac{n \cdot \sum (x - \bar{x})^4}{\left(\sum (x - \bar{x})^2\right)^2}.$$

Example 2.6.2 Continued: For the above example, the sample distribution's mode is wider than a normal distribution, and the tails are lighter, emphasized by the fact that the sample kurtosis is 2.19 (the normal distribution kurtosis is 3). The sample skewness is 0.00391, demonstrating that this sample is nearly symmetric.

Example 2.6.5 For our next example, our data are uniformly distributed. Again the boxplot is perfectly symmetric, but because the distribution has heavy tails, the box width (the inter-quartile range) takes up a larger proportion of the range. {1.2, 1.6, 2.0, 2.4, 2.8, 3.2, 3.6, 4.0, 4.4, 4.8, 5.2, 5.6, 6.0, 6.4, 6.8}. The five point summary is: *minimum* = 1.2, *lower quartile* = 2.4, *median* = 4.0, *upper quartile* = 5.6,

$maximum = 6.8$. The totals, $\sum (x - \bar{x})^2$, $\sum (x - \bar{x})^3$, and $\sum (x - \bar{x})^4$, use the mean $\bar{x} = 4$, and are computed below.

x	$(x - \bar{x})^2$	$(x - \bar{x})^3$	$(x - \bar{x})^4$
1.2	7.84	−21.952	61.4656
1.6	5.76	−13.824	33.1776
2.0	4.00	−8.000	16.0000
2.4	2.56	−4.096	6.5536
2.8	1.44	−1.728	2.0736
3.2	0.64	−0.512	0.4096
3.6	0.16	−0.064	0.0256
4.0	0.00	0.000	0.0000
4.4	0.16	0.064	0.0256
4.8	0.64	0.512	0.4096
5.2	1.44	1.728	2.0736
5.6	2.56	4.096	6.5536
6.0	4.00	8.000	16.0000
6.4	5.76	13.824	33.1776
6.8	7.84	21.952	61.4656
Σ	44.8	0.000	239.4112

The totals are:

$$\sum (x - \bar{x})^2 = 44.8000,$$

$$\sum (x - \bar{x})^3 = 0.0000,$$

$$\sum (x - \bar{x})^4 = 239.4112.$$

From these we compute the skewness.

$$skewness = \frac{\sqrt{n} \cdot \sum (x - \bar{x})^3}{\left(\sum (x - \bar{x})^2 \right)^{3/2}}$$

$$= \frac{\sqrt{15} \cdot 0.0000}{(44.8000)^{3/2}}$$

$$= 0$$

Next, we compute the sample kurtosis.

$$kurtosis = \frac{15 \cdot 239.4112}{(44.8000)^2}$$

$$\approx 1.79$$

The flatness of the distribution is indicated by the wide box plot in Figure 2.24, and emphasized by the fact that the sample *kurtosis* is 1.79, much less than the standard value of 3, indicating a data set without tails or outliers, and in this case, no peak at all.

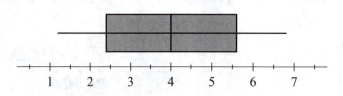

Figure 2.24: The boxplot of data values in Example 2.6.5.

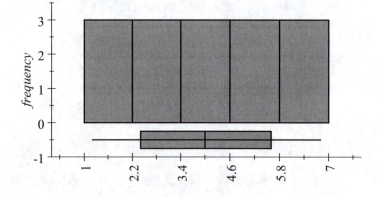

Figure 2.25: A histogram and boxplot corresponding to the data given in Example 2.6.5.

The histogram is plotted in Figure 2.25, with the boxplot.

Example 2.6.6 Our next data set is skewed to the right. {1.2, 1.6, 1.8, 1.9, 1.9, 2.0, 2.2, 2.6, 3.0, 3.5, 4.0, 4.8, 5.6, 6.6, 7.6}. The five point summary is: *minimum*= 1.2, *lower quartile* = 1.9, *median* = 2.6, *upper quartile* = 4.8, *maximum* = 7.6. Notice the box is shifted to the left of the range, and that the line in the box representing the median is also shifted to the left. The skewness of this sample is 0.925, further indicating a positively skewed (to the right) sample. The boxplot for these values is plotted in Figure 2.26.

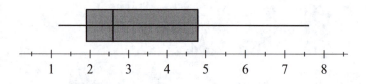

Figure 2.26: The boxplot of data values given in Example 2.6.6.

The mean of these values is $\bar{x} = 3.35$.

The totals used in the skewness and kurtosis are computed in the table below.

x	$(x-\bar{x})^2$	$(x-\bar{x})^3$	$(x-\bar{x})^4$
1.2	4.6368	−9.9847	21.5003
1.6	3.0742	−5.3901	9.4506
1.8	2.4128	−3.7480	5.8218
1.9	2.1122	−3.0697	4.4613
1.9	2.1122	−3.0697	4.4613
2.0	1.8315	−2.4786	3.3544
2.2	1.3302	−1.5341	1.7694
2.6	0.5675	−0.4275	0.3221
3.0	0.1248	−0.0441	0.0156
3.5	0.0215	0.0032	0.0005
4.0	0.4182	0.2704	0.1749
4.8	2.0928	3.0276	4.3800
5.6	5.0475	11.3401	25.4774
6.6	10.5408	34.2226	111.1094
7.6	18.0342	76.5851	325.2316
Σ	54.3573	95.7026	517.5304

The totals are

$$\sum (x - \bar{x})^2 = 54.3573,$$

$$\sum (x - \bar{x})^3 = 95.7026,$$

$$\sum (x - \bar{x})^4 = 517.5304.$$

From these we compute the skewness.

$$
\begin{aligned}
skewness &= \frac{\sqrt{n} \cdot \sum (x - \bar{x})^3}{\left(\sum (x - \bar{x})^2 \right)^{3/2}} \\
&= \frac{\sqrt{15} \cdot 95.7026}{(54.3573)^{3/2}} \\
&\approx 0.925
\end{aligned}
$$

Next, we compute the sample kurtosis.

$$
\begin{aligned}
kurtosis &= \frac{15 \cdot 517.5304}{(54.3573)^2} \\
&\approx 2.627
\end{aligned}
$$

The *kurtosis* is closer to 3, indicating a peak that exists, with a longer tail, but the *skewness* is positive, indicating that the distribution of values is skewed to the right.

The histogram and boxplot for Example 2.6.6 are plotted together in Figure 2.27.

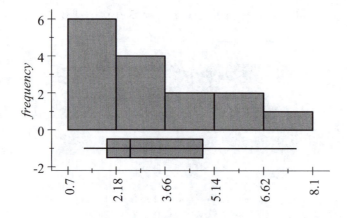

Figure 2.27: The histogram and boxplot for Example 2.6.6.

Example 2.6.7 This next collection of values is skewed to the left: {1.4, 2.4, 3.4, 4.2, 5.0, 5.5, 6.0, 6.4, 6.8, 7.0, 7.1, 7.1, 7.2, 7.4, 7.8}. The five point summary is: *minimum* = 1.4, *lower quartile* = 4.2, *median* = 6.4, *upper quartile* = 7.1, *maximum* = 7.8. The sample mean is $\bar{x} = 5.65$, and this is used to compute the totals below.

$$\sum \left(x - \bar{x}\right)^2 = 54.3573$$
$$\sum \left(x - \bar{x}\right)^3 = -95.7026,$$
$$\sum \left(x - \bar{x}\right)^4 = 517.5304.$$

From these we compute the sample *skewness* and *kurtosis*.

$$skewness = \frac{\sqrt{n} \cdot \sum \left(x - \bar{x}\right)^3}{\left(\sum \left(x - \bar{x}\right)^2\right)^{3/2}}$$
$$= \frac{\sqrt{15} \cdot (-95.7026)}{(54.3573)^{3/2}}$$
$$\approx -0.925$$

Next, we compute the sample kurtosis.

$$kurtosis = \frac{15 \cdot 517.5304}{(54.3573)^2}$$
$$\approx 2.627$$

The kurtosis is the same here as in Example 2.6.6, *kurtosis* ≈ 2.627. With equal kurtoses, the peaks of the distributions of values are the same shape here as in the last example, but the skewnesses are reversed. The *skewness* of this sample is −0.925, indicating a negatively skewed (to the left) distribution.

The boxplot of this sample is plotted in Figure 2.28.

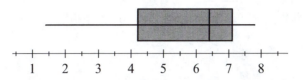

Figure 2.28: The boxplot for Example 2.6.7.

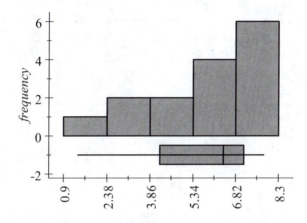

Figure 2.29: The histogram and boxplot for Example 2.6.7.

Once more, the histogram and boxplot appear together in Figure 2.29.

Example 2.6.8 Here we examine the speed of light data from Earnest Dorsey's classic experiments conducted in between Mt. Baldy and Mt. Wilson in Southern California.

				Speed of Light Estimates					
299.62	299.65	299.72	299.72	299.72	299.74	299.74	299.74	299.75	299.76
299.76	299.76	299.76	299.76	299.77	299.78	299.78	299.79	299.79	299.79
299.80	299.80	299.80	299.80	299.80	299.81	299.81	299.81	299.81	299.81
299.81	299.81	299.81	299.81	299.81	299.82	299.82	299.83	299.83	299.84
299.84	299.84	299.84	299.84	299.84	299.84	299.84	299.85	299.85	299.85
299.85	299.85	299.85	299.85	299.85	299.86	299.86	299.86	299.87	299.87
299.87	299.87	299.88	299.88	299.88	299.88	299.88	299.88	299.88	299.88
299.88	299.88	299.89	299.89	299.89	299.90	299.90	299.91	299.91	299.92
299.93	299.93	299.94	299.94	299.94	299.95	299.95	299.95	299.96	299.96
299.96	299.96	299.97	299.98	299.98	299.98	300.00	300.00	300.00	300.07

The five point summary for this sample is *minimum* = 299.62, *lower quartile* = 299.805, *median* = 299.85, *upper quartile* = 299.895, *maximum* = 300.07. The boxplot is roughly symmetric, and the box is small in length compared to the range. This is emphasized by the fact that the *kurtosis* for this sample is 3.264, indicating that the peak is *slightly* higher and narrower, with heavier tails, than the normal standard of 3, and the *skewness* is -0.018, indicating that the sample data are quite symmetric.

The boxplot for the speed of light data is plotted in Figure 2.30.

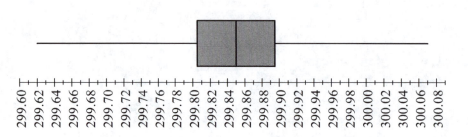

Figure 2.30: The boxplot for Example 2.6.8.

The boxplot is again plotted with its corresponding histogram in Figure 2.31.

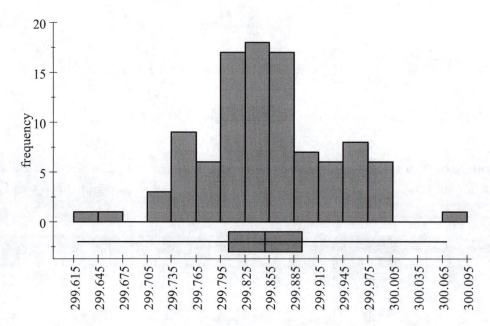

Figure 2.31: The histogram and boxplot for the speed of light estimates in Example 2.6.8.

2.6.3 Outliers

We have referred repeatedly to *outliers*, but have to this point given no objective means for determining which values are truly outliers. A standard that is often used is as follows.

Definition 2.6.9 Any value that lies more than 1.5 inter-quartile ranges from an outer edge of the box in the boxplot is an *outlier*. Thus, x is an outlier if

$$x > Q_3 + 1.5 \cdot IQR$$

or

$$x < Q_1 - 1.5 \cdot IQR.$$

An outlier is *extreme* if it is more than 3 inter-quartile ranges from an edge of the box. Therefore, x is an extreme outlier if

$$x > Q_3 + 3 \cdot IQR$$

or

$$x < Q_1 - 3 \cdot IQR.$$

Example 2.6.8 Continued: Returning to the speed of light data, the inter-quartile range is $IQR = Q_3 - Q_1 = 299.895 - 299.805 = 0.09$. For the upper limit we compute

$$\begin{aligned}
x &> Q_3 + 1.5 \cdot IQR \\
&= 299.895 + 1.5 \cdot 0.09 \\
&= 300.03.
\end{aligned}$$

For the lower limit we compute

$$\begin{aligned}
x &< Q_1 - 1.5 \cdot IQR \\
&= 299.805 - 1.5 \cdot 0.09 \\
&= 299.67.
\end{aligned}$$

So any point outside of the range from 299.67 to 300.03 would be an outlier. On the left, there are two outliers, 299.62 and 299.65, and on the right there is only 300.07.

To determine extreme outliers, we compute as follows.

$$\begin{aligned}
x &> Q_3 + 3 \cdot IQR \\
&= 299.895 + 3 \cdot 0.09 \\
&= 300.165
\end{aligned}$$

The lower limit is next.

$$\begin{aligned}
x &< Q_1 - 3 \cdot IQR \\
&= 299.805 - 3 \cdot 0.09 \\
&= 299.535
\end{aligned}$$

Any value outside of the range from 299.535 to 300.165 would be an extreme outlier. There are no values outside of this range, so our data set has no extreme outliers.

2.6.4   Enhanced Boxplots

Sometimes we want our boxplots to display outliers in a data when they exist. This is an understandable need, because we often wish to see what our data set would look like without outliers. When low outliers exist, (less than $Q_1 - 1.5 \cdot IQR$) we will start the left end of the whisker at the value, $Q_1 - 1.5 \cdot IQR$, which defines the low outliers. When high outliers exist, (greater than $Q_3 + 1.5 \cdot IQR$) we will terminate the right end of the whisker at the value, $Q_3 + 1.5 \cdot IQR$, which defines the high outliers. Outliers are plotted beyond the whiskers as asterisks.

Example 2.6.8 Continued: Recalling the speed of light data, we determined that any point outside of the range from $Q_1 - 1.5 \cdot IQR = 299.67$ to $Q_3 + 1.5 \cdot IQR = 300.03$ would be an outlier. There were two low outliers, 299.62 and 299.65, and only one high outlier, 300.07. The enhanced boxplot's whisker extends from 299.67 to 300.03, with the three outliers plotted using asterisks, as shown in Figure 2.32.

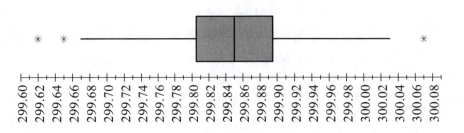

Figure 2.32: The enhanced boxplot identifies outliers with asterisks.

2.6.5   Boxplots and Normal Distributions of Values

Next, we examine the appearance of a boxplot from a population which is perfect in its normality. In Figure 2.33, an enhanced boxplot is plotted above a normal bell curve. The whiskers on the boxplot extend from $Q_1 - 1.5 \cdot IQR$ to $Q_3 + 1.5 \cdot IQR$. In such a distribution of values, $24.65\% + 50\% + 24.65\% = 99.3\%$ of all values are not characterized as outliers, and thus 0.7% of values are outliers. As we have mentioned, the interquartile range, the length of the boxplot's box, is about 25% of the range of values in such a distribution, excluding outliers. The first quartile is always 0.67 standard deviations below the mean, and the third quartile is always 0.67 standard deviations above the mean.

2.6.6   Comparing Samples with Parallel Boxplots

In Section 2.2 we used stacked dotplots to compare the distributions of two samples. In this section, we would like to refine this process. Here, we will compare sections by the same criteria (from Procedure 2.2.1 on page 59), but the visual tool we use here is a graph with parallel boxplots, both sharing the same coordinate system. The boxplot contains more precise information about the distribution of values in a data set, and we are therefore able to make more certain judgements and inferences.

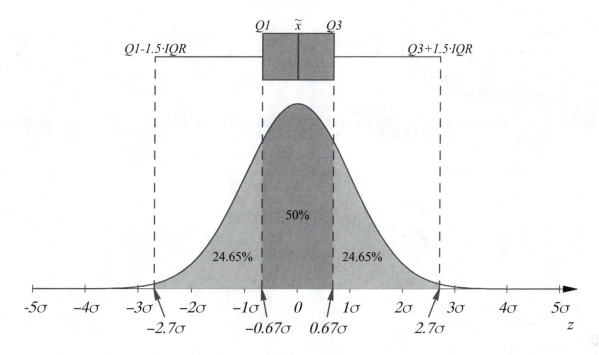

Figure 2.33: An enhanced boxplot is plotted above the bell-curve of a perfectly normal distribution of values, excluding outliers.

Keep in mind, however, that while the boxplot is a more precise tool for making comparisons, we are still making our inferences based on sample data, and thus no conclusion can be absolutely certain. The wording of any conclusions based on sample data should reflect this uncertainty. Also, since boxplots use the median as a measure of center, it may be best to make claims regarding the median. When the distributions of values are moderately symmetric, the median and mean values should be similar, and in such cases claims can be made regarding the mean of the distributions.

When considering the strength of evidence against a claim, consider not only the distance between the median lines of the boxplots, but the separation between the boxes which represent the middle 50% range of values (from Q_1 to Q_2). The wider the separation between the boxes, the stronger the evidence in any claim regarding the difference between the medians.

Example 2.6.10 Below are heights (in cm) for randomly selected adult females and for randomly selected adult males.

Females: {148.0, 152.2, 155.2, 152.7, 155.1, 155.5, 156.8, 158.4, 158.3, 156.2, 158.6, 159.9, 164.2, 157.9, 159.8, 160.9, 162.3, 158.0, 159.8, 156.9, 162.4, 160.2, 166.9, 167.2, 167.6, 164.0, 170.4, 175.9, 176.4, 184.1}

Males: {167.7, 164.1, 159.8, 165.4, 166.6, 171.2, 170.6, 168.6, 168.5, 172.4, 168.0, 172.7, 169.7, 176.3, 174.1, 172.8, 177.6, 178.9, 177.0, 175.5, 179.4, 174.8, 175.5, 182.2, 178.3,

179.1, 176.3, 178.4, 177.1, 189.1}

In Figure 2.34 we plot the stacked boxplots for these samples.

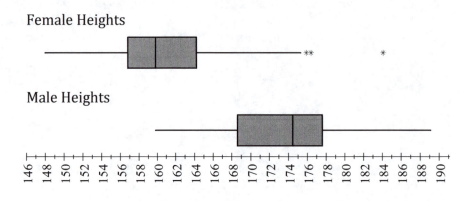

Figure 2.34: The boxplots for Example 2.6.10.

The five point summaries for these samples are below.
Females: {148.0, 156.8, 159.8, 164.2, 184.1}
Males: {159.8, 168.6, 174.45, 177.6, 189.1}

The elements of our comparison, from Procedure 2.2.1, are included below.

Step I.   **Hypothesis:** As before, we hypothesize that the females have a smaller median height than the males.

Step II.   **Summary:** Here we see that the median summary values are as follows. For the females, the median is 159.8 cm, and for males the median is 174.45 cm.

Step III.   **Shift:** The distribution of female heights is clearly shifted to the left of the males.

Step IV.   **Overlap:** Clearly the distributions overlap here, but the middle 50% regions of each sample do not overlap. The medians are therefore separated by a large distance.

Step V.   **Spread:** The spread of female heights is larger if we take their outliers into account. Without the outliers, the ranges are quite similar.

Step VI.   **Sample Size:** Both samples are of size 30. These are small, and this smallness weakens our inferences.

Step VII.   **Meaning in Context:** Both samples contain heights, in centimeters, of sample members. The context of this problem is simple, and what we observe in the sample data is logical – most men are taller than most women.

Step VIII.   **Extreme Cases:** The sample of female heights has three outliers on the right. Because the centers of these distributions are measured using medians, these outliers will not have much influence over them as averages.

Step IX. **Inferences:** The sample data seem to support our hypothesis. We claimed that the median height for males is greater than the median height for females, and this is what we observe. The inferences are weakened by the small sample sizes, but the fact that the middle 50% regions do not overlap adds strength to our conclusions.

2.6.7   Homework

For each of the problems below, give the 5-point summary, and draw an enhanced boxplot. Compute the sample *skewness* and *kurtosis*, classifying the distribution as *positively skewed*, *negatively skewed*, or *symmetric*. Classify the distribution as *platykurtic*, *mesokurtic*, or *leptokurtic*. Note whether there are any *outliers* or *extreme outliers*.

1. The following are yearly income levels for twenty randomly selected Alaska residents for the year 1999 (selected randomly from the 2000 U.S. Census):

   {9000, 10600, 60000, 80000, 5000, 2600, 14800, 4500, 45000, 23000, 21300, 50000, 17000, 49000, 85000, 38000, 64000, 25000, 23000, 55000}

2. The following are randomly selected National League baseball salaries from 1994.

   {$437,500, $500,000, $2,450,000, $350,000, $109,000, $135,000, $275,000, $846,667, $200,000, $3,583,333, $225,000, $165,000, $145,000, $350,000, $650,000, $3,500,000, $180,000, $3,125,000, $3,750,000, $112,000, $915,000, $109,000, $2,916,667, $600,000, $109,000, $135,000, $4,750,000, $4,300,000, $115,000, $1,375,000, $130,000, $525,000, $300,000, $700,000, $225,000, $109,000, $109,000, $1,100,000, $2,700,000, $2,150,000, $155,000, $250,000, $1,500,000, $192,500, $109,000, $3,333,334, $6,300,000, $109,000, $187,500, $1,008,334, $1,800,000, $140,000, $109,000, $172,500, $125,000, $130,000, $2,316,667, $3,250,000, $175,000, $575,000, $695,000, $2,575,000}

3. The following are average Math SAT scores for students admitted to 67 random colleges and universities in the United States.

   {524, 652, 507, 451, 500, 634, 750, 368, 597, 520, 621, 470, 526, 505, 550, 590, 320, 556, 482, 474, 537, 446, 400, 468, 561, 529, 520, 533, 485, 428, 567, 442, 501, 497, 470, 513, 489, 490, 455, 608, 480, 541, 458, 420, 438, 580, 516, 540, 450, 545, 465, 572, 680, 433, 530, 440, 576, 413, 552, 510, 444, 390, 478, 486, 495, 460}

4. The following values are hours worked per week for randomly selected students at Mt. San Antonio College in Walnut, California.

   { 15, 20, 40, 16, 20, 25, 19, 40, 40, 20, 20, 40, 0, 40, 35, 30, 35, 25, 29, 0, 0, 40, 16, 0, 40, 30, 25, 16, 35, 0 }.

5. The following are grade point averages for the same students whose work hours are listed above.

   { 2.7, 1.7, 2.0, 3.7, 3.1, 4.0, 3.4, 3.7, 2.5, 2.8, 3.0, 3.6, 3.3, 3.2, 3.1, 3.2, 2.6, 2.8, 2.0, 4.0, 2.7, 3.9, 3.1, 2.5, 3.7, 2.9, 3.5, 3.7, 2.2, 3.8 }

6. Below are a collection of randomly selected unemployment rates gathered from random months ranging from 1948 to 2004.

{ 7.9, 5.2, 5.5, 4.1, 10.8, 5.6, 3.1, 5.1, 6.7, 6.3, 4.6, 6.0, 8.9, 5.8, 4.8, 3.5, 3.7, 6.1, 5.4, 5.0, 7.4, 3.9, 4.4, 7.6, 2.5, 6.5, 7.1, 5.6, 5.7, 5.4, 4.2, 7.3, 6.9, 3.8, 5.9 }

7. The following are average electricity prices, in dollars per $500 \, kW \cdot h$, in the United States from 1979 to 1996.

{ 25.33, 29.83, 34.19, 37.20, 38.30, 40.06, 39.99, 40.88, 40.93, 40.66, 41.62, 42.54, 44.03, 45.82, 48.02, 48.69, 49.30, 48.93 }

8. The following are numbers of pregnancies to single teens in 1990 by state.

{ 5932, 588, 4801, 3370, 30511, 3242, 2915, 929, 1610, 13817, 9453, 1136, 674, 16069, 6398, 2090, 2182, 3780, 8808, 1146, 6193, 4919, 9416, 3353, 6039, 5995, 761, 1375, 838, 706, 7898, 2468, 19465, 7897, 481, 12980, 3589, 2384, 12349, 943, 5534, 702, 5983, 19333, 1207, 475, 6175, 4193, 1751, 5000, 455 }

9. In the 2000 U.S. presidential elections, the inaccuracy of voting technology became the subject of worldwide interest. In Florida, the two front-runner candidates, George W. Bush and Al Gore, were so close in the tally that votes had to be recounted by hand to verify the totals. In the process of recounting, it was found that the most unreliable voting machine was the *Votomatic*, with the most under-counted (votes placed but not counted) and over-counted (votes counted that were never placed) votes of any other technology. The under-counted votes for 15 different Florida counties are listed below.

{ 4946, 2070, 5090, 466, 5431, 1044, 1975, 2410, 10570, 634, 10134, 1763, 4240, 1846, 596 }

10. The over-counted votes for the same *Votomatic* machines are listed below.

{ 7826, 1134, 21855, 520, 3640, 790, 2531, 890, 17833, 1039, 19218, 2124, 4258, 994, 170 }

11. The following are the number of electoral votes for each state in the U.S.

{ 9, 3, 10, 6, 55, 9, 7, 3, 3, 27, 15, 4, 4, 21, 11, 7, 6, 8, 9, 4, 10, 12, 17, 10, 6, 11, 3, 5, 5, 4, 15, 5, 31, 15, 3, 20, 7, 7, 21, 4, 8, 3, 11, 34, 5, 3, 13, 11, 5, 10, 3 }

Sketch stacked boxplots over a common coordinate system, and use Procedure 2.2.1 on page 59 to compare the following samples. Please address the following: hypothesis, summary, shift, overlap, spread, sample size, meaning in context, extreme cases, and inferences suggested by the graphs.

12. The following are average tip rates for a random selection of female and male food servers. Compare the samples.

    Females: {14%, 10%, 14%, 12%, 15%, 13%, 13%, 13%, 14%, 14%, 13%, 12%, 17%, 12%, 12%, 14%, 15%, 12%, 11%, 14%, 13%, 13%, 8%, 7%}

    Males: {10%, 11%, 12%, 13%, 11%, 11%, 12%, 14%, 15%, 10%}

13. The following are grade point averages for smoking and non-smoking community college students. Compare the samples.

    Smokers: {2.0, 2.0, 2.0, 2.0, 2.1, 2.2, 2.4, 2.4, 2.4, 2.4, 2.5, 2.5, 2.5, 2.5, 2.5, 2.5, 2.6, 2.6, 2.6, 2.6, 2.6, 2.6, 2.6, 2.7, 2.7, 2.7, 2.7, 2.7, 2.8, 2.8, 2.8, 2.8, 2.8, 2.8, 2.8, 2.8, 2.9, 2.9, 2.9, 3.0, 3.0, 3.0, 3.0, 3.1, 3.1, 3.1, 3.2, 3.2, 3.2, 3.2, 3.2, 3.3, 3.3, 3.3, 3.4, 3.4, 3.4, 3.4, 3.5, 3.5, 3.5, 3.5, 3.6, 3.6, 3.6, 3.6, 3.6, 3.7, 3.7, 3.7, 3.7, 3.8, 3.8, 3.8, 3.8, 3.8, 3.9, 4.0, 4.0}

    Non-smokers: {2.3, 2.6, 2.6, 2.7, 2.7, 2.7, 2.8, 2.8, 2.8, 2.8, 2.8, 2.9, 2.9, 2.9, 3.0, 3.0, 3.0, 3.0, 3.0, 3.0, 3.0, 3.0, 3.1, 3.1, 3.2, 3.2, 3.2, 3.3, 3.3, 3.3, 3.3, 3.3, 3.3, 3.4, 3.4, 3.4, 3.4, 3.4, 3.5, 3.5, 3.5, 3.5, 3.5, 3.6, 3.6, 3.6, 3.6, 3.7, 3.7, 3.7, 3.7, 3.7, 3.7, 3.7, 3.7, 3.8, 3.8, 3.8, 3.8, 3.8, 3.8, 3.8, 3.8, 3.9, 3.9, 3.9, 3.9, 3.9, 3.9, 3.9, 3.9, 3.9, 3.9, 3.9, 4.0, 4.0, 4.0, 4.0, 4.0, 4.0, 4.0}

14. Below are foot lengths (in cm) for male and female adults. Compare the samples.

    Male feet: {24.4, 24.3, 25.4, 25.8, 26.3, 24.7, 25.1, 24.6, 26.1, 26.8, 25.1, 25.8, 26.4, 25.7, 25.6, 26.2, 26.2, 26.8, 26.4, 26.8, 26.3, 26.4, 24.5, 25.6, 25.7, 24.9, 27.0, 27.8, 27.6, 27.6}

    Female feet: {18.2, 20.4, 21.5, 20.4, 21.1, 19.1, 20.6, 20.4, 21.7, 21.6, 23.0, 20.9, 20.0, 21.8, 21.0, 22.3, 22.7, 22.8, 21.8, 23.7, 21.3, 22.7, 22.1, 22.6, 22.0, 21.6, 22.8, 22.1, 24.2, 23.7}

15. Below are lipoprotein levels for subjects who eat corn flakes every morning for breakfast, and also for subjects who eat oat bran (Pagano & Gauvreau, 1993, p. 252). Compare the samples.

    Corn Flakes: {4.6, 6.4, 5.4, 4.5, 4.0, 3.8, 5.0, 4.3, 3.8, 4.6, 5.4, 3.9, 2.3, 4.2}

    Oat Bran: {3.8, 5.6, 5.9, 4.8, 3.7, 3.0, 4.4, 3.7, 3.5, 3.8, 5.3, 3.7, 1.8, 4.1}

16. When college class locations are inconvenient for students, does student success decline? The following samples include student success rates for two different populations teachers – those whose classrooms are in a convenient location, and those whose classrooms are in an inconvenient location. Compare the samples.
Convenient: {37%, 51%, 47%, 61%, 63%, 59%, 32%, 70%, 46%, 44%, 68%, 68%, 56%, 89%, 76%, 32%, 57%, 61%, 79%}

Inconvenient: {42%, 66%, 46%, 57%, 73%, 63%, 39%, 66%, 64%, 55%, 61%, 34%, 36%, 44%, 25%, 44%, 30%, 44%, 69%}

17. Are grocery prices higher in more affluent neighborhoods? Thirty-one randomly selected products are purchased in two grocery stores – first in an affluent neighborhood, then again with the same products in a non-affluent neighborhood. Both stores are members of the same grocery store franchise. Compare the samples.

Affluent: {$3.00, $2.80, $3.30, $3.00, $1.80, $1.40, $4.70, $3.80, $1.80, $2.10, $1.10, $1.10, $2.80, $1.90, $0.90, $0.80, $1.40, $2.30, $1.80, $1.90, $0.20, $2.40, $1.30, $3.00, $2.80, $1.30, $1.70, $0.90, $0.70, $1.70, $1.80}

Non-Affluent: {$3.00, $2.80, $3.30, $3.10, $2.00, $1.10, $4.40, $3.30, $1.60, $1.80, $0.90, $0.90, $2.60, $1.30, $0.90, $0.80, $1.40, $2.10, $1.60, $1.90, $0.30, $2.00, $1.00, $3.30, $3.00, $1.40, $1.60, $0.90, $0.70, $1.70, $1.60}

18. Data suggest that the average *eye gain ratio* of people with schizophrenia is significantly different from the average for people without this disorder. An *eye gain ratio* is the ratio of an observer's angular eye velocity to the angular velocity of an object that the eye is following.

Compare the gain ratios of the eyes of schizophrenic people to the control group provided in Statcato (available at www.statcato.org), or using the Excel spreadsheets at http://math.mtsac.edu/statistics/excel.

Part III
Probability

Chapter 3

Basic Rules of Probability

As we proceed toward the last step in the process of statistical analysis, *making statistical inferences* (Figure 3.1), we must introduce the basic concepts of *probability*. When we make statistical inferences, they are rooted in probability, so it is important that we have a basic grasp on probability theory.

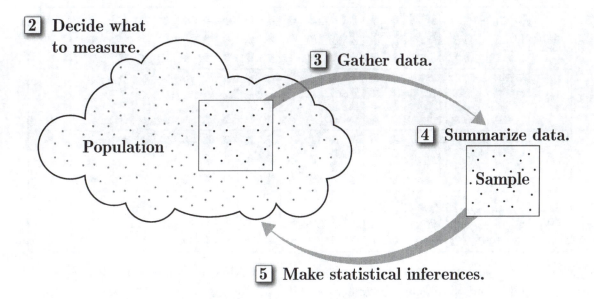

Figure 3.1: The process of statistical analysis.

Our work in probability theory proceeds as follows. In Chapter 3, we introduce the basic concepts of probability theory. In Chapters 4 and 5 we learn about discrete and continuous probability distributions. After this, we will be ready to begin making statistical inferences.

3.1   Introduction to Probability

In this section, we begin computing probabilities using the most basic formulas. In this chapter, we will have several rules to learn, but they are often unnecessary – for our purposes, all that we usually need is a basic understanding of the simplest rules, which we will cover soon.

First, we begin with a few terms which we will need.

Definition 3.1.1 An experiment will be any *trial* which may have any number of possible outcomes.

Definition 3.1.2 An event is just one of the many *possible* outcomes of an experiment. Events are often denoted by capital letters S, F, A, B, C, etc.

Example 3.1.3 Our first example will use a poker deck with no jokers, which includes a total of 52 cards. There are four suits in a poker deck: *diamonds*, *clubs*, *hearts*, and *spades*. Each suit has 13 cards: Ace, 2, 3, 4, 5, 6, 7, 8, 9, 10, Jack, Queen, and King. An image of an entire poker deck is given in Figure 3.2.

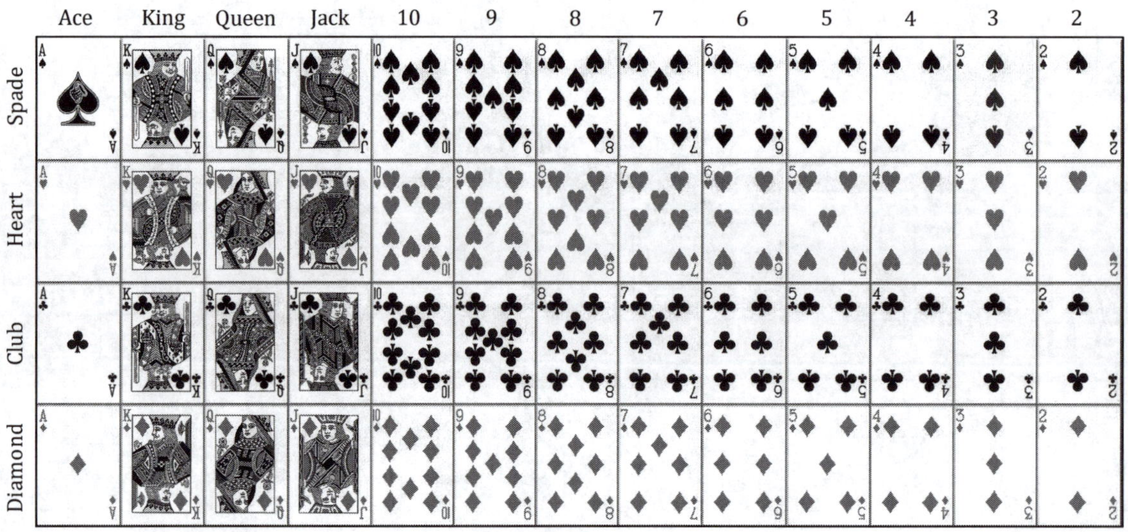

Figure 3.2: A poker deck of cards. The deck contains 52 cards, composed of 13 spades, 13 hearts, 13 clubs, and 13 diamonds. (*Image compiled by the author from 52 card images in the public domain under Wikimedia Commons licensing*).

When we say that the deck is *shuffled*, we mean that the cards are put into random order, so that if we pick a card, it may be any card at all, and each has an equal likelihood of being picked. The selection is therefore *random*.

If our experiment is to take a card from a shuffled deck, this is a *trial*. It is a trial which has many possible outcomes, or *events*. The card could be the *queen of spades*, or the card could be a *heart*. These are both outcomes which may happen, but of course, there are many others.

The way we compute probability depends partly on the number of ways an event can happen. For instance, there is only one queen of spades in the deck, so this outcome can happen in only one way. Whenever an event can happen in only one way, we call the event *simple*. There are many hearts in the deck, so the *heart* outcome may happen in many ways. Whenever an event can happen in more than one way, the event is called *compound*.

Definition 3.1.4 An event is *simple* when it can happen in only one way.

Definition 3.1.5 An event is *compound* when it can happen in more than one way.

Having given the definition of a simple event, we may proceed to define the *sample space* in a probability experiment.

Definition 3.1.6 The *sample space* for a given experiment is the set of all possible *simple* events in the experiment.

For the experiment where we pick a single card from a shuffled poker deck, the sample space is the collection of the 52 individual cards. No other simple outcomes are possible.

3.1.1　The Probability of an Event

We will denote the *probability* of event S by $P(S)$. For any event, S, the probability of S will always be between 0 and 1, inclusive. That is, for any event, S, $0 \leqslant P(S) \leqslant 1$. When $P(S)$ is near zero, S is unlikely, and when $P(S) = 0$, S is impossible. When $P(S)$ is near 1, then S is likely, but when $P(S) = 1$, then event S *must happen*. If $P(S) = 0.5$, then S is just as likely to happen as to not happen.

How we compute $P(S)$ is the matter which we must address next. There are, in some cases, ways to compute $P(S)$ exactly, but unfortunately, many times the best we can do is to estimate this value. Generally speaking, in a finite sample space, when we know that all simple outcomes are equally likely, we can compute $P(S)$ exactly. Unfortunately, it is unusual, especially when working with sample data, for all outcomes to be equally likely. When this is the case, the best we can do is to estimate probabilities using sample data and relative frequencies – but such estimates will always contain sampling error.

Definition 3.1.7 *The Classical Approach to Computing Probabilities*

In an experiment where each of N possible simple outcomes are equally likely, if event S can happen in X simple ways, then

$$P(S) = \frac{X}{N}.$$

It is important to note that, when working with data sets to compute this probability, this probability is only exact if the data includes the entire population. In this case, X is the number of times that S occurs in the entire population, while N is the finite population size, and in this case $P(S)$ will be called the population proportion (which will be denoted by$p$).

Example 3.1.3 Continued: Referring to our experiment where we draw one card randomly from a shuffled deck, each outcome is equally likely, so there are $N = 52$ simple outcomes. If event $S = $ *"the card is a heart"*, we note that there are $X = 13$ ways to draw a heart from the deck. Thus,

$$\begin{aligned} P(S) &= \frac{13}{52} \\ &= 0.250 \\ &= 25.0\%. \end{aligned}$$

We need a rule for rounding probabilities, proportions, and percents. Our rule requires that the value contains 3 *significant* digits. For our purposes it is sufficient to say that all digits are significant except *leading zeros*. Trailing zeros are significant, even though they do not alter the value of the number. They are significant because they tell us the value of a known digit in its respective position. So $13/52 = 0.25$ is rounded incorrectly, because the leading zero is not significant. But $0.25 = 0.250$, and rounding $13/52$ as 0.250 is correct, because the trailing zero is significant.

Definition 3.1.8 *Rounding Rule for Probabilities, Proportions, and Percentages*
When rounding a proportion, percent, or probability, keep at least 3 significant digits, assuming that all digits are significant *except* leading zeros.

As another example, if we define event $B = $ "the card is a queen", we still have $N = 52$, but now since there are only $X = 4$ queens in the deck, and so $P(B) = 4/52 = 0.0769$. Notice the two leading zeros are not significant, so this rounding includes only the minimum of 3 significant digits.

When it is not clear that all outcomes are equally likely, we will estimate probabilities using sample data and relative frequencies. This is often what we must do, even though such an estimate always includes sampling error. When we estimate probabilities in this way, we will use the relative frequencies which were discussed in Section 2.1.

Definition 3.1.9 Estimating Probabilities using Relative Frequencies
In a sample of size n, representing an experiment with n trials, if event S happens x times, then

$$P(S) \approx \frac{x}{n}.$$

This estimate includes sampling error, but as we will see, the error decreases as sample size increases (this principle is known as the *law of large numbers*). The true value of $P(S)$ is the value that the relative frequency approaches as n becomes indefinitely large. Another name for this relative frequency is the *sample proportion* (which will be denoted by $\hat{p} = \frac{x}{n}$).

It should be noted that some populations are not *finite*. For example, if we want to know the probability that a particular coin lands heads, we must compute the proportion of all possible coin tosses which would land heads. But there is no limit to the number of times a coin can be tossed, and thus the population is, in this sense, *infinite*. When populations are infinite, we can only estimate probabilities using relative frequencies, understanding that the true probabilities cannot be known, but that relative frequencies approach these exact probabilities as sample size, n, increases indefinitely.

Example 3.1.10 Mathematician John Kerrich tossed a coin 10,000 times while interned in a prison camp in Denmark during World War I. At various stages of the experiment, the relative frequency would climb or fall below the theoretical probability of 0.5, but as the number of tosses increased, the relative frequency tended to vary less and stay near 0.5, or 50 percent. (Only in a prison camp would a mathematician toss a coin 10,000 times and record the results!) Overall he observed 5067 heads out of these 10,000 tosses.

As we have suggested, in an application like this, there is no limit to the number of tosses which may be counted as part of the sample. Because of this, the population from which we are sampling is effectively infinite. No number of tosses could result in a population probability. Our experiment is the tossing of John Kerrich's coin, and we will compute the sample probability of $S = $ "coin lands heads" as a relative frequency.

John Kerrich conducted our experiment $n = 10,000$ times, and in these trials, the event S occurred 5067 times. With these we estimate $P(S)$ below.

$$\begin{aligned}
P(S) &\approx \frac{x}{n} \\
&= \frac{5067}{10000} \\
&= 0.5067 \\
&\approx 50.7\%
\end{aligned}$$

It is interesting that Kerrich's coin tosses did not yield a proportion that was closer to 50%. Remember that the tosses were a large sample, but a sample nevertheless, and therefore subject to sampling error. The estimate of the probability of heads was 50.7%, which seems to suggest that the coin may be biased toward heads. Should the true proportion be closer to 50%? That is a question that we cannot answer!

3.1.2   Probability and Two-Way Tables

Frequently, when gathering information for a sample, we gather more than one observation from each member. When two observations are made per subject, we have called the data *bivariate*. We have already used scatterplots to summarize bivariate quantitative data. In this section we will consider the case where categorical (qualitative) data are gathered in pairs. To summarize such bivariate categorical data, we tally observations in a two-way, or contingency, table.

The following example involves data gathered from 45 randomly selected community college students. From each student two observations were gathered: their gender and whether they routinely recycle. The results are recorded below.

Gender	Recycle?	Gender	Recycle?	Gender	Recycle?
Male	No	Male	Yes	Female	No
Female	Yes	Male	No	Male	Yes
Male	No	Male	Yes	Female	Yes
Male	Yes	Male	Yes	Female	Yes
Male	Yes	Female	Yes	Female	Yes
Female	No	Female	No	Female	No
Female	No	Male	No	Male	Yes
Female	No	Male	No	Female	No
Male	Yes	Female	No	Male	Yes
Female	No	Male	Yes	Male	Yes
Female	No	Female	Yes	Female	No
Female	No	Female	No	Female	No
Male	No	Female	No	Male	Yes
Female	Yes	Female	Yes	Male	No
Female	No	Male	Yes	Female	Yes

Each response is categorical with two possible outcomes. The two-way table uses row headings for the categories of the first observation, and column headings for the categories of the second observation. Each cell in the table corresponds to a pair of possible observations, indicated by the row and column headings, and in each cell, corresponding frequencies are tallied.

Gender	Yes	No
Male	IIII IIII III	IIII IIII
Female	IIII II	IIII IIII IIII I

Once the tallies are made, row and column totals are often computed, with a grand total given in the lower right cell (the grand total is the total of row totals or column totals).

Gender	Yes	No	Totals
Male	13	9	22
Female	7	16	23
Totals	20	25	45

Two-way tables are useful for determining whether the two variables under consideration are dependent upon one another. For instance, we might be interested in the probability that a person recycles, given that she is female. To compute this we assume that we are picking from the females, of which there are 23, and from these we see that 7 recycle. Thus,

$$P\,(recycle \text{ given that she is } female) = \frac{7}{23}$$
$$\approx 0.304$$
$$= 30.4\%.$$

In this problem, we only used 23 for the denominator because we restricted the number of outcomes to females only. The obvious comparison to make is to the probability that a person recycles, given that he is male. In the sample, there are 22 males, and of these, 13 recycle. The probability is therefore

$$P\,(recycle \text{ given that he is } male) = \frac{13}{22}$$
$$\approx 0.591$$
$$= 59.1\%.$$

Remember that these are proportions of sample data only, and we must be careful before making inferences about an entire population from a sample of only 45 people (23 females and 22 males). It may be that the population of males recycles at a higher rate than the population of females, but it is very difficult to know for sure from this small sample.

Both of the last probabilities computed above are examples of *conditional probabilities*. Conditional probabilities provide information about the outcome of the experiment that alters the probability. In the first case, we were given the condition that the person selected was female. With this new knowledge, there were only 23 outcomes. In the second case, we were told the condition that the person was male, and this limited the outcomes to 22.

An *unconditional probability* would provide no conditions at all. We would pick a person randomly from the sample – any person at all. For this example, an unconditional probability might be the probability that any randomly selected person recycles. There are 20 people who recycle, out of 45 people all together, and from these, we compute our unconditional probability.

$$P\left(recycle\right) = \frac{20}{45}$$
$$\approx 0.444$$
$$= 44.4\%$$

Example 3.1.11 The article "Relationship of Health Behaviors to Alcohol and Cigarette Use by College Students " (*J. of College Student Development* (1992)) included frequencies on drinking levels for a random sample of male and female students.

Drinking Level	$A = male$	$B = female$	Totals
$C = None$	140	186	326
$D = Low$	478	661	1139
$E = Moderate$	300	173	473
$F = High$	63	16	79
Totals	981	1036	2017

To each row and column heading, capital letters A through F have been assigned, representing each of these possible events, respectively.

Notice that the grand total in the lower right cell is the sample size, $n = 2017$. Our experiment involves randomly selecting a college student. Let $E = $ "*student drinks moderately.*" There were $x = 473$ students in this category. From this, we estimate that $P\left(E\right) \approx x/n = 473/2017$, which rounds to $0.235 = 23.5\%$.

Next, we wish to estimate, for a single student, the probability that the student is *male* and has a *high* drinking level. There were 63 males in this drinking category, so

$$P\left(A \ \& \ F\right) = \frac{63}{2017}$$
$$= 0.0312$$
$$= 3.12\%.$$

The probability that the person is a female and has a high drinking level is

$$P\left(B \ \& \ F\right) = \frac{16}{2017}$$
$$= 0.00793$$
$$= 0.793\%.$$

If we wanted to compute the probability of A or F, that a student is male or has a high drinking level, then we would have a bit of counting to do. First we need to count the male students, for a total of 981, and then we need to count those students who had high drinking levels that were not male. That would be the females with high drinking levels – another 16 students. Altogether we count $x = 981 + 16 = 997$. So,

$$\begin{aligned} P\left(A \text{ or } F\right) &\approx \frac{997}{2017} \\ &= 0.494 \\ &= 49.4\%. \end{aligned}$$

Each of these estimates includes sampling error, and all along we assume that the sample is a simple random sample.

We have covered the simplest of rules for computing probability. It is important to note again that in most cases these simple rules are sufficient for many problems, and should be used whenever possible. Unfortunately, there will be times when the more advanced formulas become necessary, and so we proceed.

3.1.3 Homework

In a recent study by Mt. San Antonio College honors student, Jacqueline Sun, 36 students were asked if they believe that global warming is real, and if they recycle regularly. The global warming question allowed 3 responses: Yes, No, and Maybe. The recycling question allowed Yes and No responses only. These observations are given below.

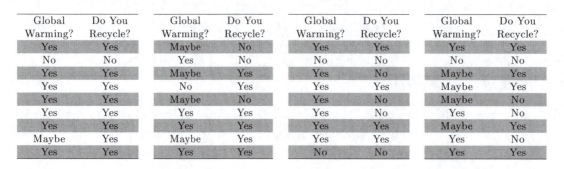

Global Warming?	Do You Recycle?	Global Warming?	Do You Recycle?	Global Warming?	Do You Recycle?	Global Warming?	Do You Recycle?
Yes	Yes	Maybe	No	Yes	Yes	Yes	Yes
No	No	Yes	No	No	No	No	No
Yes	Yes	Maybe	Yes	Yes	No	Maybe	Yes
Yes	Yes	No	Yes	Yes	Yes	Maybe	Yes
Yes	Yes	Maybe	No	Yes	No	Maybe	No
Yes	Yes	Yes	Yes	Yes	No	Yes	No
Maybe	Yes	Maybe	Yes	Yes	Yes	Maybe	Yes
Yes	Yes	Yes	Yes	No	No	Yes	No
						Yes	Yes

1. Construct a two-way table with row headings, Yes and No, for the recycle question. The column headings should be Yes, Maybe and No for the global warming question.

From the two-way table constructed in Problem 1, suppose that one person is selected randomly. Define the events $R = recycle$, and $G = global \ warming$, and answer the following questions.

2. Compute $P(R)$.

3. Compute $P(G)$.

4. Compute $P(R \ \& \ G)$.

5. Compute $P(R \ or \ G)$.

6. The events, R and G, are called *independent* if $P(R \ \& \ G) = P(R) \cdot P(G)$. Do these events appear to be independent?

In the article, "Attitudes About Marijuana and Political Views" in Psychological Reports, 1973, pp. 1051 – 1054, the following marijuana usage level frequencies were reported:

Political Views	$A = Never$	$B = Rarely$	$C = Frequently$	Totals
$D = Liberal$	479	173	119	771
$E = Conservative$	214	47	15	276
$F = Other$	172	45	85	302
Totals	865	265	219	1349

Altogether, 1,349 people were surveyed. Please use the table above to estimate probabilities for problems 7 through 12, using relative frequencies.

7. $P(A) =$ **10.** $P(D) =$

8. $P(B) =$ **11.** $P(A \ \& \ D) =$

9. $P(A \ \& \ B) =$ **12.** $P(A \ \& \ F) =$

13. Suppose a student guesses randomly on a multiple choice question, with five possible answers. What is the probability of guessing correctly?

14. If two quarters are tossed simultaneously, the sample space consists of all possible outcomes of *heads* and *tails* on the two coins. List the sample space, then give the probability that at least one quarter lands *heads*.

15. For a particular commuter train, an observer noted that out of 300 rides to work from a particular stop, 278 arrivals were *on-time*. Using relative frequencies, estimate the probability that any random arrival for this commute is on-time. What would be the probability that a commute is not on-time?

16. Suppose a couple plans on having two children. Assume that boys and girls are equally likely. What are the possible outcomes for their genders? What is the probability of having one girl and one boy?

The following data were gathered by Mt. San Antonio College honors student Helentina Pang, regarding genders of sample members and whether respective members have ever been in a car accident.

Gender	$Y = Yes$	$N = No$
$F = Female$	15	15
$M = Male$	20	10

Please answer the following problems regarding the above sample data.

17. $P(Y) =$ **23.** $P(Y \ \& \ M) =$

18. $P(N) =$ **24.** $P(Y \ \text{or} \ M) =$

19. $P(F) =$ **25.** $P(N \ \& \ F) =$

20. $P(M) =$ **26.** $P(N \ \text{or} \ F) =$

21. $P(Y \ \& \ F) =$ **27.** $P(N \ \& \ M) =$

22. $P(Y \ \text{or} \ F) =$ **28.** $P(N \ \text{or} \ M) =$

3.2   Rules for Computing Probabilities

3.2.1   The Addition Rules

In the previous section of this text, we concluded with an example where we computed $P(A \text{ or } F)$ using relative frequencies. We would like to compute this again, but this time we will develop the idea into formulas which will become the *addition rules*.

Example 3.2.1 Below are the data which involved a random sample of 2017 college students grouped by gender and drinking levels.

Drinking Level	$A = Male$	$B = Female$	Totals
$C = None$	140	186	326
$D = Low$ (1-7)	478	661	1139
$E = Moderate$ (8-24)	300	173	473
$F = High$ (25 or more)	63	16	79
Totals	981	1036	2017

Suppose we wish to estimate $P(A \text{ or } F)$ again, but this time by adding 981 *males* to 79 people whose drinking level is *high*. In doing so, we must note that we have counted the 63 *males* in the *high* category twice, so we will subtract this once from the total.

$$\text{Thus,} P(A \text{ or } F) \approx \frac{981+79-63}{2017} = \frac{997}{2017}$$

Next, we develop the first addition rule formula:

$$
\begin{aligned}
P(A \text{ or } F) &\approx \frac{981 + 79 - 63}{2017} \\
&= \frac{981}{2017} + \frac{79}{2017} - \frac{63}{2017} \\
&= P(A) + P(F) - P(A \ \& \ F)
\end{aligned}
$$

Thus, we have our formula, but note – we have already computed this same probability in the previous section of this text without the use of the addition rule. The rule required some care, realizing that we had to subtract out those people who were in both A and F categories simultaneously. Whenever it is possible for two events to happen simultaneously, in a single trial, we say that A and F are *not mutually exclusive*.
In general, the addition rule for non-mutually exclusive events is:

$$P(A \text{ or } B) = P(A) + P(B) - P(A \ \& \ B),$$

where the formula is an approximation when relative frequencies from sample data are used.

Using the classical approach, let us return to our shuffled poker deck of cards with no jokers, picking one card at random. Suppose $Q =$ "card is a queen" and $H =$ "card is a heart". To compute $P(Q \text{ or } H)$ we simply note that, out of 52 cards, there are 4 queens, 13 hearts, and one card that is both, so

$$
\begin{aligned}
P(Q \text{ or } H) &= P(Q) + P(H) - P(Q \text{ \& } H) \\
&= \frac{4}{52} + \frac{13}{52} - \frac{1}{52} \\
&= \frac{16}{52} \\
&= 0.308 \\
&= 30.8\%.
\end{aligned}
$$

Both of the previous examples had something in common – in each case there was the possibility of both events happening simultaneously (for the 63 males with high drinking levels, or the queen of hearts), so in each case the events were *not mutually exclusive*.

There are many cases, when computing $P(A \text{ or } B)$ that events A and B cannot both happen simultaneously. In such a case, the events are called *mutually exclusive*, and since there is no possibility of counting outcomes twice, we do not need to subtract out *P(A & B)* in the computation. The formula for the addition rule becomes

$$
P(A \text{ or } B) = P(A) + P(B)
$$

when A and B are mutually exclusive. The formula is an approximation when relative frequencies from sample data are used.

Returning to Example 3.2.1, recall that out of 2017 students, 473 were in the $E = $"*moderate*" category, while 79 were $F = $"*high*" category. Since no person could be in both categories, they are mutually exclusive, and thus,

$$
\begin{aligned}
P(E \text{ or } F) &= P(E) + P(F) \\
&\approx \frac{473}{2017} + \frac{79}{2017} \\
&= \frac{552}{2017} \\
&= 0.274 \\
&= 27.4\%
\end{aligned}
$$

Example 3.2.2 Refer again to the shuffled poker deck with no jokers from the previous section of this text. We again choose a single card at random, and we define $H = $"*hearts*" and $S = $"*spades*". Note that no card could be both a heart and a spade, so these events are mutually exclusive. Note also that there are 13 hearts and 13 spades out of 52 cards total, and with this we compute:

$$P(H \text{ or } S) = P(H) + P(S)$$
$$= \frac{13}{52} + \frac{13}{52}$$
$$= \frac{26}{52}$$
$$= 0.500$$
$$= 50.0\%.$$

We summarize the addition rules below.

Theorem 3.2.3 The Addition Rules

If A and B are not mutually exclusive, then $P(A \text{ or } B) = P(A) + P(B) - P(A \ \& \ B)$.

If A and B are mutually exclusive, then $P(A \text{ or } B) = P(A) + P(B)$.

3.2.2   The Rule of Complements

The next rule, the *rule of complements*, is an example of the addition rule for two special mutually exclusive events. First, for any event, S, we define the event S *complement*, denoted by $\bar{S}$, to be the event, $\bar{S} = not(S)$. So $\bar{S}$ includes every possible simple outcome that is not included in S. We should note that since S and $\bar{S}$ have no simple events in common, S and $\bar{S}$ are mutually exclusive, so $P(S \text{ or } \bar{S}) = P(S) + P(\bar{S})$. And since every simple outcome is included by either S or $\bar{S}$, one of these must occur. Thus,

$$1 = P(S \text{ or } \bar{S}) = P(S) + P(\bar{S}).$$

This is the rule of complements, which can be expressed in any of the following equivalent forms.

Theorem 3.2.4 If events S and $\overline{S}$ are complements, the rule of complements may be stated in the equivalent forms below.

$$P(S) + P(\bar{S}) = 1$$
$$P(S) = 1 - P(\bar{S})$$
$$P(\bar{S}) = 1 - P(S)$$

Applications of this can be very complicated or very simple. For our purposes, simple applications will suffice. The main usefulness of the rule of complements is as follows. Sometimes an event, S, may be rather complicated compared to its complement. When this is the case, it may be easier to compute the probability of S *indirectly* using the rule: $P(S) = 1 - P(\bar{S})$.

Example 3.2.1 Continued: Referring again to our table relating drinking levels and genders for college students, let S = "*student drinks at least one alcoholic beverage per week or is male.*" Now, that is rather complicated, given that there are many ways that this could happen. If we consider the complement of this event, we see that $\bar{S}$= "*student has no drinks per week and is female.*" If we consult the table, it is easy to see that there are 186 females in this category, out of 2017 students. So, using relative frequencies, we estimate $P(\bar{S}) \approx 186/2017$. That is good, but what we want is $P(S)$, which we estimate using the rule of complements.

$$
\begin{aligned}
P(S) &= 1 - P(\bar{S}) \\
&\approx 1 - \frac{186}{2017} \\
&= 1 - 0.0922 \\
&= 0.908 \\
&= 90.8\%.
\end{aligned}
$$

Example 3.2.5 Returning to the shuffled poker deck with no jokers, if we let A = "*card is a Diamond or Heart or Spade*", then $\bar{A}$= "*card is a Club*". Obviously, computing $\bar{A}$ is easier, and since there are 13 clubs out of 52 cards, $P(\bar{A}) = 13/52 = 0.250$, and thus:

$$
P(A) = 1 - P(\bar{A}) = 1 - 0.250 = 0.750 = 75.0\%.
$$

The rule of complements is the simplest of the rules, and will be the one which we use the most.

3.2.3 The Multiplication Rules

The multiplication rules are for computing $P(A \ \& \ B)$. Again, note that we have already computed this kind of probability several times already, without any special rules. In fact, for the examples given thus far, the use of multiplication rules greatly complicates the process, as will be seen. Still, there are problems which cannot be done easily without the multiplication rules, and so we must discuss them. In addition, the multiplication rules will become very important in section 10.2 when we develop a test for independence.

Definition 3.2.6 Events A and B are *independent* if the occurrence of one does not affect the probability of the other.

Sometimes independence is easy to determine, but at other times it is much more difficult to determine.

Example 3.2.7 The *heads* and *tails* sides of a coin are depicted in Figure 3.3. As an example, if we consider an experiment where we toss two coins, let $A =$"*first coin lands heads*", and $B =$"*second coin lands heads*", it is easy to believe that if the first coin lands heads, that this does not affect the probability that the second coin will land heads. If that is true, then these events truly are independent.

Figure 3.3: The side of the coin with an engraved head is called *heads*. The reverse side is always called *tails*. (*Photo in the public domain by the U.S. Treasury.*)

The question is, how do we compute the probability of A and B? One method we might use would be to list all simple outcomes in the sample space: { *Heads & Heads, Heads & Tails, Tails and Heads, Tails & Tails* }. In only one of these four possible outcomes do we get A and B both happening, so

$$P(A \ \& \ B) = \frac{1}{4} = 0.250 = 25.0\%,$$

but,

$$P(A \ \& \ B) = \frac{1}{4} = \frac{1}{2} \cdot \frac{1}{2} = P(A) \cdot P(B),$$

and this is, in fact, the first multiplication rule.

The reasoning behind the multiplication rule goes like this. Suppose in one experiment there are n simple outcomes, and that event A can happen in x simple ways. Next, suppose that in another experiment, there are m simple outcomes, and that event B can happen in y simple ways. If the two experiments are combined into one, the first question is how many simple outcomes are there? For each of the n possible simple outcomes of the first experiment, there are m things that can happen in the second experiment. Because of this, there are $n \cdot m$ outcomes in the combined experiment. And since for each of x ways that event A can happen there are y ways that B can happen, there are a total of $x \cdot y$ ways that A and B can happen together. This means that

$$P(A \ \& \ B) = \frac{\text{Ways } A \ \& \ B \text{ can happen together}}{\text{Total outcomes of combined experiments}} = \frac{x \cdot y}{n \cdot m} = \frac{x}{n} \cdot \frac{y}{m} = P(A) \cdot P(B).$$

Thus, we have our first multiplication rule:

Theorem 3.2.8 If A and B are independent, then $P(A \ \& \ B) = P(A) \cdot P(B)$.

Example 3.2.9 As an example, we will pick two cards from a shuffled 52 card poker deck *with replacement.* This means that we will replace the first card and reshuffle the deck before picking the second card, so there is a chance that the first card will be picked twice. Because of this, the likelihood of picking any card on the second draw is not influenced by what happened with the first card – so these outcomes are independent. Next, we define $A = $"*first card is a heart*" and $B = $"*second card is a heart*", noting that on each draw there are 13 hearts in the deck, out of 52 cards. Thus,

$$P(A \ \& \ B) = P(A) \cdot P(B) = \frac{13}{52} \cdot \frac{13}{52} = \frac{1}{4} \cdot \frac{1}{4} = \frac{1}{16} = .0625 = 6.25\%.$$

The problem becomes slightly more complicated if we consider events A and B that are *dependent.*

Definition 3.2.10 Events A and B are *dependent* if the occurrence of one effects the likelihood of the other.

This time we will conduct the same experiment, but *without replacement,* so we will not replace the first card before drawing the second card. The first probability, $P(A)$ is unchanged as there are again 13 hearts out of 52 cards. The difficulty begins when we consider the second card. We want to compute the probability that the second card is a heart, but we do not know whether the first card will be a heart or not. We do not know how many hearts are in the deck as the second card is drawn – are there 12 or 13 hearts out of the 51 cards that remain? This is what makes A and B dependent: the probability of B *depends* on whether A is true or not. This question is answered when we understand what the event $A \ \& \ B$ implies – that A must happen for $A \ \& \ B$ to happen. To compute $P(A \ \& \ B)$, we multiply the probability of A times the probability of B given that A has occurred. When we compute the probability of event B, given that event A has occurred, we are computing a *conditional probability*, and in this case it is denoted by $P(B|A)$. So, in our example, on the second card, we are to assume that the first was a heart, and there are therefore only 12 hearts left, out of 51.

Definition 3.2.11 The *conditional probability*, $P(B|A)$, is the probability of event B, given that event A is true.

The multiplication rule can now be updated to allow for dependent events.

Theorem 3.2.12 If A and B are *dependent*, then $P(A \ \& \ B) = P(A) \cdot P(B|A)$

Using the last experiment where we draw two cards from a poker deck of 52 cards without replacement, where $A =$ "*first card is a heart*" and $B =$ "*second card is a heart*", we compute

$$
\begin{aligned}
P\left(A \ \& \ B\right) &= P\left(A\right) \cdot P\left(B|A\right) \\
&= \frac{13}{52} \cdot \frac{12}{51} \\
&= \frac{1}{4} \cdot \frac{4}{17} \\
&= \frac{1}{17} \\
&= 0.0558 \\
&= 5.58\%.
\end{aligned}
$$

The multiplication rules are summarized below.

Theorem 3.2.13 The Multiplication Rules may be stated as follows.
If A and B are independent,

$$
P\left(A \ \& \ B\right) = P\left(A\right) \cdot P\left(B\right).
$$

If A and B are dependent, then

$$
P\left(A \ \& \ B\right) = P\left(A\right) \cdot P\left(B|A\right).
$$

If we apply the multiplication rules to categories from sample data using relative frequencies, the computations are more confusing, and it is often difficult to determine if events are dependent or independent. Generally speaking, whenever it is not clear whether events are dependent or independent, you should always assume that they are dependent, and this will give the correct answer even if they are independent. But, also speaking generally, whenever sample data are given and frequencies are provided, using a multiplication rule just complicates problems and is more formula than is really necessary.

Example 3.2.1 Continued: The data for this example are again provided below. Remember that each row and column has been assigned letters A through F as we refer to them below.

Drinking Level	$A = Male$	$B = Female$	Totals
$C = None$	140	186	326
$D = Low$ (1-7)	478	661	1139
$E = Moderate$ (8-24)	300	173	473
$F = High$ (25 or more)	63	16	79
Totals	981	1036	2017

Suppose we wish to compute $P(A \ \& \ F)$. It is clear that $A \ \& \ F$ occurred 63 times out of 2017 members of the sample, so $P\left(A \ \& \ F\right) \approx \frac{63}{2017} = 3.12\%$. Computing with the direct approach is easy, and it gave us a perfectly correct answer. Now, in using a multiplication rule, we will get the same answer, but the process is considerably more difficult. First we must decide whether drinking behavior is at all influenced by, or is dependent on, gender. We might consider *guessing* the answer to this question, but guessing is usually a bad idea. As mentioned previously, however, whenever we assume that the events are dependent, we will get the right answer, even if the events turn out to be independent. So we will assume that the events are dependent, and compute $P(A \& F) = P(A) \cdot P(F|A)$. Now, $P(A) = 981/2017$, and that is easy enough. But now we must compute $P(F|A)$. So we want the probability of $F = $ "*high drinking level*", given that the person was $A = $ "*male.*" Given that the person was male, we must understand that this limits the possibilities, both in the ways that F can happen, but also in the total number of possible outcomes. In fact, given that the person is male, we are only considering the data in the table that is shaded grey, the "male" column. So, given that the person is male, there are only 63 ways that F can happen, out of a total of only 981 possible outcomes. So $P(F|A) = 63/981$, and finally we apply the multiplication rule,

$$
\begin{aligned}
P(A \& F) &= P(A) \cdot P(F|A) \\
&\approx \frac{981}{2017} \cdot \frac{63}{981} \\
&= \frac{981}{2017} \cdot \frac{63}{981} \\
&= \frac{63}{2017} \\
&= 0.0312 \\
&= 3.12\%.
\end{aligned}
$$

Of course, this is the same answer which we arrived at originally, but obviously it required a bit more thought and computation when using the multiplication rule. Thus, when computing $P(A \& B)$ from sample data, it is recommended that you just count up the number of times that $A \& B$ both happened, and divide by the sample size, avoiding the multiplication rule altogether.

3.2.4 Complements and Compound Events

The most complicated type of probability that we will compute is the probability of an event which can occur in many ways. In computing such a probability, we use the addition rule, adding terms to the probability for each of the mutually exclusive ways that the event can occur. Evaluating these probabilities *directly* is often tedious, but the good news is that this tedium can be avoided when the *complement* of our event is simpler. When this is the case, the problem becomes easier if we evaluate the probability *indirectly* using the rule of complements.

Example 3.2.14 As an example, consider an experiment where we toss a coin 5 times. For each toss let $H = $ "*heads*" and $T = $ "*tails.*" The event which we wish to consider is A

= "*at least one toss lands heads.*" Computing $P(A)$ *directly* is not easy because getting at least one "*heads*" can happen in 5 different ways (where we get 1, 2, 3, 4 or 5 heads). The best approach is to compute $P(A)$ *indirectly* by applying the rule of complements. The key is in noting that if A = "*at least one toss lands heads*", then the complement of A is the event where we do not get at least one "*heads*", meaning we get all "*tails*". Thus,

$$\bar{A} = \text{``no heads''} = \text{``all tails''} = T\&T\&T\&T\&T.$$

Given that $P(T) = 0.5$, and that each coin toss is independent of all others, we compute $P(A)$ as follows.

$$\begin{aligned} P(A) &= 1 - P\left(\bar{A}\right) \\ &= 1 - P\left(T\&T\&T\&T\&T\right) \\ &= 1 - 0.5 \cdot 0.5 \cdot 0.5 \cdot 0.5 \cdot 0.5 \\ &= 1 - 0.5^5 \\ &= 1 - 0.03125 \\ &= 0.96875. \end{aligned}$$

Therefore the probability of getting at least one "*heads*" in five tosses is 96.875%.

3.2.5 Homework

The table below is based on "Ignoring a covariate: An example of Simpson's Paradox" by Appleton, D.R. French, J.M. and Vanderpump, M.P (1996, American Statistician, 50, 340-341). In 1972-1994 a one-in-six survey of the electoral roll, largely concerned with thyroid disease and heart disease was carried out in Wichkham, a mixed urban and rural district near Newcastle upon Tyne, in the UK. Twenty years later, a follow-up study was conducted to see which study members were still alive.

Here are the results for a sample of randomly selected females aged 65 to 74. Assuming 7425 women were involved, the observed frequencies are as follows.

Smoking Status	$A = Dead$	$B = Alive$	Totals
$C = Smokers$	1305	315	1620
$D = Non\text{-}smokers$	4545	1260	5805
Totals	5850	1575	7425

Use the table above to answer questions 1 – 16. Use the relative frequency approach.

1. $P(A) =$

2. $P(B) =$

3. $P(C) =$

4. $P(D) =$

5. $P(A \& B) =$

6. $P(A \text{ or } B) =$

7. $P(A \& C) =$

8. $P(A \text{ or } C) =$

9. $P(C|A) =$

10. $P(A|C) =$

11. $P(B \& D) =$

12. $P(B \text{ or } D) =$

13. $P(D|B) =$

14. $P(B|D) =$

15. $P(\bar{A}) =$

16. $P(\bar{D}) =$

One way to test whether events are independent is to use the multiplication rule for independent events. If A and B are independent then $P(A \& B) = P(A) \cdot P(B)$. Refer to the table above to answer the following questions.

17. Do A and C appear to be independent?

18. Do B and D appear to be independent?

The following data give game rating preferences by gender for randomly selected college students. These data were gathered by Sean Meshkin, honors student at Mt. San Antonio College.

Gender	Rated-E	Rated-T	Rated-M
$L = Male$	7	12	15
$F = Female$	5	17	5

Please answer the following problems regarding the above data.

19. $P(M \ \& \ L) =$

20. $P(M \text{ or } L) =$

21. $P(M|L) =$

22. $P(L|M) =$

23. $P(L \ \& \ F) =$

24. $P(M \text{ or } F) =$

25. $P(M|F) =$

26. $P(F|M) =$

27. $P(\bar{E}) =$

28. $P(\bar{F}) =$

The following give frequencies of grades by number of units attempted for randomly selected sample members. These data were gathered by Lily Bai, honors student at Mt. San Antonio College.

Units Attempted	Grade of A	Grade of B	C or Lower
$D = 0 - 12 \ Units$	1	3	1
$E = 12 - 13 \ Units$	7	10	4
$F = More \ than \ 16 \ Units$	7	3	1

Please answer the following problems regarding the above data.

29. $P(A|F) =$

30. $P(C|F) =$

31. $P(A|D) =$

32. $P(C|D) =$

33. $P(A \ \& \ F) =$

34. $P(A \text{ or } F) =$

35. $P(A \ \& \ D) =$

36. $P(A \text{ or } D) =$

37. $P(\bar{E}) =$

38. $P(\bar{B}) =$

Suppose that a card is drawn randomly from a shuffled poker deck with no jokers. Let A = "card is a jack", B = "card is a king", and C = "card is a club." Compute the following probabilities.

39. $P(A \text{ or } B) =$ **41.** $P(A \text{ or } C) =$

40. $P(A \And B) =$ **42.** $P(A \And C) =$

Suppose that we draw two cards from a shuffled poker deck with no jokers. Let A = "first card is a jack", and B = "second card is a jack".

43. Compute $P(A \And B)$ if the cards are drawn with replacement.

44. Compute $P(A \And B)$ if the cards are drawn without replacement.

Suppose that a couple plans to have three children, and that boys and girls are equally likely. List all of the possible outcomes for the genders of the three children, and use this list to answer the following questions.

45. What is the probability of having no girls?

46. What is the probability of having one girl?

47. What is the probability of having two girls?

48. What is the probability of having three girls?

49. What is the probability of having at least one girl?

Answer the following questions. Many of the events are compound, and thus their probabilities may be easier to evaluate using the rule of complements.

50. Suppose that four men are selected for a police line-up, where three people will try to identify one of them as the perpetrator of a particular crime that they witnessed. What is the probability that they identify the same person if they are all guessing randomly?

51. Suppose that you do not trust your alarm clock, so instead of having just one, you buy three and set them all every night. Suppose also that each has a 95% chance of ringing in the morning. What is the chance that your system fails, and that none ring?

52. Using the same alarm clock system from the previous problem, what is the chance that your system does not fail, so that at least one rings?

53. Given that 49.8% of the world's population is female, suppose that you select three people randomly (with replacement). What is the probability that all three are female? Would your answer be different if we did not sample with replacement?

54. If three people with different last names walk through a door, in how many different orders can they enter? What is the probability that they come in so that their last names come in alphabetic order?

55. A particular lottery asks participants to pick three numbers from 1 to 50. If a participant picks the right three numbers, he or she wins. What is the probability of winning?

Chapter 4

Discrete Probability Distributions

4.1 Discrete Probability Distributions

As we move toward inferential statistics which begin in Chapter 6, we must move into our most important application of probability, the probability distribution. Each application that we study will involve some probability distribution, and for us, that will normally mean looking up a number or two in a probability distribution table.

Probability distributions are always defined in the context of a random variable. A random variable, x, will be a quantity, representing every possible value measurable in some population. This is different from values actually measured in some sample. This distinction leads more advanced books to often use a capital X for a random variable, while using a lower case x for particular values of the random variable. For simplicity's sake, we will use lower case x in both cases, trusting that the context of any discussion will clarify to which form we are referring.

There are two types of random variables – discrete and continuous. As a reminder on the meanings of these terms, discrete quantitative data are numerical values where measurements may be ordered consecutively, and gaps exist between consecutive measurements where no measurements are possible. Continuous quantitative data are numerical values where, within a certain range, any value is possible.

Definition 4.1.1 A *random variable*, x, is a numerical and quantitative value, representing each possible measurement from a population. When the population measurements are *discrete*, x is a *discrete random variable*. When the measurements are continuous, x is a *continuous random variable*.

In this chapter, we will study discrete random variables, and in Chapter 5, we will study continuous random variables.

Example 4.1.2 As a simple example, suppose we randomly select 5 people (with replacement), and define x to be the number of females in the sample. The values that x

may assume are $\{0, 1, 2, 3, 4, 5\}$. This variable is a *discrete* random variable, because *gaps* exist between consecutive values where no measurement is possible.

Next, we define a *probability distribution* for a discrete variable. A *discrete probability distribution* is a *rule* by which we associate, to each value of the discrete variable, x, a probability, $P(x)$. This *rule* may be a formula, or a table of values.

Returning to our example where x represents the number of females in a random sample of 5 people, the probability distribution rule will be a table, where one column lists each possible x, and the next column gives the corresponding probability, $P(x)$. From this, there are a few things to note. First of all, the sum of all probabilities in the discrete distribution is equal to 1. This is because one of the mutually exclusive outcomes must occur, so that the sum of all the probabilities is one.

x	$P(x)$
0	0.03188
1	0.15813
2	0.31374
3	0.31124
4	0.15438
5	0.03063
Sum:	1.00000

Second, we note that each probability is between 0 and 1, inclusive. This is always true with any probability, but we are noting it particularly with respect to a random variable.

In considering the various possibilities for the value of this random variable, it seems reasonable to assume that if this experiment was conducted many times, there would be a mean number of females, but the actual outcomes would vary. Since it is possible for x to deviate from this mean, there should be a value for the standard deviation as well. Wanting to know the mean and standard deviation of the possible outcomes for this experiment seems natural enough, but how shall we compute them, given that we have no sample data? Is this even possible?

4.1.1   Mean, Variance and Standard Deviation

It *is* possible to compute the mean, standard deviation, and variance of a random variable from its probability distribution, and formulas for these are developed in the optional section at the end of this chapter. In the context of a probability distribution, the mean of the random variable is a population mean since it comes from *exact* probabilities, and not from sample data. In addition, in this context the mean is again called the *expected value* of the random variable, x, denoted by $\mu = E(x)$. In addition, the variance of the population is still denoted $\sigma^2 = Var(x)$, and the standard deviation is $\sigma = \sqrt{\sigma^2} = \sqrt{Var(x)}$. The formulas for the mean and variance are

$$\mu = E(x) = \sum x \cdot P(x)$$

and

$$\sigma^2 = Var(x) = \sum (x - \mu)^2 \cdot P(x),$$

which we algebraically simplify to

$$\sigma^2 = Var(x) = \sum x^2 \cdot P(x) - \mu^2.$$

The formulas for population mean (expected value), variance and standard deviation for a discrete probability distribution are summarized below.

Definition 4.1.3 The mean, variance, and standard deviation of a discrete probability distribution are given below.

$$\mu = E(x) = \sum x \cdot P(x)$$
$$\sigma^2 = Var(x) = \sum x^2 \cdot P(x) - \mu^2$$
$$\sigma = \sqrt{Var(x)} = \sqrt{\sum x^2 \cdot P(x) - \mu^2}$$

Example 4.1.2 Continued: Returning to our randomly selected sample of 5 people, where x represents the number of females, we will compute the mean, variance, and standard deviation of x. Note that these computations require two sums: $\sum x \cdot P(x)$ and $\sum x^2 \cdot P(x)$. To compute these we will add two columns to our table, with $x \cdot P(x)$ and $x^2 \cdot P(x)$ as their headings.

x	$P(x)$	$x \cdot P(x)$	$x^2 \cdot P(x)$
0	0.03188	0.0000000	0.0000000
1	0.15813	0.1581300	0.1581300
2	0.31374	0.6274799	1.2549598
3	0.31124	0.9337201	2.8011604
4	0.15438	0.6175201	2.4700803
5	0.03063	0.1531499	0.7657495
Sum:	1.00000	2.4900000	7.4500800

Both of these columns have been summed, and we see that $\sum x \cdot P(x) = 2.49$ and that $\sum x^2 \cdot P(x) = 7.45008$. Using these values we apply our formulas for the mean, variance, and standard deviation. First we compute the mean.

$$\mu = E(x) = \sum x \cdot P(x)$$
$$= 2.49$$

Next, we compute the variance.

$$\begin{aligned}
\sigma^2 &= Var\,(x) \\
&= \sum x^2 \cdot P\,(x) - \mu^2 \\
&= 7.45008 - 2.49^2 \\
&= 1.24998
\end{aligned}$$

The standard deviation is the square root of variance.

$$\begin{aligned}
\sigma &= \sqrt{\sigma^2} \\
&= \sqrt{\sum x^2 \cdot P\,(x) - \mu^2} \\
&= \sqrt{1.24998} \\
&= 1.118.
\end{aligned}$$

Notice that, in all of these computations, we resisted rounding values until the computations were finished.

To summarize what we have learned about discrete probabilities, we collect all of our facts together below.

Theorem 4.1.4 Rules for Discrete Probability Distributions

1. $0 \leqslant P\,(x) \leqslant 1$ for all x.

2. $\sum P\,(x) = 1$

3. $\mu = E\,(x) = \sum x \cdot P\,(x)$

4. $\sigma^2 = Var\,(x) = \sum x^2 \cdot P\,(x) - \mu^2$

5. $\sigma = \sqrt{\sigma^2} = \sqrt{\sum x^2 \cdot P\,(x) - \mu^2}$

4.1.2   Unusual or Significant Measurements

Once again, we revisit the idea of unusualness, now in the context of a probability distribution. Our rules are still appropriate, but they are in need some refinement.

There is no problem with taking the z-score of any x value, now that we can compute the mean and standard deviation. Referring to the last example, where $\mu = 2.49$, and $\sigma = 1.118$, we might ask whether observing $x = 4$ females in a random sample of five people would be an unusually high number. The z-score can answer this, so we compute $z = \frac{x-\mu}{\sigma} = \frac{4-2.49}{1.118} = 1.35$. This indicates that the value, $x = 4$, is only 1.35 standard deviations above average. Normally, we like to see about 2 standard deviations from average before we call it significant, so in this case, 4 females in a random group of 5 is not really unusual or significant.

We have mentioned another way of determining if an event is unusual or significant, and that is by computing the probability of that event. When it comes to random variables, we refine this slightly below. The probabilities which we will use for determining significance will be known as *p-values*.

Procedure 4.1.5 We may use *p-values* to determine significance for a given value of a random variable as follows.

- A value $x = a$ is *significantly high* if the *p-value* $= P(x \geqslant a)$ is small (less than around 5%).

- A value $x = a$ is *significantly low* if the *p-value* $= P(x \leqslant a)$ is small (less than around 5%).

Either way, we will consider an event to be statistically significant when the corresponding *p-value* satisfies the inequality below.

$$p\text{-}value \lesssim 0.05$$

The smaller the probability, the more significant the event is. If either probability is less than 1%, we might call the event *highly* significant. If the probability is slightly more than 5%, we might consider the event *moderately* significant. In Chapter 7, the p-value is used extensively for statistical inference.

Let us go back to our example. If we compute the p-value for $x = 4$ to determine if this value is significantly high, we get

$$P(x \geqslant 4) = P(x = 4) + P(x = 5) = 0.154 + 0.0306 = 0.184.$$

There is an 18.4% chance of measuring a value for x that is at least 4. When this happens, we will not consider it significant, as this will happen at random rather frequently. Thus again, observing 4 women in a random group of 5 is not significant. If we ask whether 5 women in a random group of 5 is significantly high, the answer is yes, because the p-value is

$$p\text{-}value = P(x \geqslant 5) = P(x = 5) = .0306 = 3.06\%.$$

This is unusual and significant when it happens.

4.1.3   Probability and Area

In order to visualize the way that probabilities are distributed, we can plot the probabilities in a discrete probability distribution using a histogram, in the same way that we graphed a histogram for a frequency table. Usually, when working with a discrete variable (typically with distances between consecutive measurements all equal to one), we will graph this histogram using a class width equal to 1, with each value of x individually represented. The areas of the bars will be important to us. Each bar has area equal to the width times the height.

x	$P(x)$
0	0.000001
1	0.000018
2	0.000170
3	0.001028
4	0.004403
5	0.014204
6	0.035795
7	0.072164
8	0.118209
9	0.158877
10	0.176169
11	0.161440
12	0.122052
13	0.075712
14	0.038160
15	0.015387
16	0.004847
17	0.001150
18	0.000193
19	0.000020
20	0.000001

The histogram for this discrete distribution is plotted in Figure 4.1.

The widths are equal to 1, and the heights will be probabilities for each x, so that each bar's area is 1 times its corresponding probability. Thus, area of each bar is equal to its corresponding probability. Since the sum of all probabilities is equal to one, the total area is equal to one. The problem of computing probability could then be thought of as a geometry problem, where we are merely finding out areas of geometric regions on the graph of a probability distribution. This is precisely how we will compute probabilities in Chapter 5 as we move on to continuous random variables.

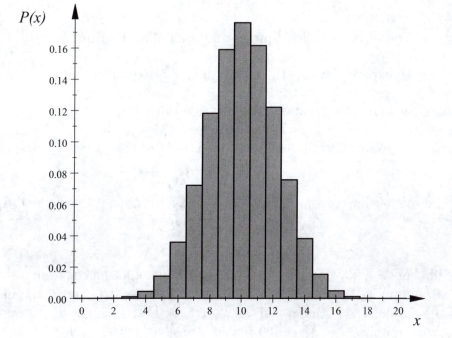

Figure 4.1: The histogram of a discrete probability distribution. The total area of all bars is equal to one.

4.1.4 Homework

For the entire world's population, approximately 14% of all people speak Chinese as their native language. Suppose four people are randomly selected from the Earth's population, and define x to be the number of these who speak Chinese as their native language. The probability distribution associated with this random variable is given to the right.

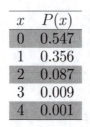

x	$P(x)$
0	0.547
1	0.356
2	0.087
3	0.009
4	0.001

1. Verify that these probabilities represent a probability distribution.

2. Calculate the expected value (the mean) of this distribution.

3. Calculate the variance and standard deviation of this distribution.

4. Compute the p-value corresponding to $x = 3$ to determine if this value is significantly high.

5. Compute the p-value corresponding to $x = 0$ to determine if this value is significantly low.

In a game of chance, a player selects two cards without replacement from a shuffled deck of 52 cards, containing 13 hearts. A bet of $160 is required to play. If the player draws no hearts the player wins nothing, and loses the $160 bet. If the player draws one heart, the player wins nothing, but the $160 bet is returned. If the player draws two hearts, the player wins $1600 (the bet is not returned). The probabilities of these outcomes are: $P(\text{no hearts}) = 56\%$; $P(\text{one heart}) = 38\%$. $P(\text{two hearts}) = 6\%$.

6. Letting x represent net winnings (winnings minus the $160 bet), construct the probability distribution for x, verifying that this is a probability distribution.

7. If the game is played again and again, what would be the average net winnings?

8. What would be the variance and standard deviation of these winnings?

Suppose an investor has two investments that she is considering, and wants to pick the one that will make the most money (on average). In the first investment, she must invest $5000. From this there is a 60% chance that she will earn $20,000 (so her net return would be $15,000), but there is a 40% chance that she will lose her $5000 investment. In the next investment, she must invest $4000, and there is an 80% chance that she will earn $10,000 (so her net return would be $6000), but there is a 20% chance that she would lose her $4000 investment.

9. What is the expected value of net earnings for the first investment?

10. What is the expected value of net earnings for the second investment?

11. Which investment should the investor pick?

Suppose a family has three children, and that boys and girls are equally likely. Let the random variable, x, represent the number of boys.

12. List the sample space – the collection of all possible outcomes. This list will have 8 possibilities.

13. Construct the probability distribution for x.

14. Compute the expected value for x.

15. Compute the variance and standard deviation for x.

4.2   Properties of Expected Value and Variance

Among the more important numerical summaries in elementary statistics are the arithmetic mean, the variance, and the standard deviation. There are other measures of center and spread, but we focus upon these primarily because of their properties.

The theory upon which elementary statistics is built relies heavily upon the properties of expected value and variance. Without those properties, the formulas of statistics would be much more complicated. Among the more important properties of expected value and variance are the properties

$$E(x \pm y) = E(x) \pm E(y),$$

and

$$Var(x \pm y) = Var(x) + Var(y),$$

where x and y are independent random variables. Here we refer to x and y as *independent* in the same sense that we referred to events as independent in Chapter 3.

4.2.1   Combinations of Random Variables

We now assume that x and y, measured from two different population, are independent random variables. To understand the properties which will be important to us, we need to discuss what is meant by the new *random variables* $x - y$, and $x + y$.

When we refer to the *variable* $x - y$ we are referring to the collection of all possible differences $x - y$, where x is from a first population of size N, and y is from a second population of size M. Since each of the N values of x can be paired with M values of y, there are $N \cdot M$ pairings of x and y, so there are $N \cdot M$ values of the variable $x - y$, and the same is true for $x + y$. We wish to compute the mean (the *expected value*) of the values $x - y$. This would be denoted $E(x - y)$, and the expected value of $x + y$ would be $E(x + y)$. The variance of these two random variables would be $Var(x - y)$, and $Var(x + y)$, respectively. The following properties of expected value and variance will be proven in Section 4.6.

When we refer to the variable kx, we mean that every value, x, in a given population is multiplied by a single constant, k, to form a completely new collection of values.

4.2.2   The Algebra of Expected Value and Variance

The properties of expected value and variance, along with the operations described which we may perform on random variables create a new type of algebra, whose properties are described below.

Theorem 4.2.1 Properties of Expected Value and Variance
Assume x and y are independent random variables.

1. $E(x + y) = E(x) + E(y)$

2. $E(x - y) = E(x) - E(y)$

3. $E(x \cdot y) = E(x) \cdot E(y)$

4. $E(kx) = k \cdot E(x)$, where k is any constant

5. $E(x^2) = E(x)^2 + Var(x)$

6. $Var(x + y) = Var(x) + Var(y)$

7. $Var(x - y) = Var(x) + Var(y)$

8. $Var(kx) = k^2 \cdot Var(x)$

We now look at a few examples which will allow us to become familiar with this new algebra.

Example 4.2.2 As an example, we will consider two small populations of values x: $\{7, 8, 8, 9\}$ and y: $\{1, 3, 3, 5\}$. The descriptions of these two populations follow.

For the first population of values x, the mean is $\mu_x = E(x) = 8$, and the variance is $\sigma_x^2 = Var(x) = 0.5$. A histogram of the population is in Figure 4.2.

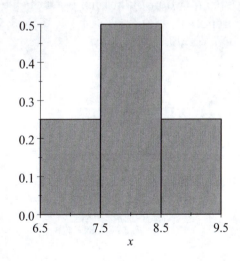

Figure 4.2: The frequency distribution of the population $x : \{7, 8, 8, 9\}$.

For our second population of values, y, the mean is $\mu_y = E(y) = 3$, and the variance is $\sigma_y^2 = Var(y) = 2.0$. A histogram of this population is plotted in Figure 4.3. Notice the larger variance accompanying the population with a wider range of values.

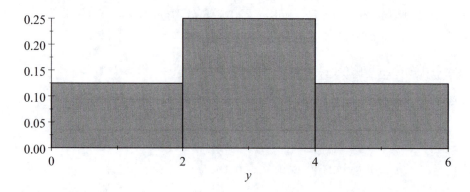

Figure 4.3: The frequency distribution of the population $y : \{1, 3, 3, 5\}$.

Also notice that an attempt has been made, with only four values, for each population to look symmetric and bell-shaped. Now, these populations are too small to satisfy the empirical rule, so they are not really normal, but still there are some similarities to normal populations.

Next, we derive the population of differences $x - y$. Again, we do this by subtracting every y value from each x. Our differences are as follows:

$\{ 7 - 1 = 6, 7 - 3 = 4, 7 - 3 = 4, 7 - 5 = 2, 8 - 1 = 7, 8 - 3 = 5, 8 - 3 = 5, 8 - 5 = 3, 8 - 1 = 7, 8 - 3 = 5, 8 - 3 = 5, 8 - 5 = 3, 9 - 1 = 8, 9 - 3 = 6, 9 - 3 = 6, 9 - 5 = 4 \}$. There are sixteen differences from two populations, both of size 4, because $4 \cdot 4 = 16$. We have our population of differences: $x - y$: $\{6, 4, 4, 2, 7, 5, 5, 3, 7, 5, 5, 3, 8, 6, 6, 4\}$. These are summarized in the frequency table below.

$x - y$	$freq$
2	1
3	2
4	3
5	4
6	3
7	2
8	1

Note the somewhat bell-shaped nature of the distribution of differences.

This has happened because the x and y distributions were somewhat bell-shaped originally.

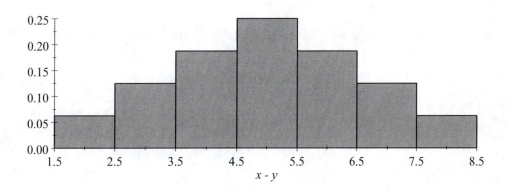

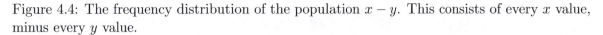

Figure 4.4: The frequency distribution of the population $x - y$. This consists of every x value, minus every y value.

The mean of all differences, $x - y$, is $\mu_{x-y} = E\left(x - y\right) = 5$, and the variance is $\sigma^2_{x-y} = Var\left(x - y\right) = 2.5$. The point to notice is that

$$
\begin{aligned}
E\left(x - y\right) &= \mu_{x-y} \\
&= 5 \\
&= 8 - 3 \\
&= \mu_x - \mu_y \\
&= E\left(x\right) - E\left(y\right),
\end{aligned}
$$

and that

$$
\begin{aligned}
Var\left(x - y\right) &= \sigma^2_{x-y} \\
&= 2.5 \\
&= 0.5 + 2.0 \\
&= \sigma^2_x + \sigma^2_y \\
&= Var\left(x\right) + Var\left(y\right),
\end{aligned}
$$

which is some confirmation of properties two and seven, stated above.

The importance of the shape of the $x - y$ distribution cannot be overstated. We will make much use of this, and state this formally as the principal below. In addition, the distribution of $k \cdot x$ is given as well.

Theorem 4.2.3 When x and y are normally distributed, then the distributions of $x - y$ and $x + y$ are also normally distributed. Additionally, when x is normally distributed, then the distribution of kx is normal as well (if $k \neq 0$).

Example 4.2.4 Here we consider again the population of x values, $\{7, 8, 8, 9\}$, from Example 4.2.2. As mentioned, the mean of these values is $\mu_x = E\left(x\right) = 8$, and the variance is $\sigma^2_x = Var\left(x\right) = 0.5$.

From these values we would like to construct the population of values $3x$, where we multiply each x by 3.

$$3x = \{21, 24, 24, 27\}$$

The mean of these values is

$$
\begin{aligned}
\mu_{3x} &= E\left(3x\right) \\
&= 3E\left(x\right) \\
&= 3 \cdot 8 \\
&= 24.
\end{aligned}
$$

The variance of the values is

$$
\begin{aligned}
\sigma_{3x}^2 &= Var\left(3x\right) \\
&= 3^2 Var\left(x\right) \\
&= 9 \cdot 0.5 \\
&= 4.5
\end{aligned}
$$

Note that the distribution of the $3x$ values is still bell-shaped as demonstrated in Figure 4.5.

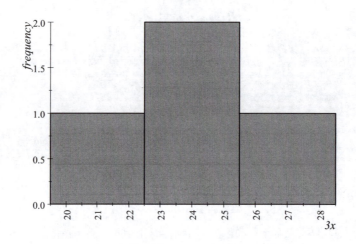

Figure 4.5: The frequency histogram of the population $3x$ from Example 4.2.4.

Example 4.2.5 We consider one more time the population of x values $\{7, 8, 8, 9\}$ with mean $\mu_x = E\left(x\right) = 8$, and the variance is $\sigma_x^2 = Var\left(x\right) = 0.5$, as well as the population of y values $\{1, 3, 3, 5\}$ with mean $\mu_y = E\left(y\right) = 3$, and the variance is $\sigma_y^2 = Var\left(y\right) = 2.0$.

From these populations, we construct the population, $2x + 4y$. The values of $2x$ are

$\{14, 16, 16, 18\}$, while the values of $4y$ are $\{4, 12, 12, 20\}$. To create $2x + 4y$, we add each value of $2x$ to each value of $4y$.

$$\left\{\begin{array}{cccc} 14 + 4, & 14 + 12, & 14 + 12, & 14 + 20 \\ 16 + 4, & 16 + 12, & 16 + 12, & 16 + 20 \\ 16 + 4, & 16 + 12, & 16 + 12, & 16 + 20 \\ 18 + 4, & 18 + 12, & 18 + 12, & 18 + 20 \end{array}\right\}$$

The resulting population is $2x + 4y$.

$$\left\{\begin{array}{cccc} 18 & 26 & 26 & 34 \\ 20 & 28 & 28 & 36 \\ 20 & 28 & 28 & 36 \\ 22 & 30 & 30 & 38 \end{array}\right\}$$

The mean of $2x + 4y$ may be computed as follows.

$$\begin{aligned} \mu_{2x+4y} &= E\left(2x + 4y\right) \\ &= E\left(2x\right) + E\left(4y\right) \\ &= 2E\left(x\right) + 4E\left(y\right) \\ &= 2 \cdot 8 + 4 \cdot 3 \\ &= 28 \end{aligned}$$

The variance of $2x + 4y$ is next.

$$\begin{aligned} \sigma^2_{2x+4y} &= Var\left(2x + 4y\right) \\ &= Var\left(2x\right) + Var\left(4y\right) \\ &= 2^2 Var\left(x\right) + 4^2 Var\left(y\right) \\ &= 4 \cdot 0.5 + 16 \cdot 2.0 \\ &= 34 \end{aligned}$$

4.2.3 Homework

The following questions refer to the small populations below.

The first population is x: $\{15, 16, 12, 14\}$, with mean $\mu_x = E(x)$, and variance $\sigma_x^2 = Var(x)$. The second population is y: $\{5, 2, 9, 4\}$, with mean $\mu_y = E(y)$, and variance $\sigma_y^2 = Var(y)$.

1. Compute the mean $\mu_x = E(x)$, and variance $\sigma_x^2 = Var(x)$ of population 1, using the statistics functions on your calculator. When computing the standard deviation, make sure to compute the population, not sample, standard deviation. Square this to get the variance.

2. Compute the mean $\mu_y = E(y)$, and variance $\sigma_y^2 = Var(y)$ of population 2, using the statistics functions on your calculator. When computing the standard deviation make sure to compute the population, not sample, standard deviation. Square this to get the variance.

3. Using properties of expected value and variance, give the value of $\mu_{x-y} = E(x-y)$ and $\sigma_{x-y}^2 = Var(x-y)$.

4. Compute the population of all differences $x - y$. Compute the mean and variance of this population and compare to your answers in Problem 3 above.

Suppose that the mean of all x values is $\mu_x = 7$ and the standard deviation is $\sigma_x = 2$. Suppose also that the mean of all y values is $\mu_y = 4$ and the standard deviation is $\sigma_y = 3$. Use the properties of expected value and variance to answer the following questions.

5. Give the mean, variance, and standard deviation of the population $3x - 2y$.

6. Give the mean, variance, and standard deviation of the population $\frac{1}{2}x + \frac{3}{2}y$.

7. Give the mean, variance, and standard deviation of the population $\frac{x+y}{2}$.

8. Give the mean, variance, and standard deviation of the population $\frac{2x+3y}{5}$.

9. Give the mean of all x^2 values.

10. Give the mean of all y^2 values.

11. The variable $x + x + x$ represents the sum of three random observations from the population of x values. The mean of these is $\overline{x} = \frac{x+x+x}{3}$. Give the mean, $\mu_{\overline{x}}$, and standard deviation, $\sigma_{\overline{x}}$, of all possible $\overline{x}$ values.

12. The variable $y + y + y$ represents the sum of three random observations from the population of y values. The mean of these is $\overline{y} = \frac{y+y+y}{3}$. Give the mean and standard deviation of all possible $\overline{y}$ values.

Answer the following questions.

13. Suppose $n = 5$ values, x, are randomly sampled from a population with mean, μ, and standard deviation, σ. The mean of these 5 values is $\overline{x} = \frac{x+x+x+x+x}{5}$. Consider the population of all possible $\overline{x}$ values which could result. Find the mean, $\mu_{\overline{x}}$, and standard deviation, $\sigma_{\overline{x}}$, of this population in terms of μ and σ.

14. Suppose $n = 6$ values, y, are randomly sampled from a population with mean, μ, and standard deviation, σ. The mean of these y values is $\overline{y} = \frac{y+y+y+y+y+y}{6}$. Consider the population of all possible $\overline{y}$ values which could result. Find the mean, $\mu_{\overline{y}}$, and standard deviation, $\sigma_{\overline{y}}$, of this population in terms of μ and σ.

15. Suppose n values, x, are randomly sampled from a population with mean, μ, and standard deviation, σ. The mean of these n values is $\overline{x}$. Consider the population of all possible $\overline{x}$ values which could result. Use the results of problems 13 and 14 to guess the formula for the mean, $\mu_{\overline{x}}$, and standard deviation, $\sigma_{\overline{x}}$, of this population in terms of μ and σ.

16. From the population of values, $\{4, 5, 6, 7\}$, gather all samples of size $n = 2$ (sampling with replacement) and compute a sample mean, $\overline{x}$. Draw a dotplot of the resulting collection of $\overline{x}$ values, and give a description of the shape of the distribution. Use the result of problem 15 to give the mean and standard deviation of the sample means.

17. From the population of values, $\{2, 3, 4, 5, 6\}$, gather all samples of size $n = 2$ (sampling with replacement) and compute a sample mean, $\overline{x}$. Draw a dotplot of the resulting collection of $\overline{x}$ values, and give a description of the shape of the distribution. Use the result of problem 15 to give the mean and standard deviation of the sample means.

4.3 Moments, Skewness and Kurtosis

In Chapter 2 we learned that the arithmetic mean is analogous to center of mass for a physical object. In this section we clarify this idea, give a physical analogy to variance, define moments, and introduce two new statistics and their respective parameters.

First, we define the objects we are describing. The statistical object is a probability distribution with its corresponding histogram. The analogous physical object is a system of discrete masses distributed along a line segment. Each object in the system is at position x, and has mass $P(x)$, and these are each analogous to values x with probability $P(x)$. In fact, the most correct term for the probability distribution function, $P(x)$, is the *probability mass function*, emphasizing the analogy between probability and mass.

4.3.1 Moments

After the analogies between probability and mass are established, we define the concept of a *moment* for each.

Definition 4.3.1 For the discrete values or positions x, with respective probabilities or masses $P(x)$, define the kth *moment* of the system about the value $x = a$ as

$$\mu_k = \sum (x - a)^k P(x)$$
$$= E\left((x - a)^k\right), \text{ for } k = 0, 1, \dots$$

Moments are always computed with respect to some point of reference, $x = a$, but usually a is equal to either zero (the *origin* on the number line) or to the mean of the data set.

With $k = 0$, we compute the 0th moment about $x = \mu$,

$$\mu_0 = E\left((x - \mu)^0\right)$$
$$= \sum (x - \mu)^0 P(x)$$
$$= \sum P(x),$$

which is the sum of all probabilities (which is always equal to one) for a probability distribution, or it is the total mass of the physical system. So the 0th moment, μ_0 is equal to 1 for a probability distribution, but $\mu_0 = total\ mass$ for a physical system of discrete masses.

With $k = 1$, we can compute the 1st moment about $x = \mu$,

$$\mu_1 = E\left((x - \mu)^1\right)$$
$$= \sum (x - \mu)^1 P(x),$$

but since this always equals zero, it is not very interesting. As an alternate, we might consider the first moment about $x = 0$, which gives

$$
\begin{aligned}
\mu_1 &= E\left((x - 0)^1\right) \\
&= \sum (x - 0)^1 P(x) \\
&= \sum x \cdot P(x) \\
&= \mu,
\end{aligned}
$$

the population mean. For the physical system, we must divide the first moment about $x = 0$ by the 0th moment to compute the center of mass, $\mu = \mu_1/\mu_0$. Since $\mu_0 = 1$ for a probability distribution, this formula works for either application.

With $k = 2$, we compute the 2nd moment about $x = \mu$,

$$
\begin{aligned}
\mu_2 &= E\left((x - \mu)^2\right) \\
&= \sum (x - \mu)^2 P(x).
\end{aligned}
$$

This is the original formula given for the variance of a discrete probability distribution, before we simplified it. For our physical system of discrete masses, this is the *moment of inertia* of the system.

Inertia is an important quantity in physics and engineering. Inertia is a quantity that describes the reluctance of an object with mass to have its velocity or angular velocity changed. Objects with mass that are at rest like to stay at rest, but once they are moving they do not like to have this velocity changed. *The measure of this kind of inertia is the object's mass*, which is the object's 0th moment about the center of mass. When such an object is in motion, it has kinetic energy, and this energy is directly proportional to its mass.

Objects with mass that spin have another kind of inertia which relates to how the object's mass is distributed about the axis of rotation. If the mass is gathered very close to this axis, then the object is easy to spin – it has little rotational inertia. If this object's mass is distributed far from the axis of rotation, then it will exhibit a greater reluctance to spin about this axis – it has greater rotational inertia. This type of inertia is called the *moment of inertia*, and it is the object's 2nd moment about the center of mass. When such an object spins, it again has kinetic energy, but this time its energy is directly proportional to its moment of inertia. The moment of inertia about the center of mass, μ, of a discrete system of masses at positions x, with mass $P(x)$ is computed by the formula $\mu_2 = \sum (x - \mu)^2 P(x)$. We note immediately that this is exactly the formula for the population variance, and so while we interpret this second moment to be a measurement of inertia for an object as it rotates about the center of mass, it may also be interpreted as a measure of how widely the discrete masses are distributed about the center of mass.

There are two additional moments which are useful in statistics, whose physical analogies are not often considered important – the third and fourth moments.

The third and fourth moments about the mean are, respectively, given below.

$$\mu_3 = E\left((x-\mu)^3\right)$$
$$= \sum (x-\mu)^3 P(x),$$

and

$$\mu_4 = E\left((x-\mu)^4\right)$$
$$= \sum (x-\mu)^4 P(x).$$

These values are of indirect importance, as they are used in the computations of *skewness* and *kurtosis*. In each case, these moments are divided by the second moment raised to a power which makes the resulting measure independent of any unit of measurement.

4.3.2   Skewness and Kurtosis

Definition 4.3.2 The *population skewness* is given by

$$skewness = \frac{\mu_3}{\mu_2^{3/2}}$$
$$= \frac{\mu_3}{\sigma^3}$$
$$= \frac{\sum (x-\mu)^3 P(x)}{\sigma^3}.$$

Definition 4.3.3 Pearson's *population kurtosis* is given by

$$kurtosis = \frac{\mu_4}{\mu_2^2}$$
$$= \frac{\mu_4}{\sigma^4}$$
$$= \frac{\sum (x-\mu)^4 P(x)}{\sigma^4}.$$

Analogous to the interpretations for sample skewness, the *skewness* of a population is a measure of how skewed the population is. When a population is skewed to the right, the *skewness* is positive, while a population skewed to the left has negative *skewness*. Populations that are nearly symmetric with respect to the mean have *skewness* close to zero, while populations that are perfectly symmetric with respect to the mean have zero *skewness*.

Like sample kurtosis, the population *Pearson kurtosis* (named after the great statistician Karl Pearson) measures the peakedness of the population. For a distribution which has a lower, wider peak and light tails (allowing few outliers) the *kurtosis* is less than 3, but as the peak narrows and rises and the tails get heavier (allowing more outliers), the *kurtosis* grows. Normal or bell-shaped populations have a *kurtosis* of 3, are called *mesokurtic*, and this is the standard by which other bell-shaped distributions are compared. When a distribution's *kurtosis* is greater than 3, it has a high, narrow peak and heavy tails, and is called *leptokurtic*. When a distribution's *kurtosis* is less than 3, its peak is lower and wider with light tails, and is called *platykurtic*. In all cases the *kurtosis* is greater than or equal to one.

Many texts use *Fisher's kurtosis* (named after the great statistician Ronald Fisher), denoted $kurt(x)$. This is simply Pearson's kurtosis minus three. The purpose is to center the kurtosis at zero, and thus giving a normal population a kurtosis of zero. Fisher's kurtosis has excellent algebraic properties, such as $kurt(x + y) = kurt(x) + kurt(y)$ (when x and y are independent and have equal variances), and $kurt(k \cdot x) = kurt(x)$. We will not use Fisher's kurtosis in this text.

Example 4.3.4 Suppose we toss an unbiased coin 6 times, and let x represent the number of tails. The values for x range from 0 tails to 6 tails. The probability distribution is given below, and the histogram is plotted in Figure 4.6.

x	$P(x)$
0	0.0156
1	0.0938
2	0.2344
3	0.3125
4	0.2344
5	0.0938
6	0.0156

In the next section of this text we will learn how to construct the distribution for this sort of problem, which we will refer to as a *binomial experiment*.

In order to compute the population skewness and kurtosis, we must first compute the mean and standard deviation. To do this we compute the sums $\sum x \cdot P(x)$ and $\sum x^2 \cdot P(x)$.

x	$P(x)$	$x \cdot P(x)$	$x^2 \cdot P(x)$
0	0.0156	0.00000	0.00000
1	0.0938	0.09375	0.09375
2	0.2344	0.46875	0.93750
3	0.3125	0.93750	2.81250
4	0.2344	0.93750	3.75000
5	0.0938	0.46875	2.34375
6	0.0156	0.09375	0.56250
$\sum$	1.0000	3.0000	10.5000

The mean is therefore $\mu = \sum x \cdot P(x) = 3.0000$, and the standard deviation is

$$\sigma = \sqrt{\sum x^2 P(x) - \mu^2} = \sqrt{10.5000 - 3.0000^2} = \sqrt{1.5} = 1.2247.$$

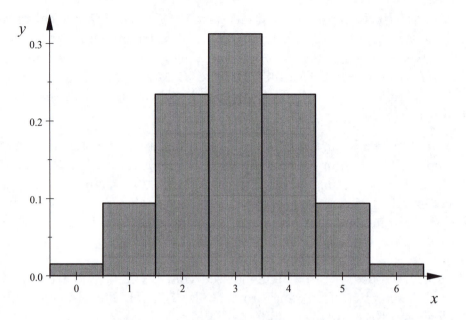

Figure 4.6: The distribution of the number of *tails* in six coin tosses.

Next, we compute the third and fourth moments about the mean.

x	$P(x)$	$(x - \mu)^3 \cdot P(x)$	$(x - \mu)^4 \cdot P(x)$
0	0.0156	−0.421875	1.265625
1	0.0938	−0.750000	1.500000
2	0.2344	−0.234375	0.234375
3	0.3125	0.000000	0.000000
4	0.2344	0.234375	0.234375
5	0.0938	0.750000	1.500000
6	0.0156	0.421875	1.265625
$\sum$	1.0000	0.000000	6.000000

The third moment about the mean is

$$\mu_3 = \sum (x - \mu)^3 \cdot P(x) = 0.000000,$$

and the fourth moment about the mean is

$$\mu_4 = \sum (x - \mu)^4 \cdot P(x) = 6.000000.$$

From these we compute skewness and kurtosis.

$$Skewness = \frac{\mu_3}{\sigma^3} = \frac{0.000000}{1.2247^3} = 0.0000;$$
$$Kurtosis = \frac{\mu_4}{\sigma^4} = \frac{6.000000}{1.2247^4} = 2.6667.$$

We see that this distribution is perfectly symmetric with *skewness* equal to zero, and is slightly *platykurtic* with *kurtosis* equal to 2.6667, slightly less than the normal standard of 3.

When working with sample data, it may be desirable to compute sample skewness and sample kurtosis. This is done with the computational formulas for the 2nd, 3rd, and 4th moments below. These formulas come directly from their respective parameter formulas above, but where μ is replaced by $\bar{x}$, and $P(x)$ is replaced by the relative frequency $\frac{f}{n}$.

Definition 4.3.5 The sample 2nd, 3rd, and 4th moments about the sample mean are:

$$m_2 = \frac{\sum f \cdot (x - \bar{x})^2}{n},$$

$$m_3 = \frac{\sum f \cdot (x - \bar{x})^3}{n},$$

and

$$m_4 = \frac{\sum f \cdot (x - \bar{x})^4}{n}.$$

With these, the sample skewness and kurtosis formulas are:

$$skewness = \frac{m_3}{m_2^{3/2}};$$

and

$$kurtosis = \frac{m_4}{m_2^2}.$$

The sample size is $n = \Sigma f$. When no frequencies are tallied, the frequencies are all $f = 1$. When all substitutions are made we derive the formulas given for sample skewness and kurtosis in Chapter 2.

Example 4.3.6 For the sample $\{1, 3, 3, 5\}$ we will compute the *skewness* and *kurtosis*. The mean of this sample is $\bar{x} = (1 + 3 + 3 + 5)/4 = 12/4 = 3$, and the sample size is $n = 4$.

x	f	$(x - \bar{x})$	$f(x - \bar{x})^2$	$f(x - \bar{x})^3$	$f(x - \bar{x})^4$
1	1	-2	4	-8	16
3	2	0	0	0	0
5	1	2	4	8	16
Σ	4	—	8	0	32

The 2nd moment is:

$$m_2 = \frac{\sum f \cdot (x - \bar{x})^2}{n} = \frac{8}{4} = 2.$$

The 3rd, and 4th moments of this sample are:

$$m_3 = \frac{\sum f \cdot (x - \bar{x})^3}{n} = \frac{0}{4} = 0,$$

and

$$m_4 = \frac{\sum f \cdot (x - \bar{x})^4}{n} = \frac{32}{4} = 8.$$

Thus, we may compute

$$skewness = \frac{m_3}{m_2^{3/2}} = \frac{0}{2^{3/2}} = 0,$$

and

$$kurtosis = \frac{m_4}{m_2^2} = \frac{8}{2^2} = \frac{8}{4} = 2.$$

The frequency histogram is plotted in Figure 4.7.

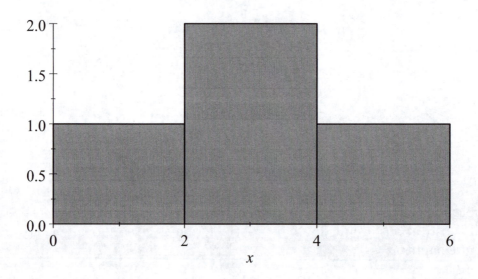

Figure 4.7: The histogram corresponding to the values in Example 4.3.6.

Visually, the graph is perfectly symmetric with respect to the mean, and this is proven by the fact that the *skewness* is zero. The *kurtosis* is 2, and because this is less than the standard value of 3, the distribution has a wider peak and shorter tails than the normal distribution and is thus classified as *platykurtic*.

4.3.3   Homework

Answer the following questions.

1. A deck of cards has 52 cards, and 13 of these are hearts. Three cards are chosen randomly *without replacement*. We define the random variable, x, to be the number of hearts drawn. The probability distribution for x is given below. Compute the 2nd, 3rd, and 4th moments, and the population skewness and kurtosis, classifying the distribution as skewed left, skewed right, or symmetric and as platykurtic, leptokurtic, or mesokurtic.

x	$P(x)$
0	0.413
1	0.436
2	0.138
3	0.013

2. For the entire world's population, approximately 14% of all people speak Chinese as their native language. Suppose four people are randomly selected from the Earth's population, and define x to be the number of these who speak Chinese as their native language. The probability distribution associated with this random variable is given below.

x	$P(x)$
0	0.547
1	0.356
2	0.087
3	0.009
4	0.001

Compute the population skewness and kurtosis, classifying the distribution as skewed left, skewed right, or symmetric and as platykurtic, leptokurtic, or mesokurtic.

3. For a family with 5 children we define x to be the number of girls. Statistically it is estimated that 49.7% of all children born are girls. Given this, the probability distribution for x is constructed below.

x	$P(x)$
0	0.0325
1	0.1600
2	0.3150
3	0.3100
4	0.1525
5	0.0300

Compute the population skewness and kurtosis, classifying the distribution as skewed left, skewed right, or symmetric and as platykurtic, leptokurtic, or mesokurtic.

4. A deck of cards has 52 cards, and 13 of these are hearts. Three cards are chosen randomly *with replacement*. We define the random variable, x, to be the number of hearts drawn. The probability distribution for x is given below. Compute the 2nd, 3rd, and 4th moments, and the population skewness and kurtosis, classifying the distribution as skewed left, skewed right, or symmetric and as platykurtic, leptokurtic, or mesokurtic.

x	$P(x)$
0	0.422
1	0.422
2	0.141
3	0.016

For each of the problems below, compute the sample 2nd, 3rd, 4th moments about the mean, along with the sample skewness and kurtosis. Classify each sample as *mesokurtic*, *leptokurtic*, or *platykurtic*. Also, classify each sample as *positively* or *negatively skewed*.

5. The following values are ages of college students who ride the bus to the college.
{ 23, 19, 19, 19, 20, 20, 22, 21, 20, 18, 22 }.

6. The following values are hours spent on Facebook each week by college students.
{ 12, 8, 4, 23, 10, 9, 20, 1, 5, 100 }.

7. The following are average hours that female college students sleep each night.
{ 7.0, 7.5, 8.0, 3.0, 9.0, 6.0, 4.5, 9.0, 6.5, 4.5 }

8. The following are average hours that male college students sleep each night.
{ 6.0, 4.0, 7.5, 6.5, 7.0, 4.0, 5.0, 7.0, 3.0, 5.5 }

4.4   Binomial Experiments

Now we move into a topic that will have long lasting ramifications in the inferential statistics chapters which lie ahead. The *binomial experiment* goes to the heart of the analysis of population and sample proportions. As mentioned previously, a population proportion, p, for a finite population of trials, is the number of times, X, a particular event, S, occurs in a population, divided by the population size, N. That is, $p = X/N$. Because of this, the population proportion is exactly equal to the probability that S occurs.

With each trial, either S occurs, or it does not. We will call the event S "*success*", and we will define $F =$ "*failure*" to be the complement of S ($F = \bar{S}$), and define $q = P(F) = P(\bar{S}) = 1 - P(S) = 1 - p$ by the rule of complements. A *binomial experiment* is one with n trials, each trial having two outcomes, S or F. We will define the random variable, x to be the number of times that S occurs in these n trials. Remember also that we previously defined a sample proportion as $\hat{p} = x/n$ where x is the number of times event S occurs in a sample of n trials. Later, we will use the sample proportion, $\hat{p}$, as an estimate of the population proportion, p.

4.4.1   The Binomial Experiment

Definition 4.4.1 A *Binomial Experiment* consists of the following:

- n independent trials

- Each trial has exactly two outcomes: $S =$ "success" and $F =$ "failure"

- $p = P(S)$, and $q = P(F) = 1 - p$, with p & q remaining constant throughout the experiment.

- The random variable, x, will represent the possible number of successes in n trials.

Example 4.4.2 Consider a random sample of $n = 5$ people. We will define a success as$S =$ "*female*" and $F =$ "*male*". Now the CIA reports that 49.8% of the world population is female, so $p = P(S) = 0.498$, and

$$
\begin{aligned}
q &= P(F) \\
&= 1 - p \\
&= 1 - 0.498 \\
&= 0.502.
\end{aligned}
$$

The random variable, x, represents the number of females out of this group.

Our goal will be to find a formula which will give the probability for each value of x

from 0 to n. In doing so, we will determine the probability distribution rule for the random variable x in the binomial experiment. To do this, we will first attempt to compute a specific probability where $x = 2$ and $n = 5$, using the rules from Chapter 3. We will do this by listing all of the ways to get $x = 2$ successes in $n = 5$ trials. Noting that two successes must be accompanied by three failures, we compute $P(x = 2)$ by listing every combination of two S events and three F events.

P(2 successes)

=P(SSFFF or SFSFF or FSSFF or FSFSF or FFSSF or FFSFS or FFFSS or SFFFS or SFFSF or FSFFS)

=P(SSFFF)+P(SFSFF)+P(FSSFF)+P(FSFSF)+P(FFSSF)+P(FFSFS)+P(FFFSS)+P(SFFFS)+P(SFFSF)+P(FSFFS)

=ppqqq+pqpqq+qppqq+qpqpq+qqppq+qqpqp+qqqpp+pqqqp+pqqpq+qpqqp

=$10p^2q^3$

Thus,

$$P(x = 2) = 10p^2q^3.$$

If we analyze this result, we understand that the 10 represents the number of ways to get $x = 2$ successes in $n = 5$ trials. We will need a way to count this without actually listing the ways individually, as this can be complicated! We are raising p to the 2nd power because p is the probability of success, and we want $x = 2$ successes. We raise q to the 3rd power because in $n = 5$ trials with $x = 2$ successes, there will be $n - 2 = 3$ failures, and q is the probability of failure.

Our probability could be rewritten as

$$P(x \text{ successes}) = 10p^x q^{n-x}.$$

But how do we count the 10 ways to do get 2 successes in 5 trials without actually listing each of the ways? The answer lies in a field of mathematics called *combinatorics*, which is concerned with *methods of counting.*

One formula that is used in combinatorics is the *combination* or *"choose"* formula,

$$_nC_x = \frac{n!}{x!\,(n - x)!}.$$

This formula counts the ways to get x successes in n trials. To understand this, we need to understand factorial notation $n! = n(n - 1)(n - 2) \cdots 3 \cdot 2 \cdot 1$. For example, $5! = 5 \cdot 4 \cdot 3 \cdot 2 \cdot 1 = 120$.

In addition, we define $0! = 1$. Applying the choose formula to our binomial experiment where $n = 5$ and $x = 2$,

$$
\begin{aligned}
_5C_2 &= \frac{5!}{2!\,(5-2)!} \\
&= \frac{5!}{2! \cdot 3!} \\
&= \frac{5 \cdot 4 \cdot \cancel{3} \cdot \cancel{2} \cdot \cancel{1}}{2 \cdot 1 \cdot \cancel{3} \cdot \cancel{2} \cdot \cancel{1}} \\
&= \frac{5 \cdot \overset{2}{\cancel{4}}}{\cancel{2} \cdot 1} \\
&= \frac{5 \cdot 2}{1} \\
&= 10
\end{aligned}
$$

Notice that $_5C_2 = 10$, which is exactly the number of ways that we listed to get $x = 2$ successes in $n = 5$ trials. Notice also that the fraction for $_5C_2$ reduced to a whole number. This will always happen, and it is important to note that when computing $_nC_x$ it is best to either use a built in function on a scientific calculator, or to simplify the expression by hand, canceling all terms in the denominator to compute the answer. It is best to avoid using the factorial button on a calculator, as answers often get rounded due to the large magnitude of the result.

With $_5C_2 = 10$ with $n = 5$ and $x = 2$, we can update our formula:

$$
\begin{aligned}
P(x\ successes) &= 10 p^x q^{n-x} \\
&= {}_nC_x\, p^x q^{n-x} \\
&= \frac{n!}{x!\,(n-x)!} p^x q^{n-x}.
\end{aligned}
$$

Finally, our probability has been expressed in a form that may be used for any binomial experiment.

Theorem 4.4.3 The Binomial Probability Distribution formula is:

$$
\begin{aligned}
P(x) &= {}_nC_x\, p^x\, q^{n-x} \\
&= \frac{n!}{x!\,(n-x)!} p^x\, q^{n-x},
\end{aligned}
$$

where $P(x)$ represents the probability of x successes in n trials in a binomial experiment.

If we apply this theorem to our original example, where $n = 5$, $x = 2$, $p = 0.498$,

$q = 0.502$, then

$$
\begin{aligned}
P\left(x = 2\right) &= \frac{n!}{x!\,(n-x)!}p^x q^{n-x}\\[6pt]
&= \frac{5!}{2!\,(5-2)!}0.498^2 \cdot 0.502^{5-2}\\[6pt]
&= 10 \cdot 0.498^2 \cdot 0.502^3\\[6pt]
&= 0.314\\[6pt]
&= 31.4\%.
\end{aligned}
$$

Example 4.4.4 Approximately 82% of the Earth's people over the age of 15 can read. Suppose that we randomly select 28 people over 15 years of age from the Earth's population. We would like to know the probability that exactly 23 out of these 28 of these people can read.

We begin by defining what success and failure are. Since we are counting the number of people out of this sample who can read, we define success as $S = $ "*can read*" and failure as $F = $ "*cannot read*". Since 82% of the world's population can read, we know that $p = P(S) = 0.82$, and thus $q = 1 - p = 0.18$. We want to compute

$$P(23 \; successes) = P(x = 23).$$

First we list the values for each variable used in the binomial probability formula: $n = 28$, $p = 0.82$, $q = 0.18$, $x = 23$. Next, we compute $_nC_x = \,_{28}C_{23}$.

$$
\begin{aligned}
_{28}C_{23} &= \frac{28!}{23!\cdot(28-23)!}\\[10pt]
&= \frac{28\cdot27\cdot26\cdot25\cdot24\cdot\cancel{23}\cdot\cancel{22}\cdots\cancel{1}}{(\cancel{23}\cdot\cancel{22}\cdots\cancel{1})\cdot(5\cdot4\cdot3\cdot2\cdot1)}\\[10pt]
&= \frac{28\cdot27\cdot26\cdot25\cdot24}{5\cdot4\cdot3\cdot2\cdot1}\\[10pt]
&= \frac{14\cdot9\cdot26\cdot5\cdot6}{1}\\[10pt]
&= 98,280.
\end{aligned}
$$

This tells us that there are $98,280$ ways to get 23 successes in 28 trials. It is good that we have the $_nC_x$ formula, and do not need to list each of the 98,280 of these ways! Now we can compute $P(x = 23)$.

$$
\begin{aligned}
P\left(x = 23\right) &= \,_nC_x\,p^x\,q^{n-x}\\[6pt]
&= \,_{28}C_{23}\cdot 0.82^{23}\cdot 0.18^{28-23}\\[6pt]
&= 98,280\cdot 0.82^{23}\cdot 0.18^5\\[6pt]
&= 0.1934\\[6pt]
&\approx 19.34\%
\end{aligned}
$$

Below, the entire probability distribution, associating a probability for each x from 0 to 28 is listed, and plotted in Figure 4.8.

x	$P(x)$
0	0.000000000000000000001
1	0.000000000000000000179
2	0.000000000000000011021
3	0.000000000000000435109
4	0.000000000000012388520
5	0.000000000000270895647
6	0.000000000004730640658
7	0.000000000067730759890
8	0.000000000809947003689
9	0.000000008199463494140
10	0.000000070970911799279
11	0.000000529055887958261
12	0.000003414369943582480
13	0.000019143817803334200
14	0.000093440063087702700
15	0.000397293305276603000
16	0.001470537025780760000
17	0.004728785729961640000
18	0.013164705951930300000
19	0.031564499650826900000
20	0.064707224284195200000
21	0.112296135265693000000
22	0.162772680915424000000
23	0.193439997609634000000
24	0.183588886620255000000
25	0.133815899580986000000
26	0.070339126702825700000
27	0.023735836994369200000
28	0.003861783003052120000

Notice that while the bell in the distribution is shifted to the right, there is still a fair amount of symmetry to the bell shape, with a single mode and probabilities nearly vanishing on each side. The distribution appears somewhat *normal*. This is quite common for binomial distributions – they are nearly normal under certain conditions which we will describe later.

4.4.2 Mean and Standard Deviation

All discrete probability distributions have a mean and standard deviation, and each may be computed by the formulas in section 4.1. The binomial distribution is no exception, but in this particular case the formulas for mean and standard deviation simplify a great

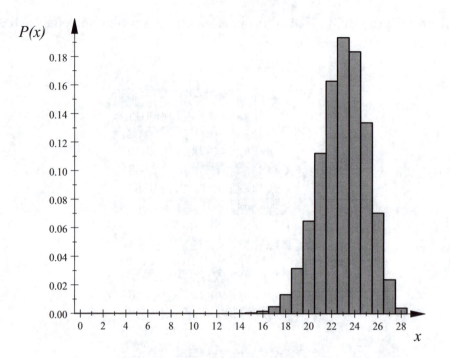

Figure 4.8: The binomial distribution for $n = 28$ and $p = 0.82$.

deal. There is a proof for the following formulas in the optional section at the end of this chapter.

Theorem 4.4.5 The mean and standard deviation for the binomial distribution are given below.

$$\mu = np$$
$$\sigma = \sqrt{npq}$$

Proof. To prove the formulas for the mean and standard deviation of a binomial experiment, we begin with the simplest case – a binomial experiment with $n = 1$ trial. If x represents the number of successes in this experiment, then the values for x are $x = 0$ and $x = 1$. The probability that $x = 0$ is q, as zero successes implies one failure. The probability that $x = 1$ is p, the probability of success. In the following table, we give the probability distribution, with the relevant totals.

x	$P(x)$	$xP(x)$	$x^2P(x)$
0	q	0	0
1	p	p	p
$\sum$	1	p	p

We see that $\mu = E(x) = \sum x P(x) = p$, and $\sum x^2 P(x) = p$ as well. With these we may compute

$$\begin{aligned}
\sigma &= \sqrt{\sum x^2 P(x) - \mu^2} \\
&= \sqrt{p - p^2} \\
&= \sqrt{p(1-p)} \\
&= \sqrt{pq}
\end{aligned}$$

The variance follows directly.

$$\sigma^2 = Var(x) = pq$$

For binomial experiments with $n > 1$ trials, it is easiest to consider each trial as a single trial experiment with expected value p and variance pq. The total number of successes, x, is the sum of the number of successes in each individual trial, and the mean of this sum is the sum of the n means, each equal to p. That sum is

$$\mu = np,$$

the mean of the overall experiment. The variance of the sum is the sum of n variances, each equal to pq. This sum is $\sigma^2 = Var(x) = npq$. We apply a square root for the standard deviation,

$$\sigma = \sqrt{npq}.$$

■

If we apply these formulas to our last example, we see that the mean number of people over 15 who can read in a random sample of 28 people is $\mu = np = 28 \cdot 0.82 = 22.96$, while the standard deviation is: $\sigma = \sqrt{npq} = \sqrt{28 \cdot 0.82 \cdot 0.18} = 2.033$.

Once we have the mean and standard deviation we can look at individual scores and determine when they are unusual or significant. For instance, would 27 readers in a random group of 28 people over 15 years of age be unusual? To answer this, we simply compute the z-score, $z = \frac{x-\mu}{\sigma} = \frac{27-22.96}{2.033} = 1.99$ which tells us that 27 is 1.99 standard deviations above average. This is close enough to the required two standard deviations for unusualness, so we will say that getting 27 readers is significant in a random group of 28. This may not *seem* unusual, being in a country with a very high literacy rate, but we must remember that, worldwide, literacy rates are much lower than they are here at home.

Example 4.4.4 Continued: If we wish to determine significance by looking at probabilities, the basic rule is that we should compute the probability of measuring a value that is at least as extreme as the one we have measured. As mentioned previously, this computation is called a p-value. To determine whether $x = 27$ is significantly high, we compute using the probabilities listed earlier in this example.

$$
\begin{aligned}
P(x \geqslant 27) &= P(27 \text{ or } 28) \\
&= P(27) + P(28) \\
&= 0.0237 + 0.00386 \\
&= 0.0276 \\
&= 2.76\%
\end{aligned}
$$

Using our rule that an event with less than a 5% chance is significant when it happens, this is again significant.

4.4.3   Skewness and Kurtosis

The *skewness* and *kurtosis* for the binomial distribution are:

$$
skewness = \frac{q-p}{\sqrt{npq}},
$$
$$
kurtosis = 3 + \frac{1-6pq}{npq}.
$$

We see that only when $p = q = 0.5$, is the skewness is zero with a symmetric distribution. Still, even when $p \neq q$, the skewness approaches zero as n increases, and the distribution approaches symmetry in the process. Also, as n increases, the kurtosis approaches the standard value of 3, which is the kurtosis for a normal, bell-shaped, distribution.

4.4.4   The History of the Binomial Probability Distribution

The Binomial Theorem from algebra gives a formula for the terms that come about after raising a binomial to the nth power. The Binomial Theorem is ancient, and various forms have been known for thousands of years. Ancient Indian and Chinese mathematicians knew it in one form or another. An Islamic mathematician named al-Karaji (1029) gave the coefficients for the expansion for exponents up to $n = 5$ (which form Pascal's Triangle). Newton gave a special version that allowed negative exponents and included infinitely many terms.

The simplest version of the formula given in the Binomial Theorem is:

$$
(p+q)^n = \sum_{x=0}^{n} {}_nC_x \cdot p^x q^{n-x} = \sum_{x=0}^{n} \frac{n!}{x!\,(n-x)!} \cdot p^x q^{n-x}.
$$

Notice the terms in the sum in the binomial expansion of $(p+q)^n$. They are exactly equal to the formula which we developed for the binomial probabilities, $P(x) = \frac{n!}{x!(n-x)!} \cdot p^x q^{n-x}$. That is quite a coincidence!

The Swiss mathematician Jacob Bernoulli was the first to have discovered that the terms in the binomial expansion for $(p + q)^n$ give the probabilities for the binomial distribution. In a paper published after his death in 1713, he stated that the probability of x successes in n trials with probabilities for success and failure equal to p and q respectively is equal to the xth term in the expansion described by the binomial theorem.

Figure 4.9: A Swiss postage stamp from 1994 featuring a portrait of mathematician Jacob Bernoulli. The portrait was painted by Bernoulli's brother Nicholas. The stamp also features Bernoulli's law of large numbers, which states that the sample mean approaches the population mean with increasing sample size.

The binomial theorem also gives us proof that the sum of the binomial probabilities is equal to one:

$$\sum_{x=0}^{n} P(x) = \sum_{x=0}^{n} \frac{n!}{x!(n-x)!} \cdot p^x q^{n-x} = (p + q)^n = 1^n = 1.$$

4.4.5   Homework

Answer the following questions.

1. Suppose $n = 10$ and $x = 4$. Compute $_nC_x$.

2. Compute $P(4)$ for a binomial experiment with $n = 10$ and $p = 0.30$.

3. Compute the mean, standard deviation, skewness and kurtosis of the random variable, x, in a binomial experiment where $n = 10$ and $p = 0.30$.

4. In a binomial experiment with $n = 10$ and $p = 0.30$, would $x = 9$ be unusually high? Use a *p-value* to answer this. As a reminder, the *p-value* corresponding to $x = 9$ is the probability of measuring an x that is at least as extreme as $x = 9$. So this is equal to $P(x \geqslant 9) = P(9 \text{ or } 10) = P(9) + P(10)$.

5. Suppose $n = 20$ and $x = 13$. Compute $_nC_x$. You may need to do this using the built-in $_nC_x$ function on your scientific calculator.

6. Compute $P(13)$ in a binomial experiment with $n = 20$ and $p = 0.45$.

7. Compute the mean, standard deviation, skewness, and kurtosis of the random variable, x, in a binomial experiment where $n = 20$ and $p = 0.45$.

8. With $n = 20$ and $p = 0.45$, would $x = 5$ be an unusually low measurement in a binomial experiment? Use a z-score to answer this.

9. Approximately 5% of the world's population speaks English as their native language. If we randomly select 10 people from anywhere in the world, what is the probability that exactly one of them speaks English natively?

10. Airlines routinely overbook flights to help lower the number of empty seats on their airplanes in order to increase profits. In the year 2005, the no-show rate was estimated to be 12%, with 88% of passengers with tickets actually showing up to take the flight. Suppose an airplane has 25 seats, and that the airline has sold 27 tickets. What is the probability that there will not be enough seats?

11. Referring to the no-show rates from Problem 10, out of 27 tickets sold, what are the mean and standard deviation for the number of people that show up to take the flight?

12. Suppose the president has a 60% approval rating among voters, and that 12 voters are randomly selected, and asked if they approve of the president. What are the mean and standard deviation of the number of approvals?

13. Referring to the approval rate in Problem 12, when 12 voters are selected, what is the probability that 8 of them approve of the president?

14. In 1936 the famous statistician Ronald A. Fisher challenged one of Gregor Mendel's famous experiments in pea genetics by saying that, in this experiment, the 6,022 yellow peas observed in a sample of 8,023 peas was much too close to the mean number of peas, and that there was only one chance in ten of being this close.

   a. Given that the yellow peas in this experiment had a 75% chance of occurring, what would the mean number of peas have been?

   b. What is the standard deviation?

   c. How many standard deviations from the mean is the value that Mendel reported?

   d. Would you consider the 6,022 observed peas to be "too close" to the expected value? In fact, this closeness occurred in all seven of Mendel's pea experiments. In each, the observed values were remarkably close to the expected value.

   e. Do you feel Mendel fabricated his numbers?

We may never know. One thing is for certain, Ronald Fisher was not afraid of confrontation, as we shall learn in Chapter 11.

15. In a weather report for a certain town, there was reported to be a 20% chance of rain for the days Monday through Friday. Assuming that these rainy days are independent of one another (even though they are likely to be *dependent*), compute the probability that it will rain at least one of these days. (Consider using the rule of complements to simplify this!)

16. What is the expected value for the number of rainy days in the previous problem?

4.5   The Poisson Distribution

To provide another example of a discrete probability distribution which is commonly encountered, we now turn our attention to the Poisson probability distribution. The Poisson distribution is unusual in that it is cannot be derived through the normal rules of probability, as we did with the binomial distribution. The Poisson distribution is actually quite closely related to the binomial, because the Poisson probabilities are derived by allowing the number of trials to increase indefinitely on a binomial experiment. The resulting probability distribution is $P(x) = \frac{e^{-\mu} \cdot \mu^x}{x!}$, for $x = 0, 1, 2, \ldots$. This formula is derived from the binomial probability distribution formula in the optional section at the end of this chapter. The number e is an irrational number approximately equal to 2.718281828, and is defined to be the value which the expression $(1 + 1/k)^k$ approaches as k increases indefinitely.

4.5.1   The Poisson Distribution

In the Poisson distribution, the random variable x is used to measure the number of times a particular event, S, occurs over some time period or spatial region. Event S should be an unlikely event which may occur many times due to the largeness of the population under investigation. There is no limit to the number of times that S can occur in the experiment, and this is what distinguishes the Poisson distribution from the binomial where x can be no larger than n. The mean of the distribution is μ, a parameter which needs to be estimated experimentally. We will prove that the variance of the Poisson is also μ, and that the sum of the probabilities is equal to 1. For the present, we summarize below.

Definition 4.5.1 The *Poisson Distribution* gives probabilities for the discrete random variable, x, which measures the number of times that event S occurs randomly in some finite interval of time or space. The occurrences of S must be independent and uniformly distributed over the relevant interval in time or space. The probabilities of each x are given by the formula below, where μ is the mean number of occurrences of S over this interval.

$$P(x) = \frac{e^{-\mu} \cdot \mu^x}{x!}$$

Example 4.5.2 As an example of an experiment whose discrete random variable is likely to be distributed according to the Poisson distribution, consider a particular hospital which, on average, delivers 6.4 babies per day. If we let x represent the number of babies born on a randomly selected day at this hospital, x, will have a Poisson distribution, and its probability distribution function is $P(x) = \frac{e^{-6.4} \cdot 6.4^x}{x!}$. Suppose we wish to compute the probability that less than 10 babies are born on any given day at this hospital. We must compute $P(x)$ for all x values ranging from zero to nine.

The first three are computed below.

x	$P(x) = \frac{e^{-6.4} \cdot 6.4^x}{x!}$
0	$P(0) = \frac{e^{-6.4} \cdot 6.4^0}{0!} = 0.001662$
1	$P(1) = \frac{e^{-6.4} \cdot 6.4^1}{1!} = 0.01064$
2	$P(2) = \frac{e^{-6.4} \cdot 6.4^2}{2!} = 0.03403$
3	$P(3) = \frac{e^{-6.4} \cdot 6.4^3}{3!} = 0.07259$

The process continues, with the table to the below summarizing these probabilities for x values ranging from zero to nine, along with their sum.

x	$P(x)$
0	0.001661557
1	0.010633967
2	0.034028693
3	0.072594545
4	0.116151272
5	0.148673628
6	0.158585203
7	0.144992186
8	0.115993749
9	0.082484444
Σ	0.885799243

The probability in which we are interested is the sum: $P(0 \leqslant x \leqslant 9) = \sum\limits_{x=0}^{9} P(x) = 0.8858$.

In Figure 4.10 is a histogram representing this particular Poisson distribution for x ranging from 0 to 25.

The distribution carries on indefinitely, but as is seen from the graph, probabilities become nearly zero quickly. Also from the graph, the Poisson distribution is skewed to the right, but becomes more bell-shaped with increasing values of μ. Note the highest bar associated with the value of x which is closest to μ.

Example 4.5.3 In the classic experiment in radioactivity conducted by Rutherford, Geiger, and Bateman in 1910, the number of alpha particles emitted by a film of radioactive polonium were counted over 2608 time intervals of 1/8 second each.

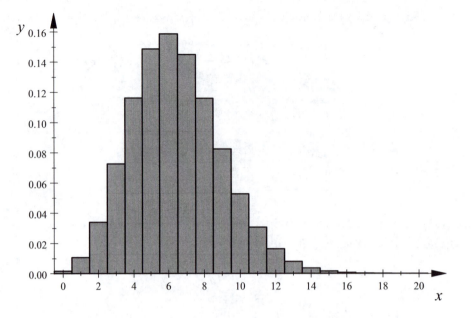

Figure 4.10: An example of a discrete Poisson Distribution.

Results are summarized below, where O represents the number of time intervals out of the 2608 in which x particles were emitted.

x = Number of alpha particles	O = Observed Number of intervals where x particles were emitted
0	57
1	203
2	383
3	525
4	532
5	408
6	273
7	139
8	45
9	27
10	10
11	4
12	0
13	1
14	1
Over 14	0
Total	2608

Altogether, 10,097 particles were emitted amongst the 2608 time intervals. On average, the mean number of particles emitted in each time interval is

$$\mu = \frac{\text{total particles emitted}}{\text{number of intervals}} = \frac{10,097}{2,608} = 3.8716$$

particles per time interval. It is also possible to compute the variance, using the variance formula for frequency tables (using the O values as frequencies),

$$s^2 = \frac{\sum f \cdot x^2 - \frac{1}{n}\left(\sum f \cdot x\right)^2}{n-1}.$$

Applying this, we get a variance of 3.6962. Remember that the variance is equal to the mean in the Poisson distribution, and we see that this is nearly so with our sample data.

Using the mean $\mu = 3.8716$, we are able to construct the Poisson probability distribution function,

$$P(x) = \frac{e^{-\mu} \cdot \mu^x}{x!}$$
$$= \frac{e^{-3.8716} \cdot 3.8716^x}{x!}.$$

If we wish to know the probability of a particular value of x, say $x = 5$, we will evaluate the function

$$P(5) = \frac{e^{-3.8716} \cdot 3.8716^5}{5!}$$
$$= \frac{0.020825 \cdot 869.866}{120}$$
$$= 0.1510$$
$$= 15.10\%.$$

The question now becomes whether our probabilities lead us to expected frequencies which are close to the frequencies actually observed. For example, the probability that $x = 5$ is 15.10%, so out of 2608 time intervals, we expected 15.10% of 2608 intervals to register $x = 5$ alpha particles. Computing, we get $0.1510 \cdot 2608 = 393.69$ expected alpha particles, and in fact 5 particles were observed 408 times out of 2608, which is quite close to what was expected. The entire distribution of probabilities is given in Table 4.1, along with observed and expected frequencies.

Clearly the observed frequencies are different from the expected frequencies, but that is to be expected due to the random nature of the observations. In Chapter 10 we will develop a test for determining if the observed frequencies are within a reasonable distance from the expected frequencies, and that the observed deviations are simply due to sampling error & chance fluctuation.

x = Number of alpha particles	$P(x) = \frac{e^{-3.8716} \cdot 3.8716^x}{x!}$	O = Observed intervals (x particles)	E = Expected intervals (x particles)
0	0.02083	57	54.31
1	0.08063	203	210.28
2	0.15608	383	407.06
3	0.20142	525	525.31
4	0.19496	532	508.44
5	0.15096	408	393.69
6	0.09741	273	254.03
7	0.05387	139	140.50
8	0.02607	45	67.99
9	0.01122	27	29.25
10	0.00434	10	11.32
11	0.00153	4	3.99
12	0.00049	0	1.29
13	0.00015	1	0.38
14	0.00004	1	0.11
Over 14	0.00001	0	0.04
Total	1.00000	2608	2608

Table 4.1: The distribution of probabilities, with observed and expected frequencies for Example 4.5.3.

4.5.2 Skewness and Kurtosis

The skewness and kurtosis for the Poisson distribution are:

$$skewness = \frac{1}{\sqrt{\mu}},$$

and

$$kurtosis = 3 + \frac{1}{\mu}.$$

We see that the skewness for the Poisson distribution is never zero, and thus the distribution is never symmetric. Still, the skewness approaches zero as μ increases, and the distribution approaches symmetry in the process. Also, as μ increases, the kurtosis approaches the standard value of 3, which is the kurtosis for a normal, bell-shaped, distribution.

4.5.3 History of the Poisson Distribution

The Poisson distribution is named after Siméon Denis Poisson (1781 – 1840), who gave its formula in a book published in 1830. It does not appear that Poisson was the first to discover it, though he made it famous. It seems that Abraham De Moivre (1667 - 1754) had developed the equivalent of the Poisson distribution in 1712, long before Poisson published his version in 1830 (De Moivre was also the first known person to use the normal probability distribution which we will discuss in the next chapter). To that point, the Poisson distribution's properties had not been determined, but they were eventually revealed by L. J. Bortkiewicz in the article *Das Gesetz der kleinen Zahlen* (1898); which includes the famous example of cavalrymen being killed by the kick of a horse (see this section's homework problems for this). The Poisson distribution was largely forgotten for years, until rediscovered by statistics genius William *Student* Gosset in 1907, and it has been widely known since then.

4.5.4   Homework

Answer the following questions.

William *Student* Gosset was a pivotal character in the history of statistics, developing Student's *t*-distribution which we will use frequently beginning in Chapter 6 of this text. In England the Poisson distribution was largely unknown, but Gosset rediscovered it in his research in 1907, and demonstrated how it occurs naturally in scientific experiments. Gosset worked for a brewery in England, and was naturally concerned with yeast as it relates to the making of alcohol. In one famous example, Student provided a frequency distribution of yeast cells divided into 400 square containers of equal size. In this experiment, 1,872 yeast cells in a solution were divided amongst the 400 squares, with each square receiving equal amounts of solution. Next, microscopes were used to count the number of yeast cells in each square container, with the observed frequencies given below.

x = Number of yeast cells	Number of Containers with x cells
0	0
1	20
2	43
3	53
4	86
5	70
6	54
7	37
8	18
9	18
10	10
11	5
12	2
13	2
14	0
15	0
16	0
Total	400

In analyzing these frequencies, *Student* noted that they closely follow the frequencies expected according to the Poisson distribution. Compute these expected frequencies by first constructing the Poisson distribution corresponding to the values of x above.

1. With 1,872 yeast cells counted, divided amongst 400 containers, what is the mean number of cells per container?

2. Give the formula for this Poisson distribution, using the mean computed in Problem 1 above.

3. For each x in the table above, compute $P(x)$ using the formula from Problem 2 above.

4. Using the probabilities computed in Problem 3 above, compute expected frequencies for each value of x by multiplying $P(x)$ by the total number of containers, which was 400. Do you agree with Student that the data fit the Poisson distribution closely?

Another classic example of data which are nearly distributed according to the Poisson distribution comes from a little book published in 1898 by Ladislaus von Bortkiewicz. This book was the first to show the Poisson distribution accurately predicting frequencies of random events. Among various examples of data the most extensive involved numbers of deaths annually by horse kicks in 14 Prussian army corps between 1875 and 1894. (14 army corps for 20 years totals 280 units of time, each tracking a single corps over a single year).

x = Number of Deaths per time unit	Observed Frequency
0	144
1	91
2	32
3	11
4	2
5 and over	0
Total	280

5. In this time, a total of 196 deaths occurred, spread amongst 280 time units. What is the mean number of deaths per time unit?

6. Give the formula for this Poisson distribution, using the mean computed in Problem 5 above.

7. For each x in the table above, compute $P(x)$ using the formula developed in Problem 6 above. To compute the probability for the "5 and over" category, use the rule of complements.

8. Using the probabilities computed in Problem 7 above, compute expected frequencies for each value of x by multiplying $P(x)$ by the total number of time units, which was 280. Do the data appear fit the Poisson distribution closely?

In 1964, Mosteller and Wallace divided James Madison's historical issues of the Federalist Papers into 262 blocks of text, each containing approximately 200 words. In each of these blocks, the number of occurrences of the word *may* were counted. In 156 of the blocks, the word *may* never occurred, while the most occurrences of *may* were 6, which happened in only one block.

Altogether, the word *may* occurs 172 times. The entire distribution is given below. (Data from *Applied Bayesian and Classical Inference: The Case of the Federalist Papers*, by Mosteller and Wallace, New York: Springer-Verlag, 1984).

Occurrences of the word *may*.	Observed Frequency
0	156
1	63
2	29
3	8
4	4
5	1
6 and over	1
Total	262

9. What is the mean number of occurrences of *may* per block of text?

10. Give the formula for this particular Poisson distribution.

11. For each x in the table above, compute $P(x)$ using the formula developed in Problem 10 above. To compute the probability for the "6 and over" category, use the rule of complements.

12. Using the probabilities computed in Problem 11 above, compute expected frequencies for each value of x by multiplying $P(x)$ by the total number of word blocks, which was 256. Do the data fit the Poisson distribution closely?

4.6 Proofs of Theorems

4.6.1 Formulas for Expected Value and Variance

Recall a few items from Chapter 2. If we are computing the mean and standard deviation of a finite population using frequencies, the formulas are: $\mu = \frac{\sum fx}{N}$, and $\sigma^2 = \frac{\sum f \cdot x^2 - \frac{1}{N} \cdot (\sum f \cdot x)^2}{N}$, where each x in the population occurs f times, and $N = \sum f$. Now, recalling that we may compute probabilities for finite populations exactly using relative frequencies for population data so that $P(x) = \frac{f}{N}$, our formula for the mean becomes $\mu = \frac{\sum fx}{N} = \sum x \cdot \frac{f}{N} = \sum x \cdot P(x)$, and the variance becomes

$$
\begin{aligned}
Var(x) &= \sigma^2 \\
&= \frac{\sum f \cdot x^2 - \frac{1}{N} \cdot (\sum f \cdot x)^2}{N} \\
&= \frac{\sum f \cdot x^2}{N} - \frac{\frac{1}{N} \cdot (\sum f \cdot x)^2}{N} \\
&= \sum \frac{f}{N} \cdot x^2 - \frac{1}{N^2} \cdot \left(\sum f \cdot x \right)^2 \\
&= \sum P(x) \cdot x^2 - \left(\frac{\sum f \cdot x}{N} \right)^2 \\
&= \sum x^2 \cdot P(x) - \left(\sum x \cdot \frac{f}{N} \right)^2 \\
&= \sum x^2 \cdot P(x) - \left(\sum x \cdot P(x) \right)^2 \\
&= \sum x^2 \cdot P(x) - \mu^2 \\
&= E(x^2) - E(x)^2
\end{aligned}
$$

While this proof applies to finite populations, it is also true for populations of indeterminate size. The proof uses relative frequencies from samples, and using the law of large numbers, the relative frequencies approach corresponding probabilities as n increases.

In the context of probability distributions, we give special names to the mean and variance of a random variable, x. The mean is denoted $\mu = E(x)$, which stands for the expected value of x. The variance of the random variable, x, is denoted $\sigma^2 = Var(x)$.

We summarize these findings below.

Summary 4.6.1 The formulas for the mean, variance, and standard deviation of a discrete probability distribution are summarized below.

$$
\mu = E(x) = \sum x \cdot P(x),
$$

$$
\sigma^2 = Var(x) = \sum x^2 \cdot P(x) - \mu^2,
$$

and

$$
\sigma = \sqrt{\sum x^2 \cdot P(x) - \mu^2}.
$$

To establish the fact that the variance formulas $Var(x) = \sum x^2 \cdot P(x) - \mu^2$ and $Var(x) = \sum (x - \mu)^2 \cdot P(x)$ are equivalent, we simplify as follows:

$$\begin{aligned} Var(x) &= \sum (x - \mu)^2 \cdot P(x) \\ &= \sum \left(x^2 \cdot P(x) - 2x \cdot \mu \cdot P(x) + \mu^2 P(x) \right) \\ &= \sum x^2 P(x) - 2\mu \sum x \cdot P(x) + \mu^2 \sum P(x) \\ &= \sum x^2 P(x) - 2\mu \cdot \mu + \mu^2 \cdot 1 \\ &= \sum x^2 P(x) - 2\mu^2 + \mu^2 \\ &= \sum x^2 P(x) - \mu^2. \end{aligned}$$

4.6.2 Proofs of Properties of Expected Value and Variance

First, a few of the properties of the summation operator, $\sum$, will be given. They have been mentioned previously, but an additional property has been added.

Theorem 4.6.2 Properties of Sums

1. $\sum (x \pm y) = \sum x \pm \sum y$

2. $\sum k \cdot x = k \cdot \sum x$, where k is constant.

3. $\sum_{i=1}^{n} k = n \cdot k$, where k is constant.

Next, a reminder of the properties of expected value and variance.

Theorem 4.6.3 Properties of Expected Value and Variance
Assume that x & y are independent random variables.

1. $E(x + y) = E(x) + E(y)$

2. $E(x - y) = E(x) - E(y)$

3. $E(x \cdot y) = E(x) \cdot E(y)$, where $x \cdot y$ represents each x times each y.

4. $E(k \cdot x) = k \cdot E(x)$, where k is any constant.

5. $E(x^2) = E(x)^2 + Var(x)$

6. $Var(x + y) = Var(x) + Var(y)$

7. $Var(x - y) = Var(x) + Var(y)$

8. $Var(k \cdot x) = k^2 \cdot Var(x)$

Below, we prove the formulas for expected value and variance, as they will be required in discussions which will arise later.

First, we will prove that $E(x + y) = E(x) + E(y)$. Recall what is meant by $x + y$. For the random variables x and y, with probability distributions $P(x)$ and $P(y)$ respectively, when we refer to $x + y$, we mean a new random variable which involves the sum of every value for x to every value for y. What we need to know to compute $E(x + y)$ is the probability distribution for $x + y$. Assume that x and y are independent, and that we are going to choose an (x, y) pair randomly so that we can compute $x + y$. The probability of choosing this particular pair randomly is $P(x\&y)$, and since x and y are independent, $P(x\&y) = P(x) \cdot P(y)$. Now, to randomly choose a particular $x + y$ value, one must choose an (x, y) pair first, and thus the probability of such an $x + y$ value would be the probability of choosing the pair, so, $P(x + y) = P(x\&y) = P(x) \cdot P(y)$. To compute the expected value of $x + y$, we need to sum each $x + y$ times its associated probability, $P(x) \cdot P(y)$, and we need to sum in a way so that every x gets paired with every y. The easiest way to do this is to first sum for a particular x, letting y vary, and then over this sum, we sum again but now letting x vary.

First, we prove the first property of expected value of two independent variables, $E(x + y) = E(x) + E(y)$.

$$
\begin{aligned}
E(x + y) &= \sum_x \sum_y (x + y) \cdot P(x + y) \\
&= \sum_x \sum_y (x + y) \cdot P(x) \cdot P(y) \\
&= \sum_x \sum_y x \cdot P(x) \cdot P(y) + \sum_x \sum_y y \cdot P(x) \cdot P(y) \\
&= \sum_x x \cdot P(x) \cdot \sum_y P(y) + \sum_x P(x) \cdot \sum_y y \cdot P(y) \\
&= E(x) \cdot 1 + 1 \cdot E(y) \\
&= E(x) + E(y)
\end{aligned}
$$

The proof of the second property, $E(x - y) = E(x) - E(y)$, is nearly identical to the first

The third property is relatively easy.

$$
\begin{aligned}
E(x \cdot y) &= \sum_x \sum_y (x \cdot y) P(x \cdot y) \\
&= \sum_x \sum_y x \cdot y \cdot P(x) \cdot P(y) \\
&= \sum_x x \cdot P(x) \cdot \sum_y y \cdot P(y) \\
&= E(x) \cdot E(y)
\end{aligned}
$$

The fourth property is trivial:

$$E\left(k \cdot x\right) = \sum_x k \cdot x \cdot P\left(x\right) = k \cdot \sum_x x \cdot P\left(x\right) = k \cdot E\left(x\right).$$

The fifth property comes from the formula for variance of a discrete probability distribution: $\sigma^2 = Var\left(x\right) = \sum x^2 \cdot P\left(x\right) - \mu^2$. This could be rewritten as $Var\left(x\right) = \sum x^2 \cdot P\left(x\right) - E\left(x\right)^2$, or as $\sum x^2 \cdot P\left(x\right) = E\left(x\right)^2 + Var\left(x\right)$. But the expression on the left of this is the expected value of x^2, so this becomes $E\left(x^2\right) = E\left(x\right)^2 + Var\left(x\right)$.

The sixth and seventh properties have nearly identical proofs, so we will just prove the seventh for $Var\left(x - y\right)$. The proof uses the fifth property, $E\left(x^2\right) = E\left(x\right)^2 + Var\left(x\right)$, but rearranged as $Var\left(x\right) = E\left(x^2\right) - E\left(x\right)^2$.

$$
\begin{aligned}
Var\left(x - y\right) &= E\left(\left(x - y\right)^2\right) - \left(E\left(x - y\right)\right)^2 \\
&= E\left(x^2 - 2xy + y\right) - \left(E\left(x\right) - E\left(y\right)\right)^2 \\
&= E\left(x^2\right) - 2E\left(xy\right) + E\left(y^2\right) - \left(\left(E\left(x\right)\right)^2 - 2 \cdot E\left(x\right) \cdot E\left(y\right) + \left(E\left(y\right)\right)^2\right) \\
&= E\left(x^2\right) - \left(E\left(x\right)\right)^2 + E\left(y^2\right) - \left(E\left(y\right)\right)^2 + \left(2 \cdot E\left(x\right) \cdot E\left(y\right) - 2 \cdot E\left(xy\right)\right) \\
&= Var\left(x\right) + Var\left(y\right) + \left(2 \cdot E\left(x\right) \cdot E\left(y\right) - 2E\left(x\right) \cdot E\left(y\right)\right) \\
&= Var\left(x\right) + Var\left(y\right)
\end{aligned}
$$

The eighth property also uses the fifth, rearranged as $Var\left(x\right) = E\left(x^2\right) - E\left(x\right)^2$:

$$
\begin{aligned}
Var\left(k \cdot x\right) &= E\left(\left(k \cdot x\right)^2\right) - E\left(k \cdot x\right)^2 \\
&= E\left(k^2 \cdot x^2\right) - \left(E\left(k \cdot x\right)\right)^2 \\
&= k^2 \cdot E\left(x^2\right) - \left(k \cdot E\left(x\right)\right)^2 \\
&= k^2 \cdot E\left(x^2\right) - k^2 \cdot E\left(x\right)^2 \\
&= k^2 \cdot \left(E\left(x^2\right) - E\left(x\right)^2\right) \\
&= k^2 \cdot Var\left(x\right)
\end{aligned}
$$

4.6.3   Sample Variance as an Unbiased Estimator of σ^2

One important consequence of the properties of expected value and variance is the ability to prove that the sample variance, s^2, is an unbiased estimator of the population variance. σ^2. This is stated formally below, along with a proof which uses the properties just developed.

Theorem 4.6.4 The sample standard deviation is an unbiased estimator of the population variance.

Proof. Now we prove that s^2 is an unbiased estimator of σ^2. This is proven by showing that $E\left(s^2\right) = \sigma^2$.

$$E\left(s^2\right) = E\left(\frac{\Sigma x^2 - \frac{1}{n}\left(\Sigma x\right)^2}{n-1}\right) \qquad \text{Def. of } s^2.$$

$$= \frac{1}{n-1}\left(E\left(\Sigma x^2\right) - \frac{1}{n}E\left(\left(\Sigma x\right)^2\right)\right) \qquad \text{Prop 2, 4}$$

$$= \frac{1}{n-1}\left(n \cdot E\left(x^2\right) - \frac{1}{n}\left(\left(E\left(\Sigma x\right)\right)^2 + Var\left(\Sigma x\right)\right)\right) \qquad \text{Prop 1, 5}$$

$$= \frac{1}{n-1}\left(n \cdot \left(\left(E\left(x\right)\right)^2 + Var\left(x\right)\right) - \frac{1}{n}\left(\left(\Sigma E\left(x\right)\right)^2 + \Sigma Var\left(x\right)\right)\right) \qquad \text{Prop 1,5,6}$$

$$= \frac{1}{n-1}\left(n\mu^2 + n\sigma^2 - \frac{1}{n}\left(\left(n\mu\right)^2 + n\sigma^2\right)\right) \qquad \text{Prop 6}$$

$$= \frac{1}{n-1}\left(n\mu^2 + n\sigma^2 - n\mu^2 - \sigma^2\right) \qquad \text{Simplify}$$

$$= \frac{1}{(n-1)}\left(n-1\right)\sigma^2 \qquad \text{Simplify}$$

$$= \sigma^2 \qquad \text{Simplify}$$

■

4.6.4   Binomial Experiments

Development of Formulas for Mean and Standard Deviation

First we show that the mean of a binomial distribution with n trials is equal to $n \cdot p$.

We begin by considering a binomial experiment with $n = 1$ trial, where p is the probability of success, and $q = 1 - p$. For this trivial case, we will use X to represent the number of successes, and since there is only one trial, the only possible values for X are 0 and 1. We will denote the mean and variance as μ_X and σ_X^2, and compute them using the formulas from section 4.1: $\mu_X = E(X) = \Sigma X \cdot P(X)$ and $\sigma_X^2 = \sum X^2 \cdot P\left(X\right) - \mu^2$. The necessary totals are computed in the table below.

X	$P(X)$	$X \cdot P(X)$	$X^2 \cdot P(X)$
0	q	0	0
1	p	p	p
Σ	1	p	p

So we see that

$$\mu_X = \Sigma X \cdot P(X) = p,$$

and

$$\begin{aligned}
\sigma_X^2 &= Var(X) \\
&= \sum X^2 \cdot P(X) - \mu^2 \\
&= p - p^2 \\
&= p \cdot (1 - p) \\
&= pq.
\end{aligned}$$

To determine the mean and variance for the case when there are n trials, we can consider each trial as its own binomial experiment as above, with X representing the number of successes in a single trial, and x representing the total number of successes in n trials. From this we know that $x = \sum X$.

For the mean of the entire experiment, we use the properties of expected value.

$$\begin{aligned}
\mu &= E(x) \\
&= E\left(\sum X\right) \\
&= \sum E(x) \\
&= \sum \mu_X \\
&= \sum p \\
&= np.
\end{aligned}$$

For the variance of the entire experiment we use the properties of variance.

$$\begin{aligned}
\sigma &= Var(x) \\
&= Var\left(\sum X\right) \\
&= \sum Var(X) \\
&= \sum \sigma_X^2 \\
&= \sum pq \\
&= npq
\end{aligned}$$

Thus, the standard deviation of the binomial experiment is $\sigma = \sqrt{npq}$.

4.6.5 Deriving the Poisson Distribution

Our last task is to derive the formula for the Poisson Distribution. The Poisson Distribution is derived from the binomial by allowing the number of trials, n, to approach infinity. To denote this idea, when we wish to say that a value, n, is approaching infinity we will write $n \to \infty$.

Theorem 4.6.5 As k approaches infinity $(k \to \infty)$, the expression $(1 + 1/k)^k$ approaches $e = 2.71828\ldots$. We denote this by writing $(1 + 1/k)^k \to e$. For example,

$$(1 + 1/100,000)^{100,000} \approx 2.71827.$$

Theorem 4.6.6 As n approaches infinity, $(1 + r/n)^n$ approaches e^r. To demonstrate this is, define $k = \frac{n}{r}$, we see that as n approaches infinity, k approaches infinity as well, and

$$
\begin{aligned}
(1 + r/n)^n &= \left(\left(1 + \frac{1}{(n/r)} \right)^{(n/r)} \right)^r \\
&= \left((1 + 1/k)^k \right)^r \\
&\to e^r.
\end{aligned}
$$

Theorem 4.6.7 As n approaches infinity, the ratio $(n - a)/n$ approaches 1 for any fixed value of a. For example, suppose $n = 100,000$ and $a = 495$. Then $(n - a)/n \approx 0.995$ which is very close to 1, and this gets continually closer to 1 as n increases.

Theorem 4.6.8 The combination formula, $_nC_x$, may be rearranged as given below.

$$
\begin{aligned}
_nC_x &= \frac{n!}{x! \cdot (n-x)!} = \frac{n \cdot (n-1) \cdot (n-2) \cdots (n-x+1) \cdot (n-x) \cdot (n-x-1) \cdots 3 \cdot 2 \cdot 1}{x! \cdot (n-x) \cdot (n-x-1) \cdots 3 \cdot 2 \cdot 1} \\
&= \frac{n \cdot (n-1) \cdot (n-2) \cdots (n-x+1)}{x!}
\end{aligned}
$$

We will use this fact below, noting that the numerator has exactly x factors.

Now we have enough information to show that the binomial distribution approaches the Poisson as n approaches infinity. We begin with the binomial formula, immediately replacing p by μ/n (recalling that $\mu = np$), and q by $1 - p = 1 - \mu/n$.

$$
\begin{aligned}
P(x) &= _nC_x \cdot p^x \cdot q^{n-x} = _nC_x \cdot \left(\frac{\mu}{n} \right)^x \cdot \left(1 - \frac{\mu}{n} \right)^{n-x} \\
&= \frac{n \cdot (n-1) \cdot (n-2) \cdots (n-x+1)}{x!} \cdot \frac{\mu^x}{n^x} \cdot \left(1 - \frac{\mu}{n} \right)^n \cdot \left(1 - \frac{\mu}{n} \right)^{-x} \\
&= \frac{n \cdot (n-1) \cdot (n-2) \cdots (n-x+1)}{n^x} \cdot \frac{\mu^x}{x!} \cdot \left(1 + \frac{-\mu}{n} \right)^n \cdot \left(1 - \frac{\mu}{n} \right)^{-x} \\
&= \frac{n}{n} \cdot \frac{(n-1)}{n} \cdot \frac{(n-2)}{n} \cdots \frac{(n-x+1)}{n} \cdot \frac{\mu^x}{x!} \cdot \left(1 + \frac{-\mu}{n} \right)^n \cdot \left(\frac{n-\mu}{n} \right)^{-x}
\end{aligned}
$$

We are nearly done. By Remark 4.6.8 above, as $n \to \infty$, each of the fractions in the product:

$$\frac{n}{n} \cdot \frac{(n-1)}{n} \cdot \frac{(n-2)}{n} \cdots \frac{(n-x+1)}{n}$$

approach one. For the same reason, the fraction $\frac{n-\mu}{n}$ also approaches 1. By Remark 4.6.6, the expression $\left(1 + \frac{-\mu}{n}\right)^n$ approaches $e^{-\mu}$ as $n \to \infty$. So, our binomial formula becomes:

$$\begin{aligned}
P(x) &= {}_nC_x \cdot p^x \cdot q^{n-x} \\
&= \frac{n}{n} \cdot \frac{(n-1)}{n} \cdot \frac{(n-2)}{n} \cdots \frac{(n-x+1)}{n} \cdot \frac{\mu^x}{x!} \cdot \left(1 + \frac{-\mu}{n}\right)^n \cdot \left(\frac{n-\mu}{n}\right)^{-x} \\
&\to 1 \cdot 1 \cdot 1 \cdots 1 \cdot \frac{\mu^x}{x!} \cdot e^{-\mu} \cdot 1^{-x} \\
&= \frac{e^{-\mu} \cdot \mu^x}{x!}
\end{aligned}$$

Thus, as $n \to \infty$, our formula becomes $P(x) = \frac{e^{-\mu} \cdot \mu^x}{x!}$, which is the formula for the Poisson Distribution.

Theorem 4.6.9 The Sum of All Poisson Probabilities is Equal to One

The proof of this statement uses a fact from calculus:

$$e^k = \sum_{x=0}^{\infty} \frac{k^x}{x!}.$$

Using this, we sum:

$$\sum_{x=0}^{\infty} P(x) = \sum_{x=0}^{\infty} \frac{e^{-\mu} \cdot \mu^x}{x!} = e^{-\mu} \cdot \sum_{x=0}^{\infty} \frac{\mu^x}{x!} = e^{-\mu} \cdot e^{\mu} = e^0 = 1.$$

Theorem 4.6.10 The variance of the Poisson Distribution is exactly equal to the mean. To prove this, the variance of the Poisson distribution is derived from the variance of the binomial distribution, where we substitute $\mu = np$, and $q = 1 - p = 1 - \mu/n$. We then let $n \to \infty$. Note that as $n \to \infty$, $\mu/n \to 0$.

$$\begin{aligned}
Var(x) &= npq \\
&= \mu(1 - \mu/n) \\
&\to \mu \cdot (1 - 0) \\
&= \mu.
\end{aligned}$$

4.6.6 Skewness and Kurtosis

The skewness and kurtosis for the Poisson distribution are derived from the binomial distribution, where we allow n to approach infinity. In the development, we make the following substitutions: $\mu = np$, so $p = \frac{\mu}{n}$, and $q = 1 - p = 1 - \frac{\mu}{n}$. The development for skewness is below.

$$
\begin{aligned}
skewness &= \frac{q - p}{\sqrt{npq}} \\[2mm]
&= \frac{1 - \frac{\mu}{n} - \frac{\mu}{n}}{\sqrt{\mu\left(1 - \frac{\mu}{n}\right)}} \\[2mm]
&= \frac{1 - \frac{2\mu}{n}}{\sqrt{\mu - \frac{\mu^2}{n}}} \\[2mm]
&\to \frac{1 - 0}{\sqrt{\mu - 0}} \\[2mm]
&= \frac{1}{\sqrt{\mu}}.
\end{aligned}
$$

The kurtosis follows next.

$$
\begin{aligned}
kurtosis &= 3 + \frac{1 - 6pq}{npq} \\[2mm]
&= 3 + \frac{1 - 6 \cdot \frac{\mu}{n} \cdot \left(1 - \frac{\mu}{n}\right)}{\mu\left(1 - \frac{\mu}{n}\right)} \\[2mm]
&= 3 + \frac{1 - 6\frac{\mu}{n} + \frac{6\mu^2}{n^2}}{\mu - \frac{\mu^2}{n}} \\[2mm]
&\to 3 + \frac{1 - 6 \cdot 0 + 0}{\mu - 0} \\[2mm]
&= 3 + \frac{1}{\mu}.
\end{aligned}
$$

Chapter 5

Continuous Probability Distributions

5.1   Continuous Random Variables

We have mentioned previously that, for continuous random variables, the problem of computing probability would become a geometric problem of computing areas. It is now time to begin this process.

First, we define the probability density function. It is important to understand the distinction between a probability density function, and probability itself. The *probability density function* is a mathematical function whose graph is, typically, a smooth curve, and whose total area underneath is equal to one. Areas under the curve, corresponding to intervals for the random variable, represent the probability that the random variable is observed to lie within that range.

With continuous random variables, it is impossible to measure values exactly, since exact values have an infinite number of places after the decimal. So when we investigate probabilities, these always relate to a range of values rather than to a single value for x. Thus, the probability that x lies between values a and b will be written as $P(a \leqslant x \leqslant b)$, and such a region is depicted by the shaded area in the Figure 5.1.

Again, note that the probability density function does not directly give probabilities – it gives the height of the distribution, but the areas under the curve are the probabilities.

For all but the simplest distributions, calculus is required to compute these areas. As this course does not have a calculus prerequisite, we will be looking up these areas on tables which are included in the appendix of this text.

5.1.1   The Continuous Uniform Distribution

We begin with the very simplest continuous probability distribution, called the *uniform distribution*. The uniform distribution has a horizontal probability density function, and

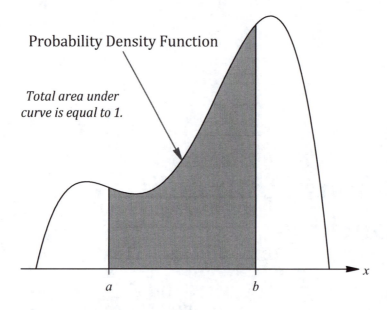

Probability Density Function

Total area under curve is equal to 1.

Figure 5.1: *The shaded region represents* $P(a \leqslant x \leqslant b)$.

values of the random variable, x, range from some minimum value, a, to a maximum value, b.

The range is the maximum value minus the minimum value, so $range = b - a$. The total area for any probability distribution is equal to 1, and because the distribution is rectangular in shape, the area is equal to the base times the height. The total area is $1 = base \cdot height$, and thus $height = \frac{1}{base} = \frac{1}{b-a}$. So the equation of the probability density function is $y = \frac{1}{b-a}$, where $a \leqslant x \leqslant b$. This density function is plotted in Figure 5.2.

Uniform Probability Distribution

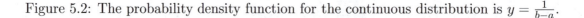

Figure 5.2: The probability density function for the continuous distribution is $y = \frac{1}{b-a}$.

Figure 5.2 gives a general view of a continuous uniform distribution of values x that range from a to b, with probability density function $y = \frac{1}{b-a}$. To compute the probability that x is between two values, say c and d, denoted by $P(c \leqslant x \leqslant d)$, we need only compute the area of this rectangular region, which is shaded in the illustration. This area is equal to the base, $d - c$, times the height, $\frac{1}{b-a}$. Thus,

$$P(c \leqslant x \leqslant d) = \frac{d-c}{b-a}.$$

Example 5.1.1 As an example, let us consider the distribution of voltages output by a particular brand of battery. These values range from 0.00 to 1.47 volts, and with an assumed uniform distribution. The range is $1.47 - 0.00 = 1.47$. The probability density function is the height of the rectangle, $y = \frac{1}{range} = \frac{1}{1.47} \approx 0.68$, and is plotted in Figure 5.3.

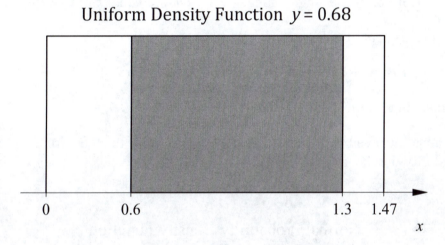

Figure 5.3: The area shaded under this uniform distribution represents $P(0.6 \leq x \leq 1.3)$.

Suppose we want to compute the probability that x lies between 0.6 and 1.3. This is represented by the area of the shaded rectangle in the figure above. The base of this rectangle is $1.3 - 0.6 = 0.7$, and the height is $y = 0.68$. The area represents the desired probability, so and we compute this as the base times the height, so, $P(0.6 \leqslant x \leqslant 1.3) = 0.7 \cdot 0.68 = 0.476$.

5.1.2 The Normal Distribution

We have begun with the uniform distribution because it is easy to explain and computations are simple, but unfortunately, we will not consider it again. The problem is that very few applications that are of interest to us involve uniformly distributed random variables. The most important distribution to us will be the distribution with the shape that we have seen again and again with discrete variables, the *bell* or *normal* shape, but now the shape will be applied to a continuous random variable. The main difference is that, instead of a bell-shaped histogram, our bell-shape needs to be a smooth curve.

Whenever a population has a normal distribution, we will simply call the population *normal.* Most often we will not be able to prove that populations are normal, but if evidence leads us to believe that the population may be normal, we will say something like *"the population appears to be approximately normal."*

For our normal distribution, we have certain requirements that we have seen in the past that we would like to continue to see with the continuous distribution. These requirements are as follows:

1. The distribution is symmetric with *skewness* equal to *zero.*

2. The *mean* is at the axis of symmetry;

3. The distribution has a single mode, with *kurtosis* equal to *three.*

4. The *mode* is equal to the *mean;*

5. The *total area* under the probability density function is 1;

6. The distribution must satisfy the *empirical rule;*

7. The range of possible values is infinite, but as values get far from the mean, the probability density must approach zero.

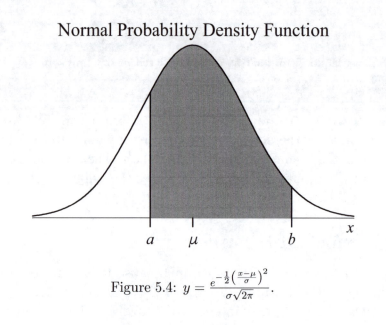

Figure 5.4: $y = \dfrac{e^{-\frac{1}{2}\left(\frac{x-\mu}{\sigma}\right)^2}}{\sigma\sqrt{2\pi}}.$

The normal probability density function is depicted in Figure 5.4. The shaded area represents $P(a \leqslant x \leqslant b)$.

These rules are our requirements, which are based on nothing more than experience. In searching for a probability density function that satisfies this, statisticians identified the *normal probability density function*

$$y = \frac{e^{-\frac{1}{2}\left(\frac{x-\mu}{\sigma}\right)^2}}{\sigma\sqrt{2\pi}},$$

as a density function that satisfies all of the requirements outlined above. This formula may look intimidating, but the good news is that it is so difficult that we do not need to work with it directly. We will find areas under this curve by first performing a transformation to *standardize* x values that are relevant to us, then we will look-up these standardized values in a table.

Note the symmetry, and that the mode is the mean, μ. What is harder to see is that the distribution goes forever in both directions, but what we do see is that the probability density is nearly zero when values stray a little way from the mean (about 3 standard deviations). What cannot be seen is that the total area is equal to one (we need to use calculus to prove this), and that the distribution satisfies the empirical rule (we will demonstrate this in Section 5.2).

The problem with making a table for looking up normal probabilities is that every normal population has a different mean and standard deviation. Because of this, the areas associated with a given range of values will vary depending on the population. This problem is resolved by *standardizing*.

Example 5.1.2 We would like to demonstrate how standardizing works, but to do so, we will return to a small population with a discrete random variable, x, that has a roughly bell-shaped distribution. The values in this population, along with their associated frequencies, are listed in the table below.

x	f	fx	fx^2
0	1	0	0
1	9	9	9
2	23	46	92
3	31	93	279
4	23	92	368
5	9	45	225
6	1	6	36
Σ	97	291	1009

The population mean is

$$\begin{aligned}
\mu &= \frac{\sum fx}{N} \\
&= \frac{291}{97} \\
&= 3,
\end{aligned}$$

and the population standard deviation is

$$\sigma = \sqrt{\frac{\sum fx^2 - \frac{1}{N}\left(\sum fx\right)^2}{N}}$$

$$= \sqrt{\frac{1009 - \frac{1}{97}\left(291\right)^2}{97}}$$

$$= \sqrt{\frac{136}{97}}$$

$$= \sqrt{1.402}$$

$$= 1.184.$$

We standardize by computing z-scores of each x in the table above. Starting with $x = 0$, the z-score is

$$z = \frac{x - \mu}{\sigma} = \frac{0 - 3}{1.184} = -2.534.$$

Next with $x = 1$, the z-score is

$$z = \frac{x - \mu}{\sigma} = \frac{1 - 3}{1.184} = -1.689.$$

Continuing, we have the population of z-scores, noting that the frequencies on the z-scores are the same as those for the values of x.

z	f	fz	fz^2
-2.534	1	-2.53	6.42
-1.689	9	-15.20	25.67
-0.845	23	-19.44	16.42
0.000	31	0.00	0.00
0.845	23	19.44	16.42
1.689	9	15.20	25.67
2.534	1	2.53	6.42
Σ	97	0.00	97.00

Notice first, that because the frequencies have not changed, that the distribution of the z-scores is the same as the distribution of the values of x – roughly normal. Next we compute the mean and standard deviation which we will denote by μ_z and σ_z respectively.

$$\mu_z = \frac{\sum fz}{N} = \frac{0}{97} = 0,$$

and

$$\sigma_z = \sqrt{\frac{\sum fz^2 - \frac{1}{N}\left(\sum fz\right)^2}{N}}$$

$$= \sqrt{\frac{97.00 - \frac{1}{97}0^2}{97}}$$

$$= \sqrt{\frac{97.00}{97}}$$

$$= \sqrt{1.000}$$

$$= 1.000.$$

So, $\mu_z = 0$ and $\sigma_z = 1$. We call this process standardizing because when we take z-scores of an entire population, we always end up with a collection of z-scores whose mean and standard deviation are $\mu_z = 0$ and $\sigma_z = 1$, respectively. Also, if the original population is normal, then the z-scores are normal. So when normal populations are standardized, they all become identical in terms of their shape, center, and variation. When this is done, the distribution is called the *standard normal distribution* of z-scores.

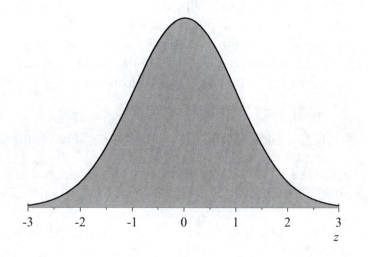

Figure 5.5: The Standard Normal Density Function.

The probability density function for the *normal distribution*,

$$y = \frac{e^{-\frac{1}{2}\left(\frac{x-\mu}{\sigma}\right)^2}}{\sigma\sqrt{2\pi}},$$

becomes the density function for the *standard normal distribution*,

$$y = \frac{e^{-\frac{z^2}{2}}}{\sqrt{2\pi}},$$

where we replace the random variable x with the random variable z, μ with $\mu_z = 0$, and σ with $\sigma_z = 1$. The graph of the *standard* normal density function is given in Figure 5.5, and in our next section we will learn to look-up areas under the standard normal curve.

5.1.3   Skewness and Kurtosis

The normal distribution is perfectly symmetric and thus its skewness is zero. We have stated repeatedly that the kurtosis of the normal distribution is equal to 3. This is the standard value to which all other distribution kurtoses are compared. We summarize below.

Theorem 5.1.3 The skewness and kurtosis of the normal distribution are:

$$skewness = 0,$$

and

$$kurtosis = 3.$$

5.1.4   The History of the Normal Distribution

The first person known to use the normal distribution was Abraham De Moivre, an 18th century statistician. As a consultant to gamblers, De Moivre was called upon to compute probabilities of different outcomes in gambling games. In this effort, De Moivre frequently used the binomial formula for many applications, but as we have seen, this formula can be difficult to use when n, the number of trials, gets large. In an effort to simplify these computations, De Moivre developed the formula for the normal distribution as an approximation to the values given by the binomial formula.

The great astronomer Galileo noticed that errors in measurements tended to be bell-shaped in their distribution, and following in his footsteps came Gauss.

In 1809, Carl Friedrich Gauss (1777-1855) independently developed the density function for the normal distribution in *Theoria Motus Corporum Coelestium in Sectionibus Conicis Solem Ambientum* (The Theory of the Motion of Heavenly Bodies moving around the Sun in Conic Sections). Gauss' development was used to describe the distribution of errors in astronomical observations. This is one case of many when important mathematical developments were created to help humanity in understanding the mysteries of the planets and the universe.

In Gauss' book, the density function appeared in a barely recognizable form. The image below is scanned from its pages.

$$\varphi \Delta = \frac{h}{\sqrt{\pi}} e^{-hh\Delta\Delta}$$

Gauss' Normal Density Function.

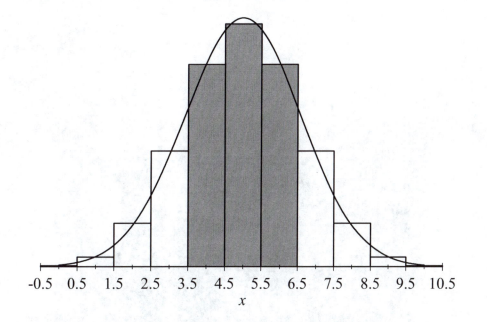

Figure 5.6: The Binomial Distribution is represented by the bars on the histogram, the normal distribution is the smooth curve.

In Gauss' density function, Δ represents the error from the mean, which we may regard as $x - \mu$, $\Delta\Delta$ is simply $\Delta^2 = (x - \mu)^2$, and h (in English literature on statistics, h, or its reciprocal, were referred to as the *modulus*) may be regarded as the reciprocal of $\sigma\sqrt{2}$, (the idea of standard deviation was not introduced until 1894 by Karl Pearson), and thus in Gauss' notation, $hh = h^2 = \frac{1}{2\sigma^2}$. Updating our notation, Gauss' density function becomes:

$$\phi\left(\Delta\right) = \frac{h}{\sqrt{\pi}}e^{-h^2\Delta^2} = \frac{1}{\sigma\sqrt{2\pi}}e^{-\frac{1}{2}\left(\frac{x-\mu}{\sigma}\right)^2},$$

which is the density function for the non-standard normal distribution.

Gauss used the normal distribution to describe the distribution of errors in measurements, and this is quite similar to how we will apply the distribution as well. Later we will apply the normal distribution as De Moivre did. The 1756 edition of De Moivre's *The Doctrine of Chance* contained his normal approximation of binomial probabilities, and in Chapter 6 we will follow in De Moivre's footsteps as we learn to do exactly what he did for exactly the same purpose – the result will be a theory for studying the distribution of sample proportions.

Figure 5.7: Portrait of French mathematician Abraham de Moivre (1667-1754). Engraving by J. Faber of a painting by J. Highmore.

Figure 5.8: A rare portrait of Gauss on the terrace of the Göttingen observatory, with the *heliometer* invented by him. Lithography of Eduard Ritmüller, after 1814 (the year when this heliometer was produced).

5.1.5   Homework

Suppose that a randomly selected used battery outputs voltages from 0 volts to 3 volts, uniformly distributed, depending on how much it has been used. Let x = *random battery voltage*.

1. Graph the probability distribution for x, stating the height of the density function.

2. Given that the mean of a uniform distribution is the midpoint of its range, state the formula for the mean.

3. Compute $P(x > 1.3)$.

4. Compute $P(x < 2.5)$.

5. Compute $P(0.8 < x < 2.7)$.

6. What is the mean of this distribution?

5.2   The Standard Normal Distribution

Next, we must learn the skill of finding areas under the standard normal distribution of z-scores. To do this, we will use a table in the appendix, Table A1.

Table A1 is an example of a *cumulative* probability distribution. When we look up a value, a, in this table, it returns the probability $P(z < a)$. The table has one section for negative z-scores, and another for positive z-scores. To look up $P(z < a)$ in Table A1, we round our z-score to the nearest one-hundredth. Next, we go to the negative or positive section of the table as appropriate, and in the first column find the row containing our z-score to the tenths place, then along the top row we find the column containing the one-hundredths place of our z-score.

At the intersection of the respective row and column, we find $P(z < a)$.

Example 5.2.1 As an example, let us look up the probability $P(z < 1.35)$. To do this, we look in the positive z-score section of the table, and in the first column we find $z = 1.3$. Next, on the top row, we find .05, the one-hundredths place, (since $1.3 + .05 = 1.35$ is our desired z-score). A small version of the positive section of the table, portrayed below, shows the intersection of the relevant row and column, containing the value 0.9115. This means that $P(z < 1.35) = 0.9115$.

Positive z-Scores

z	.00	.01	.02	.03	.04	.05	.06	.07	.08	.09
1.0	.8413	.8438	.8461	.8485	.8508	.8531	.8554	.8577	.8599	.8621
1.1	.8643	.8665	.8686	.8708	.8729	.8749	.8770	.8790	.8810	.8830
1.2	.8849	.8869	.8888	.8907	.8925	.8944	.8962	.8980	.8997	.9015
1.3	.9032	.9049	.9066	.9082	.9099	.9115	.9131	.9147	.9162	.9177
1.4	.9192	.9207	.9222	.9236	.9251	.9265	.9279	.9292	.9306	.9319
1.5	.9332	.9345	.9357	.9370	.9382	.9394	.9406	.9418	.9429	.9441

The shaded region in Figure 5.9 shades the appropriate region under the normal density function.

Example 5.2.2 As another example, suppose we wish to compute $P(z < -2.68)$. We simply go to the negative section of the table and look up as highlighted below. First we look for the row with the heading $z = -2.6$, then the column with heading .08 (since $-2.6 - 0.8 = -2.68$), and we see the value 0.0037. This means that $P(z < -2.68) = 0.0037$.

Negative z-Scores

z	.00	.01	.02	.03	.04	.05	.06	.07	.08	.09
−2.8	.0026	.0025	.0024	.0023	.0023	.0022	.0021	.0021	.0020	.0019
−2.7	.0035	.0034	.0033	.0032	.0031	.0030	.0029	.0028	.0027	.0026
−2.6	.0047	.0045	.0044	.0043	.0041	.0040	.0039	.0038	.0037	.0036
−2.5	.0062	.0060	.0059	.0057	.0055	.0054	.0052	.0051	.0049	.0048
−2.4	.0082	.0080	.0078	.0075	.0073	.0071	.0069	.0068	.0066	.0064

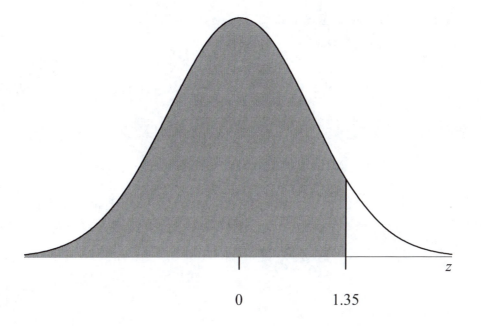

Figure 5.9: The area to the left of $z = 1.35$ under the normal density function is shaded in the figure above.

There is a matter which needs to be considered, having to do with notation and probabilities of measuring an exact value for the continuous random variable, z. The question begins in considering the difference between the probabilities $P(z \leqslant a)$, and $P(z < a)$. Geometrically these represent the areas under the standard normal probability density function for areas to the left of $z = a$. The only difference is that the first includes the boundary line $z = a$, while the second does not. Including or excluding the boundary line does not affect the area of the region, because lines have no area, and thus, the probabilities are the same. We are saying this with regard to z-scores, but in fact, it is true for any continuous random variable.

Theorem 5.2.3 For any continuous random variable, x,

$$P\left(x \leqslant a\right) = P\left(x < a\right),$$
$$P\left(x \geqslant a\right) = P\left(x > a\right),$$

and this is because $P\left(x = a\right) = 0$. For this reason, for a continuous variable only,

$$P\left(x > a\right) = 1 - P\left(x < a\right)$$

Of course, there are times when we need to compute probabilities for different kinds of ranges than those given directly in the Table A1. For example, we might wonder what is the probability that z is greater than some specific value. To compute this, we use the rule of complements.

Example 5.2.4 Referring to Example 5.2.1, we know that

$$P(z < 1.35) = 0.9115.$$

If we wish to compute the probability, $P(z > 1.35)$, we use the rule of complements:

$$
\begin{aligned}
P(z > 1.35) &= 1 - P(z < 1.35) \\
&= 1 - 0.9115 \\
&= 0.0885.
\end{aligned}
$$

We also determined that $P(z < -2.68) = 0.0037$. Using this, we can compute $P(z > -2.68)$

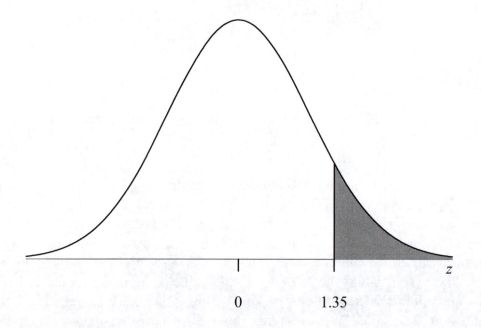

Figure 5.10: The region shaded corresponds to the probability in Example 5.2.4.

by the same principle. Thus,

$$
\begin{aligned}
P(z > -2.68) &= 1 - P(z < -2.68) \\
&= 1 - 0.0037 \\
&= 0.9963.
\end{aligned}
$$

The next type of probability is associated with an interval of finite length, and while easier to visualize, it is slightly harder to compute.

Example 5.2.5 Suppose we want to compute $P(0.47 < z < 1.85)$. This is more difficult, as Table A1 only gives areas to the left of whatever z-score we look up. To compute $P(0.47 < z < 1.85)$ we must first look up the area to the left of $z = 1.85$, then look up and subtract the area to the left of $z = 0.47$. This will give the area between $z = 0.47$ and $z = 1.85$. Thus, we compute $P(0.47 < z < 1.85)$ by subtracting $P(0.47 < z < 1.85) = P(z < 1.85) - P(z < 0.47)$. The illustration below portrays these ideas.

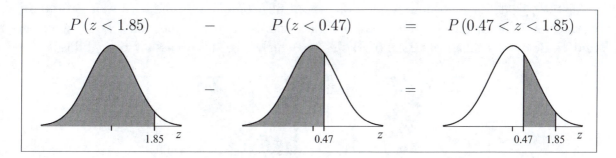

First, we look up $P(z < 1.85)$.

z	.00	.01	.02	.03	.04	.05	.06	.07	.08	.09
1.7	.9554	.9564	.9573	.9582	.9591	.9599	.9608	.9616	.9625	.9633
1.8	.9641	.9649	.9656	.9664	.9671	.9678	.9686	.9693	.9699	.9706
1.9	.9713	.9719	.9726	.9732	.9738	.9744	.9750	.9756	.9761	.9767

Above, we see the area to the left of $z = 1.85$ is 0.9678, and below we see that the area to the left of $z = 0.47$ is 0.6808.

z	.00	.01	.02	.03	.04	.05	.06	.07	.08	.09
0.3	.6179	.6217	.6255	.6293	.6331	.6368	.6406	.6443	.6480	.6517
0.4	.6554	.6591	.6628	.6664	.6700	.6736	.6772	.6808	.6844	.6879
0.5	.6915	.6950	.6985	.7019	.7054	.7088	.7123	.7157	.7190	.7224

Thus,

$$P(0.47 < z < 1.85) = P(z < 1.85) - P(z < 0.47)$$
$$= 0.9678 - 0.6808$$
$$= 0.2870.$$

Summary 5.2.6 Below, we summarize the three cases for obtaining probabilities under the standard normal distribution.

1. To compute $P(z < a)$, look up $z = a$ in Table A1. The corresponding area is $P(z < a)$.

2. To compute $P(z > a)$, look up $z = a$ in Table A1, and compute: $P(z > a) = 1 - P(z < a)$.

3. To compute $P(a < z < b)$, look up $z = a$ and $z = b$ in Table A1, and compute $P(a < z < b) = P(z < b) - P(z < a)$.

5.2.1   The Empirical Rule and The Normal Distribution

Using the rules derived above, we would like to show that the probability density chosen for the standard normal distribution satisfies the empirical rule. The first part of the rule states that, for a normal population, about 68% of all values lie within one standard deviation of the mean. That means that the z-scores of such values must be between -1 and 1. Below are small portions of Table A1, which give the necessary probabilities.

z	.00	.01
0.9	.8159	.8186
1.0	**.8413**	.8438
1.1	.8643	.8665

z	.00	.01
-1.1	.1357	.1335
-1.0	**.1587**	.1562
-0.9	.1841	.1814

The probability is computed next.

$$P(-1 < z < 1) = P(z < 1.00) - P(z < -1.00)$$
$$= 0.8413 - 0.1587$$
$$= 0.6826$$
$$\approx 68\%$$

The probability that a z-score lies between -1 and 1 is about 68%, and therefore 68% of all values in a normal population lie within one standard deviation of the mean.

The second part of the rule states that, for our normal population, about 95% of all values should lie within two standard deviations of the mean, so the z-scores of these values would lie between -2 and 2. First we find $z = 2.00$ and $z = -2.00$ in Table A1.

z	.00	.01
1.9	.9713	.9719
2.0	**.9772**	.9778
2.1	.9821	.9826

z	.00	.01
-2.1	.0179	.0174
-2.0	**.0228**	.0222
-1.9	.0287	.0281

Computing the associated probability, we proceed as follows.

$$P(-2 < z < 2) = P(z < 2.00) - P(z < -2.00)$$
$$= 0.9772 - 0.0228$$
$$= 0.9544$$
$$\approx 95\%$$

About 95% of all values should lie within two standard deviations from the mean.

Lastly, the third part of the rule states that, for this normal population, about 99.7% of all values should lie within three standard deviations of the mean. Thus, the z-scores of these values should lie between -3 and 3. Again, we find $z = 3.00$ and $z = -3.00$ in Table A1.

z	**.00**	.01
2.9	.9981	.9982
3.0	**.9987**	.9987
3.1	.9990	.9991

z	**.00**	.01
-3.1	.0010	.0009
-3.0	**.0013**	.0013
-2.9	.0019	.0018

The probability that z lies between -3 and 3 is computed next.

$$P\left(-3 < z < 3\right) = P\left(z < 3.00\right) - P\left(z < -3.00\right)$$
$$= 0.9987 - 0.0013$$
$$= 0.9974$$
$$\approx 99.7\%$$

About 99.7% of the population should lie within three standard deviations from the mean, as required.

5.2.2   Finding z-Scores When Given Probabilities

Another process that is of equal importance to looking up probabilities based on given z-scores, using Table A1, is the reverse of this process. Given a probability or area under the standard normal density function, we would like to find a z-score that acts as a boundary for this region. We will restrict such regions to those that are only bounded on one side. When a region is bounded only on the right, we will call it a *left tail*, whereas a region bounded only on the left will be called a *right tail*. The z-scores associated with left or right tails will be called *critical values*.

It is important to note that $P\left(z < 0\right) = 0.5$, since $z = 0$ is the center of symmetry and the total area under the density function is equal to 1. The reason that this is important is that whenever a left tail has an area less than 0.5, the associated z-score will be negative (see Figure 5.11), but if the left tail has an area greater than 0.5, the z-score will be positive (see Figure 5.12). Likewise, whenever a right tail has an area greater than 0.5, the associated z-score will be negative, but if the right tail has area less than 0.5, the z-score will be positive. In any case, we must remember that Table A1 only gives areas to the left of a given z-score, so that if we start with a right tail area, and are intending to find an associated z-score, we must find the area to the left first using the rule of complements, $(left\ area) = 1 - (right\ area)$.

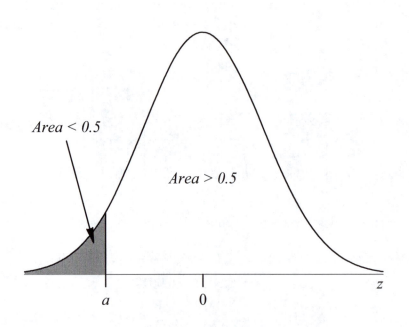

Figure 5.11: Whenever $P(z < a) < 0.5$, it will be true that $a < 0$.

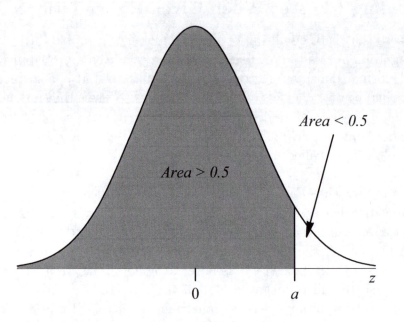

Figure 5.12: Whenever $P(z < a) > 0.5$, it will be true that $a > 0$.

As a simple example, suppose we know that the area to the left of a given z-score is 0.0192. This area is less than 0.5, so we know that the associated z-score must be negative. To find this z-score we go to the negative section of Table A1, and, scanning inside the table, we look for an area that is as close to 0.0192 as possible (rarely will we find exactly what we want). This can be a bit tedious at first, but practice is the key to getting better at it.

The cell we are looking for is highlighted below, corresponding to a z-score of $z = -2.07$.

z	.00	.01	.02	.03	.04	.05	.06	.07	.08	.09
−2.2	.0139	.0136	.0132	.0129	.0125	.0122	.0119	.0116	.0113	.0110
−2.1	.0179	.0174	.0170	.0166	.0162	.0158	.0154	.0150	.0146	.0143
−2.0	.0228	.0222	.0217	.0212	.0207	.0202	.0197	.0192	.0188	.0183
−1.9	.0287	.0281	.0274	.0268	.0262	.0256	.0250	.0244	.0239	.0233

Now suppose that we want the z-score associated with a right tail of 0.01. Remembering that right tail areas are not used in Table A1, we compute the area to the left using (*left area*) =
$1 - $ (*right area*) $= 1 - 0.01 = 0.99$. Scanning Table A1 for the area closest to 0.99, we find an area of 0.9901, the closest to the desired value in the table.

z	.00	.01	.02	.03	.04	.05	.06	.07	.08	.09
2.2	.9861	.9864	.9868	.9871	.9875	.9878	.9881	.9884	.9887	.9890
2.3	.9893	.9896	.9898	.9901	.9904	.9906	.9909	.9911	.9913	.9916
2.4	.9918	.9920	.9922	.9925	.9927	.9929	.9931	.9932	.9934	.9936

The z-score associated with this area is $z = 2.33$.

Note, that since the probability that z is less than 2.33 is 99%, we may conclude that 99% of a normal population is less than 2.33 standard deviations above the mean, so that the 99th percentile in a normal population is always 2.33 standard deviations above average.

Since we are discussing percentiles, let us try a problem with percentiles. Keep in mind that since the area to the left of $z = 0$ is 0.50, we know that $z = 0$ is the 50th percentile for z-scores. Our next question is, what z-score represents the 86th percentile? To answer, we note that the area to the left of the 86th percentile is 0.86, which is greater than 0.50, so the z-score must be positive. We therefore look to the positive z-score section of Table A1, and scan the interior of the table for an area close to 0.86.

z	.00	.01	.02	.03	.04	.05	.06	.07	.08	.09
0.9	.8159	.8186	.8212	.8238	.8264	.8289	.8315	.8340	.8365	.8389
1.0	.8413	.8438	.8461	.8485	.8508	.8531	.8554	.8577	.8599	.8621
1.1	.8643	.8665	.8686	.8708	.8729	.8749	.8770	.8790	.8810	.8830

The closest area to 0.86 is 0.8599, and this corresponds to a z-score of $z = 1.08$, the 86th percentile.

Let us do one more example before we move on. This time we want to find the 95th percentile. By definition, this means that we want the z-score which separates the lower 95% of the population from the upper 5%. So we scan our Table A1 for an area close to 0.95.

z	.00	.01	.02	.03	.04	.05	.06	.07	.08	.09
1.5	.9332	.9345	.9357	.9370	.9382	.9394	.9406	.9418	.9429	.9441
1.6	.9452	.9463	.9474	.9484	.9495	.9505	.9515	.9525	.9535	.9545
1.7	.9554	.9564	.9573	.9582	.9591	.9599	.9608	.9616	.9625	.9633

Common values from interpolation:

z	Area
1.645	0.9500
2.576	0.9950

Confidence Level	Critical Value
0.90	1.645
0.95	1.960
0.99	2.576

In scanning the table we see two values that are equally close to 0.95. There is 0.9495 and 0.9505 – both in error by exactly 0.0005, corresponding to $z = 1.64$ and $z = 1.65$. Which shall we choose? Note the asterisk at the bottom of the table, which says for an area of 0.9500, we should use $z = 1.645$, the average of the two values found in the table. The 95th percentile z-score is $z = 1.645$. This means that in any normal population, the 95th percentile is 1.645 standard deviations above average.

5.2.3   Homework

Use Table A1 to look up the following probabilities associated with the standard normal random variable, z.

1. Compute $P(z < 0)$

2. Compute $P(z < 3.18)$

3. Compute $P(z > 2.49)$

4. Compute $P(-3.05 < z < 0.73)$

5. Compute $P(z < 4.06)$

6. Compute $P(z > 2.65)$

7. Compute $P(z < 2.65)$

8. Compute $P(z > -1.27)$

9. Compute $P(z < -1.27)$

10. Compute $P(-1.27 < z < 2.65)$

11. Compute $P(1.53 < z < 2.09)$

Use Table A1 to approximate the z-score associated with the given percentile.

12. The 94th percentile.

13. The 81st percentile.

14. The 42nd percentile.

15. The 50th percentile.

Find the approximate z-score associated with the given area under the standard normal distribution.

16. Right tail area equal to 0.01

17. Right tail area equal to 0.05.

18. Right tail area equal to 0.06

19. Right tail area equal to 0.04.

20. Left tail area equal to 0.04.

Estimate the following probabilities for large magnitude z-scores. Note the footnotes at the bottom of Table A1 for these.

21. Estimate $P(z < 5.93)$

22. Estimate $P(z > 5.93)$

23. Estimate $P(z < -11.91)$

24. Estimate $P(z > -11.91)$

5.3   The Non-Standard Normal Distribution

Now that we have mastered the standard normal distribution of z-scores, it is time to bring in applications to non-standard normal distributions. For such distributions, we return to using the continuous random variable, x, from a normal population with mean, μ, and standard deviation, σ. These values are related to z-scores by the transformation formula, $z = \frac{x-\mu}{\sigma}$. This formula is used for converting values of x to z, so that we can look up associated probabilities on Table A1. There will be times when we need to convert values of z to x, and to do this, we solve the z-score formula for x:

$$x = \mu + z\sigma.$$

Figure 5.13 depicts the transformations between x and z values (the notation $z(a)$ in the diagram is used to represent the z-score of the value a). While the distributions represent different values on the horizontal axes, the areas of the shaded regions, corresponding to a value $x = a$ on the non-standard distribution, and the value $z = z(a)$ (the z-score of a) on the standard distribution, are the same.

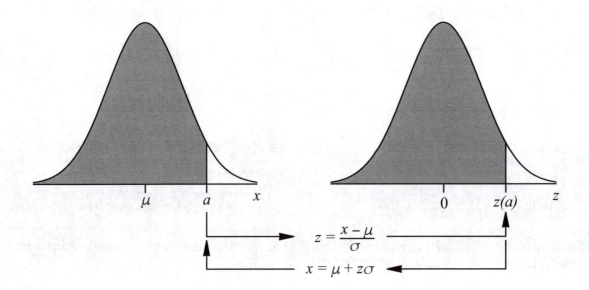

Figure 5.13: The transform from z-coordinates to and from x-coordinates.

5.3.1   Finding Probabilities Corresponding to Given Scores

Example 5.3.1 Heights of adult men are normally distributed, with a mean of 69 inches and a standard deviation of 3 inches. Use these facts to compute the probability that a randomly selected man is taller than 74 inches – that is, compute $P(x > 74)$.

To compute this, or any other probability for a normal distribution, we must always standardize our boundary points using the z-score formula. In this example, we have only one boundary value, $x = 74$, which needs to be standardized: $z = \frac{x-\mu}{\sigma} = \frac{74-69}{3} = \frac{5}{3} = 1.67$. Now we look up the probability using the methods from the previous section:

$$
\begin{aligned}
P(x > 74) &= P(z > 1.67) \\
&= 1 - P(z < 1.67) \\
&= 1 - 0.9525 \\
&= 0.0475 \\
&= 4.75\%.
\end{aligned}
$$

Now, would a height of 74 inches be unusual or significantly high? You may recall that one way we determined whether a value $x = a$ of a random variable is unusually high is by computing a p-value, $P(x > a)$. If this probability is small (less than about 5%), then the value, $x = a$, is significantly high. Since $P(x > 74) = 4.75\%$, we may conclude that such a height is significant.

Example 5.3.2 Next, let us determine the proportion of adult men whose heights are between 61 and 75 inches. First, we compute z-scores. For $x = 61$, we get

$$
z = \frac{x - \mu}{\sigma} = \frac{61 - 69}{3} = -2.67,
$$

and for $x = 75$ we get

$$
z = \frac{x - \mu}{\sigma} = \frac{75 - 69}{3} = 2.00.
$$

From this, we proceed:

$$
\begin{aligned}
P(61 < x < 75) &= P(-2.67 < z < 2.00) \\
&= P(z < 2.00) - P(z < -2.67) \\
&= 0.9772 - 0.0038 \\
&= 0.9734 \\
&= 97.34\%.
\end{aligned}
$$

We may conclude that 97.34% of adult male heights are between 61 and 75 inches.

5.3.2 Finding Scores Corresponding to Given Probabilities

There are times when we want to know a particular value of x that coincides with a known proportion or probability. For example, suppose we want to know the 99th percentile of adult male heights. From the previous section, we determined that the z-score corresponding to the 99th percentile in a normal population is $z = 2.33$. To convert this to an adult male height, we use the inverse transformation formula $x = \mu + z\sigma$. In this example, $x = \mu + z\sigma = 69 + 2.33 \cdot 3 = 75.99$, so the 99th percentile for adult male heights is almost 76 inches (six feet, four inches).

Example 5.3.3 IQ scores are normally distributed with a mean of 100, and a standard deviation of 15. Suppose a particular employer wishes only to employ people whose IQ scores are in the top 10%. What would the cutoff IQ score be for employment?

First, note that if 10% of a population is above a given value, that 90% must be below that value, so the value is the 90th percentile for the population. To solve this problem, we must first find the z-score corresponding to the 90th percentile, and then we will use the inverse transformation $x = \mu + z\sigma$ to convert this to an IQ score.

To find the 90th percentile, we scan Table A1 for the closest probability to $0.90 = 90\%$.

z	.00	.01	.02	.03	.04	.05	.06	.07	.08	.09
1.1	.8643	.8665	.8686	.8708	.8729	.8749	.8770	.8790	.8810	.8830
1.2	.8849	.8869	.8888	.8907	.8925	.8944	.8962	.8980	.8997	.9015
1.3	.9032	.9049	.9066	.9082	.9099	.9115	.9131	.9147	.9162	.9177

In scanning the table, we see that the closest probability to 0.90 is 0.8997. This corresponds to a z-score of $z = 1.28$. Whatever this cutoff IQ score is, we know that it is 1.28 standard deviations above average. To convert to an IQ score we apply the formula

$$x = \mu + z\sigma = 100 + 1.28 \cdot 15 = 119.2,$$

which we will round to 119. Thus, we have a cutoff score, and our employer will only hire applicants with an IQ of 119 or higher.

5.3.3   Homework

IQ Scores are approximately normal in their distribution, with a mean equal to 100, and a standard deviation equal to approximately 15. Let x represent a randomly selected IQ score, and compute the following probabilities. Assume that x is a *continuous* random variable.

1. $P(x < 119)$

2. $P(x < 81)$

3. $P(x > 133)$

4. $P(x > 77)$

5. $P(82 < x < 91)$

6. $P(110 < x < 139)$

Find the given IQ percentile

7. The 96th percentile

8. The 95th percentile

9. The 99th percentile

10. The 40th percentile

Heights of adult males are normally distributed with a mean of 69 inches, and a standard deviation of 3 inches. Use a p-value to determine whether the given adult male heights are significantly high or low, as indicated. Recall that $x = a$ is *significantly high* if $P(x > a)$ is small (less than about 5%), and $x = a$ is *significantly low* if $P(x < a)$ is small (again, less than about 5%).

11. Is $x = 73.5$ significantly high?

12. Is $x = 78.3$ significantly high?

13. Is $x = 60.2$ significantly low?

14. Is $x = 65.4$ significantly low?

Continuing to use the adult male heights described above, answer the following questions.

15. Give the range of heights which contains the middle 95% of all adult male heights (this range would exclude the lower 2.5% and the upper 2.5% of adult male heights).

16. What is the usual range of adult male heights, using the two standard deviation rule? Does your answer agree with the answer given on the previous problem?

17. What is the 85th percentile of adult male heights?

18. What is the 10th percentile of adult male heights?

19. What is the probability that an adult male is greater than 74.5 inches?

The mean GPA for students at a prestigious university is normally distributed with a mean of 3.463 and a standard deviation of 0.298. **Given these facts, please answer the following questions.**

20. Supposing that students who graduate with honors need a GPA of 3.5 or higher, what proportion of students at UCLA graduate with honors?

21. What is the minimum percentile for graduation with honors?

22. What GPA represents the 95th percentile?

23. Assuming that a given GPA below 2.0 results in the corresponding student being placed on academic probation, give the proportion of students who are on academic probation.

24. What GPA is the 10th percentile? Are students below the 10th percentile on academic probation?

Part IV

Inferential Statistics

The Process of Statistical Analysis

In Chapter 1, we discussed the process of statistical analysis. Each step of this process has been discussed, except the last. In Part IV of this text, we begin the process of *statistical inference*, which is summarized in Figure 5.14.

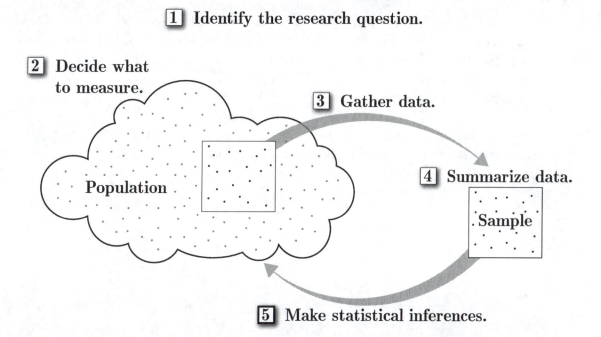

Figure 5.14: The process of statistical analysis, step 5.

Step I. Identify the Research Question

In statistical research, we ask a question that can be answered by data. Most questions that we ask have to do with a population parameter – either a proportion or a mean. The nature of the question can usually be answered by an estimate or a test regarding the hypothetical value of the parameter. In Chapter 6, we focus on estimates and their margins of error.

Step II. Decide What to Measure

When a research question is answerable by a categorical (non-numerical) response, we summarize with sample proportions. When the question is answerable with a quantitative response, we will typically summarize with a mean.

Step III. Gather Data

Samples should be gathered in a way so that they represent their populations well. When a sample is representative, it can be used to make reliable inferences. In experimental studies, the sample should be divided into blocks which are similar, so as to diminish the effects of confounding variables.

Step IV. **Summarize Data**

Categorical data are summarized by proportions and pie or ribbon charts. Quantitative data are summarized with averages and standard deviations, and dotplots, boxplots or histograms.

Step V. **Make Statistical Inferences**

When a chosen sample represents a population well, it can tell us about the population. In statistics, an inference is a somewhat reliable *guess* about a population. The inferences we make are estimates and hypothesis tests regarding unknown population parameters. These are used to answer the chosen research question.

The last step, *statistical inference*, is the focus of the remaining material in this text. We begin in Chapter 6 with *estimates and sampling distributions.*

Chapter 6

Estimates & Sampling Distributions

Now that we have a solid grasp on the use of our tables for the normal distribution, we may finally move on to our main goal: inferential statistics. In Chapter 6, our main goals are to determine the nature of the distributions of our two main statistics: the sample mean and sample proportion. Once we understand the nature of these distributions, we will be able to estimate a maximum error for these statistics, compute confidence intervals, and determine sample sizes for chosen maximum errors. From there we will move on into hypothesis testing, where we will learn to test claims that are made about unknown population parameters.

6.1   Introduction to Sampling Distributions

6.1.1   Estimates and Their Distributions

We begin our discussion by an overview of estimates. First of all, we must understand that, while a population has only one value for a particular parameter, there are many estimates of this parameter. In fact, each sample of any particular size will give a different estimate. Because of this variability, we understand that, just as values in any diverse population may vary, statistics – estimates of a parameter – will vary and have their own distribution, complete with mean and standard deviation. The distribution of a statistic is called a *sampling distribution*.

Our main tools in inferential statistics all stem from the ideas which we have developed regarding what is usual, and what is unusual. And when populations were normal, the analysis became much more refined, and centered about the z-scores of values in the population. For us to use these refinements in determining what is unusual or usual for an estimate, it would be nice to know whether the collection of all possible estimates is normal, and what is the mean and standard deviation of these estimates, so that we can talk about z-scores.

In the context of a distribution of estimates, the standard deviation of estimates is often referred to as a *standard error*, because any deviation of an estimate from the value being estimated is an error. A standard deviation is therefore a standard error.

Our goals are the following:

1. To determine the shape of the sampling distribution for a given estimate or statistic;

2. When a sampling distribution is normal, to determine the distribution's mean and standard error (or *deviation*).

We begin this process in the abstract. We assume that $\widehat{x}$ is an estimate of a population mean, μ, and that this estimate is normally distributed. Our goal is to develop a few formulas regarding the errors present in such estimates, but before we do, we discuss what it is that makes a good estimator.

6.1.2   Rules for Good Estimators

Suppose that we wish to estimate a population mean, μ. There are usually several estimators of μ. Which estimate is best? Before we can answer this, we need to define what is meant by *best*. In general, our rules for a best estimate are as follows:

1. *The best estimates come from samples which represent their populations well.*
   In our developments, we assume that all estimates are derived from *simple random samples*. Other types of samples (such as *stratified* samples) can work as well (sometimes better), but in this introductory course, we assume the random sample because it is easiest to develop.

2. *The best estimates are unbiased.*
   What makes an estimate unbiased? Simply put, an estimate is unbiased when the mean, or expected value, of the estimates is equal to the thing which is being estimated. That is, $\widehat{x}$ is an unbiased estimate if $E(\widehat{x}) = \mu$.

3. *The best unbiased estimate is that which varies the least.*
   As we will see, in some cases more than one estimator is unbiased, and when that is the case, we should pick the estimator which has the smallest variation. Choose $\widehat{x}$ as an estimator over other unbiased estimators if its standard error (i.e. its *standard deviation*) is smaller than the others. Since error increases with variability, it only makes sense to pick the estimator that varies the least.

6.1.3   The General Form of a Confidence Interval

Again, let us suppose that $\widehat{x}$ is an unbiased estimate of a population mean, μ, so $E(\widehat{x}) = \mu$. Let us also suppose that the standard error (or *deviation*) of $\widehat{x}$ is $\sigma_{\widehat{x}}$. We would like to develop what is called a confidence interval for the parameter μ.

Our development of the confidence interval begins by simply constructing a symmetric interval, with the mean at the center. This interval will represent a proportion of the distribution of estimates (the sampling distribution) called the *level of confidence*. When

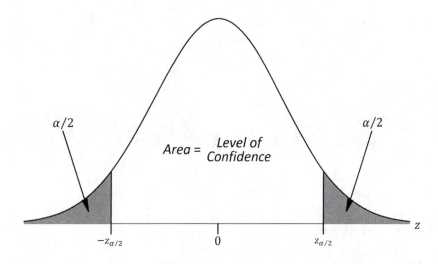

Figure 6.1: The geometry of the confidence interval on the standard normal (z) distribution.

we consider what this interval would look like, it is easiest to begin with the standardized z-scores of this distribution of estimates, as illustrated in Figure 6.1.

As pictured in Figure 6.1, we begin by centering our z-interval at zero, creating the interval symmetrically about this point, with a corresponding area equal to the level of confidence. The areas outside of this interval are the tails, right and left. By the rule of complements the total area of the tails is equal to

$$1 - (level\ of\ confidence),$$

and we denote this quantity by $\alpha = 1 - (level\ of\ confidence)$.

By symmetry, each tail has area equal to $\alpha/2$. We denote the endpoints of our symmetric interval as $\pm z_{\alpha/2}$, and will refer to these values occasionally as *critical values*. To bring this into the world of estimates, we must transform our two z endpoints into $\widehat{x}$ values using the formula $\widehat{x} = \mu + z\sigma_{\widehat{x}}$. Substituting $\pm z_{\alpha/2}$ for z gives us $\widehat{x} = \mu \pm z_{\alpha/2}\sigma_{\widehat{x}}$. These are the endpoints of the interval transformed to the distribution of estimates, $\widehat{x}$. What this means is that the proportion of estimates, $\widehat{x}$, that lie between $\widehat{x} = \mu - z_{\alpha/2}\sigma_{\widehat{x}}$ and $x = \mu + z_{\alpha/2}\sigma_{\widehat{x}}$ is equal to the level of confidence. If we denote E to be the value $E = z_{\alpha/2}\sigma_{\widehat{x}}$, then the endpoints of our interval are $\widehat{x} = \mu \pm E$, and we plot this interval with its corresponding density function in Figure 6.2.

As stated, the proportion of values, $\widehat{x}$, within this interval is equal to the level of confidence, and since μ is at the center of the interval, if $\widehat{x}$ is inside the interval, the distance from $\widehat{x}$ to μ, $|\mu - \widehat{x}|$ is less than E. Thus, for each $\widehat{x}$ inside the interval, we know that $|\mu - \widehat{x}| < E$. Using rules from algebra, this is equivalent to $-E < \mu - \widehat{x} < E$, and adding $\widehat{x}$ to each component of this inequality gives $\widehat{x} - E < \mu < \widehat{x} + E$. This is the general form of a *confidence interval*.

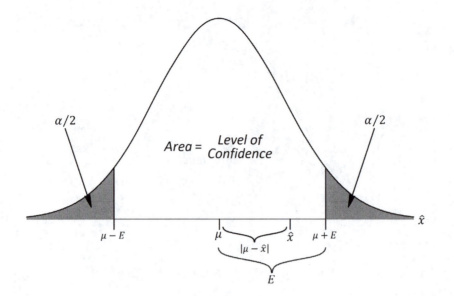

Figure 6.2: The level of confidence is the proportion of estimates for which the true error $|\mu_{\widehat{x}} - \widehat{x}|$ is less than the maximum error, E.

Theorem 6.1.1 Assuming $\widehat{x}$ is normally distributed and an unbiased estimate of μ, and that the standard error (or *deviation*) of all $\widehat{x}$ values is $\sigma_{\widehat{x}}$, the maximum error in the estimate is

$$E = z_{\alpha/2}\sigma_{\widehat{x}},$$

and the corresponding confidence interval is

$$\widehat{x} - E < \mu < \widehat{x} + E.$$

To correctly interpret the confidence interval, it should be stated that the proportion of such intervals that actually contain the mean is equal to the level of confidence.

The maximum error formula, $E = z_{\alpha/2}\sigma_{\widehat{x}}$, requires the z-score denoted as $z_{\alpha/2}$. To look up $z_{\alpha/2}$, we remember that the area to the right of this positive z-score is equal to $\alpha/2$, but that Table A1 only uses areas to the left of such a score, so we search the table for the value closest to $1 - \alpha/2$.

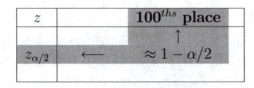

For example, let us suppose our level of confidence is 99%. This means that $\alpha = 1-0.99 = 0.01$. Now, in the table, we need to look for a value that is closest to $1 - \alpha/2 = 1 - \frac{0.01}{2} = 1-0.005 = 0.9950$. In scanning the interior of the table for this, we see two values equally close: 0.9949, and 0.9951.

z	.02	.03	.04	.05	.06	.07	.08	.09
2.4	.9922	.9925	.9927	.9929	.9931	.9932	.9934	.9936
2.5	.9941	.9943	.9945	.9946	.9948	.9949	.9951	.9952
2.6	.9956	.9957	.9959	.9960	.9961	.9962	.9963	.9964

Common values from interpolation:

z	Area
1.645	0.9500
2.576	0.9950

Confidence Level	Critical Value
0.90	1.645
0.95	1.960
0.99	2.576

A note at the bottom of the table tells us to use $z_{\alpha/2} = 2.576$ when the corresponding area is 0.9950. It also might be helpful to note the small table on the bottom right, indicating common $z_{\alpha/2}$ critical values, including the 2.576 critical value that goes with 99% confidence.

Note also, that with 90% confidence, $z_{\alpha/2} = 1.645$, and with 95% confidence, $z_{\alpha/2} = 1.960$.

As a final example, what $z_{\alpha/2}$ critical value is associated with 96% confidence? In this case,

$$\alpha = 1 - 0.96 = 0.04$$

and

$$1 - \alpha/2 = 1 - \frac{0.04}{2} = 0.98.$$

Scanning the table for a value close to 0.98 below, we see 0.9798 is closest (*not* the 0.9803 ... be careful!), and this is associated with $z_{\alpha/2} = 2.05$.

z	.00	.01	.02	.03	.04	.05	.06	.07	.08	.09
1.9	.9713	.9719	.9726	.9732	.9738	.9744	.9750	.9756	.9761	.9767
2.0	.9772	.9778	.9783	.9788	.9793	.9798	.9803	.9808	.9812	.9817
2.1	.9821	.9826	.9830	.9834	.9838	.9842	.9846	.9850	.9854	.9857

For now, all that you need is to practice looking up $z_{\alpha/2}$. The general form of a confidence interval is by definition non-specific, and not meant to be applied directly to any application. This interval will become useful as we apply it to specific sampling distributions throughout the chapter.

The modern notion of the confidence interval is a relatively new development. The first use of the modern confidence interval appears in the 1934 paper, "On the Two Different Aspects of the Representative Method," written by Jerzy Neyman (1894-1981) in *Journal of the Royal Statistical Society, 97*, 558-625 .

Definition 6.1.2 The Generalized Test Statistic

When we take the z-score of a statistic, $\widehat{x}$, that is normally distributed, the result is called a *test statistic*. The formula, $z = \frac{\widehat{x}-\mu}{\sigma}$ is the general form, but the mean and standard error will take on other forms as we proceed. The main use of a test statistic will be to judge whether a statistic is significantly different from an assumed value of the mean, μ.

6.1.4   Homework

Determine the value of $z_{\alpha/2}$ corresponding to the given level of significance.

1. 95% confidence.

4. 94% confidence.

2. 92% confidence.

5. 99% confidence.

3. 96% confidence.

6. 90% confidence.

Suppose that an estimate $\widehat{x} = 10$ of an unknown parameter, μ, has standard deviation $\sigma_{\widehat{x}} = 2$. Compute the confidence interval corresponding to the given levels of confidence. Give the correct interpretation of the result.

7. 95% confidence.

10. 94% confidence.

8. 92% confidence.

11. 99% confidence.

9. 96% confidence.

12. 90% confidence.

Answer the following questions.

13. Based on the results of the previous problems, what is the effect on the margin of error by raising the level of confidence. Is this maximum error smaller or larger?

14. What is the effect on the confidence interval by lowering the level of confidence? Is the interval narrower or wider?

15. A sample median is an unbiased estimate of its population's median, and is sometimes biased as an estimate of its population's mean. Would the sample median be a better estimate of the population mean or median?

16. The sample variance is an unbiased estimator of its population's variance. Unfortunately, taking square roots of this unbiased estimate introduces bias. For what parameter is the square root of sample variance a biased estimator?

17. To estimate the median age of all Macintosh users, a survey is conducted on a popular website for Macintosh users, yielding a sample median. Is this a good way to estimate the unknown parameter? Would it be reasonable to use the sample median to create a confidence interval for the population median?

18. Suppose that we wish to estimate the proportion of attendees at a professional football game who like the parking accommodations. To gather data, workers are placed at each entrance, and every 20th worker is surveyed. A sample proportion is computed as an estimate of the population proportion. Is this method likely to result in creating a sample that is representative of its population? Assuming that sample proportions are normal in their distribution, and unbiased in their estimate of the population proportion. Could we use the sample proportion to construct a reliable confidence interval for the population proportion?

6.2   Sampling Distributions of Sample Proportions

6.2.1   The Normal Approximation of the Binomial Distribution

Now we begin the process of developing our first sampling distribution. This begins with a review of *binomial experiments*, which give us a theoretical structure for understanding sample proportions. Our first task is to show that, in many cases, the binomial probabilities can be approximated using the normal distribution. This will quickly lead to the understanding that sample proportions have an approximately normal distribution under certain conditions. This is important because our discussion of sampling distributions in Section 6.1 assumed that the many values of a given statistic would be normally distributed.

We begin with a review of the binomial experiment.

A *binomial experiment* consists of the following:

- n independent trials (meaning we must sample *with replacement*);

- Each trial has exactly two outcomes: $S = $ "success" and $F = $ "failure;"

- $p = P(S)$, and $q = P(F) = 1 - p$, with p and q remaining constant throughout the experiment.

- The random variable, x, will represent the possible number of successes in n trials.

- With these definitions, we developed a formula for computing the probability of x successes in n trials:

$$P(x) = \ _nC_x p^x q^{n-x} = \frac{n!}{x!\,(n-x)!} p^x q^{n-x}$$

Example 6.2.1 Revisiting an example covered previously, suppose we pick $n = 12$ people randomly, and define a success to be $S = $ "*person is a woman*". Demographic data which tell us that $p = P(S) = 0.498$. Using this, we compute the probability of each value of x from 0 to 12. The probability distribution is reproduced below.

x	$P(x)$
0	0.00025612
1	0.00304896
2	0.01663565
3	0.05501032
4	0.12278698
5	0.19489375
6	0.22556428
7	0.19180025
8	0.11891998
9	0.05243218
10	0.01560432
11	0.00281454
12	0.00023268

Table 6.1: The binomial distribution for Example 6.2.1.

Figure 6.3 plots this distribution of probabilities. In the figure, the density function for the corresponding normal distribution is plotted as well. Note that the density function's curve touches the top of each bar in the histogram at the apparent center (excluding the center bar).

You may remember that, in Chapter 5, we mentioned that the normal distribution was first developed by Abraham de Moivre around 1756 as a means for approximating binomial probabilities. His reason for doing this was that the formula above can be quite difficult to use when n becomes large. We have the same difficulty to deal with as well. We have much work still to be done with binomial experiments, but the formula that we have is too complex. Our solution will be to follow de Moivre's lead, and use the normal curve to approximate these probabilities, which is the main reason that it was originally developed.

We proceed by noting first that the distribution, while not smooth nor continuous, is rather symmetric and bell-shaped, and we wonder if somehow the continuous normal distribution could be used to approximate the probabilities given by this distribution. Our goal is to compute the probability that x lies in a range of values. For our first example, we compute $P(5 \leqslant x \leqslant 7)$. This region is shaded in Figure 6.4, and a smooth bell curve is superimposed over the histogram.

As we use a continuous distribution to approximate a discrete distribution it must be noted that with this discrete variable each value for x has a class width of one. On the histogram, each bar has a width of one as well, and if the areas of the bars are to be approximated precisely, we must take this into account. The bar that represents $x = 5$ begins at $x = 4.5$, and the bar that represents $x = 7$ ends at $x = 7.5$. This correction from $5 \leqslant x \leqslant 7$ for the discrete variable to $4.5 < x < 7.5$ for the continuous variable is called the *continuity correction*, and is always necessary when we use a continuous distribution to approximate a probability for a discrete variable. Once we have made the continuity correction, we take the z-score of each interval endpoint. Of course, we need the mean and standard deviation for this:

$$\begin{aligned} \mu &= np \\ &= 12 \cdot 0.498 \\ &= 5.976, \\ \sigma &= \sqrt{npq} \\ &= \sqrt{12 \cdot 0.498 \cdot 0.502} \\ &= 1.7320. \end{aligned}$$

For $x = 4.5$, we compute

$$\begin{aligned} z &= \frac{x - \mu}{\sigma} \\ &= \frac{4.5 - 5.976}{1.7320} \\ &= -0.85. \end{aligned}$$

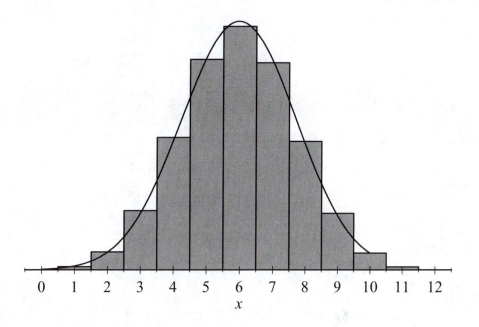

Figure 6.3: The density function of the normal distribution is superimposed over the binomial probability histogram. Note how the density function's graph touches the tops of most bars near their centers.

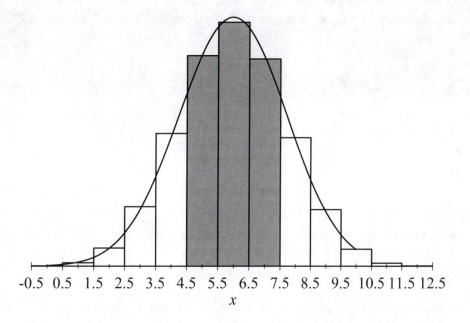

Figure 6.4: The binomial probability $P(5 \leq x \leq 7)$ must be updated to use class boundaries, as $P(4.5 \leq x \leq 7.5)$ when approximated with the normal distribution. This adjustment is called the *continuity correction*.

For $x = 7.5$, we compute

$$z = \frac{x - \mu}{\sigma}$$
$$= \frac{7.5 - 5.976}{1.7320}$$
$$= 0.88.$$

Now, we can use these in our estimate:

$$P(5 \leqslant x \leqslant 7) \approx P(4.5 < x < 7.5)$$
$$= P(-0.85 < z < 0.88)$$
$$= 0.8106 - 0.1977$$
$$= 0.6129.$$

z	.04	**.05**	.06
-0.9	.1736	.1711	.1685
-0.8	.2005	**.1977**	.1949
-0.7	.2296	.2266	.2236

z	.07	**.08**	.09
0.7	.7794	.7823	.7852
0.8	.8078	**.8106**	.8133
0.9	.8340	.8365	.8389

This is only an approximation. To get an exact value, we use the binomial distribution formula. This formula was used to produce the values in Table 6.1 on page 239, and from this we compute

$$P(5 \leqslant x \leqslant 7) = P(5 \text{ or } 6 \text{ or } 7)$$
$$= P(5) + P(6) + P(7)$$
$$= 0.1949 + 0.2256 + 0.1918$$
$$= 0.6123.$$

This answer is exact to four decimal places, but in such problems when the range of x values is large, computing an exact answer can be rather tedious. Our approximate value of 0.6129 is quite close to the exact answer, and as both values round to 61%, we will consider the approximation close enough.

There are times when the binomial distribution does not look normal. The normal distribution is symmetric, but the binomial can look quite *asymmetric*. This will happen when the mean is close to zero, as in Figure 6.5, or to n, as in Figure 6.6. Because the binomial distribution does not extend forever in both directions as the normal does, when the mean is close to one of the endpoints of the distribution's range, the tail associated to that endpoint is truncated, and the symmetry is lost.

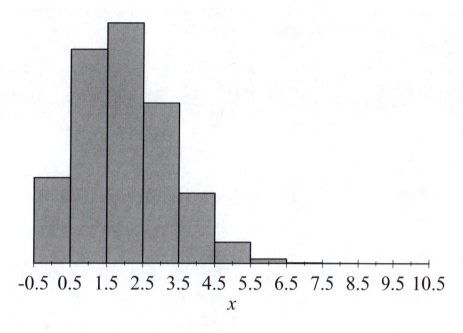

Figure 6.5: When a binomial distribution is skewed to the right, it has a shorter left tail, and does not appear *normal*.

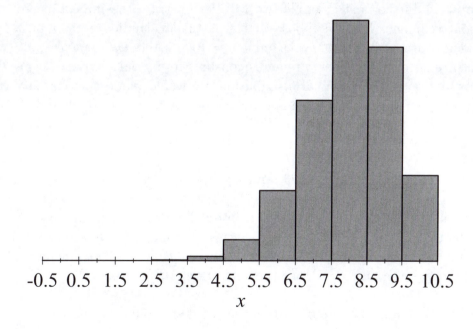

Figure 6.6: When a binomial distribution is skewed to the left, it has a shorter right tail, and does not appear *normal*.

Refer to the illustrations in Figures 6.5 and 6.6, noting the lack of symmetry. The asymmetry is due to the fact the mean of the binomial distribution is either too close to zero (on the left), or too close to n (on the right).

Our rule of thumb is simply that we want the mean to be a distance of at least 10 from 0, and at least 10 from n. And since μ is np, we need

$$np - 0 = np \geqslant 10,$$

and

$$n - np = n(1 - p)$$
$$= nq \geqslant 10.$$

Proposition 6.2.2 Rule of Thumb
The Normal Distribution may be used to approximate the Binomial Distribution when $np \geqslant 10$, and $nq \geqslant 10$

It might be noted that on the last example, $\mu = np = 5.976$, and $nq = 6.024$. The conditions for the normal approximation *were not met*, and yet, in spite of this we arrived at an approximation which was quite close to the correct value. When the conditions *are* met, the approximations tend to be *even better*.

Example 6.2.3 In the next example, we will demonstrate the importance of this approximation technique. Let us suppose that a particular candidate has a 60% approval rating among voters, and that we will sample $n = 1000$ voters *with replacement* (we must sample with replacement so that p remains constant from trial to trial). Using the normal approximation to the binomial distribution, we would like the probability that 640 or more voters give this candidate an approval. Here we have $n = 1000$, and $p = 0.60$, so

$$q = 1 - p = 1 - 0.60 = 0.40.$$

Checking $\mu = np = 1000 \cdot 0.60 = 600$, and $nq = 1000 \cdot 0.40 = 400$ we see that both of these values are at least 10, as required. Next, we wish to compute $P(x \geqslant 640)$. As before, we must adjust the endpoint of this interval as we convert the discrete variable to a continuous one. In general, this will always be done by adding or subtracting 0.5 to each endpoint of the interval in question. Here, since we want to include 640 in the interval, we must subtract 0.5, so that the correction is $P(x > 639.5)$.

All that is left is to find the z-score of this corrected endpoint. We need

$$\mu = np$$
$$= 1000 \cdot 0.60$$
$$= 600,$$

and

$$\begin{aligned}
\sigma &= \sqrt{npq} \\
&= \sqrt{1000 \cdot 0.6 \cdot 0.4} \\
&= 15.4919,
\end{aligned}$$

and so our z-score is:

$$\begin{aligned}
z &= \frac{x - \mu}{\sigma} \\
&= \frac{639.5 - 600}{15.4919} \\
&= 2.55.
\end{aligned}$$

z	.04	**.05**	.06
2.4	.9927	.9929	.9931
2.5	.9945	**.9946**	.9948
2.6	.9959	.9960	.9961

Thus, we compute

$$\begin{aligned}
P(x \geqslant 640) &= P(x > 639.5) \\
&= P(z > 2.55) \\
&= 1 - 0.9946 \\
&= 0.0054
\end{aligned}$$

It is important to recognize that this approach was quite simple compared to the exact approach, which would involve summing probabilities computed using the formula

$$P(x) = \frac{n!}{x!\,(n-x)!} p^x q^{n-x}$$

for each x from 640 to 1000. That would require a great deal of computation, which would be quick for a computer, but would involve in each step a round-off, introducing error at each of these hundreds of computations. It is difficult to say whether attempting an "exact" answer with the binomial formula would be more precise than the normal approximation introduced here. Using the computer algebra system *Maple*, we apply the binomial formula, and sum over all x from 640 to 1000, and we get $P(x \geqslant 640) = 0.00519$. This compares closely to the normal approximation 0.0054. Which value is a better approximation is a subject of a different course altogether, but it suffices to say that we should consider neither value to be exact. Still, the similarity of the two values validates them both to some extent.

6.2.2   The Sampling Distribution of Sample Proportions

Now that we understand the conditions under which the binomial random variable, x, is approximately normally distributed, we may consider the sampling distribution of sample proportions. Whenever we measure a proportion or a percentage from a sample, when sampling with replacement, we are conducting a binomial experiment. Where n has represented the number of trials, it now represents the sample size. The random variable x continues to represent the number of successes, and we define $\hat{p} = x/n$ to be the sample proportion – the proportion of trials in the sample which were successes. The collection of all possible $\hat{p}$ values from samples of size n is our sampling distribution, and we need to know all of the standard descriptors for this distribution: the shape, the mean, and the standard deviation.

In Chapter 4, we developed several properties about expected value and variance, including the following:

Theorem 6.2.4 Properties of Expected Value and Variance

1. $E(kx) = k \cdot E(x)$, where k is any constant

2. $Var(kx) = k^2 \cdot Var(x)$

3. If x is normally distributed and $k \neq 0$, then kx is normally distributed.

We know that, for a binomial experiment, x is an approximately normal random variable as long as $np \geqslant 10$ and $nq \geqslant 10$, with mean $\mu = E(x) = np$, and standard deviation $\sigma = \sqrt{npq}$, (so $Var(x) = \sigma^2 = npq$).

Now, $\hat{p} = x/n = (1/n) \cdot x$, and since $1/n \neq 0$, Property 3 above applies. Thus the collection of all possible $\hat{p} = x/n$ values is approximately normally distributed as long as $np \geqslant 10$ and $nq \geqslant 10$.

Now that we know the shape of our sampling distribution of $\hat{p}$ values, we need the mean, $\mu_{\hat{p}}$, and standard error (or *deviation*), $\sigma_{\hat{p}}$, of these $\hat{p}$ values. These develop readily using properties 1 and 2 above.

$$\begin{aligned}
\mu_{\hat{p}} &= E(\hat{p}) \\
&= E\left(\frac{x}{n}\right) \\
&= \left(\frac{1}{n}\right) \cdot E(x) \\
&= \frac{1}{n} \cdot np \\
&= p
\end{aligned}$$

The mean is thus,

$$\mu_{\hat{p}} = p.$$

To determine the standard error (or *deviation*), we begin with the variance of sample proportions, $\sigma_{\hat{p}}^2$.

$$
\begin{aligned}
\sigma_{\hat{p}}^2 &= Var\,(\hat{p}) \\
&= Var\left(\frac{x}{n}\right) \\
&= \left(\frac{1}{n}\right)^2 Var\,(x) \\
&= \frac{1}{n^2} \cdot npq \\
&= \frac{pq}{n}
\end{aligned}
$$

The standard error (or *deviation*) is the square root of the variance,

$$
\begin{aligned}
\sigma_{\hat{p}} &= \sqrt{Var\,(\hat{p})} \\
&= \sqrt{\frac{pq}{n}}.
\end{aligned}
$$

You may remember that the parameter, p, has always been the probability of success in a binomial experiment. This definition will always hold true, but we will now say this in a different way. We will now refer to p as the population proportion of successes out of the entire population of size N. So p is a population proportion, while $\hat{p}$ is a sample proportion which is an estimator of p. Note that we have shown above that $E(\hat{p}) = p$, and this is important since it tells us that $\hat{p}$ is an *unbiased point estimate* of p. We call $\hat{p}$ a *point estimate* of p, because it represents a single point on a number line. Later we will learn the confidence interval for p, and at times we will refer to it as an *interval estimate*. We summarize all that we have learned in this discussion below.

6.2.3   The Central Limit Theorem for Sample Proportions

Theorem 6.2.5 The Central Limit Theorem for Sample Proportions
In sampling from a population of binomial trials with proportion of successes equal to p, consider the sampling distribution of sample proportions – the collection of all sample proportions, $\hat{p}$, sampled with replacement from samples of size n. This sampling distribution is approximately normal when $np \geq 10$ and $nq \geq 10$. The sampling distribution has mean, $\mu_{\hat{p}}$, and standard deviation $\sigma_{\hat{p}}$. and we have proven the following facts.

1. The mean of the sampling distribution of sample proportions is given by

$$
\mu_{\hat{p}} = p.
$$

This tells us that $\hat{p}$ is an *unbiased point estimate* of p.

2. The standard deviation of the sampling distribution of sample proportions is

$$\sigma_{\hat{p}} = \sqrt{\frac{pq}{n}}.$$

Since a sample proportion is an estimate of the population proportion, and any deviation from the population proportion is error, this quantity is often referred to as the *standard error* of the sample proportions.

The above theorem assumes that samples are drawn with replacement. This is the rule, but it is also worth noting that when populations are very large, sampling with or without replacement tends to give the same results stated in the theorem above. Only with small populations is the *with replacement* requirement strict.

Now that we know all about the distribution of sample proportions – most importantly that they are approximately normally distributed – we may apply this information to all that we have learned about normal distributions. In particular, we can take a z-score of a sample proportion, which we will call a *test statistic*, and we can compute a confidence interval for a population proportion.

The test statistic is easiest. The z-score formula is

$$z = \frac{x - \mu}{\sigma},$$

and if we want to take a z-score of a $\hat{p}$ value, we must subtract the appropriate mean, $\mu_{\hat{p}} = p$, and divide by the appropriate standard error (or *deviation*),

$$\sigma_{\hat{p}} = \sqrt{\frac{p \cdot q}{n}}.$$

So our test statistic is

$$z = \frac{\hat{p} - \mu_{\hat{p}}}{\sigma_{\hat{p}}}$$

$$= \frac{\hat{p} - p}{\sqrt{\frac{pq}{n}}}.$$

Summary 6.2.6 The Test Statistic for a Sample Proportion
The z-score of a sample proportion is

$$z = \frac{\hat{p} - p}{\sqrt{\frac{pq}{n}}}.$$

This statistic measures the number of standard deviations that $\hat{p}$ is from p, and can be used to determine when a particular $\hat{p}$ is unusually high or low. Such z-scores are approximately normal in their distribution provided that $np \geq 10$, and $nq \geq 10$.

Example 6.2.3 Continued: As an example, let us return to the application where we survey 1000 voters and count the number of voters who approve of a particular candidate, and again let us assume that the population approval rate is $p = 60\% = 0.60$ (so $q = 1 - p = 1 - 0.60 = 0.40$). Suppose, in our sample, that we measure 652 approvals, so $\hat{p} = x/n = 652/1000 = 0.652$. Is this proportion an unusually or significantly high value?

To answer, we look to our test statistic, which is the z-score of the $\hat{p}$ that we have computed.

$$z = \frac{\hat{p} - p}{\sqrt{\frac{pq}{n}}}$$
$$= \frac{0.652 - 0.60}{\sqrt{\frac{0.60 \cdot 0.40}{1000}}}$$
$$= 3.36.$$

So, $\hat{p} = 0.652$ is 3.36 standard deviations above $p = 0.60$, and by the 2 standard deviation rule, this is significant.

If you recall, we also have determined significance by measuring the p-value, which is the probability of getting a value that is at least as extreme as the one we have measured.

z	.05	.06	.07
3.2	.9994	.9994	.9995
3.3	.9996	**.9996**	.9996
3.4	.9997	.9997	.9997

In this case, we the p-value is

$$P(\hat{p} > 0.652) = P(z > 3.36)$$
$$= 1 - P(z < 3.36)$$
$$= 1 - 0.9996$$
$$= 0.0004.$$

This is very small, and thus we conclude that the observed proportion, $\hat{p} = 0.652$, being very unlikely, is highly significant. It is so significant that we should consider the possibility that our old 60% approval rating is no longer valid.

This process of recognizing a significant difference between an assumed value of a parameter and a statistic from a sample is called a *hypothesis test*, and will be the primary subject of Chapter 7.

We finish this section by developing the formulas for the confidence interval for a population proportion. These formulas will be applied to examples in the next section of this text.

6.2.4   The Confidence Interval for a Population Proportion

We finish this section by developing the formulas for the confidence interval for a population proportion, and these formulas will be applied to examples in the next section of this text.

The general form of a confidence interval is summarized below.

Summary 6.2.7 Assuming $\widehat{x}$ is normally distributed, and that x is an unbiased estimate of μ, and that the standard error (or *deviation*) of all $\widehat{x}$ values is $\sigma_{\widehat{x}}$, the maximum error in the estimate is $E = z_{\alpha/2}\sigma_{\widehat{x}}$, and the corresponding confidence interval is

$$\widehat{x} - E < \mu < \widehat{x} + E.$$

Our normally distributed, *unbiased* estimate of p is $\hat{p}$, with mean $\mu_{\hat{p}} = p$, and standard error $\sigma_{\hat{p}} = \sqrt{\frac{pq}{n}}$. The maximum error formula becomes

$$E = z_{\alpha/2}\sigma_{\hat{p}}$$
$$= z_{\alpha/2}\sqrt{\frac{pq}{n}},$$

and the endpoints of the interval which were $x \pm E$ become $\hat{p} \pm E$, with $\mu_{\hat{p}} = p$ in between. So the confidence interval is

$$\hat{p} - E < \mu_{\hat{p}} < \hat{p} + E,$$

or

$$\hat{p} - E < p < \hat{p} + E.$$

The only problem is that, as we move forward, we will no longer assume that we know the value of the population proportion, p. In this case, it becomes impossible to compute

$$E = z_{\alpha/2}\sqrt{\frac{pq}{n}},$$

since it requires values for p and q. The best we can do is to use our best estimate for p, namely $\hat{p}$, and define $\hat{q} = 1 - \hat{p}$ to use in place of p and q. Thus, the formula we actually work with is

$$E = z_{\alpha/2}\sqrt{\frac{\hat{p}\hat{q}}{n}}.$$

Summary 6.2.8 The confidence interval for a population proportion is:

$$\hat{p} - E < p < \hat{p} + E,$$

where

$$E = z_{\alpha/2}\sqrt{\frac{\hat{p}\hat{q}}{n}}.$$

6.2.5   Homework

After verifying that $np \geq 10$ and $nq \geq 10$, use the normal distribution to approximate the binomial to answer the following questions.

1. For a binomial experiment with $n = 100$ and $p = 0.65$, find $P(x \geqslant 80)$.

2. Again, with the same binomial experiment as Problem 1, find $P(60 \leqslant x \leqslant 79)$.

3. For a politician with a 60% approval rate, what is the probability that, out of a sample of 1000 people, more than 65% of the sample approves of the politician?

4. Given that 49.8% of the world's population is female, what is the probability that less than 480 out of 1000 randomly selected people are female?

5. You may recall that, in 2005, the percentage of airline ticket purchasers who actually arrive to take their flight is 88%. Suppose that an airline routinely over-books their flights by 6% of the number of seats that a plane actually can seat. Thus, for an airplane with 250 seats, 265 tickets are actually sold. In such a case, what is the likelihood that the airplane does not have enough seats for the people who arrive to take the flight?

6. In 1936 the great statistician Ronald A. Fisher challenged one of Gregor Mendel's famous experiments in pea genetics by saying that, in this experiment, the 6,022 yellow peas observed in a sample of 8,023 peas was much too close to the mean number of peas, and that there was an only *one in ten* chance of being this close. Given that the yellow peas in this experiment had a 75% chance of occurring, the mean (or expected value) of the number of yellow peas out of 8023 is $\mu = 6017.25$, which we will round momentarily to 6017, with a standard deviation of $\sigma = 38.76$. The distance from the observed number of yellow peas, 6022, to the expected number of yellow peas, 6017, is 5. We wish to compute the probability of observing a number of yellow peas which is within 5 of 6017. This is a range from 6012 to 6022.

   a. What is the probability of being within 5 yellow peas from the expected value? That is, compute $P(6012 < x < 6022)$, using $\mu = 6017.25$ and $\sigma = 38.76$. Use the normal distribution to approximate the binomial distribution. The requirements for normality are met in this problem ($np = 8023 \cdot 0.75 = 6017.25 \geq 10$, and $nq = 8023 \cdot 0.25 = 2005.75 \geq 10$).

   b. Does your result agree with Fisher's claim?

   c. As mentioned previously, this closeness occurred in each of Mendel's seven famous pea experiments. Do you feel that Mendel was being honest?

Compute the z-score of the given sample proportion. Use the z-score to determine whether the sample proportion is unusually large or small.

7. Suppose a population proportion is equal to 65%. Suppose also that a sample of size 500 is drawn, and a sample proportion of 62% is observed. Is this unusually small? Use a z-score to justify your answer.

8. Given that 48.8% of the world's population is female, would it be unusual to gather a sample of size $n = 1000$ and observe a sample proportion of 52% females? Use a z-score as justification for your answer.

9. For a politician with a 60% approval rate, would it be unusual for a sample of size 4000 to yield a sample approval rate of 62%?

10. If 88% of all people who buy plane tickets actually take their flight, would it be unusual for 84% of 250 people who purchased tickets to actually take the flight?

After verifying that $np \geq 10$ and $nq \geq 10$, compute the p-value corresponding to the events described in Exercises 7 through 10.

11. Compute the p-value for the event described in Exercise 7 to determine if the observed value of $\widehat{p}$ is unusually small.

12. Compute the p-value for the event described in Exercise 8 to determine if the observed value of $\widehat{p}$ is unusually large.

13. Compute the p-value for the event described in Exercise 9 to determine if the observed value of $\widehat{p}$ is unusually large.

14. Compute the p-value for the event described in Exercise 10 to determine if the observed value of $\widehat{p}$ is unusually small.

Questions 15 through 18 discuss the sampling distribution of sample proportions for a very small population. Suppose a population is of size 4, with a population proportion $p = 75\%$. This means that 75% of the population (or three-fourths) are "successes" while the rest (one fourth) are failures. Thus, the population is: $\{S_1, S_2, S_3, F\}$. Suppose also that we wish to choose a sample from this population of size $n = 2$, to compute a sample proportion.

15. List all samples from this population of size $n = 2$. Remember that we must sample with replacement, and thus any population member may be selected as many as two times for a given sample. There will be exactly *sixteen* samples.

16. From each of the sixteen samples above, compute the sample proportion – this is the number of successes in a sample divided by the sample's size (which is equal to 2 in this case). This population of sample proportions is the *sampling distribution* of sample proportions.

17. The population of sample proportions above has mean $\mu_{\hat{p}}$, and standard error $\sigma_{\hat{p}}$. Compute this mean and standard error using the mean and *population* standard deviation functions built into your calculator.

18. Recalling that $n = 2$, $p = 0.75$, with $q = 1 - p = 0.25$, verify that $\mu_{\hat{p}} = p$, and that $\sigma_{\hat{p}} = \sqrt{\frac{pq}{n}}$.

6.3   Intervals for a Population Proportion

In the previous section, we developed a process for computing a confidence interval for a population proportion, p. In this section we apply this process, and also learn the relationship between maximum error and sample size.

Confidence intervals for a population proportion are used when a research question is being answered with categorical data – where we wish to know the value of the population proportion for a given categorical response. This response corresponds to a success in a binomial experiment.

The requirements for the confidence interval are those of the central limit theorem for sample proportions: $np \geqslant 10$ and $nq \geqslant 10$. We cannot verify these requirements, but it is understood that when n is large, these requirements are likely to be true. But, how large should n be?

We do not know the values of p and q, but we sample data will yield sample proportions, $\widehat{p}$ and $\widehat{q}$. If we use these as estimates of p and q, the requirements lead to the following. First, the requirement, $np \geq 10$ yields an approximate form.

$$
\begin{aligned}
np &\approx n\widehat{p} \\
&= n \cdot \frac{x}{n} \\
&= x \geq 10
\end{aligned}
$$

Thus, np may be approximately equal to x, the number of successes in the binomial experiment. Second, the requirement, $nq \geq 10$, can be approximated as follows.

$$
\begin{aligned}
nq &\approx n\widehat{q} \\
&= n\left(1 - \frac{x}{n}\right) \\
&= n - x \geq 10
\end{aligned}
$$

Therefore nq is approximately equal to $n - x$, the number of failures in the experiment.

Given that np and nq are unknown, our best attempt to satisfy the requirements by verifying that the number of sample successes and failures are each greater than 10.

6.3.1 Large Sample Intervals for a Population Proportion

First, we give a procedure for computing a confidence interval for a population proportion.

Procedure 6.3.1 Constructing a Large Sample Confidence Interval for a Population Proportion

Step I. Gather $\hat{p} = x/n$, n, and the level of confidence (convert percentages to proportions first!). Compute $\hat{q} = 1 - \hat{p}$, and $\alpha = 1 - $ (confidence level). Requirement: n must be large enough so that np and nq are each at least 10. This requirement is plausible if the random sample contains at least 10 successes and 10 failures.

Step II. Look up $z_{\alpha/2}$ on Table A1

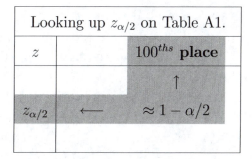

Step III. Compute the maximum error in the estimate:

$$E = z_{\alpha/2}\sqrt{\frac{\hat{p}\hat{q}}{n}}.$$

Step IV. Compute the endpoints of the confidence interval:

$$\hat{p} - E < p < \hat{p} + E.$$

Step V. As an interpretation, we may say that the proportion of such intervals that actually contains the unknown proportion, p, is equal to the level of confidence.

Example 6.3.2 Let us consider again the case where we survey 1000 voters, and find that 652 of them approve of a particular candidate. Following the steps outlined above, we will construct a 95% confidence interval for the true population proportion, p.

Step I. First, we gather data. $\hat{p} = x/n = 652/1000 = 0.652$, $n = 1000$, *level of confidence* $= 0.95$. Next we compute $\hat{q} = 1 - \hat{p} = 1 - 0.652 = 0.348$, $\alpha = 1 - $ (*confidence level*) $= 1 - 0.95 = 0.05$. Note that $n = 1000$, which seems sufficiently large, and there are many more than 10 successes and 10 failures in the sample..

Step II. Now we need to look up $z_{\alpha/2}$. First we compute $1 - \alpha/2 = 1 - \frac{0.05}{2} = 1 - 0.025 = 0.9750$. We want to scan Table A1 for an area close to 0.9750. As seen below we find $z_{\alpha/2} = 1.96$.

\multicolumn{4}{c}{Looking up $z_{\alpha/2}$ on Table A1.}
z	.05	**.06**	.07
1.8	.9678	**.9686**	.9693
1.9	**.9744**	**.9750**	.9756
2.0	.9798	.9803	.9808

Step III. Next we compute the maximum error: $E = z_{\alpha/2}\sqrt{\frac{\hat{p}\hat{q}}{n}} = 1.96 \cdot \sqrt{\frac{0.652 \cdot 0.348}{1000}} = 0.0295$. We have rounded this according to our round-off rule for proportions which requires us to keep at least three significant digits (where leading zeros are not significant).

Step IV. Now that we have the maximum error, we compute the confidence interval.

$$\hat{p} - E < p < \hat{p} + E$$
$$0.652 - 0.0295 < p < 0.652 + 0.0295$$
$$0.6225 < p < 0.6815$$
$$62.3\% < p < 68.2\%$$

Step V. To interpret this result we will say, with 95% confidence that the proportion of voters that approve of this candidate is between 62.3% and 68.2%. We might add that 95% of such intervals will actually contain the proportion in question.

Example 6.3.3 In the San Luis Obispo Tribune, an Associated Press article reported that a survey had been conducted of 750 randomly selected workers, employed full-time. Of this group, 125 stated that they had been angered by a co-worker to the point of wanting to hit them (but they did not) (Sept. 7, 1999). Construct a 96% confidence interval for the proportion of full-time workers who have wanted to hit a co-worker (but did not).

Step I. In this example, $\hat{p} = x/n = 125/750 = 0.167$, $n = 750$, *level of confidence* $= 0.96$. Next we compute $\hat{q} = 1 - \hat{p} = 1 - 0.167 = 0.833$, $\alpha = 1 - $ (*confidence level*) $= 1 - 0.96 = 0.04$. Note that $n = 750$, which seems sufficiently large. Note that there are 125 successes and 625 failures in the sample, both very much larger than 10.

Step II. Now we need to look up $z_{\alpha/2}$. First we compute $1 - \alpha/2 = 1 - 0.04/2 = 1 - 0.02 = 0.9800$. We want to scan Table A1 for an area close to 0.9800. As seen below, there are two values that are very close: 0.9798 and 0.9803. 0.9798 is closer to 0.9800, so that is what we will pick, and with it we have the associated z-score: $z_{\alpha/2} = 2.05$.

Looking up $z_{\alpha/2}$ on Table A1.			
z	.04	**.05**	.06
1.9	.9738	.9744	.9750
2.0	.9793	**.9798**	.9803
2.1	.9838	.9842	.9846

Step III. Next we compute the maximum error: $E = z_{\alpha/2}\sqrt{\frac{\hat{p}\hat{q}}{n}} = 2.05 \cdot \sqrt{\frac{0.167 \cdot 0.833}{750}} = 0.0279$. We again round to no less than least three significant digits.

Step IV. Now that we have the maximum error, we compute the confidence interval.

$$\hat{p} - E \; < \; p < \hat{p} + E$$
$$0.167 - 0.0279 \; < \; p < 0.167 + 0.0279$$
$$0.1391 \; < \; p < 0.1949$$
$$13.9\% \; < \; p < 19.5\%$$

Step V. To interpret this result we will say, with 96% certainty, that the proportion of full-time employees who have wanted to hit a co-worker in the last year, but did not, is between 13.9% and 19.5%. We might add that 96% of such intervals will actually contain the proportion in question.

6.3.2 Sample Size

Thanks to the formula for maximum error in estimating a population proportion, $E = z_{\alpha/2}\sqrt{\frac{\hat{p}\hat{q}}{n}}$, we know the relationship between sample size and maximum error. Suppose we wish to choose a maximum error that is acceptable for a particular study. We can use this formula to tell us the corresponding sample size by solving for n:

$$E = z_{\alpha/2}\sqrt{\frac{\hat{p}\hat{q}}{n}}$$
$$(E)^2 = \left(z_{\alpha/2}\sqrt{\frac{\hat{p}\hat{q}}{n}}\right)^2$$
$$E^2 = \frac{z_{\alpha/2}^2\hat{p}\hat{q}}{n}$$
$$n = \frac{z_{\alpha/2}^2\hat{p}\hat{q}}{E^2}$$

The problem with this formula is that it requires that some value for $\hat{p}$ be provided. This could be found by conducting a pilot study, whose purpose is, among other things, to give some idea of the value of a statistic like $\hat{p}$.

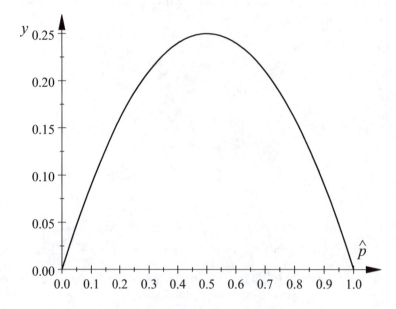

Figure 6.7: The graph of $y = \hat{p}\hat{q}$ has maximum value $\hat{p}\hat{q} = 0.25$, when $\hat{p} = 0.5$.

Another approach would be to consider that, since $0 \leqslant \hat{p} \leqslant 1$ and $0 \leqslant \hat{q} \leqslant 1$, the range of values for the product $\hat{p}\hat{q}$ is finite. If we plot

$$
\begin{aligned}
y &= \hat{p}\hat{q} \\
&= \hat{p}(1 - \hat{p}) \\
&= \hat{p} - \hat{p}^2
\end{aligned}
$$

on $\hat{p}y$-coordinate axes, the graph is a parabola, opening downward (see Figure 6.7), and the maximum value for $y = \hat{p}\hat{q}$ is $\hat{p}\hat{q} = 0.25$ when $\hat{p} = 0.5$. Thus, if we alter our formula by changing $\widehat{p}\widehat{q}$ to 0.25, we will have a sample size that is sufficiently large. The drawback of course is that this change will make the sample larger than is really necessary, and this usually translates to spending more time and money on data gathering than is necessary. Because of this, we are well advised to conduct a pilot study to find a value for $\hat{p}$, then use our original formula to enlarge the sample to whatever size is necessary. We summarize the sample size formulas below.

Summary 6.3.4 Sample Size for Estimating a Population Proportion
When we want to choose an acceptable value for the maximum error in the estimate, E, of a population proportion, we must sample at least the amount specified by one of the formulas below. If we can obtain an estimate, $\hat{p}$, from a pilot study, we should sample at least:

$$
n = \frac{z_{\alpha/2}^2 \hat{p}\hat{q}}{E^2}.
$$

If we cannot obtain an estimate, $\hat{p}$, we should sample at least:

$$n = \frac{z_{\alpha/2}^2 \, (0.25)}{E^2}.$$

The words *at least* are used to tell us to never use a value smaller than that given by the formula. Because of this, when rounding a sample size, we should always <u>round up</u> – pick the smallest integer that is greater than, or equal to, the value provided by the formula.

Example 6.3.3 Continued: Suppose, referring to the previous example, we wanted to specify a maximum error of 0.01. We will use the estimate for $\hat{p} = 0.167$ provided by this study, with the same level of confidence to lower the error to 0.01.

Because the level of confidence has not changed, our α has not changed, and thus our $z_{\alpha/2}$ has not changed. We proceed to compute the sample size

$$n = \frac{z_{\alpha/2}^2 \hat{p}\hat{q}}{E^2} = \frac{2.05^2 \cdot 0.167 \cdot 0.833}{0.01^2} = 5846.14,$$

which we must round up to $n = 5847$. That is painfully large, and may not be worth the extraordinary effort that would be required to enlarge the study to this size. The error previously was 0.0279, and after seeing what is required for a 0.01 error, the old 0.0279 does not look so bad after all!

It also might be worth pointing out that, if we use the formula $n = \frac{z_{\alpha/2}^2 (0.25)}{E^2}$, we would get

$$n = \frac{2.05^2 \, (0.25)}{0.01^2} = 10506.3$$

which we round up to $n = 10507$. Now the sample size is nearly double what it was when we had a $\hat{p}$ value from our pilot study.

6.3.3 Homework

Answer the following questions.

1. For a 96% confidence level, look up $z_{\alpha/2}$ on Table A1.

2. For a 98% confidence level, look up $z_{\alpha/2}$ on Table A1.

3. For a 92% confidence level, look up $z_{\alpha/2}$ on Table A1.

4. For the confidence interval, $.64 < p < .72$, determine $\hat{p}$ and the maximum error, E.

5. Express the confidence interval $p = .476 \pm .031$ in the form $a < p < b$.

6. Find the maximum error in the estimate of a population proportion, p, if $n = 1000$, and $\hat{p} = .524$, with 96% confidence.

7. Find the maximum error in the estimate of a population proportion, p, if $n = 563$, $x = 382$, with 98% confidence.

8. Compute the confidence interval for the population proportion, p, if $n = 200$, $x = 75$, using 99% confidence.

9. Compute the confidence interval for the population proportion, p, if $n = 921$, $\hat{p} = 0.217$, using 95% confidence.

10. Determine the sample size necessary to estimate the population proportion, p, to within a maximum error of 0.04, at 94% confidence, with $\hat{p}$ unknown.

11. Determine the sample size necessary to estimate the population proportion, p, to within a maximum error of 0.02, at 96% confidence, with $\hat{p} = 0.65$.

After verifying that the requirements are met, construct the following confidence intervals.

12. To determine an approval rate, 1000 voters are randomly selected, and of these 573 say they approve of the president. Give the 95% confidence interval for the approval rate (the proportion of approvals).

13. To determine the proportion of people in the world who are female, 890 people are randomly selected, and 440 of these are women. Give the 98% confidence interval for the proportion of the world's population who are women.

14. To determine the percentage of airline ticket purchasers who actually go to their airport to take the flight, 1265 airline tickets are sampled, and 87.6% of these represented people who actually arrived at the airport for the flight. Construct a 99% confidence interval for the proportion of tickets sold which are actually used for flying.

15. In Mendel's famous pea experiments, out of 8023 peas, exactly 6022 of these were yellow. Construct a 95% confidence interval for the proportion of such peas that are yellow.

16. Using the statistics of problem 12 as a pilot study, use the sample proportion with the same level of confidence to give the sample size necessary to bring the maximum error in the estimate of p down to 2 percentage points.

17. Using the statistics of problem 13 as a pilot study, use the sample proportion with the same level of confidence to give the sample size necessary to bring the maximum error in the estimate of p down to 1 percentage point.

18. Using the statistics of problem 14 as a pilot study, use the sample proportion with the same level of confidence to determine the sample size which would bring the maximum error in the estimate of p up to 3 percentage points.

19. Using the statistics of problem 15 as a pilot study, use the sample proportion with the same level of confidence to determine the sample size which would bring the maximum error in the estimate of p up to 2 percentage points.

20. Suppose a random sample consists of 500 Americans, of these, and that 63% of these are in favor of the death penalty for those who commit murder. Construct a 96% confidence interval for the proportion of Americans who favor the death penalty for those who commit murder.

6.4   The Sampling Distribution of Sample Means

6.4.1   The Population of all Sample Means

What we have done for sample proportions in Sections 6.2 and 6.3, we now must do for sample means. That is, for a given population of values, x, we will consider the sampling distribution of all sample means, $\bar{x}$, taken from samples of size n. Our most important goal will be to determine the shape of this distribution, and once this is settled we will need to determine the mean and standard deviation of this distribution. Along the way we will attempt to determine whether a sample mean, $\bar{x}$, is an unbiased estimator of μ. Once this is done, we will be able to develop a test statistic, or z-score, formula for a sample mean.

It will be easiest to begin by considering an example. In Figure 6.8, we plot the distribution of a small parent population of values x: $\{5, 6, 7, 8\}$. These values all occur with a frequency of one, and are dispersed at regular intervals, so the distribution is uniform, or flat.

The mean of this parent population is $\mu = 6.5$, and the standard deviation is $\sigma = 1.11803$.

Our goal is now to create the sampling distribution of all sample means, taken from samples of size $n = 3$. In doing so, we will sample with replacement. This process creates 64 samples, listed below.

$\{5,5,5\}$, $\{5,5,6\}$, $\{5,5,7\}$, $\{5,5,8\}$, $\{5,6,5\}$, $\{5,6,6\}$, $\{5,6,7\}$, $\{5,6,8\}$, $\{5,7,5\}$, $\{5,7,6\}$, $\{5,7,7\}$, $\{5,7,8\}$, $\{5,8,5\}$, $\{5,8,6\}$, $\{5,8,7\}$, $\{5,8,8\}$, $\{6,5,5\}$, $\{6,5,6\}$, $\{6,5,7\}$, $\{6,5,8\}$, $\{6,6,5\}$, $\{6,6,6\}$, $\{6,6,7\}$, $\{6,6,8\}$, $\{6,7,5\}$, $\{6,7,6\}$, $\{6,7,7\}$, $\{6,7,8\}$, $\{6,8,5\}$, $\{6,8,6\}$, $\{6,8,7\}$, $\{6,8,8\}$, $\{7,5,5\}$, $\{7,5,6\}$, $\{7,5,7\}$, $\{7,5,8\}$, $\{7,6,5\}$, $\{7,6,6\}$, $\{7,6,7\}$, $\{7,6,8\}$, $\{7,7,5\}$, $\{7,7,6\}$, $\{7,7,7\}$, $\{7,7,8\}$, $\{7,8,5\}$, $\{7,8,6\}$, $\{7,8,7\}$, $\{7,8,8\}$, $\{8,5,5\}$, $\{8,5,6\}$, $\{8,5,7\}$, $\{8,5,8\}$, $\{8,6,5\}$, $\{8,6,6\}$, $\{8,6,7\}$, $\{8,6,8\}$, $\{8,7,5\}$, $\{8,7,6\}$, $\{8,7,7\}$, $\{8,7,8\}$, $\{8,8,5\}$, $\{8,8,6\}$, $\{8,8,7\}$, $\{8,8,8\}$

From each of these, we compute a sample mean, $\bar{x}$.

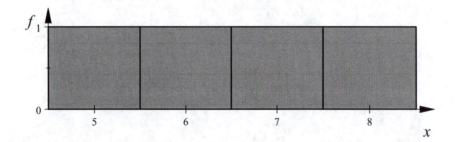

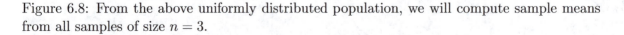

Figure 6.8: From the above uniformly distributed population, we will compute sample means from all samples of size $n = 3$.

This has been done, and the sampling distribution consisting of all 64 sample means from samples of size 3 are listed below.

$\bar{x}$: $\{5, 5\frac{1}{3}, 5\frac{2}{3}, 6, 5\frac{1}{3}, 5\frac{2}{3}, 6, 6\frac{1}{3}, 5\frac{2}{3}, 6, 6\frac{1}{3}, 6\frac{2}{3}, 6, 6\frac{1}{3}, 6\frac{2}{3}, 7, 5\frac{1}{3}, 5\frac{2}{3}, 6, 6\frac{1}{3}, 5\frac{2}{3}, 6, 6\frac{1}{3}, 6\frac{2}{3}, 6,$
$6\frac{1}{3}, 6\frac{2}{3}, 7, 6\frac{1}{3}, 6\frac{2}{3}, 7, 7\frac{1}{3}, 5\frac{2}{3}, 6, 6\frac{1}{3}, 6\frac{2}{3}, 6, 6\frac{1}{3}, 6\frac{2}{3}, 7, 6\frac{1}{3}, 6\frac{2}{3}, 7, 7\frac{1}{3}, 6\frac{2}{3}, 7, 7\frac{1}{3}, 7\frac{2}{3}, 6, 6\frac{1}{3}, 6\frac{2}{3},$
$7, 6\frac{1}{3}, 6\frac{2}{3}, 7, 7\frac{1}{3}, 6\frac{2}{3}, 7, 7\frac{1}{3}, 7\frac{2}{3}, 7, 7\frac{1}{3}, 7\frac{2}{3}, 8\}$

We wish to plot a frequency histogram, so the means are first sorted below.

$\bar{x}$: $\{5, 5\frac{1}{3}, 5\frac{1}{3}, 5\frac{1}{3}, 5\frac{2}{3}, 5\frac{2}{3}, 5\frac{2}{3}, 5\frac{2}{3}, 5\frac{2}{3}, 5\frac{2}{3}, 6, 6, 6, 6, 6, 6, 6, 6, 6, 6, 6\frac{1}{3}, 6\frac{1}{3}, 6\frac{1}{3}, 6\frac{1}{3}, 6\frac{1}{3}, 6\frac{1}{3},$
$6\frac{1}{3}, 6\frac{1}{3}, 6\frac{1}{3}, 6\frac{1}{3}, 6\frac{1}{3}, 6\frac{2}{3}, 6\frac{2}{3}, 6\frac{2}{3}, 6\frac{2}{3}, 6\frac{2}{3}, 6\frac{2}{3}, 6\frac{2}{3}, 6\frac{2}{3}, 6\frac{1}{3}, 6\frac{2}{3}, 6\frac{2}{3}, 6\frac{2}{3}, 7, 7, 7, 7, 7, 7, 7, 7,$
$7, 7, 7\frac{1}{3}, 7\frac{1}{3}, 7\frac{1}{3}, 7\frac{1}{3}, 7\frac{1}{3}, 7\frac{1}{3}, 7\frac{1}{3}, 7\frac{1}{3}, 7\frac{1}{3}, 8\}$

Remember, our goal is to compute a confidence interval for a population mean, and a test statistic for a sample mean. To do this, we must determine the mean, standard error (or *deviation*), and shape of this distribution. The histogram for this sampling distribution is given in Figure 6.9.

As is clearly visible, the distribution of sample means appears to be symmetric, and roughly bell-shaped. The *skewness* is 0, and the *kurtosis* is 2.55, which is not too different from the normal standard of 3. This is very important, because we know so much about bell-shaped distributions. What makes this really remarkable is that the population from which samples were taken was not bell-shaped. We have spent so much effort worrying about whether populations are normal or not, when in practice, it will not really matter because sample means tend to have normal distributions, even when the populations they are taken from are not normal.

If we conduct the experiment with larger sample sizes, we see a tendency toward normality (with $n = 4$ the kurtosis is 2.66, and with $n = 5$, the kurtosis is 2.73). This happens because of the fact that, even when the values of x are not near their mean, μ, the sample means, $\bar{x}$, are estimates of μ and tend to be drawn toward it, especially when sample sizes are reasonably large.

6.4.2 Mean and Standard Error

Now that we see a roughly normal sampling distribution of sample means, we would like to know the mean and standard error (or *deviation*) of these values. We use the notation $\mu_{\bar{x}}$ to represent the mean of all sample means (from samples of size n), and $\sigma_{\bar{x}}$ to represent the standard error of all of these sample means. In computing the mean of this sampling distribution, as listed above, we get $\mu_{\bar{x}} = 2.5$, while $\sigma_{\bar{x}} = 0.6455$. We would like to be able to know these values without having to compute them using the entire sampling distribution.

In comparing the mean and standard error (or *deviation*) from the sampling distribution, $\mu_{\bar{x}} = 2.5$, and $\sigma_{\bar{x}} = 0.6455$, to the mean and standard deviation of the parent population, $\mu = 2.5$, and $\sigma = 1.11803$, one might quickly observe that $\mu_{\bar{x}} = \mu$. This is convenient,

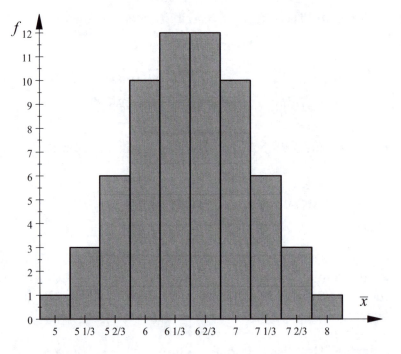

Figure 6.9: With samples of size $n = 3$, there are $4^3 = 64$ samples. The distribution is symmetric, with *kurtosis* $= 2.55$.

and in fact, it will always be the case. The proof of this is simple, but requires a reminder of the properties of expected value and variance.

Theorem 6.4.1 Properties of Expected Value and Variance

1. $E(k \cdot x) = k \cdot E(x)$, where k is any constant.

2. $E(x + y) = E(x) + E(y)$, and thus $E(\sum x) = \sum E(x)$.

3. $Var(k \cdot x) = k^2 \cdot Var(x)$

4. $Var(x + y) = Var(x) + Var(y)$, and thus $Var(\sum x) = \sum Var(x)$.

5. If x is normally distributed, then $\sum x$ is normally distributed.

The proof that $\mu_{\bar{x}} = \mu$ is relatively straight forward.

$$
\begin{aligned}
\mu_{\bar{x}} &= E(\bar{x}) \\
&= E\left(\frac{\sum x}{n}\right) \\
&= \frac{1}{n} \cdot E\left(\sum x\right) \\
&= \frac{1}{n} \sum E(x) \\
&= \frac{1}{n} \sum \mu \\
&= \frac{1}{n} \cdot n \cdot \mu \\
&= \mu
\end{aligned}
$$

It is important to note that, in this line of logic, we see that $E(\bar{x}) = \mu$. So the mean, or expected value, of the sample means is equal to the population mean which they are estimating. This tells us the important fact that $\bar{x}$ is an unbiased point estimate of μ. We refer to $\bar{x}$ as a *point estimate* of μ, as it represents a single point on a number line. Later, we will introduce confidence intervals for μ, and these will be referred to as interval estimates for μ.

The difficulty arises when we notice that $\sigma_{\bar{x}} \neq \sigma$, and yet we still need a way to determine $\sigma_{\bar{x}}$ without actually having to compute it from the sampling distribution directly. The key comes in the proof below which looks at the variances $\sigma_{\bar{x}}^2$ and σ^2.

$$
\begin{aligned}
\sigma_{\bar{x}}^2 &= Var(\bar{x}) \\
&= Var\left(\frac{\sum x}{n}\right) \\
&= \frac{1}{n^2} Var\left(\sum x\right) \\
&= \frac{1}{n^2} \sum Var(x) \\
&= \frac{1}{n^2} \sum \sigma^2 \\
&= \frac{1}{n^2} \cdot n \cdot \sigma^2 \\
&= \frac{\sigma^2}{n}
\end{aligned}
$$

In the step above which uses $Var\left(\sum x\right) = \sum Var(x)$, we emphasize that this is only valid when the x values are independent, and this is only true if sampling is done *with replacement*.

So $\sigma_{\bar{x}}^2 = \frac{\sigma^2}{n}$, and taking square roots of each side gives $\sigma_{\bar{x}} = \frac{\sigma}{\sqrt{n}}$.

If we apply this to our sampling distribution, where $\sigma = 1.11803$, and the sample sizes were all $n = 3$, we can compute $\sigma_{\bar{x}} = \frac{\sigma}{\sqrt{n}} = \frac{1.11803}{\sqrt{3}} = 0.6455$. This is exactly the same value for $\sigma_{\bar{x}}$ obtained above by directly computing the standard error (or *deviation*) of the sampling distribution, but this time we did it without consulting the distribution at all.

6.4.3 The Central Limit Theorem for Sample Means

We have done all that we needed to do with respect to the sampling distribution of all sample means from samples of size n. We summarize our findings below.

Theorem 6.4.2 The Central Limit Theorem for Sample Means
In sampling from a parent population whose mean and standard deviation are μ and σ respectively, consider the sampling distribution of sample means – the collection of all sample means, $\bar{x}$, from samples of size n, where samples are drawn with replacement. The sampling distribution of sample means is approximately normal as long as either $n > 30$, or the parent population is normal. This sampling distribution has mean, $\mu_{\bar{x}}$, and standard error $\sigma_{\bar{x}}$, and the following facts can be proven:

1. The mean of the sampling distribution of sample means is given by

$$\mu_{\bar{x}} = \mu.$$

   This tells us that $\bar{x}$ is an *unbiased point estimate* of μ.

2. The standard deviation of the sampling distribution of sample means is $\sigma_{\bar{x}} = \frac{\sigma}{\sqrt{n}}$. Since a sample mean is an estimate of the population mean, and any deviation from the population mean is error, this quantity is often referred to as the *standard error* of the sample means.

The Central Limit Theorem assumes that samples are drawn with replacement. This is the requirement, but when populations are very large, the theorem is approximately true even when sampling is done without replacement. Only when populations are small, and we sample without replacement, do we need to alter the theorem. In such cases the formula for $\sigma_{\bar{x}}$ changes to

$$\sigma_{\bar{x}} = \frac{\sigma}{\sqrt{n}} \cdot \sqrt{\frac{N-n}{N-1}},$$

where N is the population size. This formula is rarely used. Whenever the sample size is less than 5% of the population size, the factor

$$\sqrt{\frac{N-n}{N-1}}$$

is so close to 1 that it has very little effect on the result, and is usually ignored.

As we have implied, when sample size increases, the distribution of sample means approaches a continuous normal distribution. Figures 6.10, 6.11, 6.12, and 6.13 demonstrate

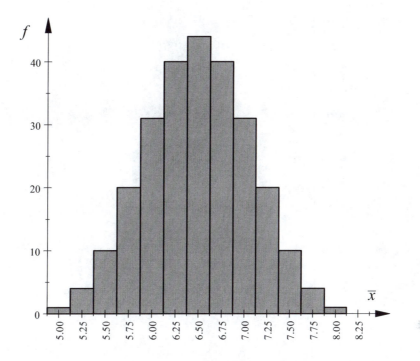

Figure 6.10: With samples of size $n = 4$, there are $4^4 = 256$ samples. The distribution is symmetric, with *kurtosis* = 2.66.

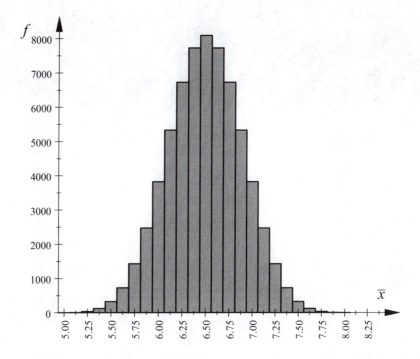

Figure 6.11: With samples of size $n = 8$, there are $4^8 = 65\,536$ samples. The distribution is symmetric, with *kurtosis* = 2.83. Distances between sample means are $\frac{1}{8}$.

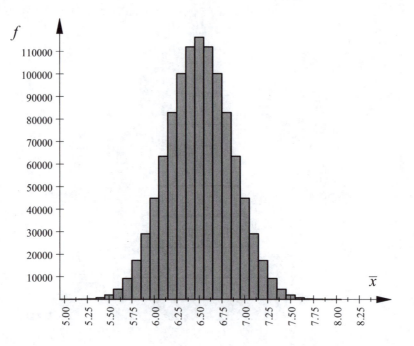

Figure 6.12: With samples of size $n = 10$, there are $4^{10} = 1\,048\,576$ samples. The distribution is symmetric, with *kurtosis* $= 2.86$. Distances between sample means are $\frac{1}{10}$.

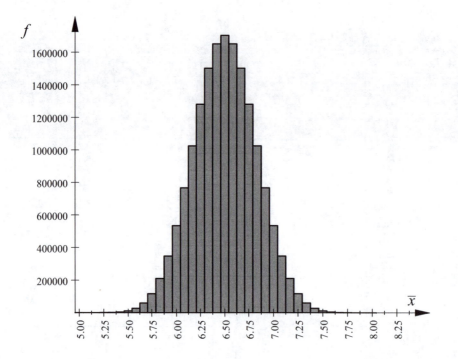

Figure 6.13: With samples of size $n = 12$, there are $4^{12} = 16\,777\,216$ samples. The distribution is symmetric, with *kurtosis* $= 2.89$. Distances between sample means are $\frac{1}{12}$, so the distribution is approaching continuity.

this progression toward normality as samples from the population $\{5, 6, 7, 8\}$ increase from $n = 4$ to $n = 12$.

In Figure 6.13, the sample means from samples of size $n = 12$ are summarized. If we plot the corresponding distribution of discrete probabilities, the graph is the same, but the total area of the rectangles is 1. The probability distribution is plotted in Figure 6.14 with the corresponding normal distribution density function. The graph is symmetric, and the *kurtosis* is equal to 2.89, nearly equal to the normal standard of 3. The two graphs align quite nicely.

Now that we have all of the main descriptors for this sampling distribution, we may develop formulas for the confidence interval for μ, and the test statistic for $\bar{x}$.

The general form for confidence intervals is $x - E < \mu < x + E$, where x is an unbiased estimate of μ, normally distributed, and $E = z_{\alpha/2}\sigma$. If we substitute $\bar{x}$ for x, then we must replace μ with the proper mean, $\mu_{\bar{x}}$, and σ with the proper standard error, $\sigma_{\bar{x}}$. The confidence interval becomes

$$\bar{x} - E < \mu_{\bar{x}} < \bar{x} + E,$$

but since $\mu_{\bar{x}} = \mu$, this becomes

$$\bar{x} - E < \mu < \bar{x} + E,$$

and the error formula becomes

$$E = z_{\alpha/2}\sigma_{\bar{x}}$$
$$= z_{\alpha/2}\frac{\sigma}{\sqrt{n}},$$

since

$$\sigma_{\bar{x}} = \frac{\sigma}{\sqrt{n}}.$$

We summarize this below, but wait until the next section before providing examples.

Summary 6.4.3 The Confidence Interval for a Population Mean is:

$$\bar{x} - E < \mu < \bar{x} + E,$$

where

$$E = z_{\alpha/2}\frac{\sigma}{\sqrt{n}}.$$

And using the substitutions described above, the z-score becomes a test statistic:

$$z = \frac{\bar{x} - \mu_{\bar{x}}}{\sigma_{\bar{x}}}$$
$$= \frac{\bar{x} - \mu}{\sigma/\sqrt{n}}.$$

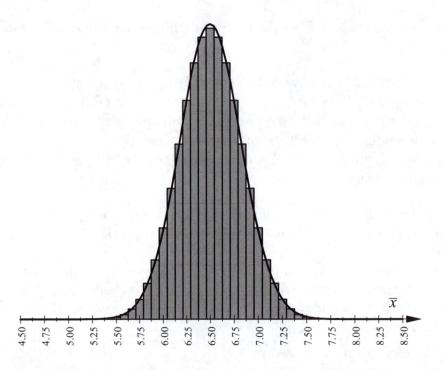

Figure 6.14: The probability distribution of sample means, $\overline{x}$, from all samples of size $n = 12$, drawn with replacement from the population $\{5, 6, 7, 8\}$. The distribution is symmetric, with *kurtosis* = 2.89. The corresponding density function for the normal distribution is superimposed over the histogram.

Summary 6.4.4 The Test Statistic for a Sample Mean
The z-score of a sample mean is

$$z = \frac{\bar{x} - \mu}{\sigma/\sqrt{n}}.$$

As before, we use test statistics to determine when a value, in this case a sample mean, is unusually high or low.

Example 6.4.5 The mean IQ score is equal to 100, and the standard deviation is 15. Suppose 65 randomly selected adults attend a course on logic and reasoning. After the course, they all take an IQ test and a mean IQ score of $\bar{x} = 107.64$ is the result. Is this sample mean significantly greater than the population mean of $\mu = 100$?

To determine this, we compute the z-score of our sample mean:

$$\begin{aligned}
z &= \frac{\bar{x} - \mu}{\sigma/\sqrt{n}} \\
&= \frac{107.64 - 100}{15/\sqrt{65}} \\
&= 4.11.
\end{aligned}$$

So the sample mean is 4.11 standard deviations above the population mean. That seems rather significant. A more sophisticated measure of significance is the *p-value*, which is, as mentioned previously, the probability of measuring a value at least as extreme as the one we have. In this case, using Table A1, the *p-value* is

$$
\begin{aligned}
p\text{-}value &= P\left(\bar{x} > 107.64\right) \\
&= P\left(z > 4.11\right) \\
&= 1 - P\left(z < 4.11\right) \\
&= 1 - 0.9999 \\
&= 0.0001.
\end{aligned}
$$

Note that at the bottom of Table A1 the words, *For values of z greater than 4.09, use 0.9999 for the area.* This is what we have done. Now, when probabilities are very small for events that have occurred, the event is *statistically significant*. This particular *p*-value indicates that the probability of exceeding this sample mean is 0.0001, which is very, very small. We conclude that this is highly statistically significant, and that the logic and reasoning course appears to be effective in raising the average test score.

Once again, we have used a *p-value* to determine when a difference is significant between an assumed population parameter, in this case μ, and an associated statistic, in this case $\bar{x}$. Again, we have determined that the difference was significant, and that the treatment (the logic course) was effective. In doing so, we have conducted another hypothesis test, albeit in a rather informal setting. As mentioned previously, hypothesis tests will be examined closely beginning in Chapter 7.

6.4.4   Homework

For problems 1 through 4, consider the population of values $x : \{1, 2, 3, 4\}$, whose mean is $\mu = 2.5$, and whose standard deviation is $\sigma = 1.1180$. This distribution is uniformly distributed, with all outcomes equally likely.

1. List all samples from this population of size $n = 2$. Remember that we must sample with replacement, so that any population member may be selected as many as two times. This means that there will be sixteen such samples.

2. From each of the sixteen samples above, compute the sample mean, $\bar{x}$. This population of sample means is the sampling distribution of sample means.

3. The population of sample means above has mean $\mu_{\bar{x}}$ and standard error (or *deviation*) $\sigma_{\bar{x}}$. Compute this mean and standard error using the mean and population standard deviation functions built into your calculator.

4. Recalling that $n = 2$, $\mu = 2.5$, and $\sigma = 1.1180$, verify that $\mu_{\bar{x}} = \mu$, and that $\sigma_{\bar{x}} = \sigma/\sqrt{n}$.

For problems 5 & 6, use the test statistic $z = \frac{\bar{x} - \mu}{\sigma/\sqrt{n}}$ to compute the z-score of the given sample mean, then look up the appropriate probability.

5. The mean height for men in Sweden is 5.709 feet, with a standard deviation of 0.197 feet. Suppose we select 5 men from Sweden randomly and compute a sample mean. What is the probability that this mean is greater than 6 feet? (Source: L. J. Launer National Institute of Public Health and Environmental Protection, RIVM/CCM, PO Box 1, 3720 BA, Bilthoven, The Netherlands).

6. The mean height for women in Sweden is 5.282 feet, with a standard deviation of 0.164 feet. Suppose we select 5 women from Sweden randomly and compute a sample mean. What is the probability that this mean is greater than 6 feet? (Source: L. J. Launer National Institute of Public Health and Environmental Protection, RIVM/CCM, PO Box 1, 3720 BA, Bilthoven, The Netherlands).

6.5 Intervals for a Population Mean (σ Known)

In the previous section, we developed formulas for the confidence interval for a population mean. Confidence intervals for a population mean are used when a research question is being answered with quantitative data – where we wish to know the value of the population's primary summary value, its mean, μ.

The unfortunate fact about this confidence interval is that the formula for the error requires a value that we are unlikely to know – the population standard deviation, σ. If we do not know the value of the population mean, we certainly would not know the population standard deviation. Even so, in this section we proceed with the assumption that we *do* know σ, and using this, construct confidence intervals for μ. In the section that follows, we will rectify this situation by introducing a method for computing a confidence interval for μ when σ is *unknown*.

The requirement for the procedure in this section (other than σ being known) is the requirement of the Central Limit Theorem for sample means – that is, either $n > 30$, or the population is normal. This procedure is *robust*, meaning that even when the requirements of this procedure are not perfectly true, the results still tend to be reliable.

6.5.1 Confidence Intervals for a Population Mean (σ Known)

Procedure 6.5.1 Constructing a Confidence Interval for a Population Mean (σ Known)

Step I. Verify that $n > 30$, or the population is normal. Gather $\bar{x}$, σ, n, and the level of confidence. Compute $\alpha = 1 - (\text{confidence level})$.

Step II. Look up $z_{\alpha/2}$ on Table A1.

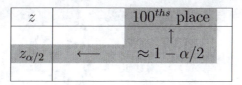

Step III. Compute the maximum error in the estimate:

$$E = z_{\alpha/2}\frac{\sigma}{\sqrt{n}}.$$

Step IV. Compute the endpoints of the confidence interval:

$$\bar{x} - E < \mu < \bar{x} + E$$

Step V. As an interpretation, we may say that the proportion of such intervals that actually contain the unknown mean, μ, is equal to the level of confidence.

Example 6.5.2 In 2005 a random sample of 20 counties in California yielded a mean of 17.2 days where air quality was unhealthy for sensitive individuals, with an assumed population standard deviation of 16.3. Construct a 94% confidence interval for the mean number of days in California where air quality was unhealthy for sensitive individuals. An analysis of the data shows that the data appear to be normal (which is required since the sample size is less than 30).

We proceed as follows.

Step I. We have assumed that our population is normal. $\bar{x} = 17.2$, $\sigma = 16.3$, $n = 20$, & the level of confidence is 0.94. Also, $\alpha = 1 - (level\ of\ confidence) = 1 - 0.94 = 0.06$.

Step II. We compute $1 - \alpha/2 = 1 - 0.06/2 = 0.97$, and so we seek 0.97 in Table A1. The closest area to 0.97 is 0.9699, which corresponds to $z_{\alpha/2} = 1.88$.

Looking up $z_{\alpha/2}$ on Table A1.			
z	.07	**.08**	.09
1.7	.9616	.9625	.9633
1.8	.9693	**.9699**	.9706
1.9	.9756	.9761	.9767

Step III. The maximum error is $E = z_{\alpha/2}\frac{\sigma}{\sqrt{n}} = 1.88 \cdot \frac{16.3}{\sqrt{20}} = 6.85$.

Step IV. The confidence interval is last:

$$\bar{x} - E\ <\ \mu < \bar{x} + E$$
$$17.2 - 6.85\ <\ \mu < 17.2 + 6.85$$
$$10.35\ <\ \mu < 24.05$$

Step V. To conclude, we will say with 94% confidence that the mean number of days where air quality was unhealthy in California is between 10.35 and 24.05. 94% of such intervals will actually contain the true mean.

Example 6.5.3 Returning again to the Galapagos Island elevations, we have a mean of 412.24 feet, from a sample of size $n = 25$, with an assumed population standard deviation of $\sigma = 447.60$ feet. We will construct a 99% confidence interval for the mean elevation. Again, since the sample size is less than 30, we assume the population is normal.

Step I. We have assumed that the population is normal. $\bar{x} = 412.24$, $\sigma = 447.60$, $n = 25$, and the level of confidence is 99% = 0.99. Also, $\alpha = 1 - (level\ of\ confidence) = 1 - 0.99 = 0.01$.

Step II. We can save steps on finding $z_{\alpha/2}$ if we note the few special cases at the bottom of Table A1. The small table reproduced there shows that, for a confidence level of 99% = 0.99, the critical value is $z_{\alpha/2} = 2.576$.

Conf. Level	Critical Value
0.90	1.645
0.95	1.960
0.99	**2.576**

Step III. The maximum error is:

$$E = z_{\alpha/2}\frac{\sigma}{\sqrt{n}}$$
$$= 2.576 \cdot \frac{447.60}{\sqrt{25}}$$
$$= 230.60.$$

Step IV. And the confidence interval:

$$\bar{x} - E < \mu < \bar{x} + E$$
$$412.24 - 230.60 < \mu < 412.24 + 230.60$$
$$181.64 < \mu < 642.84$$

Step V. We are 99% confident that the mean elevation of the Galapagos Islands is between 181.64 and 642.84.

6.5.2 Sample Size When Estimating a Population Mean

The maximum error formula, $E = z_{\alpha/2}\frac{\sigma}{\sqrt{n}}$, for a sample mean is the relationship between maximum error and sample size. For a given sample size, the formula returns the maximum error. If we wish to release control over sample size, and gain control over the maximum error, we should solve the equation for the sample size:

$$E = z_{\alpha/2}\frac{\sigma}{\sqrt{n}}$$
$$\sqrt{n} = \frac{z_{\alpha/2} \cdot \sigma}{E}$$
$$n = \left(\frac{z_{\alpha/2} \cdot \sigma}{E}\right)^2$$

Using this formula, we may determine the sample size associated with a chosen value for E. Once again, we have a formula that requires a value, namely σ, that we almost certainly will not know, so we are forced to estimate σ by using a sample standard deviation from a pilot study.

Summary 6.5.4 Sample Size When Estimating a Population Mean
If we want to estimate a population mean with an error that is less than a chosen maximum error, E, we must sample at least:

$$n = \left(\frac{z_{\alpha/2} \cdot \sigma}{E}\right)^2,$$

where we round this value up to a whole number, greater than or equal to this value, and σ is estimated using a sample standard deviation from a pilot study.

Example 6.5.2 Continued: Suppose we wish to lower our maximum error from 6.5.2 to 3 days – still with 94% confidence. We continue to estimate that $\sigma = 16.3$. With the confidence level unchanged, our value for $z_{\alpha/2}$ remains unchanged, $z_{\alpha/2} = 1.88$.

$$n = \left(\frac{z_{\alpha/2} \cdot \sigma}{E}\right)^2 = \left(\frac{1.88 \cdot 16.3}{3}\right)^2 = 104.3,$$

We *round this up* to $n = 105$.

6.5.3 Homework

In Table A1, look up $z_{\alpha/2}$ corresponding to the given level of confidence.

1. 97% confidence

2. 93% confidence

In problems 3 through 5, compute confidence intervals for a population mean, using the given data.

3. Confidence level: 95%, $\bar{x} = 13.54$, $\sigma = 3.02$, $n = 45$.

4. Confidence level: 98%, $\bar{x} = 105.32$, $\sigma = 12.13$, $n = 200$.

5. Confidence level: 94%, $\bar{x} = 3.2$, $\sigma = 0.061$, $n = 15$, the sample appears to come from a normal population.

In problems 6 and 7, compute the minimum sample size corresponding to the given maximum error.

6. Confidence level: 90%, $\sigma = 0.218$, $E = 0.09$.

7. Confidence level: 98%, $\sigma = 0.218$, $E = 0.09$.

Compute the confidence interval or sample size as requested.

8. For the 100 speed of light in air measurements discussed previously,, the mean is 299.8524 thousands of kilometers per second, with an assumed population standard deviation of 0.07901. Compute the 99% confidence interval for the population mean of such measurements. The speed of light in air is currently estimated at 299.7244 thousand kilometers per second. Does your confidence interval contain this true (*supposed*) population mean? Can you explain any disagreement?

9. The mean elevation of 25 randomly selected (with replacement) Galapagos Islands is 412.24, with an assumed population standard deviation equal to 447.61. Assuming that this sample is measured from a normal population, construct a 94% confidence interval for the mean elevation of all of the Galapagos Islands.

10. The mean of 52 randomly selected donation amounts to a charitable organization is $1037.54, with an assumed population standard deviation of $283.06. Compute the 95% confidence interval for the mean of all contributions to this organization.

11. Using the statistics of problem #8 as a pilot study, use the given standard deviation with the same level of confidence to give the sample size necessary to bring the maximum error in the estimate of μ down to 0.01 thousand kilometers per second.

12. Using the statistics of problem #9 as a pilot study, use the given standard deviation with the same level of confidence to give the sample size necessary to bring the maximum error in the estimate of μ down to 150.

13. Using the statistics of problem #10 as a pilot study, use the given standard deviation with the same level of confidence to give the sample size necessary to bring the maximum error in the estimate of μ down to $60.00.

14. Below statistics are provided for the mean study hours per week for 32 randomly selected statistics students, along with an assumed population standard deviation. Compute a 94% confidence interval for the mean of all such students. These values were gathered by Mt. San Antonio College honors student Jessica Yang.

Statistics Students	
n	32
Sample Mean	2.519
Population Standard Deviation	1.359

15. Below statistics are provided for the mean and standard deviation of study hours per week for 32 randomly selected calculus students. Compute a 95% confidence interval for the mean of all such students. These values were gathered by Mt. San Antonio College honors student Jessica Yang.

Calculus Students	
n	32
Sample Mean	3.625
Population Standard Deviation	1.431

16. From a random sample of 99 UCLA college students, a mean GPA of 3.463 is observed, with an assumed population standard deviation of 0.298. Construct a 98% confidence interval for the mean GPA of students at UCLA.

17. From a random sample of 99 university students, the mean number of college football games attended in a given season was 5.03, with an assumed population standard deviation of 2.58. Construct a 97% confidence interval for the mean number of college football games attended per season.

6.6 Intervals for a Population Mean (σ Unknown)

In the previous section, we gave a method for computing a confidence interval for a population mean, where we required that the population standard deviation, σ, was known. This assumption is, of course, not likely to be true, especially in the context where we are computing a confidence interval for an unknown population mean, μ. If μ is unknown, there is really no possibility that σ is known, so we are forced to estimate σ by using a sample standard deviation, s. This will be our approach, but what may not be obvious is the fact that we can no longer use Table A1 in this case.

When we first introduced the idea of standardizing a normal population of values, x, we stated that taking z-scores gives a new population of z-scores, normally distributed, where the mean is always zero, and the standard deviation is always one. This is true whenever we take z-scores, including when we *standardize* a sample mean with the z-score (or *test statistic*)

$$z = \frac{\bar{x} - \mu}{\sigma/\sqrt{n}}.$$

Using this formula, if we take the z-score of all sample means taken from samples of size n, we create a normal population whose mean is zero, and whose standard deviation is one. The problem is, again, that we are unlikely to know the value for σ, and so we use a sample standard deviation, s, to estimate σ.

In practice, the test statistic is generally computed as

$$\frac{\bar{x} - \mu}{s/\sqrt{n}},$$

and it would seem like the problem is solved ... but unfortunately it is not. When computing

$$z = \frac{\bar{x} - \mu}{\sigma/\sqrt{n}},$$

this statistic only can assume a single value for a given value of $\bar{x}$, because μ, σ, and n are constants. But when we use s in place of σ, there may be many samples whose mean is equal to $\bar{x}$, all with different sample standard deviations, s. So for a given $\bar{x}$, the statistic

$$\frac{\bar{x} - \mu}{s/\sqrt{n}}$$

will vary, and because of this added variability, the standard deviation of its varying values is greater than the standard deviation of

$$z = \frac{\bar{x} - \mu}{\sigma/\sqrt{n}}$$

which is equal to one. In fact, as the sample size gets smaller, the standard deviation of the values of the statistic

$$\frac{\bar{x} - \mu}{s/\sqrt{n}}$$

gets larger. This is all to say that the statistic,

$$\frac{\bar{x} - \mu}{s/\sqrt{n}},$$

does not behave like a z-score, whose standard deviation is always one.

A solution to this problem was found in the probability distribution developed by statistician William Gosset (whose pen name was Student, pictured in Figure 6.15), an employee of Guinness Brewery in Dublin. In 1908, through mathematical deduction and empirical data, Gosset published a probability distribution, which became known as Student's t-distribution.

Figure 6.15: William "Student" Gosset, the developer of *Student's* t-distribution. (Photograph is from Gosset's obituary in the *Annals of Eugenics*, now in the public domain).

This distribution was later adapted by Ronald A. Fisher to describe the distribution of the statistic described above, known as the t-test statistic,

$$t = \frac{\bar{x} - \mu}{s/\sqrt{n}}.$$

We do not refer to the t statistic as a standardized variable (as we do exclusively with z-scores). Instead, we refer to t values as *studentized*, after the pen name of Gosset. The distribution is known as Student's and not Gosset's, because Gosset's employer, Guinness Brewery, did not allow employees to publish because they were afraid that employees would publish the company's trade secrets. Gosset's distribution was not originally understood particularly well. Initially, the degrees of freedom issue was misunderstood. This was resolved, and the tables of values refined by the great statistician Ronald A. Fisher. We will learn more about R. A. Fisher in Chapter 11.

The t-distribution is symmetric with respect to the mean, which is equal to zero, and the standard deviation is greater than 1. The standard deviation decreases, approaching 1, as the degrees of freedom increases. A reminder of the meaning of degrees of freedom is in order. In a sample of size n, there are n deviations from the mean $x - \bar{x}$. We mentioned that the sum of the n deviations from the mean is always equal to zero, so that the sum of the first $n - 1$ deviations from the mean can be any value, but the last deviation must be the negative of this value, so that the sum of all n deviations is equal to zero.

Because of this, we say that a sample standard deviation has $n - 1$ degrees of freedom, and as stated, the standard deviation of values

$$t = \frac{\bar{x} - \mu}{s/\sqrt{n}}$$

decreases as the degrees of freedom increase, approaching a standard deviation of one. As degrees of freedom increase, the t-critical values approach corresponding z-critical values. We use degrees of freedom, not sample size, to measure the variability in this statistic because the added variability due to using a sample standard deviation in the denominator is caused by the number of deviations that vary randomly. The last deviation does not vary randomly, so it does not contribute to the random variation of the statistic.

In practical terms, all that we need to remember is that when σ is unknown, we must look-up $t_{\alpha/2}$ in Table A2, instead of $z_{\alpha/2}$ (which we have been using when σ was known). It is important to note that the last row of Table A2 gives z critical values for commonly used α values, so the use of Table A1 for looking-up critical z values is usually unnecessary. After this, we only need to concern ourselves with the two following facts. First, when computing a confidence interval when σ is unknown, we will adjust our error formula to use $t_{\alpha/2}$ instead of $z_{\alpha/2}$, and s instead of σ:

$$E = t_{\alpha/2} \frac{s}{\sqrt{n}}.$$

Second, when computing a test statistic for a sample mean when σ is unknown, we use

$$t = \frac{\bar{x} - \mu}{s/\sqrt{n}}.$$

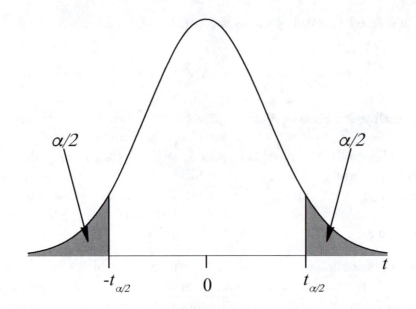

Figure 6.16: The geometry of the confidence interval on Student's t distribution.

To use Table A2 for a confidence interval, we locate the row with the appropriate degrees of freedom, $n - 1$, and find α under the *Two-Tail Application* heading (confidence intervals divide α by two, distributing it over two tails). For example, suppose our level of confidence is 95%, so $\alpha = 1 - 0.95 = 0.05$, and suppose that $n = 27$, so that we have $n - 1 = 26$ degrees of freedom. To look up $t_{\alpha/2}$ we find the column heading for $\alpha = 0.05$ under *Two-Tail Applications*, and we look for a row heading of $df = n - 1 = 26$, and these correspond to $t_{\alpha/2} = 2.056$.

	Two Tail Applications					
df	$\alpha = .01$	$\alpha = .02$	$\alpha = .05$	$\alpha = .10$	$\alpha = .20$	$\alpha = .40$
25	2.787	2.485	2.060	1.708	1.316	.856
26	2.779	2.479	**2.056**	1.706	1.315	.856
27	2.771	2.473	2.052	1.703	1.314	.855

The procedure below is used for computing a confidence interval for a population mean when σ is unknown. The procedure requires that either the sample size $n > 30$, or the population is normal. This procedure is robust, so if these rules are not exactly true, the confidence interval is still likely to be true.

6.6.1   Confidence Intervals for a Population Mean (σ Unknown)

Procedure 6.6.1 Constructing a Confidence Interval for a Population Mean (σ Unknown)

   Step I. Verify that $n > 30$, or the population is normal. Gather $\bar{x}$, s, n, and the level of confidence. Additionally, we must compute $\alpha = 1 - (\textit{confidence level})$.

   Step II. Look up $t_{\alpha/2}$ on Table A2 – make sure to find α under Two-Tail Applications.

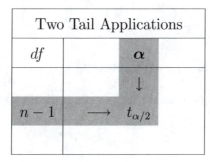

   Step III. Compute the maximum error in the estimate:

$$E = t_{\alpha/2}\frac{s}{\sqrt{n}}.$$

   Step IV. Compute the endpoints of the confidence interval:

$$\bar{x} - E < \mu < \bar{x} + E.$$

   Step V. As an interpretation, we should say that the proportion of such intervals that actually contain the unknown mean, μ, is equal to the level of confidence.

Example 6.6.2 We return to the $n = 52$ weekly contributions to a charitable organization discussed in Chapter 2. From this sample we computed a mean contribution of $\bar{x} = 1037.54$, and a sample standard deviation of $s = 283.06$. Here the sample size is greater than 30, so normality is not required (even though the histogram did appear roughly bell-shaped). We would like to compute a 99% confidence interval for the population mean.

We proceed through the steps as follows.

   Step I. We have already verified that $n > 30$. $\bar{x} = 1037.54$, $s = 283.06$, $n = 52$ and the *level of confidence* $= 0.99$. Using this, we compute $\alpha = 1 - 0.99 = 0.01$.

Step II. From Table A2 we look up $t_{\alpha/2}$. There is no $df = 51$, so we will pick the closest with 50 degrees of freedom. So we will use $t_{\alpha/2} = 2.678$.

Two Tail Applications		
df	$\alpha = .01$	$\alpha = .02$
49	2.680	2.405
50	**2.678**	2.403
75	2.643	2.377

Step III. The maximum error in the estimate is: $E = t_{\alpha/2}\frac{s}{\sqrt{n}} = 2.678 \cdot \frac{283.06}{\sqrt{52}} = 105.12$.

Step IV. Now we compute the endpoints of the confidence interval:

$$\bar{x} - E < \mu < \bar{x} + E$$

$$1037.54 - 105.12 < \mu < 1037.54 + 105.12$$

$$932.42 < \mu < 1142.66$$

Step V. We may interpret this by saying that we have 99% confidence that the mean is between 932.42 and 1142.66, and 99% of such intervals actually contain the mean.

Example 6.6.3 Previously, we gave the miles per gallon measurements for ten round-trip commutes in a Toyota Prius Hybrid vehicle from Lake Arrowhead to Walnut, California. The miles per gallon measurements are: {46.2, 44.9, 47.0, 46.3, 45.6, 46.0, 44.6, 45.9, 46.3, 45.6}. The mean of these values is 45.84, and the standard deviation is 0.70. We will assume that these values are measured from a normal population. We will compute a 95% confidence interval for the mean miles per gallon.

The steps are as follows.

Step I. This time n is not greater than 30, but we are assuming the population is normal, so we may proceed. $\bar{x} = 45.84$, $s = 0.70$, $n = 10$ and the *level of confidence* $= 0.95$. Also, $\alpha = 1 - 0.95 = 0.05$

Step II. This time $df = n - 1 = 9$, and we look up $t_{\alpha/2}$ on Table A2, where we find $t_{\alpha/2} = 2.262$.

		Two Tail Applications	
df	$\alpha = .02$	$\boldsymbol{\alpha = .05}$	$\alpha = .10$
8	2.896	2.306	1.860
9	2.821	**2.262**	1.833
10	2.764	2.228	1.812

Step III. The maximum error in the estimate is: $E = t_{\alpha/2}\frac{s}{\sqrt{n}} = 2.262 \cdot \frac{0.70}{\sqrt{10}} = 0.501$.

Step IV. Now we compute the endpoints of the confidence interval:

$$\bar{x} - E < \mu < \bar{x} + E$$

$$45.84 - 0.501 < \mu < 45.84 + 0.501$$

$$45.34 < \mu < 46.34$$

Step V. We may interpret this by saying that we have 95% confidence that the mean is between 45.34 miles per gallon and 46.34 miles per gallon, and 95% of such intervals actually contain the mean.

6.6.2 Sample Size

In the previous section, we used the formula $n = \left(\frac{z_{\alpha/2} \cdot \sigma}{E}\right)^2$ to determine the sample size necessary to estimate a population mean to within a given maximum error, E. It is very unlikely that we will know a value for σ, but even so, we will not update this formula with a $t_{\alpha/2}$ value, as looking this up requires a sample size, n, which has not been determined yet. We will continue to use $z_{\alpha/2}$ with the hope that the resulting sample size will be large enough so that $t_{\alpha/2}$ is approximately equal to $z_{\alpha/2}$.

6.6.3   Homework

Look up $t_{\alpha/2}$ corresponding to the given level of confidence and sample size.

1. 94% confidence level; $n = 38$.

2. 98% confidence level; $n = 51$.

Compute the maximum error in the estimate of the population mean. Assume populations corresponding to small sample sizes are normal.

3. 96% confidence level; $n = 11$; $s = 4.36$.

4. 92% confidence level; $n = 76$; $s = 151.27$.

Compute the confidence interval for the population mean using the given information. Assume populations corresponding to small sample sizes are normal.

5. 99% confidence level; $n = 27$; $\bar{x} = 117.83$; $s = 45.21$.

6. 95% confidence level; $n = 37$; $\bar{x} = 1.15$; $s = 0.045$.

Compute the requested confidence intervals.

7. From a sample consisting of lifetimes (in thousands of cycles) of 101 rectangular strips of aluminum subjected to repeated alternating stress at 21,000 psi, 18 cycles per second, the mean is 1399.75, with a standard deviation of 389.31. (These values are from the Journal of the American Statistical Association, 1958, p. 159). Compute the 98% confidence interval for the mean lifetime of all such rectangular strips.

8. A random sample of 30 students at Mt. San Antonio College in Walnut, California, gathered hours worked per week, yielded a mean of 23.7 hours, with a sample standard deviation of 13.8 hours. Assuming that these values were sampled from a normal population, compute the 94% confidence interval for the mean number of hours worked by students at this college.

9. The data which follow are from an article on peanut butter from *Consumer Reports* (Sept. 1990), which reported the following scores for various brands: {22, 30, 30, 36, 39, 40, 40, 41, 44, 45, 50, 50, 53, 56, 56, 56, 62, 65, 68}. Assuming these values are sampled from a normal population, compute the 95% confidence interval for the population mean. (You will need to compute the sample mean and standard deviation of these values first!)

10. The following data are overall grades for your instructor's statistics students gathered from Fall, 2005. { 59.3, 60.2, 68.4, 69.1, 70.8, 71.7, 72.3, 73.9, 75.1, 76.1, 76.1, 80.1, 84.9, 85.3, 86, 88.1, 88.3, 93, 94, 98.6 }. Assuming that these values were sampled from a normal population, compute the 98% confidence interval for the mean overall grade for your instructor's statistics students.

11. The sample below includes heights for 30 randomly selected male college students. Assuming that these values were sampled from a normal population, construct a 96% confidence interval for the mean height of all male college students.

Height for Males in Centimeters					
167.7	171.2	168.0	172.8	179.4	179.1
164.1	170.6	172.7	177.6	174.8	176.3
159.8	168.6	169.7	178.9	175.5	178.4
165.4	168.5	176.3	177.0	182.2	177.1
166.6	172.4	174.1	175.5	178.3	189.1

12. Below are hours spent on the internet per week for 31 randomly selected male college students. Construct a 96% confidence interval for the mean hours spent on the internet per week for all male college students. These values were gathered by Mt. San Antonio College honors student Warren Hsiao.

Time on Internet – Males						
10	42	54	73	34	54	25
40	42	59	98	36	33	95
42	36	51	45	45	66	64
23	32	105	47	43	21	16
	25		55		70	

13. Below are hours spent on the internet per week for 31 randomly selected female college students. Construct a 96% confidence interval for the mean hours spent on the internet per week for all female college students. These values were gathered by Mt. San Antonio College honors student Warren Hsiao.

Time on Internet – Females						
76	27	41	7	48	12	89
25	42	23	44	54	65	54
15	21	32	57	25	37	24
7	47	8	21	25	41	78
	31		42		52	

14. Based on your findings in the previous two problems, is it plausible that male and female college students spend, on average, the same amount of time on the internet per week?

Chapter 7

Single Parameter Hypothesis Tests

In Chapter 1, we discussed *the process of statistical analysis*, summarized in Figure 7.1. Each step of this process has been covered, but we have only partially discussed the last, *statistical inference*.

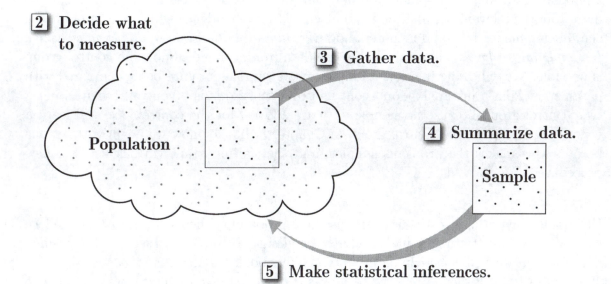

Figure 7.1: The process of statistical analysis.

As we continue with the last step, *statistical inference*, we use this chapter to introduce *hypothesis testing*. With a hypothesis test, we use sample data to address a *research question* regarding the validity of a claim about a population parameter.

Several of the ideas used in hypothesis tests have been discussed already, namely the *p-value*, the *test statistic*, and the idea of *unusualness* or *significance*. These ideas will be used throughout the remainder of the text, as we apply them to various applications.

Recall the concept of the sampling distribution. There are many possible estimates of any population parameter. Remember also that the estimates which we are primarily concerned with, sample proportions and sample means, are approximately normal in their distributions as long as the conditions of their respective central limit theorems are satisfied. This distribution tells us that estimates of population parameters tend to be wrong, and yet we are in the position of using these erroneous estimates to make decisions about population parameters. In an experiment, we introduce a treatment, designed to bring about some change. *The question becomes whether, after applying a treatment, a fluctuation in a statistic is a true response to the treatment, or merely due to natural sample fluctuation because of population variability and sampling error.* The hypothesis test is a tool for separating sampling error from true significant change, and our key tool for this separation is the *p-value*. We will deem a change significant when it is unlikely to have happened by chance. The *p-value* measures this likelihood, and when it is very small, we will consider the change significant. How small should the *p-value* be? Normally we like the *p-value* to be less than around 5% (our old familiar rule), but this can vary, and we leave this to researchers and their readers to decide.

The tests in this chapter all involve a claim about a single population parameter – either a population mean, or a population proportion. The test of the claim will require sample data from the relevant population, in an attempt to give evidence which may support the claim being made. It should be noted, however, *that all of the tests in this chapter have a fundamental weakness* – since they only use data from a single sample, no control group is possible. We will bring in a few examples where sample data are being compared with parameters from a different population, and in a sense this is a control, but it is not a true control group because no sample data are allowed for the control. We will learn a way to implement a true control group in Chapter 8, but for now we begin without one because this makes the problems simpler, and that is important when learning a new procedure.

Historically, hypothesis testing is nearly as old as probability theory itself, but the modern form of hypothesis testing is due largely to the genius of Ronald A. Fisher, who outlined much of what we outline here in his ground breaking book *Statistical Methods for Research Workers* in 1925. Fisher's preferred term was *tests of significance*, and it was Fisher who coined the terms *null hypothesis* and *level of significance* which we will learn soon. Other statisticians contributed to the theory as well, including William *Student* Gosset, and Karl Pearson's son, Egon Pearson.

We will learn more about Ronald A. Fisher and Karl Pearson in later chapters, but for now we must learn the processes created by these people, which have become universally accepted standard methods of research. In this learning process, much of the language we will introduce may be unfamiliar and awkward. Regardless, these terms are the language of statistical inference which is spoken by researchers everywhere, so we must learn them as well.

7.1 Hypothesis Testing – An Overview

In this section we will summarize the basic steps and concepts in hypothesis testing, without actually conducting any complete tests. The concepts will apply to all tests covered in Chapters 7 and 8. All tests that we cover in these chapters will involve means or proportions. After that, we will study more advanced applications, but still the essential structure of the tests will remain the same.

Let us begin our discussion on the steps involved in a hypothesis test involving a single population parameter.

7.1.1 Single Parameter Hypothesis Tests

Step I. **Gather Sample Data and Verify that Requirements are Met**

In all hypothesis testing procedures, it is assumed that we are making inferences by using simple random samples. Simple random samples tend to represent their populations well, and produce good estimates of their corresponding parameters. When samples are not representative of their populations, they cannot be used to make reliable inferences.

Once we are assured of the quality of our sample we can proceed with the hypothesis test. From our sample data, we will compute a test statistic. A test statistic requires the basic summary values from our sample – when studying proportions we will require a sample proportion and sample size, and for means we will require a sample mean, a standard deviation, and a sample size for each sample that we consider. The requirements for each test are those for the approximate normality of the underlying sampling distributions. For proportions we will require that np and nq are each greater than 10. For means, we require the sample size to be greater than 30, or that the parent population is normal.

Hypothesis tests also require a level of significance. The *level of significance* can be interpreted in several ways. Previously, we had a rule that stated that an event is *significant* when it happens and has a probability less than 5%. With this rule, 5% is our level of significance. We now will allow this level to vary. Our level of significance will still represent the probability of an unusual event, and it should be small, but we will not define precisely what *small* means. Generally, it would be less than 10%, but even so, this should be up to the researcher to decide.

Step II. **Identify the Claims**

The word *hypothesis* means *assumption*. In statistical hypothesis testing, we will always make an assumption about one or more population parameters. In chapter seven, we will make an assumption, which we will call the *null hypothesis*, denoted by H_0, about the value of a population parameter – either a population proportion, a population mean. This is the assumption which we will be testing. Every null hypothesis will be countered with another hypothesis called the *alternative hypothesis*, denoted by H_1, which we will support if we reject the assumption made in the null hypothesis.

The process we will always follow will be to assume that the null hypothesis is true. We will thus compute a test statistic using sample data and the assumed value of a population parameter to see how many standard deviations the statistic is from the assumed value of the associated parameter. When the number of standard deviations is large (we will define precisely how large), we will reject the null hypothesis, in favor of the alternative.

The null hypothesis will always assume a value for a population parameter. H_1 is an alternative to the null hypothesis, H_0, and is thus a statement that is contrary to it. In this chapter, the null hypothesis is always $H_0 : p = value$ for proportions, or $H_0 : \mu = value$ for means. The alternative hypothesis has three forms for each parameter, so for proportions, $H_1 : p < value$, $H_1 : p > value$, or $H_1 : p \neq value$. The choice of these would depend on whatever claim we are being asked to test. For means, the possible alternative hypotheses are $H_1 : \mu < value$, $H_1 : \mu > value$, or $H_1 : \mu \neq value$.

There are three types of tests, and the type is determined by the alternative hypothesis, H_1. When the alternative hypothesis claims that the parameter is greater than the assumed value in H_0, we will be looking for a statistic that is greater than the assumed parameter, and in such a case the test statistic will be positive, and we will call this significant if it lies deep within the right tail of the appropriate distribution. This is a *right-tail* test.

When the alternative hypothesis claims that the parameter is less than the assumed value in H_0, we will be looking for a statistic that is less than the assumed parameter, and in such a case the test statistic will be negative, and we will call this significant if it lies deep within the left tail of the appropriate distribution. This is a *left-tail test*.

When the alternative hypothesis claims that the parameter is not equal to the assumed value in H_0, we will be looking for a statistic that is different than the assumed parameter, and in such a case the test statistic may be positive *or* negative, but we will call this significant if it lies deep within either the left tail or the right tail of the appropriate distribution. This is a *two-tail test*.

Step III. **Compute the Test Statistic (T.S.)**

In Chapter 6, we introduced the *test statistics* (abbreviated *T.S.*) which we will use in this chapter. In each case, they measure the number of standard errors (or *deviations*) which the statistic in question lies from the assumed parameter.

To summarize, the test statistics for single sample tests are as follows. When testing a claim regarding a single population proportion, the test statistic is:

$$T.S.: z = \frac{\hat{p} - p}{\sqrt{\frac{pq}{n}}},$$

and it measures the number of standard errors that the sample proportion lies from the assumed population proportion.

When testing a claim regarding a single population mean, there are two possibilities for the test statistic: when σ is known, the test statistic is

$$T.S.: z = \frac{\bar{x} - \mu}{\sigma/\sqrt{n}},$$

but when σ is unknown we will use the test statistic

$$T.S.: t = \frac{\bar{x} - \mu}{s/\sqrt{n}}.$$

The first measures the number of standard errors that the sample mean lies from the population mean, while the second estimates this quantity.

Step IV. **Look Up the Critical Value(s) (C.V.)**

We have always said that *two* standard errors from the mean, as measured by a z- (or estimated by a t-) score is significant. This will roughly agree with our new, more refined approach, but we will now allow this number to vary, and the exact value will depend on the level of significance.

The number of standard errors required, above or below the assumed parameter, will now be called a *critical value*, abbreviated *C.V.* All along our critical value has been 2. The way we will now determine our critical value will now depend on the value for α, and whether we are conducting a right-, left-, or two-tail test. In a right-tail test, the area to the right of the positive critical value is α, as depicted in Figure 7.2. In a left-tail test, the area to the left of the negative critical value is α, as shown in Figure 7.3. In a two-tail test, the areas outside of the two critical values are each $\alpha/2$. Together the left and right tails form the rejection region, whose total area is α, as pictured in Figure 7.4.

In each type of test, there is a rejection region, corresponding to the shaded regions in the diagrams above. In each case, the total area of the rejection region is equal to α. We will be rejecting the null hypothesis whenever the test statistic lands in the rejection region. Note that for a two-tail test, each tail has area equal to $\alpha/2$. This is so that the total area of the rejection region is still equal to α, consistent with the other two test types. This is important because we will want the probability of a rejection (when the null hypothesis

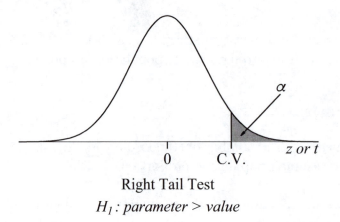

Right Tail Test

$H_1 : parameter > value$

Figure 7.2: For a right-tail test, the area to the right of the critical value is equal to α. This region is called the *rejection region*.

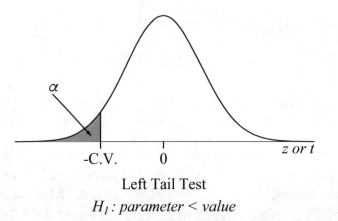

Left Tail Test

$H_1 : parameter < value$

Figure 7.3: For a left-tail test, the area to the left of the critical value is equal to α. This region is the *rejection region*.

is true) to be equal to α, the area of each rejection region.

Whether conducting a z-test or a t-test, we should remember that Table A2 will give positive z- or t-critical values for most levels of significance commonly used. The z-critical values are in the bottom row which is labeled *Large* (z). We only use t-critical values when working with a mean where a population standard deviation is unknown, and we look it up under the appropriate degrees of freedom. If we require a negative critical value, just change the sign of the corresponding positive critical value.

Step V. **Look up the p-Value**

We have defined *p-values* previously. In this context, we define the *p-value* to be the probability of measuring a test statistic that is at least as extreme as the one we have measured. How this principle is applied depends on the type of test we are conducting. For a right-tail test, we find the area to the right of

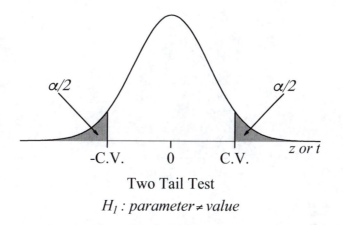

Two Tail Test

$H_1 : parameter \neq value$

Figure 7.4: Two-tail tests have a positive and a negative critical value. The areas beyond each critical value have area $\alpha/2$. These form the *rejection region*.

the test statistic as depicted in Figure 7.5. For a left-tail test, we find the area

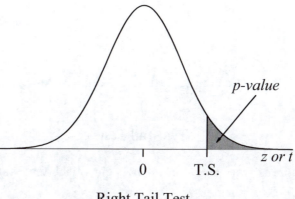

Right Tail Test

Figure 7.5: The *p-value* is the area to the right of the test statistic (T.S.).

to the left of the test statistic as depicted in Figure 7.6. The two-tail test is trickier – we must double the area of the smaller tail that is defined by the test statistic. To simplify this, we will double the area to the right of the absolute value of the test statistic, as depicted in Figure 7.7. This is consistent with the treatment for α on a two-tail test, where it is twice the area of either tail.

When using the z-distribution, we will look-up *p-values* in Table A1. When using the t-distribution, we will need a new table, as Table A2 is used for critical values only, and finding *p-values* precisely is impossible. We will use Table A3 for this, and Table A3 is similar to Table A1, giving areas to the left of critical values.

The mechanical process for computing a *p-value* is as follows. For a right-tail test, look up the test statistic in Table A1 (or A3 for t-tests), and use $p\text{-}value = 1 - (area\ in\ table)$. For a left-tail test, look up the test statistic in

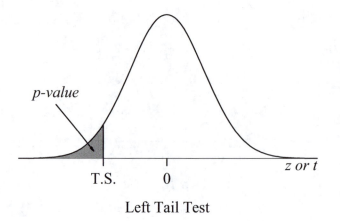

Figure 7.6: The *p-value* is the area to the left of the test statistic (T.S.).

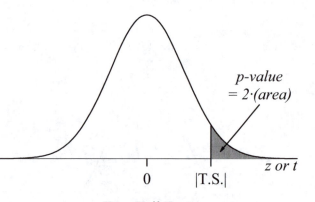

Figure 7.7: The *p-value* is twice the area to the right of the test statistic's absolute value.

Table A1 (or A3), and the area you find there is the *p-value*. For a two-tail test, look up the absolute value of the test statistic in Table A1 (or A3) and $p\text{-}value = 2 \cdot (1 - area\ in\ table)$.

As we have said, the *p-value* is the probability of observing a test statistic that is at least as extreme as the one we have, and the smaller its value, the more significant the difference between the assumed parameter and the observed statistic. The simplest rule for making a decision about the null hypothesis is that whenever the *p-value* is less than α (the level of significance), the deviation is significant, and so the null-hypothesis is probably not true, and should be rejected. While simple, this rule for decision making is rather crude, as sometimes the *p-value* can be quite close to α, and in such a case an absolute yes or no may not be appropriate. In such a case, a simple maybe is probably better. In reporting the results of research, it is often considered best to report the *p-value* only, and leave the decision about its meaning to the reader. This is typical in research, and therefore it becomes important for the reader to have a basic ability to interpret the *p-value*. In summary, smaller *p-values* indicate more significant deviations of the statistic from the assumed value of the parameter.

To help you in developing a basic intuition for interpreting *p-values*, the table below is given. These interpretations are not absolute, and vary from discipline to discipline.

p-Value Range	Observed Significance
p-value < 0.01	Highly Significant
$0.01 \leqslant$ *p-value* < 0.05	Significant
$0.05 \leqslant$ *p-value* < 0.10	Moderately Significant
p-value $\geqslant 0.10$	Insignificant

Step VI. **Make a Conclusion**

Finally, we need to make a decision about the claim that was made. The simplest rule for making a decision is based on the *p-value*. If the *p-value* is less than α, then we will reject the null hypothesis in favor of the alternative hypothesis. Whenever this happens, the test statistic, when compared to the critical value, will be in the rejection region and, again, we will reject the null hypothesis. If these do not happen, we will not accept the null-hypothesis, but instead we will fail to reject it. This is subtle, and can be confusing. The reason is due to the fact that we do not prove the null hypothesis – we just assume it to be true, and then see if the data disagree with that assumption.

Because sample data always lead to sampling error, we must understand that such error can cause us to make wrong decisions. Thus, rejecting the null-hypothesis does not automatically make it false, and failing to reject it does not automatically make it true. We assumed it was true, but failing to reject this does not prove that it is true. Because of this, it is important that the wording of our conclusions be chosen carefully. The following guidelines may be helpful in choosing the wording of conclusions to hypothesis tests.

Summary 7.1.1 Conclusions for Hypothesis Tests

If we reject the null hypothesis, we may say that the data are in support of the alternative hypothesis, and do not support the null hypothesis. If we fail to reject the null hypothesis, we will say that the data do not give sufficient reason to reject the null hypothesis, but are not in support of the alternative hypothesis.

Conclusions are a required step for a hypothesis test, and they should be written using language that anyone who has a basic understanding of averages and proportions can understand. For the more statistically sophisticated reader, it is nice to mention the *p-value* and what kind of significance it indicates, as outlined under the *p-value* interpretations in step #6 above.

7.1.2   Significant Difference Versus Sampling Error

Often, we can be surprised by the outcome of a hypothesis test, especially when the sample data seem to favor the claim that has been made. For example, there are times when our *original claim* will say that the population mean is greater than some value, and our sample mean may be greater than this value. Yet, sometimes we will conclude that the data do not support the original claim. The reason for this is that, while the sample mean was larger than the assumed value, it was not significant enough to be considered anything more than sampling error due to population variability.

When we conduct a hypothesis test, we are making an effort to separate fluctuations in our statistics which are due to error and variability from true significant change that may be due to a real response to some meaningful treatment, or an assumption about a population parameter that is simply not true. When we see that a statistic favors the claim that we wish to prove, we should be careful not to make hasty conclusions, keeping in mind that these values contain error due to population variability. The hypothesis test is our tool for objectively distinguishing sampling error from truly significant differences.

7.1.3   Errors in Hypothesis Test Conclusions

Whatever the outcome of a hypothesis test, we are always aware that sample data may lead to wrong conclusions, because of the variability in a given statistic's sampling distribution. It is important for us to understand this, and the types of errors that may be made. The table below demonstrates the ways that we can make right and wrong decisions, and gives names to the types of errors.

	We rejected H_0	We failed to reject H_0
H_0 is true	Type I Error	Correct Decision
H_0 is false	Correct Decision	Type II Error

Thus, a *Type I Error* is rejecting a true null hypothesis, while a *Type II Error* is failing to reject a false null hypothesis. Suppose our data lead us to make a Type I Error, so the value for the parameter which was assumed is actually correct. The only way to reject the null hypothesis would be for the test statistic to land in the rejection region, whose area is α. Since this area is equal to the probability of this event, the probability of a Type I Error, when H_0 is true, is equal to α.

The probability of a Type II Error is more difficult to compute, because it depends on the value of the unknown parameter (either the population proportion, p, or mean, μ). Still, we denote the probability of a Type II Error by the Greek letter β (*beta*).

7.1.4 Power

The *power* of a hypothesis test is the probability of rejecting the null hypothesis. When H_0 is false, the probability of *not* rejecting it is β, so in this case, the power would be $1-\beta$. Since β is generally unknown (its value depends on the unknown population parameter that is being tested), the power of a test is not often known. Still, the relationships between the factors that effect the test can give information regarding how they influence the power.

The main factors which influence the power of a test are as follows. First, the discrepancy between the assumed value of the parameter in question and the true value of that parameter affects the power of the test. The larger the discrepancy between these values, the higher the power of the test. Secondly, the value for α will influence the power. Larger α values increase power. Finally, the sample size can influence the power of a test. Larger sample sizes correspond to higher power.

Generally speaking, a hypothesis test has good properties when the probabilities of Type I and Type II errors are small. When the null hypothesis is false and β is small, the test should result in a rejection of the null hypothesis, and so the power of the test $(1-\beta)$ will be close to one. With sufficient sample size, the hypothesis tests covered in this text have power close to 1, as long as their conditions are met.

While we will not compute β, it is important to understand the relationship between α and β. We have the ability to pick any value for α, but we must understand that different values for α can yield different decisions to our test. Picking a large value for α makes the rejection region larger, and thus makes it more likely that we will reject the null hypothesis, but less likely that we will fail to reject the null hypothesis. If the null hypothesis is false, and we fail to reject it, we have made a Type II Error, and picking a larger α has decreased the chance of this, so β has *decreased*. By a similar argument, picking a smaller value for α causes β to *increase*. We cannot determine β directly as we can determine α, but we can influence its value indirectly.

There is another way to control the probability of a Type II Error, β, which is preferrable to manipulating α. It is not difficult to demonstrate that increasing the sample size, n, can dramatically decrease β (larger n gives a narrower distribution making it more likely for a test statistic to land in a tail). As β decreases, and the power of the test, $1-\beta$, increases. Larger sample size is the best way to improve the power of a hypothesis test.

7.1.5 Choosing α

With all of this in mind, when designing a statistical study, remember to use the largest sample size that is reasonable. This will increase the power of the test, and decrease β. After this, when choosing α, the best rule of thumb is to pick the largest α that can be tolerated, remembering that α is the probability of a Type I Error. Doing so will give the smallest value of β that corresponds to a tolerable value of α.

Summary 7.1.2 When choosing a value for α, consider the consequences of the Type I or Type II errors. With these in mind, choose the largest value of α that can be tolerated as the probability of the Type I Error.

7.1.6   Real-World Significance vs. Statistical Significance

Hypothesis tests are all about distinguishing significant change or difference from sampling error. What they cannot separate is statistical significance from real world significance. There are times when an improvement, which is due to an experimental treatment, is statistically significant but is still so small that it may have little or no impact on those population members who are receiving it. This will happen when the maximum error in our estimates is small, and improvements are unlikely to be due to this error (as indicated by the *p-value*). This is very objective, and that is good, but if no change can be detected *subjectively* by population members, what has been accomplished? In summary we must remember that once statistical significance has been determined, there are still questions to be answered about real-world significance.

7.1.7   Sampling Frame and Applicability of Results

Even when a statistical study is done well, it is important to understand that its results may only apply to a limited population. A medical study may attempt to discover information regarding a new treatment for women, but if the sample only includes American women, then results may only be applicable to the population of American women, who face different health issues than non-American women. While the target population may be all women, the sample actually represents a subset of this population, American women, which we refer to as the study's *sampling frame*.

Definition 7.1.3 The *sampling frame* of a given study is the population subset which is best represented by the sample. It is not always valid to infer that a study's results are valid outside of its sampling frame.

When we conduct a poll of U.S. citizens by calling random phone numbers, then, while the population of interest is U.S. citizens, the sampling frame consists only of those citizens with phones. Many members of this population will not be represented by this sample, so the study's results may not apply to them.

7.1.8 Homework

Identify the null and alternative hypotheses, and state whether the test is right- left- or two-tailed.

1. More than 2 out of 3 dentists choose Zest.

2. It is not true that the average age of the top 200 tennis players is equal to 25.

3. The tires on new cars last much less, on average, than 40,000 miles. This leads me to believe that car companies sell cars with cheap tires!

Identify the appropriate test statistic formula.

4. More than 2 out of 3 dentists choose Zest. We have a sample with 559 observations.

5. It is not true that the average age of the top 200 tennis players is equal to 25. In this analysis, we will use a sample standard deviation, because the population standard deviation is unknown. We have 32 observations in our sample data to test this claim.

6. The tires on new cars last much less on average than 40,000 miles. We assume that the population standard deviation is known to be 5,000 miles, and we have a sample with 50 observations.

Look up the critical value corresponding to the type of test and level of significance. Critical values for the z-distribution are in the bottom row of Table A2.

7. 1% level of significance, z-distribution, right-tail test.

8. 4% level of significance, z-distribution, left-tail test.

9. 5% level of significance, z-distribution, two-tail test.

Look up the critical value corresponding to the type of test and level of significance. Critical values for the t-distribution are in the appropriate degrees of freedom (df) row in Table A2. For left-tail critical values, use the negative of the critical value listed in the table.

10. 1% level of significance, t-distribution, right-tail test, $df = 7$.

11. 4% level of significance, t-distribution, left-tail test, $df = 21$.

12. 5% level of significance, t-distribution, two-tail test, $df = 54$.

For the given test statistic and type of test, look up the associated z-distribution p-value (Table A1), and state whether this indicates a significant difference from the assumed population parameter.

13. Test statistic: $z = 2.67$, right-tail test.

14. Test statistic: $z = 2.67$, left-tail test.

15. Test statistic: $z = 2.67$, two-tail test.

16. Test statistic: $z = -1.84$, right-tail test.

17. Test statistic: $z = -1.84$, left-tail test.

18. Test statistic: $z = -1.84$, two-tail test.

State the appropriate conclusion for the hypothesis test. Use the p-value to indicate how significant the sample data are.

19. Claim: More than 2 out of 3 dentists choose Zest. p-Value $= 0.045$. Significance level, $\alpha = 0.05$.

20. It is not true that the average age of the top 200 tennis players is equal to 25. p-Value $= 0.063$. Significance level, $\alpha = 0.05$.

21. The tires on new cars last less, on average, than 40,000 miles. p-Value $= 0.015$. Significance level, $\alpha = 0.01$.

State the type of error (Type I or II) that has been made for the following claims.

22. Claim: More than 2 out of 3 dentists choose Zest. Sample data support the claim. What error is made if 2/3 of dentists actually *do* choose Zest?

23. Claim: It is not true that the average age of the top 200 tennis players is equal to 25. Sample data do not support this claim. What error is made if the claim is actually true?

24. Claim: The tires on new cars last much less on average than 40,000 miles. Sample data do not support this claim. What error is made if the claim is actually true?

Answer the following questions.

25. In a 2011 survey on the website *hunch.com,* a sample of 15,818 readers were surveyed, and 32% of those surveyed owned an iPhone. Would it be reasonable to use this sample proportion to test the claim that $\frac{1}{3}$ of all people own an iPhone?

26. In another survey of 388, 315 hunch.com website readers, about 58% of more than 90, 000 Macintosh users identified themselves as politically liberal. Would it be reasonable to use this sample proportion to test the claim that more than $\frac{1}{2}$ of Mac users are politically liberal?

27. What is the sampling frame for the study referenced in Problem 25? Would it be reasonable to apply this study's results to this sampling frame?

28. What is the sampling frame for the study referenced in Problem 26? Would it be reasonable to apply this study's results to this sampling frame?

Problems 29 through 34 refer to the following situation.

About 16% of adults are alcohol dependent. It is claimed that a larger proportion of energy drink consumers are alcohol dependent when compared to other adults. The null hypothesis $(H_0 : p = 0.16)$ claims that there is no difference in alcohol dependence rates between these groups. The alternative hypothesis $(H_1 : p > 0.16)$ claims the proportion of energy drink consumers that are are alcohol dependent is greater than 0.16.

29. Describe what must happen for a Type I Error to occur.

30. What is the consequence of the Type I Error?

31. Describe what must happen for a Type II Error to occur.

32. What is the consequence of the Type II Error?

33. Which of the two is worse, the Type I or Type II Error?

34. You have chosen the worst of the two errors. With this in mind, what value for α would be more appropriate: 0.01 or 0.05?

Problems 35 through 40 refer to the following situation.

A manufacturer of parachutes was sued in a class action because a high rate (2%) of their parachutes failed to open in practice. A new parachute was developed, and it was claimed that the failure rate is lower for the new chute. The null hypothesis $(H_0 : p = 0.02)$ claimed that the failure rate of the new chute is the same as for the old chute. The alternative hypothesis $(H_1 : p < 0.02)$ claims that the failure rate of the new chute is lower than the rate for the old chute.

35. Describe what must happen for a Type I Error to occur.

36. What is the consequence of the Type I Error?

37. Describe what must happen for a Type II Error to occur.

38. What is the consequence of the Type II Error?

39. Which of the two is worse, the Type I or Type II error?

40. You have chosen the worst of the two errors. With this in mind, what value for α would be more appropriate: 0.01 or 0.05?

7.2 Large Sample Tests for a Population Proportion

7.2.1 Claims Regarding a Single Population Proportion

Now that we have a basic understanding of the steps involved in a hypothesis test, we may proceed to our first application – testing claims regarding a single population proportion, p. The steps are summarized below.

Procedure 7.2.1 Testing Claims About a Single Population Proportion

Step I. **Gather Sample Data**

For this test, you will need $\hat{p} = x/n$, n, and the level of significance, α.

Step II. **Identify the Claims and Verify that Requirements are Met**

The null hypothesis is always $H_0 : p = value$. The alternative hypothesis is one of the following:

H_1: $p > value$, H_1: $p < value$, or H_1: $p \neq value$.

This procedure assumes a simple random sample. Such samples tend to represent their populations well. The requirements for this test are the requirements for using the normal distribution to approximate the binomial – that whatever is used for p in the null hypothesis, and with $q = 1 - p$, we must have $np \geqslant 10$, and $nq \geqslant 10$. We do not pick a hypothetical p so that the requirement is met; rather, we choose n sufficiently large so that the requirement is met. This is why we call this a *large sample test*.

Step III. **Compute the Test Statistic (T.S.)**

For this test, the test statistic is the z-score of the sample proportion,

$$T.S. : z = \frac{\hat{p} - p}{\sqrt{\frac{pq}{n}}}.$$

The formula uses p from H_0 and $q = 1 - p$.

Step IV. **Look Up the Critical Value(s) (C.V.) on Table A2.**

The type of test is determined by H_1, and the critical values are found in Table A2. These are z-critical values, so be sure to use the bottom row of the table where the z scores are found. The right tail critical value is positive as pictured in Figure 7.8. The left tail critical value is negative, as pictured in Figure 7.9. The two-tail test has a positive and negative critical value, as depicted in Figure 7.10.

Step V. **Look up the *p-Value* on Table A1**

For a right-tail test, look up the test statistic in Table A1, and use *p-value* = $1 - (area\ in\ table)$, as pictured in Figure 7.11. For a left-tail test, look up the test statistic in Table A1, and the area you find there is the *p-value*. Such a *p*-value is depicted in Figure 7.12. For a two-tail test, look up the absolute

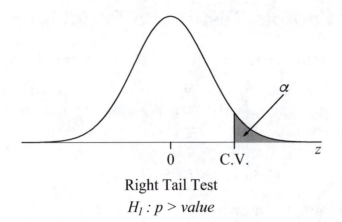

Right Tail Test

$H_1 : p > value$

Figure 7.8: For a right-tail test, the area to the right of the critical value is equal to α. This region is called the rejection region.

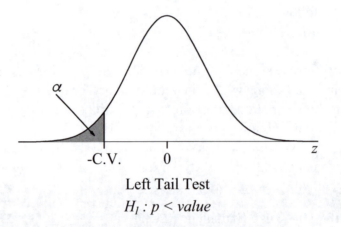

Left Tail Test

$H_1 : p < value$

Figure 7.9: For a left-tail test, the area to the left of the critical value is equal to α. This region is called the rejection region.

value of the test statistic in Table A1, and $p\text{-}value = 2 \cdot (1 - area\ in\ table)$. Figure 7.13 depicts this case.

Step VI. **Make a Conclusion**

If the $p\text{-}value$ is less than α, or equivalently, if the test statistic lies in the critical region, we reject the null hypothesis in favor of the alternative. Otherwise, we fail to reject the null hypothesis. Write a conclusion in plain language, using the rules below.

If we reject the null hypothesis, we may say that the data are in support of the alternative hypothesis, and do not support the null hypothesis. If we fail to reject the null hypothesis, we will say that the data do not give sufficient reason to reject the null hypothesis, but they are not in support of the alternative hypothesis.

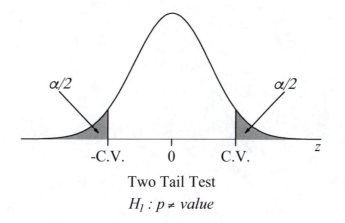

Two Tail Test

$H_1 : p \neq value$

Figure 7.10: There is a positive and a negative critical value. The areas beyond each critical value have area $\alpha/2$. These form the rejection region.

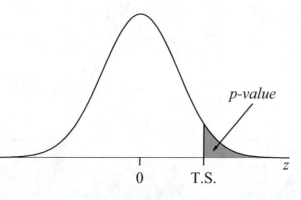

Right Tail Test

Figure 7.11: The *p-value* is the area to the right of the test statistic (T.S.).

Do not forget to mention the *p-value* and what kind of significance it indicates. Next, we will work through a few examples.

Example 7.2.2 A Right-Tail Test: We have stated several times that the proportion of people in the world who are female is 49.8%. Suppose you survey 1000 randomly selected individuals, and find 528 of these are females. Test the claim that the proportion of people who are female in the world is actually greater than 49.8% (against the claim that the proportion is equal to 49.8%), using a 1% level of significance.

Step I. **Gather Sample Data**
We will need $\hat{p} = x/n = 528/1000 = 0.528$, and $n = 1000$. The level of significance is $\alpha = 0.01$.

Step II. **Identify the Claims and Verify that Requirements are Met**
The claim we are testing is that the proportion of people who are women is greater than 49.8%. The null hypothesis must be $H_0 : p = 0.498$. The alternative hypothesis $H_1 : p > 0.498$.

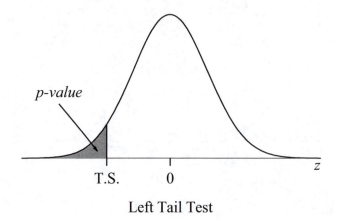

Left Tail Test

Figure 7.12: The *p-value* is the area to the left of the test statistic (T.S.).

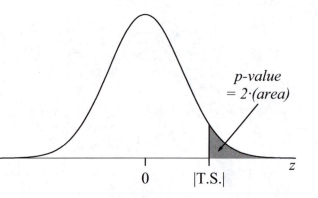

Two Tail Test

Figure 7.13: The *p-value* is twice the area to the right of the test statistic's absolute value.

For verifying requirements, $np = 1000 \cdot 0.498 = 498 \geq 10$, and $nq = 1000 \cdot 0.502 = 502 \geq 10$. So, both requirements are easily met.

Step III. **Compute the Test Statistic (T.S.)**

To compute the test statistic, we use $p = 0.498$ from the null hypothesis, and $q = 1 - p = 1 - 0.498 = 0.502$.

Thus, the test statistic is:

$$T.S. : z = \frac{\hat{p} - p}{\sqrt{\frac{pq}{n}}}$$

$$= \frac{0.528 - 0.498}{\sqrt{\frac{0.498 \cdot 0.502}{1000}}}$$

$$= 1.897.$$

Step IV. **Look Up the Critical Value(s) (C.V.) on Table A2**
The alternative hypothesis is $H_1 : p > 0.498$. If we think of the greater than symbol in this hypothesis as an arrow pointing to the right, we are reminded that this is a right-tail test with $\alpha = 0.01$.

Our test statistic is a z-score, and therefore our critical value must be a z-score as well. In Table A2, we remember that z-scores are in the bottom row, so we ignore degrees of freedom on this problem. Also, this is a one-tail test, so we look up our α in the one-tail application section of Table A2.

One Tail Applications			
df	$\alpha = .005$	$\alpha = .01$	$\alpha = .02$
1000	2.581	2.330	2.056
Large (z)	2.576	**2.326**	2.054

From Table A2, we see that the critical value is $CV : z = 2.326$. This critical value is pictured in Figure 7.14.

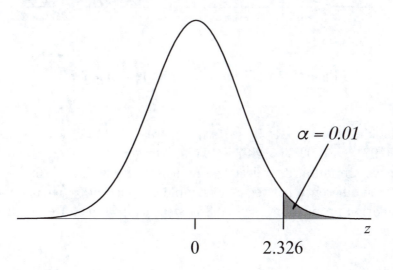

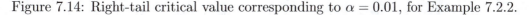

Figure 7.14: Right-tail critical value corresponding to $\alpha = 0.01$, for Example 7.2.2.

The rejection region includes all values of z greater than or equal to $z = 2.326$. Notice that our test statistic is $z = 1.897$, which does not lie in the rejection region. Because of this, *we will fail to reject the null hypothesis.*

Step V.   **Look up the *p-Value* on Table A1**

Next, we must look-up the *p-value*. This is a right-tail test, so we will look up the area associated with the test statistic $z = 1.90$ (rounded to two places due to the constraints in the table) in Table A1, and use $p\text{-}value = 1 - (area)$.

z	**.00**	.01
1.8	.9641	.9649
1.9	**.9713**	.9719
2.0	.9772	.9778

The tail on the z-distribution whose area is the p-value is pictured in Figure 7.15.

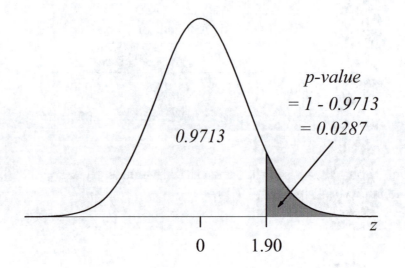

Figure 7.15: The *p-value* from Example 7.2.2.

In looking up $z = 1.90$, we see an area of 0.9713, so we have $p\text{-}value = 1 - 0.9713 = 0.0287 = 2.87\%$. This is larger than α and demonstrates again that we are failing to reject the null hypothesis. Still, this *p-value* indicates that the difference between the sample proportion $\hat{p} = 0.528$ and the assumed population proportion $p = 0.498$ is *significant*, but not enough at a 1% significance level.

Step VI. **Make a Conclusion**

Because our test statistic was not in the rejection region found in step 5, and equivalently our *p-value* was larger than α, we are failing to reject the null hypothesis that the proportion of people who are female is equal to 49.8%. The original claim was that the proportion is greater than 49.8%, but because we did not reject the null hypothesis, we conclude that the data do not support this claim. Our conclusion should be something like the following statement.

The sample data do not support the claim that more than 49.8% of the world's population is female. Still the difference between the sample proportion and the assumed population proportion was significant, with a p-value of 0.0287.

Note that we make it clear that the data are what lead us to this decision, and that our opinion has nothing to do with this.

Finally, we should take a moment to observe what we have done. We claimed that the proportion of the world's population that is female is more than 49.8%, and gave evidence by providing a random sample which was 52.8% female. This seems to support our claim, yet in our conclusion we said that the data do not support the original claim. This seems almost contradictory until we remember that all sample data contain sampling error. In reality, there is solid evidence that the world population really is 49.8% female. Even so, it would be very unlikely for a sample proportion to be exactly this amount. Remember the sampling distribution of sample proportions – the mean of sample proportions is the population proportion (which we assume is 49.8%), and because of the normality and symmetry of this distribution, half of all sample proportions are below this amount, and half are greater than this amount.

Even *with* a population proportion of 49.8%, we would expect half of all sample proportions to be greater than this amount, so our sample proportion of 52.8% does not really prove anything. This proportion had a *p-value* of 2.87%, meaning that a proportion at least as extreme as ours would occur randomly in 2.87% of all sample proportions. This is rare and significant, but not significant enough if we are using a 1% significance level. At 1% significance, we only consider a proportion significant if it has less than 1% chance of occurring, otherwise we will consider this deviation from the assumed 49.8% proportion as nothing more than *sampling error*. Any deviation that is not significant will be considered by us as being sampling error, and thus not strong enough evidence to support the alternative hypothesis.

Example 7.2.3 A Two-Tail Test: When 84 students of Mt. San Antonio College, in Walnut, California, were randomly selected, 58 were registered to vote (data provided by Nicole Hernandez, an honors student at Mt. San Antonio College). Test the claim made by a fictional college administrator that the proportion of students who are registered to vote at this college is equal to 88%, against the alternative that this is not equal to 88%, using a 6% significance level.

Step I. **Gather Sample Data**
We will need $\hat{p} = x/n = 58/84 = 0.690$, and $n = 84$. The level of significance is $\alpha = 0.06$.

Step II. **Identify the Claims and Verify that Requirements are Met**
The claim we are testing is that the proportion of students who are registered to vote at this college is equal to 88%. The null hypothesis must therefore be $H_0 : p = 0.88$. The alternative hypothesis is $H_1 : p \neq 0.88$.

In verifying requirements, $np = 84 \cdot 0.88 = 73.92 \geq 10$. Also, $q = 1 - p = 1 - 0.88 = 0.12$, so $nq = 84 \cdot 0.12 = 10.08 \geqslant 10$. Again, both requirements are met.

Step III. **Compute the Test Statistic (T.S.)**
To compute the test statistic, we use $p = 0.88$ from the null hypothesis, and $q = 1 - p = 1 - 0.498 = 0.502$.

Thus, the test statistic is:

$$T.S. : z = \frac{\hat{p} - p}{\sqrt{\frac{pq}{n}}}$$
$$= \frac{0.690 - 0.88}{\sqrt{\frac{0.88 \cdot 0.12}{84}}}$$
$$= -5.359.$$

Step IV. **Look Up the Critical Value(s) (C.V.) on Table A2**
The alternative hypothesis is $H_1 : p \neq 0.88$. This indicates a two-tail test with $\alpha = 0.06$.

Our test statistic is a z-score, and therefore our critical value must be a z-score as well. Again, in Table A2, we remember that z-scores are in the bottom row, so we ignore degrees of freedom on this problem. This problem is a two-tail test, so we look up our α in the two-tail application section of Table A2.

Two-tail Applications			
df	$\alpha = .05$	$\alpha = .06$	$\alpha = .08$
1000	1.962	1.883	1.752
Large (z)	1.960	**1.881**	1.751

From Table A2, we see a critical value of $z = 1.881$.

Since this is a two-tail test, we include this, along with its negative, so the critical values are $CV : z = \pm 1.881$. The critical values are depicted in Figure

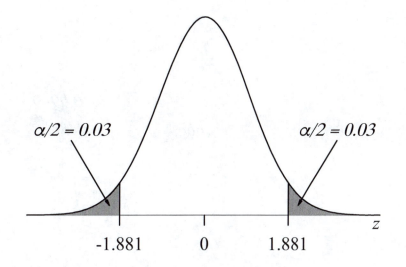

Figure 7.16: Critical values for Example 7.2.3.

7.16. The rejection region includes all values greater than or equal to 1.881, or less than −1.881. Notice that our test statistic is $z = -5.359$, which lies in the left tail of the rejection region. Because of this, *we will reject the null hypothesis.*

Step V. **Look up the p-Value on Table A1**
Next, we must look-up the *p-value*. This is a two-tail test, so we will look up the area associated with the absolute value of the test statistic, $|T.S.| = |z| = |-5.36| = 5.36$, (rounded to two places due to the constraints in the table) in Table A1. The table does not allow z-scores as high as 5.36, so we follow the footnote at the bottom of the table, assuming that 0.9999 is the associated area. With this, we compute the p-value $= 2 \cdot (1 - area)$.

The corresponding region under the normal density curve is pictured in Figure 7.17.

Thus, we have

$$p\text{-}value = 2 \cdot (1 - 0.9999)$$
$$= 2 \cdot 0.0001$$
$$= 0.0002,$$

(remember, we multiply the area of the tail times two for a two-tail *p-value*). This *p-value* is much smaller than α and demonstrates again that we will reject the null hypothesis. The difference between the sample proportion and the assumed population proportion of 88% is highly significant as indicated by the smallness of the *p-value*.

Step VI. **Make a Conclusion**
Because the *p-value* is less than α, or equivalently, because the test statistic

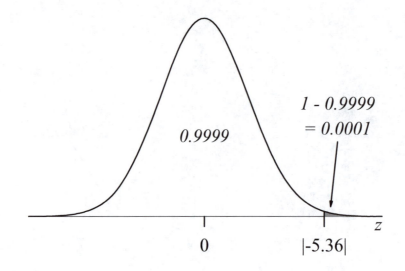

Figure 7.17: Determining the *p-value* for Example 7.2.3.

lies within the rejection region, we are rejecting the null hypothesis in favor of the alternative hypothesis which is $H_1 : p \neq 0.88$. The original claim is the same as the null hypothesis – that the population proportion is equal to 88%, and we have rejected this, so the data do not support the original claim.

The data do not support the claim made by the college administrator that the proportion of students who are registered to vote at Mt. San Antonio College is equal to 88%. The difference between the sample proportion and the assumed 88% proportion was highly significant, with a p-value of 0.0002.

This time we have rejected the null hypothesis that the population proportion is 88%. This has happened because our sample proportion was not only different from 88%, but so different that obtaining a sample proportion at least as extreme as ours had only a 0.0002 probability. This is so unlikely to have happened by chance that we are forced to conclude that the observed deviation could not be mere sampling error, but a mistaken assumption about the population proportion being 88%. We have therefore rejected this assumption.

Example 7.2.4 A Left-Tail Test: When surveying 592 randomly selected students at Mt. San Antonio College, it is found that 204 of them carpool to the college at least three times per week (these data were collected by Brianne Batson, an honors student at Mt. San Antonio College). We would like to test the claim that less than 40% of students at this college carpool to the college at least three times per week, using a 4% level of significance.

Step I. **Gather Sample Data**

We will need $\hat{p} = x/n = 204/592 = 0.345$, and $n = 592$. For this example, the level of significance is $\alpha = 0.04$.

Step II. **Identify the Claims and Verify that Requirements are Met**

The claim we are testing is that the proportion of students at this college who carpool at least 3 times per week is less than 40%. The null hypothesis must be $H_0 : p = 0.40$. The alternative hypothesis is $H_1 : p < 0.40$. For verifying requirements, $np = 592 \cdot 0.40 = 236.8 \geq 10$. Now, $q = 1 - p = 1 - 0.40 = 0.60$, so $nq = 592 \cdot 0.60 = 355.2 \geq 10$. Both requirements are met.

Step III. **Compute the Test Statistic (T.S.)**

To compute the test statistic, we use $p = 0.40$ from the null hypothesis, and $q = 0.60$. Thus, the test statistic is

$$T.S. : z = \frac{\hat{p} - p}{\sqrt{\frac{pq}{n}}}$$

$$= \frac{0.345 - 0.40}{\sqrt{\frac{0.40 \cdot 0.60}{592}}}$$

$$= -2.732.$$

Step IV. **Look Up the Critical Value(s) (C.V.) on Table A2**

The alternative hypothesis is $H_1 : p < 0.40$. If we think of the *less than* symbol in this hypothesis as an arrow pointing to the left, we are reminded that this is a left-tail test with $\alpha = 0.04$. As mentioned in previous examples, our test statistic is a z-score, and therefore our critical value must be a z-score as well. In Table A2, we remember that z-scores are in the bottom row, so we ignore degrees of freedom on this problem. Also, this is a one-tail test, so we look up our α in the one-tail application section of Table A2.

	One Tail Applications		
df	$\alpha = .03$	$\alpha = .04$	$\alpha = .05$
1000	1.883	1.752	1.646
Large (z)	1.881	**1.751**	1.645

From Table A2, we see a value of $z = 1.751$, but since this is a left-tail test, we must use the negative of this value, so we have $CV : z = -1.751$. This critical value is plotted in Figure 7.18.

The rejection region includes all values less than or equal to $z = -1.751$. Notice that our test statistic is $z = -2.732$, which lies within this rejection region. Because of this, *we will reject the null hypothesis.*

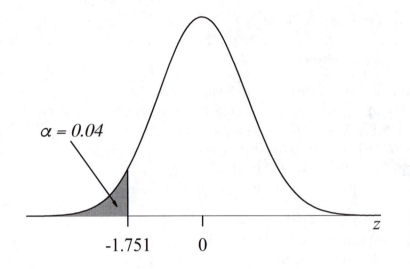

Figure 7.18: The critical value from Example 7.2.4.

Step V. Look up the p-Value on Table A1

Next, we must look-up the p-value. This is a left-tail test, so we will look up the area associated with the test statistic $z = -2.73$ (rounded to two places due to the constraints in the table) in Table A1, and since this table gives the area to the left of the test statistic, the value found in the table is the p-value.

z	.02	**.03**	.04
-2.8	.0024	.0023	.0023
-2.7	.0033	**.0032**	.0031
-2.6	.0044	.0043	.0041

In looking up $z = -2.73$, we see an area of 0.0032 and so we have p-value = 0.0032. The corresponding region on the normal distribution is shaded in Figure 7.19.

This is much smaller than α and demonstrates again that we are rejecting the null hypothesis. This p-value indicates that the difference between the sample proportion $\hat{p} = 0.345$ and the assumed population proportion $p = 0.40$ is *highly significant*.

Step VI. Make a Conclusion

We have rejected the null hypothesis in favor of the alternative hypothesis, which was $H_1 : p < 0.40$. This is our original claim, and so we decide that the data support the original claim. Our conclusion is below.

The sample data support the claim that the proportion of students who carpool to Mt. San Antonio College at least three days per week is less than 40%.

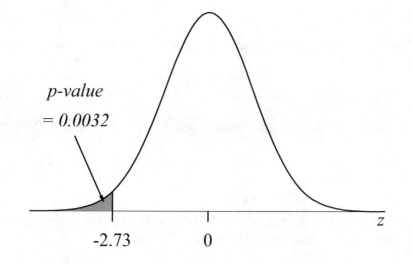

Figure 7.19: The *p-value* from Example 7.2.4.

The difference between the sample proportion and the assumed 40% proportion was highly significant, with a p-value of 0.0032.

Again we see a sample proportion which was very unlikely to have occurred by chance. We assumed that the proportion was 40%, but our sample proportion was 34.5%. This was significantly less than 40% because it only had a 0.0032 probability of occurring randomly, and thus we cannot dismiss it as sampling error. We therefore reject our assumption that the population proportion is equal to 40%.

Congratulations! We have made it through our first section of hypothesis testing. We have conducted left, right, and two-tail tests, each regarding a population proportion. The next two sections of this chapter involve tests regarding a population mean. The subject of the tests is different, but the good news is that the process is nearly identical. In mastering this topic, we find that mastery of topics to come is relatively easy.

7.2.2 Homework

Test the following claims. This process should include each step as outlined in the text.

1. Original claim: $p > 0.38$. Sample data: $x = 185$, $n = 450$. Level of significance: 5%.

2. Original claim: $p \neq 0.80$. Sample data: $x = 635$, $n = 826$. Level of significance: 4%.

3. Original claim: $p < 0.82$. Sample data: $\hat{p} = 74.2\%$, $n = 120$. Level of significance: 2%.

Determine the null and alternative hypotheses corresponding to the given claim, then use the p-value to make a conclusion.

4. In 2002, a study (by the *Women's Health Initiative*) included a large sample of post-menopausal women from the U.S. who were put on hormone replacement therapy. Of these, 1.7% had heart attacks afterward. In testing the claim that this sample proportion is greater than the 1.3% rate for other post-menopausal women, a p-value of 0.02 was observed. Assume that all test requirements are satisfied by this large random sample. Do the data support the claim (at $\alpha = 0.05$) that the heart attack rate is greater for post-menopausal women on hormone replacement therapy? Is it clear that all women should not use hormone replacement therapy?

5. In a survey of readers of the website *hunch.com*, about 58% of a large sample of Macintosh users identified themselves as politically liberal. Suppose that, in testing the claim that more than $\frac{1}{2}$ of Mac users are politically liberal, we determine a p-value of 0.0033. What is the appropriate conclusion at $\alpha = 0.01$ significance?

Test the following claims. This process should include each step as outlined in the text.

6. A politician's approval rating was 60%, but now a new poll has been conducted consisting of 850 voters, and of these, 534 approvals were given. Can you conclude, at 5% significance, that the politician's approval rating is increased above the old 60% approval rating?

7. In Mendel's pea experiments, 6,022 yellow peas were observed out of a sample of 8,023 peas. Test the claim, at 1% significance, that the proportion of these peas that were yellow is equal to 75%, against the alternative that the proportion is not 75%.

8. In 2005, the poverty rate in the United States was 12%. This represents the proportion of people whose family income is below the poverty line. In Russia, the poverty rate appears to be higher. Suppose a random sample of 200 Russian families is selected, and 36 of these have incomes below the poverty line (these figures are based on statistical data on Russia in the 2005 CIA World Factbook). Test the claim at 1% significance that the poverty rate is higher in Russia than the 12% rate for the U.S.

9. The literacy rate is defined to be the proportion of people over 15 years of age who can read. The literacy rate in the United States is 97%. This is very high, but some countries appear to be doing even better. For example, the rate in Japan appears to be *higher* than 97%. Suppose we gather a sample of 450 Japanese people over 15 years of age, and of these, 445 can read. Test the claim at 1% significance that the literacy rate in Japan is higher than the U.S. literacy rate of 97%. (The proportions from this sample are based on statistics given in the CIA World Factbook).

10. In 1972-1994 a one-in-six survey of the electoral roll, largely concerned with thyroid disease and heart disease was carried out in Wichkham, a mixed urban and rural district near Newcastle upon Tyne, in the UK. Twenty years later, a follow-up study was conducted to see which study members were still alive. The proportion of non-smoking women in the study who had died by 1994 was 78.2%. The proportion of smoking women who had died by 1994 was 80.6%. At the 5% significance level, test the claim that the proportion of women who smoke in this age bracket who are dead after 20 years of smoking is greater than 78.2%, which was the proportion observed in the sample for the non-smoking women. Assume the sample contained 1000 smoking women.

11. Success in community college mathematics courses is defined to be completion with a grade of C or better. In 2005 the overall success rate for students in mathematics courses at Mt. San Antonio College was 54.2%. In a sample of 1444 Math Pre-Algebra students, the success rate was 56.9%. Test the claim at 2% significance that the success rate for Pre-Algebra students is higher than the overall rate of 54.2%.

12. Again referring to success rates for mathematics courses at Mt. San Antonio College, the success rate for a sample of 1201 Beginning Algebra students was 52.2%. Test the claim at 5% significance that the success rate for Beginning Algebra students is lower than the overall success rate in mathematics at Mt. San Antonio College, which is 54.2%

13. The following problem is based on statistics given in "The Price of Privilege," a book written in 2006 by Madeline Levine, Ph.D. A sample of 110 children of clinically depressed mothers revealed that 67 of these developed some form of psychological illness. Test the claim that the proportion of all children of clinically depressed mothers who develop psychological illnesses is greater than 50%. Use a 5% significance level.

7.3   Tests for a Population Mean (σ Known)

7.3.1   Claims Regarding a Single Population Mean

In this section, we will test claims regarding a single population mean, μ, where the population standard deviation, σ, is known. As we have said before, it is unlikely that we would know σ if we do not know μ, and so this test, while of academic importance, is unlikely to be useful for many real applications. We cover this first, and then in the following section we will make the topic practical by assuming σ is unknown, and using s in its place. Of course, when we do this, our test statistic will no longer be normally distributed, and we will thus switch to Student's t-distribution for critical values and *p-values*. For now, however, we proceed with the dubious assumption that we know the true value for σ, and give the procedure for the hypothesis test.

Procedure 7.3.1 Testing Claims About a Single Population Mean (σ Known)

    Step I. Gather Sample Data, σ, and Verify that Requirements are Met
We need $\bar{x}$, σ (the *known* population standard deviation), n, and α, the level of significance. This procedure assumes a simple random sample. Such samples tend to represent their populations well. The requirements for this test are the requirements of the central limit theorem for sample means, which are either $n > 30$, or the population is normal.

    Step II. Identify the Claims
The null hypothesis is always H_0: $\mu = $ value. The alternative hypothesis is one of the following: H_1: $\mu > $ value, H_1: $\mu < $ value, or H_1: $\mu \neq $ value.

    Step III. Compute the Test Statistic (T.S.)
For this section, we use the T.S.: $z = \frac{\bar{x}-\mu}{\sigma/\sqrt{n}}$. The formula uses μ from H_0.

    Step IV. Look Up the Critical Value(s) (C.V.) on Table A2.
As before, the type of test is determined by H_1, and the critical values are found in the bottom row of Table A2. These are z-critical values, so be sure to use the bottom row of the table where the z scores are found. As before, the right tail test has a positive critical value as shown in Figure 7.20. The critical value for a left tail test is negative as depicted in Figure 7.21. There are two critical values for a two-tail test, one positive and one negative, as pictured in Figure 7.22.

    Step V. Look up the *p-Value* on Table A1
Use the diagrams in Figures 7.23, 7.24, and 7.25 for help in visualizing the *p-values*. For a right-tail test, look up the test statistic in Table A1, and use *p-value* $= 1 - (area\ in\ table)$. For a left-tail test, look up the test statistic in Table A1, and the area you find there is the *p-value*. For a two-tail test, look up the absolute value of the test statistic in Table A1, and use *p-value* $= 2 \cdot (1 - area\ in\ table)$.

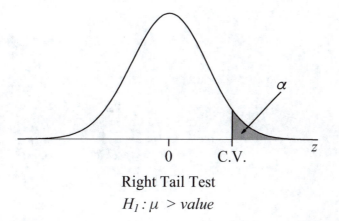

Right Tail Test

$H_1 : \mu > value$

Figure 7.20: For a right-tail test, the area to the right of the critical value is equal to α. This region is called the rejection region.

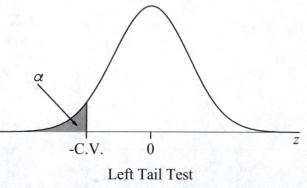

Left Tail Test

$H_1 : \mu < value$

Figure 7.21: For a left-tail test, the area to the left of the critical value is equal to α. This region is called the rejection region.

Step VI. **Make a Conclusion**

This is done as before, but for convenience we give the rules for making a conclusion below. Do not forget to mention the *p-value* and what kind of significance it indicates.

Example 7.3.2 A Right-Tail Test: The mean IQ score for adults is 100, with a known population standard deviation of $\sigma = 15$. A group of 45 college level statistics students is randomly selected, and given an IQ test. The mean score for these students is 107.6. Test the claim that the mean IQ for college level statistics students is greater than 100, using a 1% level of significance.

Step I. **Gather Sample Data, σ, and Verify that Requirements are Met**

From the problem we gather $\bar{x} = 107.6$, $\sigma = 15$, and $n = 45$. From the problem, we gather that $\alpha = 0.01$. The requirements for this test are met because $n > 30$, and σ is known.

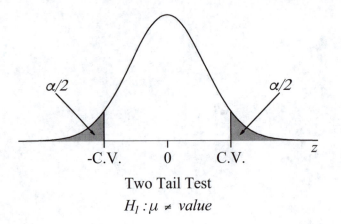

Two Tail Test

$H_1 : \mu \neq value$

Figure 7.22: On a two-tail test, there is a positive and a negative critical value. The areas beyond each critical value have area $\alpha/2$. These form the rejection region.

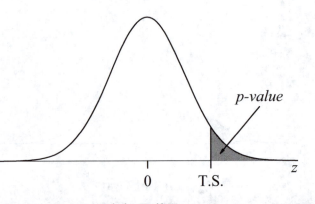

Right Tail Test

Figure 7.23: The *p-value* is the area to the right of the test statistic (T.S.).

Step II. **Identify the Claims**

The null hypothesis must be $H_0 : \mu = 100$. The alternative hypothesis is $H_1 : \mu > 100$.

Step III. **Compute the Test Statistic (T.S.)**

Remember, the formula uses μ from H_0.

$$T.S. : z = \frac{\bar{x} - \mu}{\sigma/\sqrt{n}}$$
$$= \frac{107.6 - 100}{15/\sqrt{45}}$$
$$= 3.399.$$

Step IV. **Look Up the Critical Value(s) (C.V.) on Table A2.**

Now remember that the alternative hypothesis indicates the tail for the test. Here, the alternative hypothesis is $H_1 : \mu > 100$, with an arrow pointing to the right, so this is a right-tail test, with $\alpha = 0.01$. Referring to Table A2, we

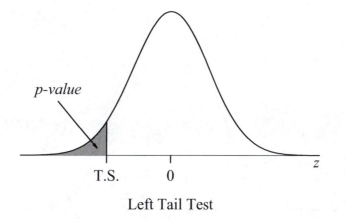

Left Tail Test

Figure 7.24: The *p-value* is the area to the left of the test statistic (T.S.).

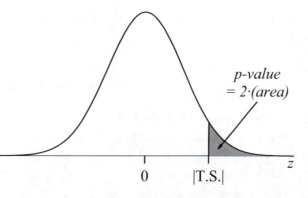

Two Tail Test

Figure 7.25: For a two-tail test, the *p-value* is twice the area to the right of the test statistic's absolute value.

remember that z-scores are in the bottom row of the table (ignoring degrees of freedom for t-scores), and we are conducting a one-tail application.

One Tail Applications			
df	$\alpha = .005$	$\boldsymbol{\alpha = .01}$	$\alpha = .02$
1000	2.581	2.330	2.056
Large (z)	2.576	**2.326**	2.054

The critical value is pictured in Figure 7.26.

From Table A2, we see in the bottom row that the critical value is $CV : z =$

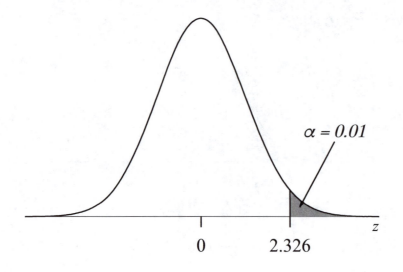

Figure 7.26: The critical value from Example 7.3.2.

2.326. Considering the test statistic $TS : z = 3.399$, we see that this is within the rejection region, so we will be rejecting the null hypothesis.

Step V. **Look up the p-Value on Table A1**

Remember that, for a right-tail test, we look up the test statistic in Table A1, and use p-value $= 1 - (area\ in\ table)$. Our test statistic is $z = 3.40$ (rounded to two places for use on Table A1), so we look this up.

z	.00	.01
3.3	.9995	.9995
3.4	**.9997**	.9997
3.5	.9998	.9998

The region on the normal distribution whose area is equal to this p-value is plotted in Figure 7.27.

The area found in Table A1 is 0.9997, so p-value $= 1 - 0.9997 = 0.0003$. This indicates a highly significant difference between the sample mean of 107.6 and the assumed population mean of 100. Because the p-value is less than α, we see again that will be rejecting the null hypothesis.

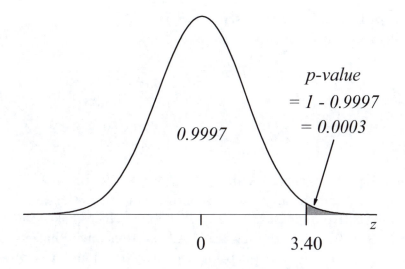

Figure 7.27: The *p-value* corresponding to a test statistic of $z = 3.40$ in Example 7.3.2.

Step VI. **Make a Conclusion**

Our original claim was that the mean IQ score for college statistics is greater than 100. This was the alternative hypothesis ($H_1 : \mu > 100$), and since we are rejecting the null hypothesis, we will favor the alternative hypothesis, which favors the original claim. Possible wording for the conclusion is below.

The data support the claim that the mean IQ score for college level statistics students is greater than 100. The difference between the sample mean and the assumed mean of 100 was highly significant, as demonstrated by the p-value of 0.0003.

Example 7.3.3 A Two-Tail Test: According to the 2005 CIA World Fact Book, in the year 2005, the mean life expectancy, worldwide, was 64.33 years, with an assumed population standard deviation of 25 years. The life expectancy in the United States was 77.71 years. Supposing that this expectancy for the United States was measured from a sample of 50 people, test the claim at 1% significance that the United States life expectancy is different from the world average.

Step I. **Gather Sample Data, σ, and Verify that Requirements are Met**

From the problem we gather $\bar{x} = 77.71$, $\sigma = 25$, and $\sigma = 25$. From the problem, we gather that $\alpha = 0.01$. The requirements for this test are met because $n > 30$, and we are assuming that σ is known.

Step II. **Identify the Claims**

The null hypothesis is $H_0 : \mu = 64.33$ (the worldwide mean). The alternative hypothesis is $H_1 : \mu \neq 64.33$.

Step III. **Compute the Test Statistic (T.S.)**

Remember, the formula uses μ from H_0.

$$T.S. : z = \frac{\bar{x} - \mu}{\sigma/\sqrt{n}}$$

$$= \frac{77.71 - 64.33}{25/\sqrt{50}}$$

$$= 3.784.$$

Step IV. **Look Up the Critical Value(s) (C.V.) on Table A2.**

Remember that the alternative hypothesis indicates the tail for the test. Here, the alternative hypothesis is $H_1 : \mu \neq 64.33$, which makes this a two-tail test, with $\alpha = 0.01$. Referring to Table A2, we remember that z-scores are in the bottom row of the table (ignoring degrees of freedom for t-scores), and we are conducting a two-tail application.

Two Tail Applications		
df	$\alpha = .01$	$\alpha = .02$
1000	2.581	2.330
Large (z)	**2.576**	2.326

From Table A2, we see in the bottom row the value $z = 2.576$.

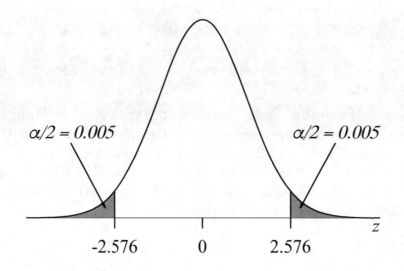

Figure 7.28: The critical values for Example 7.3.3.

This is a two-tail test, so the critical values are $CV : z = \pm 2.576$. These are plotted in Figure 7.28. Considering the test statistic, $TS : z = 3.784$, we see that this is, again, within the rejection region of the right tail, so we will be rejecting the null hypothesis.

Step V. **Look up the p-$Value$ on Table A1**

Recall that, for a two-tail test, we look up the absolute value of the test statistic in Table A1, and use p-$value = 2 \cdot (1- \ area)$. The absolute value of our test statistic is 3.78 (rounded to two places for use on Table A1), so we look this up.

z	.07	.08	.09
3.6	.9999	.9999	.9999
3.7	.9999	**.9999**	.9999
3.8	.9999	.9999	.9999

The area found in Table A1 is 0.9999, so p-$value = 2 \cdot (1 - 0.9999) = 0.0002$. The right tail area for this p-$value$ is pictured in Figure 7.29.

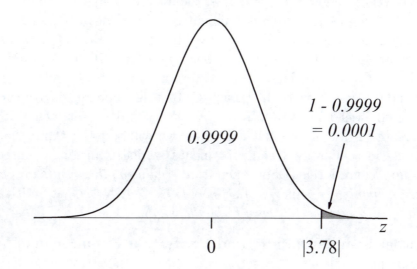

Figure 7.29: The p-value for Example 7.3.3.

This indicates a highly significant difference between the United States sample mean of 77.71 and the worldwide population mean of 64.33. Because the p-$value$ is less than α, we see again that will be rejecting the null hypothesis.

Step VI. **Make a Conclusion**

Our original claim was that United States life expectancy is different from the world average. This was the alternative hypothesis, $H_1 : \mu \neq 64.33$, and since we are rejecting the null hypothesis, we will favor the alternative hypothesis, which favors the original claim. Possible wording for the conclusion is given below.

The data support the claim that the mean life expectancy for people who live in the United States is different from the worldwide life expectancy of 64.33 years. The difference between the sample mean and the assumed mean of 64.33 years was highly significant, as demonstrated by the p-value of 0.0002.

Remember statistical significance and real-world significance are two different things. In this case, we have seen that the mean life expectancy of people who live in the United States is statistically significantly greater than the mean life expectancy of all people worldwide. In fact, the sample mean is more than 13 years larger than the population mean. Most of us would agree that 13 years longer to live, on average, is highly significant in a subjective real-world sense.

Example 7.3.4 A Left-Tail Test

At one time, it was thought that "seeding" clouds by spraying chemicals into clouds with airplanes utilizing crop-dusting equipment could significantly increase the amount of rainfall that might come from those clouds. In one experiment, data were collected from mid-1964 through 1971, where clouds were seeded in 270 storms, and sample data were gathered. For the seeded clouds, the mean rainfall was 1.53 inches with an assumed population standard deviation of 1.34 inches. In this same region, the mean rainfall for unseeded clouds was 1.74 inches. From this data, it does not seem like the seeding caused an increase in rainfall amounts. In fact, we will test the claim that the seeded clouds have a mean that is less than 1.74 inches, which is mean for unseeded clouds, using a 0.5% significance level. (Data taken from *Analyzing the results of a cloud-seeding experiment in Tasmania*, by Miller, A.J., Shaw, D.E., Veitch, L.G. & Smith, E.J. (1979), *Communications in Statistics - Theory & Methods*, vol. A8).

Step I. **Gather Sample Data, σ, and Verify that Requirements are Met**
From the problem we gather $\bar{x} = 1.53$, $\sigma = 1.34$, and $n = 270$. This time we have $\alpha = 0.005$. This is very small, but it is in our table, so there is no problem. The requirements for this test are met because $n > 30$, and σ is known.

Step II. **Identify the Claims**
The null hypothesis must be $H_0 : \mu = 1.74$. The alternative hypothesis is $H_1 : \mu < 1.74$.

Step III. **Compute the Test Statistic (T.S.)**
Remember, the formula uses μ from H_0.

$$T.S. : z = \frac{\bar{x} - \mu}{\sigma/\sqrt{n}}$$
$$= \frac{1.53 - 1.74}{1.34/\sqrt{270}}$$
$$= -2.575.$$

Step IV. **Look Up the Critical Value(s) (C.V.) on Table A2.**

Now remember that the alternative hypothesis indicates the tail for the test. Here, the alternative hypothesis is $H_1 : \mu < 1.74$, with an arrow pointing to the left, so this is a left-tail test, with $\alpha = 0.005$. Referring to Table A2, we remember that z-scores are in the bottom row of the table (ignoring degrees of freedom for t-scores), and we are conducting a one-tail application.

One Tail Applications		
df	$\alpha = .005$	$\alpha = .01$
1000	2.581	2.330
Large (z)	**2.576**	2.326

From Table A2, we see in the bottom row the value $z = 2.576$, but since this is a left-tail test, we will make this negative. Our critical value is $C.V. : z = -2.576$. Considering the test statistic, $T.S. : z = -2.575$, we see that this *very close*, but it does not fall within the rejection region, so we will be *failing to reject* the null hypothesis.

Step V. **Look up the *p-Value***

Remember that, for a left-tail test, we look up the test statistic in Table A1, and use *p-value* = (*area in table*). Our test statistic is $z = -2.57$ (rounded *down* and not up because this would push the test statistic into the rejection region where it should not be!) and so we look this up.

z	.06	.07	.08
-2.6	.0039	.0038	.0037
-2.5	.0052	**.0051**	.0049
-2.4	.0069	.0068	.0066

The area found in Table A1 is 0.0051, so *p-value* = 0.0051. This indicates a highly significant difference between the sample mean of 1.53 inches and the assumed population mean of 1.74 inches. Still, because the *p-value* is larger than α, we see again that will be failing to reject the null hypothesis. This one was close, however, and the decision would have been the reverse if α had been larger.

Step VI. **Make a Conclusion**

The null hypothesis was that the mean rainfall for seeded clouds was equal to 1.74 inches, the same as for unseeded clouds. We failed to reject this. Now, the original claim stated that the mean rainfall for seeded clouds was less than the mean rainfall for unseeded clouds, which is the alternative hypothesis, but we are not supporting the alternative hypothesis this time, so we cannot support the original claim. The conclusion is below.

The data do not support the claim (at 0.5% significance level), that the mean rainfall amount for seeded clouds is less than the mean rainfall amount for unseeded clouds which is 1.74 inches. The difference between the sample mean and the assumed mean of 1.74 inches years was highly significant, as demonstrated by the p-value of 0.0051, yet this did not meet the level of significance required by $\alpha = 0.005$.

The conclusion of the last example is rather unsatisfying because the significance required for the test statistic is extraordinarily high. We claimed that the population mean was significantly less than 1.74 inches, and observed a sample mean of 1.53 inches. The *p-value* of this statistic was 0.0051, indicating that randomly obtaining a mean at least as extreme as this is highly unlikely to have happened by chance. Still, we have not called this significant, and essentially we are dismissing the deviation of the sample mean as being nothing more than sampling error.

This is what happens when we choose a very small value for α. Remember that α is the probability of a Type I error, and sometimes we pick small α values in order to avoid the Type I error. In this case the Type I error is rejecting the null hypothesis that the mean rainfall for seeded clouds is equal to that for unseeded clouds, when this claim is actually true. Scientifically, it is not logical that the alternative hypothesis it true – that seeding clouds to *increase* rainfall should result in causing rainfall to *decrease*, and yet this is what the alternative hypothesis is claiming. Because of this, we are very reluctant to allow sample data to upset our scientific reasoning, and therefore are requiring a very significant deviation below the assumed mean before we will question the validity of our null hypothesis. This is why we have chosen such a small α value, demonstrating how applications should be considered very carefully before we decide on an α value. In this case, we consider the Type I error to be an error to avoid, so we have chosen a very small α value.

7.3.2   Homework

Test the following claims. This process should include each step as outlined in the text.

1. Original claim: $\mu > 180$. Sample data: $\bar{x} = 185.2$, $n = 450$. $\sigma = 45.7$, and $\alpha = 0.06$.

2. Original claim: $\mu \neq 43.6$. Sample data: $\bar{x} = 41.2$, $n = 20$, but sample data indicate that normality is plausible. $\sigma = 6.3$, and level of significance is 8%.

3. Original claim: $\mu < 1.59$. Sample data: $\bar{x} = 1.61$, $n = 34$, $\sigma = 0.21$, level of significance: 5%.

4. The GPAs of a random sample of 79 community college students who smoke are summarized in Figure 7.30. Does the graph give evidence that the sample is drawn from a normal population? Would it be appropriate to test the claim that the mean GPA of community college students is greater than 2.90? If so, test the claim at 1% significance, given that the sample mean is 3.01 and we assume that the population standard deviation is 0.5.

5. The foot lengths (in centimeters) of 29 randomly selected female students of Mt. San Antonio College are summarized in the histogram in Figure 7.31. Their mean is 21.8 in. Assume the population standard deviation is 1.2 in. Test the claim at 1% significance that the mean foot length for female college students is less than 22.5 cm.

Test the following claims. This process should include each step outlined in the text.

6. The mean of 52 randomly selected donation amounts to a charitable organization is $1037.54, with an assumed population standard deviation of $283.06. We understand that this mean contains sampling error, so it may not be true that the population mean is greater than $1000, but we would like to see if the sample mean is so much greater than $1000 that the differences cannot be considered nothing more than sampling error. Test this claim at a 4% level of significance and discuss whether the deviation of the sample mean above $1000 is likely to be sampling error or a truly significant difference which is not likely to be sampling error.

7. For the 100 speed of light measurements from chapter two, the mean is 299.8524 thousands of kilometers per second, with an assumed population standard deviation of 0.07901. Test the claim at 1% significance that the mean of all such speed of light (in air) estimates is equal to 299.7244 thousand kilometers per second (which is currently the best estimate for the speed of light in air), against the alternative that this is not true.

   If conducted properly, you will reject the null hypothesis for this test. Can you explain what type of error is likely to have been made (Type I or II), and why the data led us to make such an error?

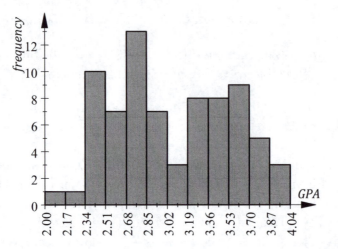

Figure 7.30: The GPAs of 79 randomly selected community college smokers.

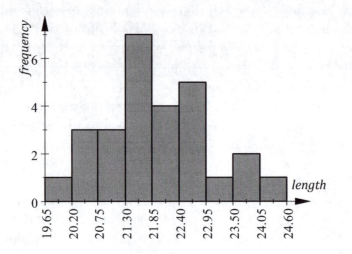

Figure 7.31: The foot lengths (in centimeters) of 29 female students of Mt. San Antonio College.

8. In a study of the length of time that students require to earn bachelor's degrees, 45 students are randomly selected and they are found to have a mean of 4.8 years (based on data from the National Center for Education Statistics). Assuming that the population standard deviation is known to be 2.1 years, test the claim at 2% significance that the average student takes more than 4 years to earn a bachelor's degree.

9. For a collection of 70 rainfall totals from Tasmania for clouds chemically seeded to increase rainfall amounts, the mean is 1.702 inches, with an assumed population standard deviation of 1.314 inches. Test the claim at 2% significance that the mean rainfall for seeded clouds in this region is less than 2 inches.

10. The journal *Environmental Concentration and Toxicology* published the article "Trace Metals in Sea Scallops" (vol. 19, pp. 326 – 1334), which gave cadmium amounts in 40 sea scallops (a shellfish eaten by humans) observed at a number of different locations in North Atlantic waters. Cadmium is used in alloys, in the electronics industry, in nickel-cadmium rechargeable batteries, and in many other applications. Cadmium poisoning leads to loss of calcium ions from the bones, leaving them brittle and easily broken. It also causes severe abdominal pain, vomiting, diarrhea, and a choking sensation. The mean cadmium level in these 40 sea scallops was 10.03 mg, with an assumed population standard deviation of 4.99 mg. Understanding that sampling error can sometimes give misleading statistics, do the data provide strong enough evidence to claim, at 1% significance, that the mean cadmium amount in sea scallops is greater than 8 mg?

7.4   Tests for a Population Mean (σ Unknown)

7.4.1   Claims Regarding a Single Population Mean

It is time to move into the more practical application, where we cannot know the population standard deviation, while testing a claim about an unknown population mean. As we learned in chapter six, the statistic

$$t = \frac{\bar{x} - \mu}{s/\sqrt{n}}$$

is distributed according to Student's t-distribution with $n-1$ degrees of freedom. This test has the same requirements as given for the normality of the sampling distribution of sample means in Chapter 6 – namely that either $n > 30$, or the population is normal. Of course, we are requiring that σ is unknown. The procedure for the test is below.

Procedure 7.4.1 Testing Claims About a Single Population Mean (σ Unknown)

Step I. **Gather Sample Data, and Verify that Requirements are Met**

We need $\bar{x}$, s, n,and α,the level of significance. This procedure assumes a simple random sample. Such samples tend to represent their populations well. The requirements for this test are the requirements of the central limit for sample means, which are either $n > 30$, or the population is normal. Also, σ should be *unknown*.

Step II. **Identify the Claims**
The null hypothesis is always $H_0 : \mu = value$. The alternative hypothesis is one of the following: $H_1 : \mu > value$, $H_1 : \mu < value$, or $H_1 : \mu \neq value$.

Step III. **Compute the Test Statistic (T.S.)**
For this section, we use the test statistic,

$$T.S. : t = \frac{\bar{x} - \mu}{s/\sqrt{n}}.$$

The formula uses μ from H_0, and note that it is a t-score and not a z-score. The critical values will be t-values as well.

Step IV. **Look Up the Critical Value(s) (C.V.) on Table A2.**
As before, the type of test is determined by H_1, and the critical values are found in Table A2. These are t-critical values, so be sure to use $n-1$ for the degrees of freedom, and *do not* go to the bottom row where the z-critical values are stored.

Step V. **Look up the *p-Value* on Table A3**
This is a new table. It is similar to Table A1, but is less precise. As with Table A1, we look up our test statistic to the nearest tenth in the first column, but now the top row is used for degrees of freedom. Like Table A1, there is

a section for negative test statistics, and another for positive test statistics. Also like Table A1, the table returns the area to the left of the test statistic. Below we demonstrate the use of Table A3, which is found in Appendix A.

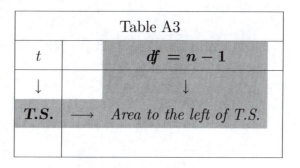

Table A3		
t		$df = n - 1$
$\downarrow$		$\downarrow$
T.S.	$\longrightarrow$	*Area to the left of T.S.*

Once you have the area, you proceed as before. For a right-tail test, look up the test statistic in Table A3, and use *p-value* $= 1 - ($*area in table*$)$. For a left-tail test, look up the test statistic in Table A3, and the area you find there is the *p-value*. For a two-tail test, look up the *absolute value* of the test statistic in Table A1, and *p-value* $= 2 \cdot (1 - $*area in table*$)$.

Step VI. **Make a Conclusion**
This is done as before – make your conclusion using non-technical language. Do not forget to mention the *p-value* for the technically minded and what kind of significance it indicates.

Example 7.4.2 A Right-Tail Test: At Mt. San Antonio College, a program called the Summer Bridge Program has been implemented with the hopes of bringing up students' grades and success rates in courses which traditionally keep some students from succeeding in college, namely Mathematics and English. For one particular math class, a group of 30 Summer Bridge math students were randomly selected, and their average grade was 3.036 (on the 4 point grading scale), with a standard deviation of 1.104. Assuming that these grades are normally distributed, and using a 1% level of significance, test the claim that the Summer Bridge math students get higher grades, on average, than the average for the non-Summer Bridge students in this math class, which is 2.074.

Step I. **Gather Sample Data, and Verify that Requirements are Met**
We gather $\bar{x} = 3.036$, $s = 1.104$, and $n = 30$. For this example, $\alpha = 0.01$. We have assumed that the population is normal.

Step II. **Identify the Claims**
The null hypothesis is $H_0 : \mu = 2.074$. The alternative hypothesis is the original claim: $H_1 : \mu > 2.074$.

Step III. **Compute the Test Statistic (T.S.).**

$$T.S. : t = \frac{\bar{x} - \mu}{s/\sqrt{n}}$$
$$= \frac{3.036 - 2.074}{1.104/\sqrt{30}}$$
$$= 4.773.$$

Step IV. **Look Up the Critical Value(s) (C.V.) on Table A2.**
This problem is a right-tail test, since $H_1 : \mu > 2.074$ is pointing to the right, with $\alpha = 0.01$. Also, we will use degrees of freedom $df = n - 1 = 30 - 1 = 29$.

	One Tail Applications		
df	$\alpha = .005$	$\alpha = .01$	$\alpha = .02$
28	2.763	2.467	2.154
29	2.756	2.462	2.150
30	2.750	2.457	2.147

From Table A2, we see that our critical value is $C.V. : t = 2.462$. This critical value is depicted in Figure 7.32. Notice that our test statistic is deep within

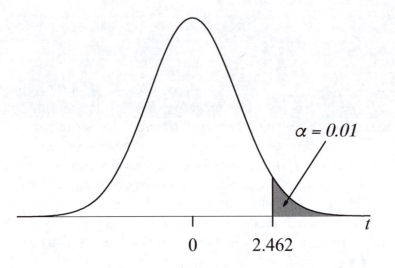

$\alpha = 0.01$

0   2.462 t

Figure 7.32: The critical value for Example 7.4.2.

the rejection region, so we will be rejecting the null hypothesis.

Step V. **Look up the p-Value on Table A3**
Our test statistic is, again $T.S. : t = 4.8$, where we have rounded this to one place after the decimal.

Remember, we have 29 degrees of freedom, as we look up the area to the left of our test statistic.

	Degrees of Freedom		
T.S.	28	$n - 1 = 29$	30
4.7	.999	.999	.999
4.8	.999	**.999**	.999
4.9	.999	.999	.999

For a right-tail test, so we use *p-value* $= 1 - (area\ in\ table) = 1 - 0.999 = 0.001$. The area used to compute the *p-value* is depicted in Figure 7.33. This indicates

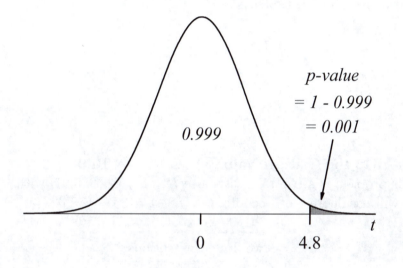

Figure 7.33: The area used to compute the *p-value* for Example 7.4.2.

that the higher mean grade for students in the Summer Bridge Program, compared to mean for non-Bridge students, is highly significant. Because the *p-value* is much smaller than α, we are reminded that we will be rejecting the null hypothesis.

Step VI. **Make a Conclusion**

We are rejecting the null hypothesis, in favor of the alternative, which was the original claim. So these sample data are in support of the original claim, and we conclude below.

The sample data support the claim that the mean grade for math students in the summer bridge program is higher than the mean for math students who do not enroll in this program. The observed difference is highly significant (with p-value = 0.0001).

Example 7.4.3 A Two-Tail Test: According to the 2000 census, the mean income level for California individuals was \$34,965.97. In Alaska, a sample of 33 individuals gave a mean income level of \$34,493.03 for the year 1999, with a sample standard deviation of \$27,118.78. Test the claim, at 10% significance, that the mean income level for individuals in Alaska is different from \$34,965.97 (the mean for California).

Step I. **Gather Sample Data, and Verify that Requirements are Met**
We gather $\bar{x} = 34,493.03$, $s = 27,118.78$, and $n = 33$. This time we have a significance level of $\alpha = 0.10$. Our sample size is greater than 30, so the test's requirements are met.

Step II. **Identify the Claims**
The null hypothesis is $H_0 : \mu = 34,965.97$. The alternative hypothesis is the original claim: $H_1 : \mu \neq 34,965.97$, so we have a two-tail test.

Step III. **Compute the Test Statistic (T.S.)**

$$
\begin{aligned}
T.S. : t &= \frac{\bar{x} - \mu}{s/\sqrt{n}} \\
&= \frac{34493.03 - 34965.97}{27118.78/\sqrt{33}} \\
&= -0.100
\end{aligned}
$$

Step IV. **Look Up the Critical Value(s) (C.V.) on Table A2.**
This problem is a two-tail test, since $H_1 : \mu \neq 34,965.97$, with $\alpha = 0.10$. Also, we will use degrees of freedom $df = n - 1 = 33 - 1 = 32$.

	Two Tail Applications		
df	$\alpha = .08$	$\alpha = .10$	$\alpha = .12$
31	1.810	1.696	1.599
32	1.808	**1.694**	1.597
33	1.806	1.692	1.596

From Table A2, we see the value $t = 1.694$. This is a two-tail test, so we will be using two critical values, $C.V. : t = \pm 1.694$. These are plotted in Figure 7.34. Notice that our test statistic is nowhere near the rejection region, so we will be failing to reject the null hypothesis.

Step V. **Look up the p-Value on Table A3**
Our test statistic is, again $T.S.t = -0.1$, where we have rounded this to one place after the decimal. Remember, we have 32 degrees of freedom, but the closest to this in this table is 30 degrees of freedom, so we will use this. Since

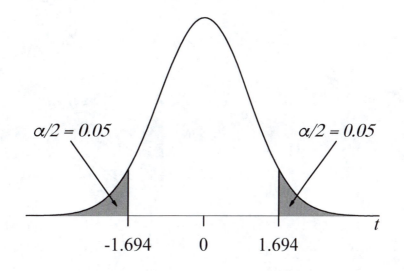

Figure 7.34: The critical values for Example 7.4.3.

this is a two-tail test, we will look up the *absolute value* of the test statistic, to find the area to its left.

	Degrees of Freedom		
T.S.	29	$n - 1 = 30$	40
0.0	.500	.500	.500
0.1	.540	**.540**	.540
0.2	.579	.579	.579

The area used to compute the *p-value* is plotted in Figure 7.35.

For the two-tail tests we use

$$p\text{-}value = 2 \cdot (1 - area)$$
$$= 2 \cdot (1 - 0.540)$$
$$= 0.920.$$

This large *p-value* indicates that the mean income for residence in Alaska in 1999 is not significantly different from the California mean. Because the *p-value* is much larger than α, we again state that we will be failing to reject the null hypothesis.

Step VI. **Make a Conclusion**
This time we have failed to reject the null hypothesis that $H_0 : \mu = 34,965.97$, which is different from the original claim that the mean was different. This means that we cannot support the original claim, and so we conclude as follows.

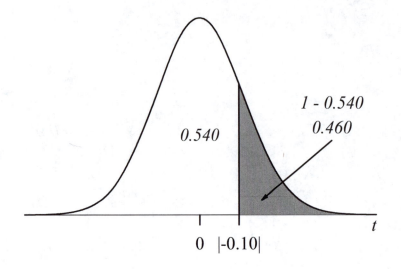

Figure 7.35: The area used to compute the *p-value* for Example 7.4.3.

The data do not support the claim that the mean income in Alaska is different from $34,965.97 (the mean for California). Any difference in sample data was insignificant, as demonstrated by the p-value of 0.920.

Remember that, though the sample data did show a difference in mean incomes, we always remember that when the difference is insignificant (as indicated by a large *p-value*), we will assume that this is due merely to sampling error.

Example 7.4.4 A Left-Tail Test: Here we go back to U.S. Census 2000 data. According to the census, the mean weight for Alaska adults is 151.79 pounds. A sample taken from 31 California residents gave a mean weight of 136.57 pounds, with a standard deviation of 36.19 pounds. Test the claim, at 1% significance, that the mean weight of California residents is less than 151.79 pounds, the mean for Alaska. (Author's note – the weights recorded in the U.S. Census look awfully small for some adults, making me wonder if some people were not telling the truth, especially in California!).

Step I. **Gather Sample Data, and Verify that Requirements are Met**
We gather $\bar{x} = 136.57$, $s = 36.19$, and $n = 31$. For this example, $\alpha = 0.01$. Since $n > 30$, the requirements are met.

Step II. **Identify the Claims**
The null hypothesis is $H_0 : \mu = 151.79$. The alternative hypothesis is the original claim: $H_1 : \mu < 151.79$.

Step III. **Compute the Test Statistic (T.S.)**

$$TS : t = \frac{\bar{x} - \mu}{s/\sqrt{n}}$$
$$= \frac{136.57 - 151.79}{36.19/\sqrt{31}}$$
$$= -2.342.$$

Step IV. **Look Up the Critical Value(s) (C.V.) on Table A2.**

This problem is a left-tail test, since $H_1 : \mu < 151.79$ is pointing to the left, with $\alpha = 0.01$. Also, we will use degrees of freedom $df = n - 1 = 31 - 1 = 30$.

	One Tail Applications		
df	$\alpha = .005$	$\alpha = .01$	$\alpha = .02$
29	2.756	2.462	2.150
30	2.750	**2.457**	2.147
31	2.744	2.453	2.144

From Table A2, we see the value $t = 2.457$, but since this is a left-tail test, we make this negative, so that our critical value is $C.V. : t = -2.457$. Notice that our test statistic is not within the rejection region, so we will be failing to reject the null hypothesis.

Step V. **Look up the p-Value on Table A3**

Our test statistic is, again $T.S.t = -2.3$, where we have rounded this to one place after the decimal. Remember, we have 30 degrees of freedom, as we look up the area to the left of our test statistic.

	Degrees of Freedom		
$T.S.$	29	$n - 1 = 30$	40
-2.2	.018	.018	.017
-2.3	.014	**.014**	.013
-2.4	.012	.011	.011

For a left-tail test we use p-value $= (area\ in\ table) = 0.014$. This indicates that the mean weight of adult Californians in the sample is significantly lower than the mean from Alaska. However, because the p-value is not smaller than α, we are reminded that we are not rejecting the null hypothesis.

Step VI. **Make a Conclusion**

The null hypothesis stated that the mean weight of Californians is equal to 151.79 pounds, and we failed to reject this, which is contrary to the original claim. We will not support the original claim in our conclusion.

The data do not support the claim that the mean weight of adult Californians is less than 151.79 pounds, the mean for adult Alaska residents. Still, the p-value of 0.014 indicated a significant difference between these means, but it did not meet the required significance level of $\alpha = 0.01$.

With this section concluded, we have completed our procedures for single sample tests which use the z and t distributions. Our next task will be to develop tests which compare parameters from two samples. In the following section we will develop a test which compares means from matched-pair samples. This test is so similar to the single parameter tests which we have covered that we will include it here with the single parameter tests. Technically, it is still a single parameter test, but the parameter is new. The mean, μ_d, represents the difference of two population means where the populations are paired value to value. This will all be explained in the following section, but for now, we conclude the sections on single sample z and t tests with a review in the form of a table on the following page.

7.4.2   Homework

Test the following claims. This process should include each step outlined in the text.

1. Original claim:$\mu > 67.9$. Sample data: $\bar{x} = 75.3$, $s = 45.7$, $n = 33$. Level of significance: 5%.

2. Original claim: $\mu \neq 114.3$. Sample data: $\bar{x} = 105.4$, $s = 9.2$, $n = 45$. Level of significance: 8%.

3. Original claim: $\mu < 1.59$. Sample data: $\bar{x} = 1.49$, $s = 0.21$, $n = 18$, but sample data indicate that normality is plausible. Level of significance: 6%.

4. The axial load of an aluminum can is the maximum weight that the sides can support before collapsing. The axial load is an important measure, because the lids are pressed onto the sides with a great deal of force. A soft drink company experimented with thinner aluminum cans, and a single batch of 75 thinner cans had a mean axial load of 269 lb, with a standard deviation of 32 lb. Using a 0.04 significance level, the claim was tested that the thinner cans have a mean axial load that is greater than 259 lb, as required by the manufacturer. The p-value for the test was 0.0034. Do the data support the claim?

5. From a sample of 26 randomly selected body temperatures of healthy adults, the mean is $97.6\,^\circ$ F, with a standard deviation of $1.08\,^\circ$ F. In order to test the claim at 1% significance that the mean temperature of healthy adults is less than $98.6\,^\circ$ F, a p-value of less than 0.0001 is computed. Do the data support the claim?

Test the following claims. This process should include each step as outlined in the text.

6. In Chapter 2, we gave the miles per gallon measurements for ten round-trip commutes in a Toyota Prius Hybrid vehicle from Lake Arrowhead to Walnut, California. The miles per gallon measurements are: {46.2, 44.9, 47.0, 46.3, 45.6, 46.0, 44.6, 45.9, 46.3, 45.6}. The mean of these values is 45.84, and the standard deviation is 0.70. We will assume that these values are measured from a normal population. Test the claim at 4% significance that the mean mileage for this commute in the Toyota Prius Hybrid is equal to 46 miles per gallon, against the claim that the mean is not equal to 46 miles per gallon.

7. The following values are hours worked per week for randomly selected students at Mt. San Antonio College in Walnut, California. { 15, 20, 40, 16, 20, 25, 19, 40, 40, 20, 20, 40, 0, 40, 35, 30, 35, 25, 29, 0, 0, 40, 16, 0, 40, 30, 25, 16, 35, 0 }. Assuming that these values are taken from a normal population, test the claim at 4% significance that students at this college work, on average, less than 28 hours per week.

8. The following are grade point averages for randomly selected students at Mt. San Antonio College in Walnut, California. { 2.7, 1.7, 2.0, 3.7, 3.1, 4.0, 3.4, 3.7, 2.5, 2.8, 3.0, 3.6, 3.3, 3.2, 3.1, 3.2, 2.6, 2.8, 2.0, 4.0, 2.7, 3.9, 3.1, 2.5, 3.7, 2.9, 3.5, 3.7, 2.2, 3.8 }. Test the claim at 1% significance that the mean G.P.A. of students at this college is greater than 2.80.

9. The data which follow are from an article on peanut butter from *Consumer Reports* (Sept. 1990), which reported the following scores for various brands: {22, 30, 30, 36, 39, 40, 40, 41, 44, 45, 50, 50, 53, 56, 56, 56, 62, 65, 68}. Assuming these values are sampled from a normal population, test the claim at 3% significance that these values are sampled from a population whose mean is less than 52.

10. The following data are overall grades for your instructor's statistics students. { 59.3, 60.2, 68.4, 69.1, 70.8, 71.7, 72.3, 73.9, 75.1, 76.1, 76.1, 80.1, 84.9, 85.3, 86, 88.1, 88.3, 93, 94, 98.6 }. Assuming that these values are sampled from a normal population, test the claim at 2% significance that the mean overall grade for your instructor's statistics students is equal to 84.0.

11. The following data are overall homework scores for your instructor's statistics students. { 33, 38, 35, 78, 85, 110, 90, 83, 78, 78, 113, 113, 85, 65, 103, 53, 118, 75, 108, 108, 25, 115, 50, 105, 70, 90, 70, 103, 85, 80, 115, 105, 113, 115, 100 }. Assuming that these values are sampled from a normal population, test the claim at 3% significance that the mean homework score for your instructor's statistics students is greater than 75.

7.5   Tests Comparing Means from Paired Samples

In this section, we begin the move toward comparing two samples where one is a treatment group and the other is a control group used as a basis for comparison. This first attempt will fall slightly short of this – we *will* compare two sets of numbers, usually measured from the same group before and after some treatment is applied. Thus, the *before treatment* measurements may be used as a basis to which we may compare the values measured after the treatment. This is not a true control, because no effort is made to control confounding due to the placebo effect, but we are beginning to move toward a true control.

Whenever we work with *before* and *after treatment* measurements, we will call the samples *dependent*, or *matched-pairs*. With matched-pair samples, it is not necessarily the case that measurements be taken before and after a treatment. It will be the case, however, that they be the same type of measurements, taken under two different conditions, and that the two samples have a natural one-to-one pairing between them.

7.5.1   Claims which Compare Means from Paired Samples

The way that we shall compare the samples is by subtracting one member of each pair from the other, to measure the difference. We will then compute the mean and standard deviation of all such differences, and using these, test a claim about the mean difference for the entire population. The test statistic will be the same as used in the previous section, but we will change the names of some of the variables.

Example 7.5.1 Prozac is a drug which is meant to elevate the moods of people who suffer from depression. In the following sample (provided by the WebStat website of the University of New England, School of Psychology), mood levels were recorded before and after taking Prozac. In these values, a higher score means a less depressed psychological state. We will label the *Before Prozac* values as x, and the *After Prozac* values as y.

$x = Before$	$y = After$
3	5
0	1
6	5
7	7
4	10
3	9
2	7
1	11
4	8

As mentioned, we are not directly interested in the *before* or *after* values. What we want to know is how the patients' moods were changed.

We measure the change by subtracting each y from its corresponding x, and calling the difference $d = x - y$.

x	y	$d = x - y$
3	5	-2
0	1	-1
6	5	1
7	7	0
4	10	-6
3	9	-6
2	7	-5
1	11	-10
4	8	-4

We will be computing the sample mean, $\bar{d}$, and standard deviation, s_d, of these difference with the usual methods. Do not let the new labels confuse you, these are the same mean and standard deviations introduced in Chapter 2, but we are calling the sample values d, instead of x.

In computing the mean and standard deviation of the sample d values, we have $\bar{d} = -3.667$, $s_d = 3.5$, and we note that the sample size (the number of differences) is $n = 9$. We will assume that the population of *differences* is normal.

Next, we would like to introduce notation for the population means of the *before* (x) measurements and the *after* (y) measurements. Let μ_1 be the *before* mean, and μ_2 be the *after* mean. Since we are computing the mean of the differences, it makes sense to define the mean of all differences to be $\mu_d = \mu_1 - \mu_2$.

Let us test a claim. Let us claim that Prozac is effective in improving the mental state of patients with depression. So we are claiming that the mean after treatment is higher than the mean before treatment, or $\mu_1 < \mu_2$. That could be our alternative hypothesis. Note that this is equivalent to saying that $\mu_1 - \mu_2 < 0$, and since $\mu_d = \mu_1 - \mu_2$, we are saying that $\mu_d < 0$, which is another form of the same alternative hypothesis. Either way, this is a left-tail test. So the hypotheses are

$$H_0 : \mu_1 = \mu_2 \text{ (i.e. } \mu_d = 0),$$

and

$$H_1 : \mu_1 < \mu_2 \text{ (i.e. } \mu_d < 0).$$

Now, our old test statistic for a sample mean, where σ is unknown, is $t = \frac{\bar{x} - \mu}{s/\sqrt{n}}$. This is what we need, but we will update it to use the correct names:

$$T.S. : t = \frac{\bar{d} - \mu_d}{s_d/\sqrt{n}}.$$

For our example, the test statistic becomes

$$T.S. : t = \frac{\bar{d} - \mu_d}{s_d/\sqrt{n}}$$
$$= \frac{-3.667 - 0}{3.5/\sqrt{9}}$$
$$= -3.143.$$

Assuming a level of significance of $\alpha = 0.01$, we look up the critical value on Table A2 with $df = n - 1 = 9 - 1 = 8$ degrees of freedom.

	One Tail Applications		
df	$\alpha = .005$	$\boldsymbol{\alpha = .01}$	$\alpha = .02$
7	3.499	2.998	2.517
8	3.355	**2.896**	2.449
9	3.250	2.821	2.398

The table reads $t = 2.896$, but this is a left-tail test, so we will use the critical value $C.V. : t = -2.896$. This critical value is pictured in Figure 7.36.

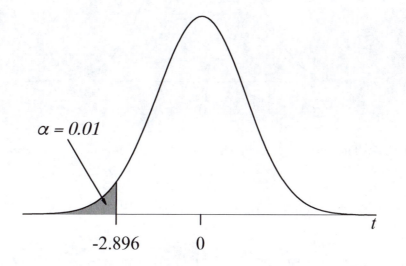

$\alpha = 0.01$

-2.896   0

Figure 7.36: The critical value for Example 7.5.1.

Note that the test statistic does lie within the rejection region, so we will be rejecting the null hypothesis.

Looking up the *p-value* for this test on Table A3, we remember to use the section for negative test statistics, with T.S.: $t = -3.143$.

Since this is a left-tail test, we use the area directly from the table as the *p-value*.

	Degrees of Freedom		
T.S.	7	8	9
−3.0	.010	.009	.008
−3.1	.009	.007	.006
−3.2	.008	.006	.005

So, the *p-value* is 0.007, which indicates that the mean difference is highly statistically significant, and it appears that the Prozac is having some statistically significant effect on the average mood of patients who use it. The shaded region in Figure 7.37 is equal to this *p-value*.

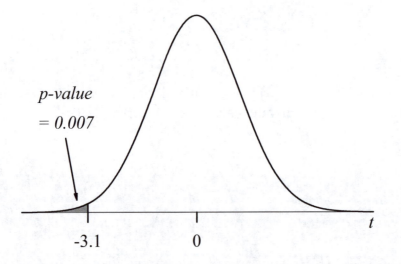

Figure 7.37: The shaded area above is equal to the *p-value* for Example 7.5.1.

Because this *p-value* is less than α, we reemphasize that the null hypothesis is to be rejected.

To conclude, *the sample data support the claim that Prozac is effective on raising the mean mood of patients with depression. The improvement was highly significant, as indicated by the p-value of 0.007.*

With a *p-value* as small as 0.0073, we have concluded that the improvement after treatment is unlikely to be due to sampling error, and that it is more likely to be a result of the treatment. Prozac does appear to bring about a statistically significant improvement in patients' moods. Whether this statistically significant improvement actually has any *real-world significance* – an improvement that patients can actually detect subjectively – we must leave to patients and their doctors to decide.

We have done all of this without a procedure, because it was easiest to explain the ideas while going through an example. Now that this is done, we will outline the process, and then conclude with another example.

Procedure 7.5.2 Testing Claims About Two Means Using Matched Pair Samples

Step I. **Gather Sample Data, and Verify that Requirements are Met**
For each x, y pair present, compute $d = x - y$, then compute $\bar{d}$, the mean of the d values, and s_d, the standard deviation of the d values. We also need n, which is the number of sample (x, y) pairs, and α, the level of significance. This procedure assumes a simple random sample. Such samples tend to represent their populations well. Additionally, it is required that either $n > 30$, or that the population of differences, d, is normal.

Step II. **Identify the Claims**
The null hypothesis is usually H_0: $\mu_d = 0$. In this case the alternative hypothesis would be one of the following: H_1: $\mu_1 > \mu_2$ (i.e. $\mu_d > 0$), H_1: $\mu_1 < \mu_2$ (i.e. $\mu_d < 0$), or H_1: $\mu_1 \neq \mu_2$ (i.e. $\mu_d \neq 0$).

Step III. **Compute the Test Statistic (T.S.)**
For this section, we use the test statistic,

$$T.S. : t = \frac{\bar{d} - \mu_d}{s_d/\sqrt{n}}.$$

The formula uses μ_d from H_0, and when the null hypothesis is $H_0 : \mu_d = 0$, the second term in the numerator vanishes.

Step IV. **Look Up the Critical Value(s) (C.V.) on Table A2.**
As before, the type of test is determined by H_1, and the critical values are found in Table A2. These are t-critical values, so be sure to use $n - 1$ for the degrees of freedom, and *do not* go to the bottom row where the z-critical values are stored.

Step V. **Look up the p-Value on Table A3**
This is done as in Section 7.4.

	Degrees of Freedom
t	$n - 1$
$\downarrow$	$\downarrow$
T.S. $\longrightarrow$	*Area to the left of T.S.*

Once you have the area, you proceed as before. For a right-tail test, look up the test statistic in Table A3, and use *p-value* = 1 − (*area in table*). For a left-tail test, look up the test statistic in Table A3, and the area you find there is the *p-value*. For a two-tail test, look up the absolute value of the test statistic in Table A1, and *p-value* = 2 · (1 − *area in table*).

Step VI. **Make a Conclusion**

The conclusions are as before. Try to make them as understandable to ordinary people as possible, and include a comment about the *p-value* for the more statistically minded people.

Example 7.5.3 A Two-Tail Test: Do physical characteristics of twin babies affect their birth order? Is it possible that the firstborn has an advantage in some way over the second born? The data below give head circumferences for 10 pairs of twins, with $x = firstborn\ head\ circumference$, and $y = second\ born\ head\ circumference$. The mean head circumference for the firstborn babies was 56.25 cm, while the mean for the second born babies was 56 cm. The firstborns have a slightly larger mean head circumference, but is this statistically significant?

Head Circumferences	
$x = Firstborn$	$y = Secondborn$
54.7	54.2
53.0	52.9
57.8	56.9
56.6	55.3
53.1	54.8
57.2	57.2
57.2	57.2
57.2	55.8
57.2	56.5
58.5	59.2

These data are naturally given in pairs, where the same measurement is being taken for the x and y values in each pair, thus this is a matched-pair application. Let us test the claim that the mean head circumference for the firstborn twin is larger than the mean head circumference for the second born twin, using a 6% level of significance, and assuming that the x and y values both come from normal populations, respectively.

Step I. **Gather Sample Data, and Verify that Requirements are Met**

The $d = x - y$ values are tabulated below, and the mean and standard deviation of the differences are $\bar{d} = 0.25$ and $s_d = 0.941$, respectively. Also $n = 10$. The level of significance is $\alpha = 0.06$.

We have assumed that the population of differences is normal.

x	y	$d = x - y$
54.7	54.2	0.5
53.0	52.9	0.1
57.8	56.9	0.9
56.6	55.3	1.3
53.1	54.8	−1.7
57.2	57.2	0.0
57.2	57.2	0.0
57.2	55.8	1.4
57.2	56.5	0.7
58.5	59.2	−0.7

Step II. Identify the Claims

The original claim is that the mean head circumference for the firstborn twin is larger than the mean head circumference for the second born twin, or $\mu_1 > \mu_2$, so this is the alternative hypothesis,

$$H_1 : \mu_d > 0 \text{ (i.e. } \mu_1 > \mu_2).$$

The null hypothesis is H_0: $\mu_d = 0$, as usual.

Step III. Compute the Test Statistic (T.S.)

This time, the null hypothesis is H_0: $\mu_d = 0$, so we compute the test statistic.

$$T.S. : t = \frac{\bar{d} - \mu_d}{s_d / \sqrt{n}}$$
$$= \frac{0.25 - 0}{0.941 / \sqrt{10}}$$
$$= 0.840.$$

Step IV. Look Up the Critical Value(s) (C.V.) on Table A2.

As usual, the type of test is determined by H_1, which, in this case is H_1: $\mu_d > 0$ (or $\mu_1 > \mu_2$), indicating a right-tail test, and the critical values are found in Table A2, under $\alpha = 0.06$, with $df = n - 1 = 10 - 1 = 9$ degrees of freedom.

	One Tail Applications		
df	$\alpha = .05$	$\alpha = .06$	$\alpha = .08$
8	1.860	1.740	1.549
9	1.833	**1.718**	1.532
10	1.812	1.700	1.518

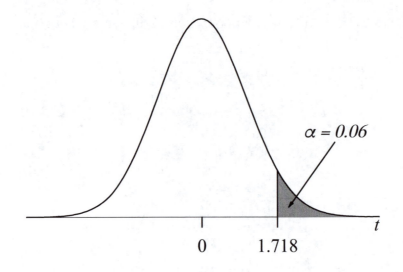

Figure 7.38: The critical value for Example 7.5.3.

From Table A2, we see that the critical value is $C.V. : t = 1.718$. This is pictured in Figure 7.38.

The test statistic did not make it into the rejection region, so we will fail to reject the null hypothesis.

Step V. **Look up the p-Value on Table A3**
Rounding our test statistic to $t = 0.8$, and under 9 degrees of freedom we look up the area to the left of the test statistic on Table A3.

	Degrees of Freedom		
$T.S.$	8	**9**	10
0.7	.748	.749	.750
0.8	.777	**.778**	.779
0.9	.803	.804	.805

The area that we find in Table A3 is 0.778, but since this is a right-tail test, we use

$$p - value = 1 - area$$
$$= 1 - 0.778$$
$$= 0.222.$$

This p-value is the area of the shaded region pictured in Figure 7.39.

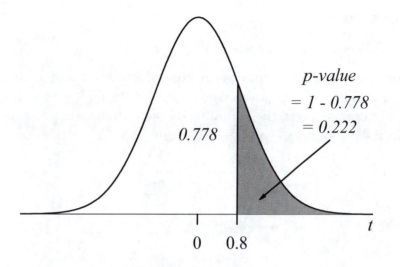

Figure 7.39: The shaded region above is equal to the *p-value* for Example 7.5.3.

This is rather large, and indicates that the difference between these two means is not significant, and emphasizes again that we will fail to reject the null hypothesis.

Step VI. **Make a Conclusion**

We have failed to reject the null hypothesis, which says that there is no difference between the mean head circumferences for first and second born twins. This is contrary to the original claim, so we will conclude that the data do not support the original claim. Here we have another case where the sample data do seem to be in support of the original claim, but the slightly larger average for the firstborn children is not significant, and thus we will dismiss it as nothing more than sampling error.

The sample data do not support the claim that the mean head circumference for the firstborn twin is larger than the mean head circumference of the second born twin. There was a difference, but this was statistically insignificant, as highlighted by a p-value of 0.222.

7.5.2 Homework

Conduct the following hypothesis tests.

1. The following are scores for an algebra test which was taken by 7 students before and after attending an algebra review seminar. Assuming that the population of differences is normal, Test the claim at 1% significance that the algebra review is effective in raising students' average test score.

Before	After
75	79
62	67
81	82
58	61
90	89
73	76
69	75

2. To determine the effect that exercise has on lactic acid levels in the blood, the authors of "A Descriptive Analysis of Elite-Level Racquetball" (Research Quarterly for Exercise and Sport, 1991, pp. 109 – 114), listed the blood lactic acid levels of 8 different players, before and after playing three games of racquetball. At the 1% level of significance, test the claim that the mean lactic acid levels are higher after exercise. Assume that the population of differences is normal.

Before	After
13	18
20	37
17	40
13	35
13	30
16	20
15	33
16	19

3. The compound mCPP is thought to be a hunger suppressant. In an experiment, nine obese men had their weight change (kg) recorded after each of two two-week periods, once when taking a placebo and once when taking mCPP. There was a two week *washout period* between measurement periods. The values are given below. Note that negative values under mCPP or Placebo indicate a weight loss. A negative difference between mCPP and placebo indicates that more weight was lost with mCPP than with the placebo. The data are summarized in the table below. Test the claim at 5% significance that, on average, more weight is lost with the mCPP appetite suppressant than with a placebo. These data are published at: http://www.stat.wisc.edu/courses/st371-larget/paired-handout.pdf.

Subject	mCPP	Placebo	Difference
1	0.0	–1.1	1.1
2	–1.1	0.5	–1.6
3	–1.6	0.5	–2.1
4	–0.3	0.0	–0.3
5	–1.1	–0.5	–0.6
6	–0.9	1.3	–2.2
7	–0.5	–1.4	0.9
8	0.7	0.0	0.7
9	–1.2	–0.8	–0.4

4. A study was completed to investigate whether oat bran lowers serum cholesterol levels in people (Pagano & Gauvreau, 1993, p. 252). Fourteen individuals were randomly assigned a diet that included either oat bran or corn flakes. After two weeks on the initial diet, serum low-density lipoprotein levels (mg/dl) were measured. Each subject was then 'crossed-over" to the alternate diet. After two-weeks on this second diet, low-density lipoprotein levels were once again recorded. The data are recorded below. Assuming the population of differences is normal, test the claim that the mean lipoprotein level is higher when patients are eating corn flakes for breakfast than when they are eating oatmeal for breakfast. Use 1% significance. These values are published at: http://www2.sjsu.edu/faculty/gerstman/StatPrimer/paired.pdf.

Corn Flakes	Oat Bran
4.61	3.84
6.42	5.57
5.40	5.85
4.54	4.80
3.98	3.68
3.82	2.96
5.01	4.41
4.34	3.72
3.80	3.49
4.56	3.84
5.35	5.26
3.89	3.73
2.25	1.84
4.24	4.14

5. In a recent study, body dissatisfaction was measured in adult women ranging in age from 20 to 84. Using a *figure rating scale*, the women were asked to select their current figure and their ideal figure from a series of silhouette drawings. The drawings were numbered from 10-90. Body dissatisfaction is the discrepancy (or difference) between the ideal and current ratings. The following table lists the means that were compared in the study. Test the claim, at 2% significance, that the mean body dissatisfaction level (or difference) is greater than 9. (Thus, on this example we are testing a null hypothesis H_0: $\mu_d = 9$.) Assume that the population of differences is normal. These data are based on a study conducted by Tiggemann, M. & Lynch, J. E. (2001). *Body image across the life span in adult women: The role of self-objectification.* Developmental Psychology, 37(2), 243-253.

Age Group	Current Figure	Ideal Figure
20 – 29	44.75	34.65
30 – 39	44.50	35.54
40 – 49	45.84	35.95
50 – 59	47.79	38.05
60 – 69	47.93	38.00
70 – 85	51.81	40.50

6. In an experiment, two eggs from each of four female frogs are collected, and one egg is put into a treatment group, while the other is put into a control group. The treatment group of eggs is injected with progesterone, while the control group is not injected. The cAMP level was then measured for each egg in each group. The table below summarizes the results. At the 5% level of significance, test the claim that the mean cAMP level for the eggs in the control group is greater than the mean cAMP level for the eggs which were injected with progesterone. Assume that the population of differences is normal.

Frog	Control	Progesterone
1	6.01	5.23
2	2.28	1.21
3	1.51	1.40
4	2.12	1.38

7. The following table gives arm spans and corresponding heights of 31 randomly selected adults. Many people claim that a person's arm span is the same as the corresponding person's height. Use these values to test the claim that the mean arm span equal to the mean height. Test this claim at 1% significance. These values were measured by Mt. San Antonio College honors student Alphonso Tran.

Arm Span (cm)	Height (cm)
170.9	172.0
175.0	177.0
177.0	173.0
177.0	176.0
178.1	178.1
183.9	180.1
188.0	188.0
188.2	186.9
185.9	183.9
188.0	182.1
188.2	181.1
188.5	192.0
194.1	193.0
196.1	183.9
199.9	185.9
167.9	168.9
150.1	156.0
152.9	159.0
156.0	162.1
157.0	159.8
159.0	162.1
159.8	154.9
161.0	159.8
161.0	162.1
162.8	167.1
162.1	170.2
165.1	166.1
170.2	170.2
170.4	167.1
173.0	184.9
199.9	176.0

8. When college class locations are inconvenient for students, does student success decline? The following sample includes 19 success rate pairs, each pair corresponding to a particular teacher who taught two identical mathematics classes in two different locations – one conveniently located, and the other inconveniently located. Under the assumption that the population of differences between these rates is normal, test the claim at 8% significance that the mean success rate is higher in classrooms which are conveniently located for students.

Teacher	Convenient	Inconvenient
1	37.3	41.9
2	51.3	65.6
3	47.2	46.1
4	60.5	57.4
5	62.5	72.6
6	59.4	62.5
7	32.4	39.4
8	70.0	65.7
9	45.7	63.6
10	44.4	54.5
11	68.4	60.8
12	67.8	34.2
13	55.8	36.1
14	88.8	43.7
15	76.4	25.0
16	32.2	43.7
17	57.3	30.2
18	60.5	44.3
19	79.3	68.5

9. Do grocery stores charge different prices in stores located in affluent neighborhoods compared to stores in less affluent neighborhoods? The following paired data give prices for a given product at two different stores – one in an affluent area and one in a non-affluent area. Test the claim that the mean price for these randomly selected products is greater in the chosen affluent area at 1% significance.

Product	Price (Affluent Area)	Price (Non-Affluent Area)
lowfat milk	2.99	2.99
nonfat milk	2.79	2.79
whole milk	3.29	3.29
apple juice	2.99	3.09
brown rice	1.79	1.99
pasta noodles	1.39	1.10
rocky road ice cream	4.69	4.43
cookie dough ice cream	3.79	3.29
carrots	1.79	1.58
toilet paper 4-pack	2.07	1.79
1 liter cola	1.09	0.89
1 liter lemon-lime soda	1.09	0.89
cheese	2.79	2.59
dozen eggs	1.89	1.29
1 lb. oranges	0.88	0.88
1 lb. bananas	0.79	0.79
broccoli	1.35	1.35
wheat bread	2.29	2.09
white bread	1.79	1.59
orange juice	1.87	1.87
cup noodles	0.20	0.25
cookies	2.39	2.00
1 gal. water	1.29	0.99
fresh meat	3.00	3.25
fresh poultry	2.79	3.00
butter	1.29	1.39
brown sugar	1.68	1.62
salt	0.88	0.88
pepper	0.69	0.69
vegetable oil	1.69	1.69
bacon	1.77	1.59

Chapter 8

Tests Comparing Two Parameters

In this chapter, our goal is to make hypothesis testing more practical. In Chapter 7, the basic concepts of hypothesis testing were introduced, but they all had a common limitation – none of them allowed for a comparison between a treatment group and an independent control group. We would like to correct this limitation here.

By the end of this chapter we will be able to compare means and proportions from treatment groups to means and proportions, respectively, from control groups. Before we can do this, we need to discuss the tool which we will use for this comparison. This tool is simply subtraction. We will subtract sample proportions first, and then sample means secondly. These differences have probability distributions, in fact sampling distributions, which we must learn to understand before we can proceed. This is how we begin in Section 8.2.

8.1   Sampling Distributions of Differences

As mentioned, our main tool for comparing two statistics – one from a treatment group and one from a control group – is subtraction. In this section, our goal is to understand the sampling distribution of differences. Once we understand this distribution, we may then begin to discuss the conditions which make one of these differences unusual or significant.

We begin with a review of the rules of expected value, variance, and differences between two random variables, all covered in Chapter 4.

The rules which we need for this discussion are given again below.

1. $E(x - y) = E(x) - E(y)$

2. $Var(x - y) = Var(x) + Var(y)$

3. If x and y are independent random variables from two respective normally distributed populations, then the population of all differences, $x - y$, is normal as well.

This is all the theory we need to develop a theory for sampling distributions of differences between two statistics.

8.1.1 The Sampling Distribution of Differences, $\hat{p}_1 - \hat{p}_2$

Let us start with two hypothetical populations, where we are measuring sample proportions from each. From each of these populations, we consider the sampling distributions of all possible sample proportions.

The Sampling Distribution of Sample Proportions, $\hat{p}_1$, from Population 1

This population includes sample proportions, $\hat{p}_1$, from samples of size n_1 which are all estimates of a population proportion, p_1. These estimates are approximately normal in their distribution provided that $n_1 p_1 \gtrsim 10$ and $n_1 q_1 \gtrsim 10$, with mean $\mu_{\hat{p}_1} = E\left(\hat{p}_1\right) = p_1$, standard error (or *deviation*), $\sigma_{\hat{p}_1} = \sqrt{\frac{p_1 q_1}{n_1}}$, and variance $Var\left(\hat{p}_1\right) = \sigma_{\hat{p}_1}^2 = \frac{p_1 q_1}{n_1}$.

The Sampling Distribution of Sample Proportions, $\hat{p}_2$, from Population 2

This population includes sample proportions, $\hat{p}_2$, from samples of size n_2 which are all estimates of a population proportion, p_2. These estimates are approximately normal in their distribution provided that $n_2 p_2 \gtrsim 10$ and $n_2 q_2 \gtrsim 10$, with mean $\mu_{\hat{p}_2} = E\left(\hat{p}_2\right) = p_2$, standard error, $\sigma_{\hat{p}_2} = \sqrt{\frac{p_2 q_2}{n_2}}$, and variance $Var\left(\hat{p}_2\right) = \sigma_{\hat{p}_2}^2 = \frac{p_2 q_2}{n_2}$.

The Sampling Distribution of Differences, $\hat{p}_1 - \hat{p}_2$

We compare two sample proportions, $\hat{p}_1$ and $\hat{p}_2$, by subtracting them. So that we may understand how these differences behave, we construct the sampling distribution of differences, $\hat{p}_1 - \hat{p}_2$. First, we address the issue of normality. We know that the difference between two normal random variables is normal, and we have given conditions above indicating when $\hat{p}_1$ and $\hat{p}_2$ are normally distributed, and under these conditions, the differences, $\hat{p}_1 - \hat{p}_2$, are normally distributed as well.

This population of differences has mean that is determined as follows.

$$
\begin{aligned}
\mu_{\hat{p}_1 - \hat{p}_2} &= E\left(\hat{p}_1 - \hat{p}_2\right) \\
&= E\left(\hat{p}_1\right) - E\left(\hat{p}_2\right) \\
&= p_1 - p_2.
\end{aligned}
$$

The variance of this population is established as follows.

$$
\begin{aligned}
\sigma_{\hat{p}_1 - \hat{p}_2}^2 &= Var\left(\hat{p}_1 - \hat{p}_2\right) \\
&= Var\left(\hat{p}_1\right) + Var\left(\hat{p}_2\right) \\
&= \frac{p_1 q_1}{n_1} + \frac{p_2 q_2}{n_2}
\end{aligned}
$$

The standard error (or *deviation*) is the square root of variance.

$$\sigma_{\hat{p}_1 - \hat{p}_2} = \sqrt{Var\left(\hat{p}_1 - \hat{p}_2\right)}$$
$$= \sqrt{\frac{p_1 q_1}{n_1} + \frac{p_2 q_2}{n_2}}$$

We finally have what we need. Our main tool for determining when a difference, $\hat{p}_1 - \hat{p}_2$, is significant is the z-score, and now that we have the mean, and the standard deviation, we are able to compute it.

$$z = \frac{(\hat{p}_1 - \hat{p}_2) - \mu_{\hat{p}_1 - \hat{p}_2}}{\sigma_{\hat{p}_1 - \hat{p}_2}}$$
$$= \frac{(\hat{p}_1 - \hat{p}_2) - (p_1 - p_2)}{\sqrt{\frac{p_1 q_1}{n_1} + \frac{p_2 q_2}{n_2}}}.$$

Theorem 8.1.1 The difference, $\hat{p}_1 - \hat{p}_2$, may be standardized using the z-score formula given below.

$$z = \frac{(\hat{p}_1 - \hat{p}_2) - (p_1 - p_2)}{\sqrt{\frac{p_1 q_1}{n_1} + \frac{p_2 q_2}{n_2}}}$$

This statistic is approximately normal in its distribution provided that $n_1 p_1 \gtrsim 10$, $n_1 q_1 \gtrsim 10$, $n_2 p_2 \gtrsim 10$ and $n_2 q_2 \gtrsim 10$. Because p_1 and p_2 are generally unknown, these conditions cannot be verified. However, there is some sense that the conditions may be met if each sample has at least 10 successes and 10 failures.

This will be the model for the first test statistic of Chapter 8. It needs some minor alteration, because we will not know all of the values required by the formula. The numerator is no problem, since $\hat{p}_1$ and $\hat{p}_2$ will come from sample data, and the difference $p_1 - p_2$ will be assumed in the null hypothesis. Most often the null hypothesis will simply be that there is no difference between the population proportions, meaning $p_1 = p_2$ and so, in the numerator, we will use $p_1 - p_2 = 0$.

The problem of evaluation is in the denominator. The denominator requires p_1, p_2, q_1, and q_2 values, but these are all unknown population parameters. Our solution will be to estimate these parameters using statistics, but how we use them will depend on the null hypothesis. We know $\hat{p}_1 = \frac{x_1}{n_1}$ and $\hat{p}_2 = \frac{x_2}{n_2}$, with $\hat{q}_1 = 1 - \hat{p}_1$ and $\hat{q}_2 = 1 - \hat{p}_2$. These could be used to estimate p_1, p_2, q_1, and q_2, but if the null hypothesis says that $p_1 = p_2$, we should not use different values for these in the denominator. Our best approach is, assuming that $p_1 = p_2$, to estimate this common proportion by pooling the samples together to compute a common proportion for both samples. This will be the total number of successes divided by the total of both sample sizes. This proportion will be called $\bar{p}$, the *pooled proportion*, which is thus given by the formula, $\bar{p} = \frac{x_1 + x_2}{n_1 + n_2}$, and we will define $\bar{q} = 1 - \bar{p}$. So, when our null hypothesis says $p_1 = p_2$, we will use the test statistic:

$$T.S. : z = \frac{(\hat{p}_1 - \hat{p}_2) - (p_1 - p_2)}{\sqrt{\frac{\bar{p}\bar{q}}{n_1} + \frac{\bar{p}\bar{q}}{n_2}}}.$$

If, however, in the unusual case that the null hypothesis assumes that the difference $p_1 - p_2 = k \neq 0$, then we may use different values for p_1 and p_2. We will thus estimate them by $\hat{p}_1$ and $\hat{p}_2$, respectively, and use $\hat{q}_1 = 1 - \hat{p}_1$ and $\hat{q}_2 = 1 - \hat{p}_2$ as estimates for q_1, and q_2. In this case the test statistic will be:

$$z = \frac{(\hat{p}_1 - \hat{p}_2) - (p_1 - p_2)}{\sqrt{\frac{\hat{p}_1 \hat{q}_1}{n_1} + \frac{\hat{p}_2 \hat{q}_2}{n_2}}}.$$

8.1.2   The Sampling Distribution of Differences, $\bar{x}_1 - \bar{x}_2$

Next, we create another sampling distribution, this time it will be for the difference of two sample means $\bar{x}_1 - \bar{x}_2$. The development is conceptually the same.

Let us start with two hypothetical populations, where we are gathering sample means from each.

The Sampling Distribution of Sample Means, $\bar{x}_1$, from Population 1

Consider the sampling distribution of sample means, $\bar{x}_1$, from samples gathered randomly from Population 1 whose mean is μ_1 and standard deviation is σ_1. These sample means are gathered from samples of size n_1 have mean $\mu_{\bar{x}_1} = E(\bar{x}_1) = \mu_1$, standard error $\sigma_{\bar{x}_1} = \frac{\sigma_1}{\sqrt{n_1}}$, and variance $\sigma_{\bar{x}_1}^2 = Var(\bar{x}_1) = \frac{\sigma_1^2}{n_1}$, and their distribution is approximately normal in provided that $n_1 > 30$ or Population 1 is normal.

The Sampling Distribution of Sample Means, $\bar{x}_2$, from Population 2

Next, consider the sampling distribution of sample means, $\bar{x}_2$, from samples gathered randomly from Population 2 whose mean is μ_2 and standard deviation is σ_2. These sample means are gathered from samples of size n_2 have mean $\mu_{\bar{x}_2} = E(\bar{x}_2) = \mu_2$, standard error $\sigma_{\bar{x}_2} = \frac{\sigma_2}{\sqrt{n_2}}$, and variance $\sigma_{\bar{x}_2}^2 = Var(\bar{x}_2) = \frac{\sigma_2^2}{n_2}$, and their distribution is approximately normal in provided that $n_2 > 30$ or Population 2 is normal.

Our goal is to compare means from samples drawn independently from two different populations. Presumably, these samples represent treatment and control groups in an experimental study. As before, our main tool for comparing means, $\bar{x}_1$ and $\bar{x}_2$, is subtracting them: $\bar{x}_1 - \bar{x}_2$. To determine when such a difference is significant, we need knowledge of the distribution of such differences. Thus we construct another sampling distribution, this time for the differences, $\bar{x}_1 - \bar{x}_2$.

The Sampling Distribution of Differences, $\bar{x}_1 - \bar{x}_2$

The distribution of differences between two sample means, $\bar{x}_1 - \bar{x}_2$, is normal under the same conditions of normality for the sampling distributions of sample means for $\bar{x}_1$ from Population 1 and $\bar{x}_2$ from Population 2. Those conditions are that Populations 1 and 2 are either normal, or the size of the samples drawn from their respective populations is greater than 30.

The mean of the distribution of differences is $\mu_{\bar{x}_1 - \bar{x}_2}$, and its value is established easily.

$$
\begin{aligned}
\mu_{\bar{x}_1 - \bar{x}_2} &= E\left(\bar{x}_1 - \bar{x}_2\right) \\
&= E\left(\bar{x}_1\right) - E\left(\bar{x}_2\right) \\
&= \mu_1 - \mu_2
\end{aligned}
$$

Of course, we need the standard deviation, but this is derived from the variance, which is determined next.

$$
\begin{aligned}
\sigma^2_{\bar{x}_1 - \bar{x}_2} &= Var\left(\bar{x}_1 - \bar{x}_2\right) \\
&= Var\left(\bar{x}_1\right) + Var\left(\bar{x}_2\right) \\
&= \frac{\sigma_1^2}{n_1} + \frac{\sigma_2^2}{n_2}
\end{aligned}
$$

The standard deviation is the last development.

$$
\begin{aligned}
\sigma_{\bar{x}_1 - \bar{x}_2} &= \sqrt{Var\left(\bar{x}_1 - \bar{x}_2\right)} \\
&= \sqrt{\frac{\sigma_1^2}{n_1} + \frac{\sigma_2^2}{n_2}}
\end{aligned}
$$

We finally have what we need. Our main tool for determining when a difference, $\hat{p}_1 - \hat{p}_2$, is significant, is the z-score, and now that we have normality, the mean, and the standard deviation, we may standardize the difference with a z-score.

$$
\begin{aligned}
z &= \frac{\left(\bar{x}_1 - \bar{x}_2\right) - \mu_{\bar{x}_1 - \bar{x}_2}}{\sigma_{\bar{x}_1 - \bar{x}_2}} \\
&= \frac{\left(\bar{x}_1 - \bar{x}_2\right) - \left(\mu_1 - \mu_2\right)}{\sqrt{\frac{\sigma_1^2}{n_1} + \frac{\sigma_2^2}{n_2}}}.
\end{aligned}
$$

Theorem 8.1.2 The difference, $\bar{x}_1 - \bar{x}_2$, is standardized with the z-score formula below.

$$
z = \frac{\left(\bar{x}_1 - \bar{x}_2\right) - \left(\mu_1 - \mu_2\right)}{\sqrt{\frac{\sigma_1^2}{n_1} + \frac{\sigma_2^2}{n_2}}}
$$

This statistic is distributed normally if either $n_1 > 30$ or Population 1 is normal, and either $n_2 > 30$ or Population 2 is normal.

This test statistic may be used to determine when the difference between two means is significant. Unfortunately, the values for σ_1 and σ_2 are unknown. We are therefore forced to use s_1 and s_2 in their places. This substitution introduces a difficulty. We know that the introduction of sample standard deviations in place of population standard deviations in the test statistic,

$$z = \frac{\overline{x} - \mu}{\sigma/\sqrt{n}},$$

makes the statistic non-normal in its distribution. The statistic is distributed according to Student's t-distribution, which allows for a varying sample standard deviation in the denominator. Our problem here is analogous, but now we have two varying sample standard deviations, s_1 and s_2, in the denominator which will replace σ_1 and σ_2. Student's distribution does not describe the variation of such a statistic, but using the appropriate degrees of freedom, the match is quite close.

Summary 8.1.3 The studentized variable,

$$t = \frac{(\overline{x}_1 - \overline{x}_2) - (\mu_1 - \mu_2)}{\sqrt{\frac{s_1^2}{n_1} + \frac{s_2^2}{n_2}}},$$

varies according to a distribution which is approximately Student's t-distribution if we use degrees of freedom $df = \frac{(V_1 + V_2)^2}{\frac{V_1^2}{n_1 - 1} + \frac{V_2^2}{n_2 - 1}}$, where $V_1 = \frac{s_1^2}{n_1}$ and $V_2 = \frac{s_2^2}{n_2}$. Here the degrees of freedom should be *truncated* (*rounded down*) when the result is not a whole number. This result is used for the *Smith-Satterthwaite test*.

8.1.3   Homework

Compute the test statistic for the difference of two proportions using the provided information. State whether the conditions for approximate normality are met, and decide informally whether the data support the rejection of the null hypothesis.

1. Sample data: $x_1 = 450$, $n_1 = 940$, $x_2 = 550$, $n_2 = 875$.
   Null hypothesis: $H_0 : p_1 = p_2$.

2. Sample data: $x_1 = 110$, $n_1 = 300$, $x_2 = 115$, $n_2 = 250$.
   Null hypothesis: $H_0 : p_1 = p_2$.

3. Sample data: $x_1 = 175$, $n_1 = 525$, $x_2 = 100$, $n_2 = 320$.
   Null hypothesis: $H_0 : p_1 - p_2 = 0.01$.

4. Sample data: $x_1 = 450$, $n_1 = 940$, $x_2 = 550$, $n_2 = 875$.
   Null hypothesis: $H_0 : p_1 - p_2 = 0.10$.

Compute the test statistic for the difference of two means using the provided information. State whether the conditions for approximate normality are met, and decide informally whether the data support the rejection of the null hypothesis.

5. Sample data: $\overline{x}_1 = 12$, $s_1 = 3$, $n_1 = 75$, $\overline{x}_2 = 13$, $s_2 = 4$, $n_2 = 95$.
   Null hypothesis: $H_0 : \mu_1 = \mu_2$.

6. Sample data: $\overline{x}_1 = 3.27$, $s_1 = 0.76$, $n_1 = 31$, $\overline{x}_2 = 2.94$, $s_2 = 0.45$, $n_2 = 56$.
   Null hypothesis: $H_0 : \mu_1 = \mu_2$.

7. Sample data: $\overline{x}_1 = 65$, $s_1 = 12$, $n_1 = 110$, $\overline{x}_2 = 57$, $s_2 = 15$, $n_2 = 182$.
   Null hypothesis: $H_0 : \mu_1 - \mu_2 = 4$.

8. Sample data: $\overline{x}_1 = 0.67$, $s_1 = 0.012$, $n_1 = 87$, $\overline{x}_2 = 0.59$, $s_2 = 0.009$, $n_2 = 79$.
   Null hypothesis: $H_0 : \mu_1 - \mu_2 = 0.077$.

8.2   Large Sample Tests for Two Proportions

8.2.1   Testing the Null Hypothesis $H_0 : p_1 - p_2 = 0$

In Section 8.1, we developed a test statistic for testing claims regarding the difference between two proportions, representing responses from two different treatments in a single population, or from two different populations altogether. The next step is to outline a procedure for testing such a claim.

Procedure 8.2.1 Testing Claims Regarding a Difference Between Two Proportions

Step I.   **Gather Sample Data and Verify that Requirements Are Met**
We will need $\hat{p}_1 = x_1/n_1$, $\hat{p}_2 = x_2/n_2$, $\bar{p} = \frac{x_1+x_2}{n_1+n_2}$, $\bar{q} = 1 - \bar{p}$, and α, the level of significance. Samples should be independent. For observational studies, we assume that samples are random. For experimental studies, sample members are assumed to be divided into two groups (experimental blocks) by random assignment. This test requires that $n_1 p_1 \gtrsim 10$, $n_1 q_1 \gtrsim 10$, $n_2 p_2 \gtrsim 10$ and $n_2 q_2 \gtrsim 10$. The best we can do to ensure this is to require that the sample sizes n_1 and n_2 be *large* (without defining exactly what this means), so that each of the requirements are met. As we have said previously, our best attempt to verify this will be check that there are at least 10 successes and failures in each sample.

Step II.   **Identify the Claims**
The null hypothesis is usually $H_0 : p_1 = p_2$. In such cases, the alternative hypothesis is one of the following: $H_1 : p_1 > p_2$, $H_1 : p_1 < p_2$, or $H_1 : p_1 \neq p_2$.

Step III.   **Compute the Test Statistic (T.S.)**
For the null hypothesis, $H_0 : p_1 = p_2$, we use the test statistic,

$$T.S. : z = \frac{(\hat{p}_1 - \hat{p}_2) - (p_1 - p_2)}{\sqrt{\frac{\bar{p}\bar{q}}{n_1} + \frac{\bar{p}\bar{q}}{n_2}}},$$

where $\bar{p} = \frac{x_1+x_2}{n_1+n_2}$, and $\bar{q} = 1 - \bar{p}$. For the expression, $p_1 - p_2$, in the numerator, the null hypothesis requires that we use $p_1 - p_2 = 0$. For other null hypotheses, see the test statistic given at the end of this section.

Step IV.   **Look Up the Critical Value(s) (C.V.) on Table A2.**
The method for looking up the critical value(s) is no different from before. Remember that, with proportions, the test statistic is normally distributed as long as the normality requirements for the sampling distribution are met. So we only use z critical values for this test, which are found in the bottom row of Table A2, and we disregard the degrees of freedom used in a t-test.

Step V.   **Look up the p-Value on Table A1**
For a right-tail test, look up the test statistic in Table A1, and use $p\text{-}value = 1 - (area\ in\ table)$. For a left-tail test, look up the test statistic in Table

A1, and the area you find there is the *p-value*. For a two-tail test, look up the absolute value of the test statistic in Table A1, and *p-value* $= 2 \cdot (1 - area\ in\ table)$.

Step VI. **Make a Conclusion**

Conclusions are made as before. When the test statistic lies in the rejection region, we reject the null hypothesis in favor of the alternative. Otherwise we fail to reject the null hypothesis. For observational studies, the conclusion refers to the difference between the population proportions. For experimental studies, the conclusion refers to the treatment, and when groups are divided by random selection, a rejection of the null hypothesis allows us to infer that the treatment is the cause for the observed difference.

Example 8.2.2 In comparing the Summer Bridge mathematics students at Mt. San Antonio College with a group of non-Summer Bridge mathematics students, two random samples were drawn and success rates were computed. Success is defined as finishing the course with a C or better. Failure is getting a D, F, or W. For the Summer Bridge sample, the success rate was 86.7% from a sample 90 students. In the other sample, the success rate was 54.8%, with 93 students involved. We would like to test the claim, at 1% significance, that the Summer Bridge student success rate is higher than that for the control group.

Step I. **Gather Sample Data and Verify that Requirements Are Met**

The Bridge group was given first, so we will use this for Sample 1, and the non-bridge group represents Sample 2. $\hat{p}_1 = x_1/n_1 = 0.867$, $n_1 = 30$, $\hat{p}_2 = x_2/n_2 = 0.548$, $n_2 = 31$. We are not directly given the values of x_1 and x_2, but we know that $x_1 = \hat{p}_1 \cdot n_1 = 0.867 \cdot 90 \approx 78$, and $x_2 = \hat{p}_2 \cdot n_2 = 0.548 \cdot 93 \approx 51$. Using these, we compute $\bar{p} = \frac{x_1 + x_2}{n_1 + n_2} = \frac{78 + 51}{90 + 93} = 0.705$, and so $\bar{q} = 1 - \bar{p} = 1 - 0.705 = 0.295$. For this example, we have chosen $\alpha = 0.01$. The samples have 78 and 51 successes respectively. Additionally, there are 12 and 42 failures, respectively. Each of these is greater than or equal to 10, as required.

Step II. **Identify the Claims**

The null hypothesis is $H_0 : p_1 = p_2$. In this case, the claim was that the Summer Bridge success rate is larger than the rate for the non-Bridge group, so the alternative hypothesis is: $H_1 : p_1 > p_2$. This will be a right-tail test.

Step III. **Compute the Test Statistic (T.S.)**

For the null hypothesis, $H_0 : p_1 = p_2$, we use

$$TS : z = \frac{(\hat{p}_1 - \hat{p}_2) - (p_1 - p_2)}{\sqrt{\frac{\bar{p}\bar{q}}{n_1} + \frac{\bar{p}\bar{q}}{n_2}}}$$

$$= \frac{(0.867 - 0.548) - 0}{\sqrt{\frac{0.705 \cdot 0.295}{90} + \frac{0.705 \cdot 0.295}{93}}}$$

$$= 4.731.$$

This tells us that the given success rates differ by 4.731 standard deviations.

Step IV. **Look Up the Critical Value(s) (C.V.) on Table A2.**
Looking at the bottom of Table A2, we see that our critical value is $C.V. : z = 2.326$.

One Tail Applications			
df	$\alpha = .005$	$\alpha = .01$	$\alpha = .02$
1000	2.581	2.330	2.056
Large (z)	2.576	**2.326**	2.054

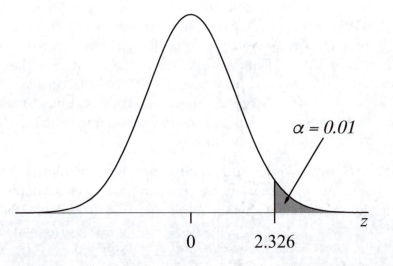

Figure 8.1: The critical value for Example 8.2.2.

The test statistic lies in the rejection region, so we must reject the null hypothesis.

Step V. **Look up the p-Value on Table A1**
This is a right-tail test, so we look up the test statistic in Table A1 (the table directs us to use 0.999 for the area), and use $p\text{-}value = 1 - (area\ in\ table)$. In this case, $p\text{-}value = 1 - 0.999 = 0.001$, which is less than α, so again, we are rejecting the null hypothesis.

Step VI. **Make a Conclusion**
The null hypothesis says that there is no difference between these success rates, and we have rejected this in favor of the alternative hypothesis, which states that the success rate is higher for the Summer Bridge students. That is the original claim, so the data support the original claim. Because the groups in

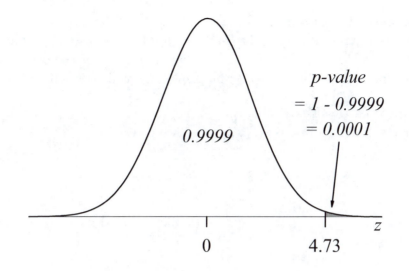

Figure 8.2: The *p-value* for Example 8.2.2.

this study were not created by random assignment, the study is observational, not experimental. The conclusion should be something similar to the following.

The data support the claim that the success rate for students in the Summer Bridge mathematics program is higher than the success rate for a control group of non-Summer Bridge students. This is emphasized by the p-value of 0.0032.

Example 8.2.3 If you were a landlord, would you rent an apartment to a person with AIDS? The authors of "Accommodating Persons with AIDS: Acceptance and Rejection in Rental Situations" (J. of Applied Social Psychology (1999): 261 – 270), chose 160 random advertisements for apartment rentals. The advertisements were separated into two groups by random selection. In the first group, 80 calls were made to landlords, asking to rent a room. It was mentioned that the renter had AIDS, and, after being released from the hospital, would need a place to live. Out of those 80 calls, 32 landlords were willing to rent the room in question. In the second group, 80 additional calls were made where no reference to AIDS was mentioned. Of those, 61 landlords were willing to rent the room in question. Test the claim at 2% significance that the proportion of rooms available to AIDS patients is lower than the proportion available to the general public.

Step I. **Gather Sample Data and Verify that Requirements Are Met**
The calls with an AIDS reference were mentioned first, so they will be Sample 1. The control group will be Sample 2. $\hat{p}_1 = x_1/n_1 = 32/80 = 0.400$, $n_1 = 80$, $\hat{p}_2 = x_2/n_2 = 61/80 = 0.763$, and $n_2 = 80$. Also, $\bar{p} = \frac{x_1+x_2}{n_1+n_2} = \frac{32+61}{80+80} = 0.581$, and so $\bar{q} = 1 - \bar{p} = 1 - 0.581 = 0.419$. For this latest example, we have chosen $\alpha = 0.02$. These samples have 32 and 61 successes, with 48 and 19 failures, respectively. Each of these is greater than or equal to 10, as required.

Step II. **Identify the Claims**

The null hypothesis is again $H_0 : p_1 = p_2$. In this case, the claim was that the proportion for AIDS patients is smaller than that for the control group, so the alternative hypothesis is: $H_1 : p_1 < p_2$. This will be a left-tail test.

Step III. **Compute the Test Statistic (T.S.)**

For the null hypothesis, $H_0 : p_1 = p_2$, we use

$$TS : z = \frac{(\hat{p}_1 - \hat{p}_2) - (p_1 - p_2)}{\sqrt{\frac{\bar{p}\bar{q}}{n_1} + \frac{\bar{p}\bar{q}}{n_2}}}$$

$$= \frac{(0.400 - 0.763) - 0}{\sqrt{\frac{0.581 \cdot 0.419}{80} + \frac{0.581 \cdot 0.419}{80}}}$$

$$= -4.653.$$

This tells us that these rates differ by −4.653 standard deviations.

Step IV. **Look Up the Critical Value(s) (C.V.) on Table A2.**

Looking at the bottom of Table A2, under 2% significance, we see the value $z = 2.054$, but noting that this is a left-tail test, we must make this negative, so our critical value is $C.V. : z = -2.054$.

One Tail Applications			
df	$\alpha = .01$	$\alpha = .02$	$\alpha = .025$
1000	2.330	2.056	1.962
Large (z)	2.326	**2.054**	1.960

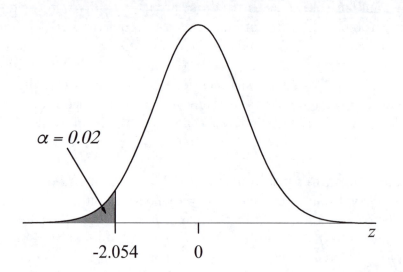

Figure 8.3: The critical value for Example 8.2.3.

The test statistic lies well within the rejection region, so we must reject the null hypothesis.

Step V. **Look up the *p-Value* on Table A1**

In searching table A1 for our test statistic, $T.S. : z = -4.652$, we note that the smallest z-score on the table is $z = -4.09$, which has an associated p-*value* of 0.0001. Because our z-score is less than -4.09, we may conclude that the true p-*value* is actually less than 0.0001 (but greater than 0). Using computer software to compute this value, we can arrive at the true p-*value*, whose value turns out to be 0.00000166. This indicates a highly significant difference between the sample proportions, and again emphasizes the strong rejection of the null hypothesis.

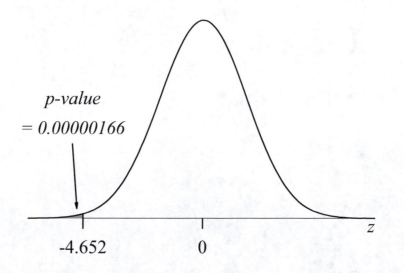

Figure 8.4: The *p-value* for Example 8.2.3.

Step VI. **Make a Conclusion**

Again, we are rejecting the null hypothesis in favor of the alternative, that the proportion of rental units available to AIDS patients is lower than the proportion for the general population. This was the original claim, so the data support the claim. Because the two groups of calls were created by random selection, this study was experimental, and a causal relationship may be inferred. The conclusion refers to the treatment below.

The data support the claim that the proportion of rental units is decreased by revealing that the tenant has AIDS. The difference between the sample proportions was highly significant, as emphasized by the small p-value of 0.00000166.

8.2.2 Testing the Null Hypothesis $H_0 : p_1 - p_2 = k \neq 0$

As mentioned in Section 8.1, when we test the more unusual null hypothesis $H_0 : p_1 - p_2 = k \neq 0$, the test statistic is different. In this case we would use

$$TS : z = \frac{(\hat{p}_1 - \hat{p}_2) - (p_1 - p_2)}{\sqrt{\frac{\hat{p}_1 \hat{q}_1}{n_1} + \frac{\hat{p}_2 \hat{q}_2}{n_2}}},$$

where the difference in the numerator $p_1 - p_2$ is assumed to be $p_1 - p_2 = k$, by the null hypothesis. We mention this here for reference, but as this application is unusual, will not provide examples.

8.2.3 Homework

Test the claims below, including each step as outlined.

1. Claim: $p_1 < p_2$; $x_1 = 45$, $n_1 = 69$, $x_2 = 57$, $n_2 = 73$, $\alpha = 0.05$.

2. Claim: $p_1 > p_2$; $x_1 = 322$, $n_1 = 511$, $x_2 = 284$, $n_2 = 496$, $\alpha = 0.02$.

3. Claim: $p_1 \neq p_2$; $x_1 = 24$, $n_1 = 71$, $x_2 = 36$, $n_2 = 144$, $\alpha = 0.08$.

4. Test the claim at 1% significance that the proportion from population 1 is not equal to the proportion from population 2. In support of this claim, sample 1 contained 148 successes in 336 trials, sample 2 contained 274 successes in 550 trials.

5. Test the claim at 5% significance that the proportion from population 1 is less than that for population 2, with the following sample data. Sample 1 contained 35 successes and 80 failures, and sample 2 contained 93 successes and 210 failures.

Answer the following questions.

6. An article claims that people who consume more than one energy drink per week are more likely to develop alcohol dependence when compared to people who do not consume energy drinks. The study included two samples – one of people who consumed more than one energy drink per week, and another who did not consume energy drinks. The samples yielded a p-value of 0.013, and the null hypothesis was rejected. Is it appropriate to conclude that energy drink consumption tends to cause people to become alcohol dependent? Give justification for your answer.

7. A study claims that teenage female depression rates are higher than teenage male depression rates. The sample included random samples of female and male teens, and a p-value of 0.045 is determined from sample data in support of the claim. Which conclusion do the data support: that gender is the cause of higher rates of depression for female teens, or more simply that depression rates are higher among female teens?

8. In a recent medical trial, a sample of volunteers was randomly divided into two groups to test the claim that the HPV vaccination side effect rate is greater than the side effect rate for a placebo. One group was given the HPV vaccination, and another was given a placebo. The difference in rates for the worst side effect (nausea) was insignificant, with p-value of 0.278. What sort of conclusion would be most appropriate: that the HPV vaccination side effect rate is no greater than the side effect rate for a placebo, or that the HPV vaccination does not cause a higher proportion of side effects when compared to a placebo.

9. On the website, *hunch.com*, a study was conducted to analyze the beliefs of *birthers* – people who believed that Obama was not a U.S. citizen (blog.hunch.com). This study was reported by national news organizations such as the Huffington Post. One survey question asked of survey participants on the Hunch website was "do you believe in

UFOs or alien visits to earth?" The difference between the sample proportions is 0.16. In testing the claim that a higher proportion of birthers believe in UFOs, the p-value is nearly zero (less than 0.0001). Do the data support the claim that the proportion of birthers who believe in UFOs is greater than for non-birthers?

Test the following claims. Conclusions should refer to populations in observational studies, and to treatments in experimental studies.

10. The following data are based on a study published by Unicef in 2006. In a sample of 752 girls (of school age or older) randomly sampled worldwide, 632 were permitted (by circumstance) to attend elementary school. In a sample of 783 boys (of school age or older) randomly sampled worldwide, 681 were permitted (by circumstance) to attend elementary school. Test the claim at 5% significance that the proportion of girls who are allowed to attend elementary school is smaller than for boys.

11. In the Mathematics Department at Mt. San Antonio College, there are two ways to complete Intermediate Algebra – this can be done by taking Math 71, a single semester approach, or by taking Math 71A and Math 71B, a two semester approach – and a student is not considered successful until completing Math 71B. In a random sample of 1007 Math 71 students, 57.1% of these people completed the course successfully (did not receive a W, D, or F). In Math 71B, with a random sample of 158 students, 57.5% of these completed Math 71B successfully. Test the claim at 6% significance that the proportion of students successfully completing Math 71 is equal to the proportion of students successfully completing Math 71B.

12. The values in this problem are, again, based on a study published by Unicef in 2006. A commonly used measure of poverty is the proportion of people in a given area who live on less than $1 a day. In a random sample of 879 people from Eastern and South Africa, 334 of these live on less than $1 a day. In sampling 1032 people from South Asia, 340 of these live on less than $1 a day. Test the claim at 4% significance that the proportion of people who live on less than $1 a day in Eastern and Southern Africa is greater than the proportion for South Asia.

13. In the Mathematics Department at Mt. San Antonio College, there are two ways to complete Beginning Algebra – this can be done by taking Math 51, a single semester approach, or by taking Math 51A and Math 51B, a two semester approach – and a student is not considered successful until completing Math 51B. In a random sample of 1086 Math 51 students, 52.1% of these people completed the course successfully (did not receive a W, D, or F). In Math 51B, with a random sample of 112 students, 59.8% of these completed Math 51B successfully. Test the claim at 4% significance that the proportion of students successfully completing Math 51 is less than the proportion of students successfully completing Math 51B.

14. In the article "Work-Related Attitudes" (Psychological Reports, 1991, pp. 443 – 450), 224 of 395 elementary school teachers said that they were very satisfied with their jobs, while 126 of 266 high school teachers said that they were very satisfied

with their jobs. Test the claim at 1% significance that the proportion of elementary school teachers who are satisfied with their jobs is greater than that proportion for high school teachers.

15. The following problem is based on statistics given in "The Price of Privilege," a book written in 2006 by Madeline Levine, Ph.D. We define a family to be *affluent* if its annual income is in the $120,000 to $160,000 range. From a random sample of 75 teenage girls from affluent families, it is learned that 17 of these suffer from clinical depression. Likewise, from a sample of 116 teenage girls only 10 suffer from clinical depression. Using this information, test the claim at 3% significance that the depression rate is higher for teenage girls from affluent families, compared to the rate for all teenage girls.

16. When college class locations are inconvenient for students, does student success decline? The following samples include success rates for two different populations of students – one where students are taking a class in a convenient location, and another where students are taking their class in an inconvenient location. Test the claim at 1% significance that the mean success rate is higher in classrooms which are conveniently located for students. These values are gathered from records in the Mathematics Department at Mt. San Antonio College.

Convenient Location		Inconvenient Location	
n	4450	n	4190
successes	2414	*successes*	2126
rate	54.25%	*rate*	50.74%

17. Recent studies suggest that drinking diet sodas can actually be accompanied by weight gain, rather than weight loss. The following data are based on a study conducted by Sharon P. Fowler, MPH, and colleagues at the University of Texas Health Science Center, San Antonio. From a sample of 1000 people who drink $\frac{1}{2}$ can of diet soda per day, 260 of these are obese. From a sample of 924 people who drink 1 can of diet soda per day, 280 of these are obese. Test the claim at 5% significance that the proportion of people who are obese is lower for people who drink $\frac{1}{2}$ can of soda per day, compared to those who drink 1 can of soda per day.

18. This problem is based on the Abecedarian (A-B-C-Darian) Project, where a group children (overwhelmingly African-American) born in the 1970s were randomly divided into two groups.

The first group (the treatment group) received quality preschool care for 6 to 8 hours per week. The second group received no preschool care, but both groups received nutritional supplements, health care, and social services to help control confounding variables. Among the treatment group of 57 children, 21 eventually attended college. Assume that, among the control group of 67 children, 10 eventually attended college.

Test the claim at 1% significance that quality preschool care increases the proportion of children who eventually attend college. (Source: Wikipedia article, *The Abecedarian Early Intervention Project.*)

19. Most teens who experience clinical depression never receive treatment, even though depression has many effective treatment options.

In a clinical trial, two groups of depressed teens were created by random assignment. The first group included 40 teens who were given the antidepressant *Fluoxetine* with psychiatric therapy. Of these, 29 improved meaningfully. In a similar group of 50 teens, placebos were given (with blinding), also with psychiatric therapy. Among this group, 16 improved meaningfully.

Test the claim at 2% significance that the proportion of teens who improve after the Fluoxetine treatment is greater than the proportion for patients who received a placebo.

8.3   Comparing Means from Independent Samples

We have developed a test statistic for making a claim about the difference between two means, and thus we outline a procedure for testing such a claim. These two means may be from two different populations, or they may be from a single population where we are considering average responses from two different treatments. We have considered the cases where both population standard deviations are known, where the test statistic is

$$z = \frac{(\bar{x}_1 - \bar{x}_2) - (\mu_1 - \mu_2)}{\sqrt{\frac{\sigma_1^2}{n_1} + \frac{\sigma_2^2}{n_2}}},$$

but given that the population standard deviations are unlikely to be known, we will disregard this possibility and only consider the case where they are unknown.

8.3.1   Comparing Means from Independent Samples

Procedure 8.3.1 Testing Claims Regarding a Difference between Two Population Means

Step I. **Gather Sample Data and Verify that Requirements Are Met**
We will need $\bar{x}_1$, s_1, n_1, $\bar{x}_2$, s_2, and n_2, and α, the level of significance. Samples should be independent and we assume that they are random. This test requires that n_1 and n_2 are both greater than 30, or that the populations are normal.

Step II. **Identify the Claims**
The null hypothesis is usually $H_0 : \mu_1 = \mu_2$. In this case, the alternative hypothesis is one of the following: $H_1 : \mu_1 > \mu_2$, $H_1 : \mu_1 < \mu_2$, or $H_1 : \mu_1 \neq \mu_2$.

Step III. **Compute the Test Statistic (T.S.)**
Here, our test statistic is

$$T.S. : t = \frac{(\bar{x}_1 - \bar{x}_2) - (\mu_1 - \mu_2)}{\sqrt{\frac{s_1^2}{n_1} + \frac{s_2^2}{n_2}}},$$

which is approximately distributed according to the t-distribution, provided that we use the degrees of freedom given in the following step below. When the null hypothesis is $H_0 : \mu_1 = \mu_2$, we must use $\mu_1 - \mu_2 = 0$ in the numerator.

Step IV. **Look Up the Critical Value(s) (C.V.) on Table A2.**
The method for looking up the critical value(s) is as usual, but we use degrees of freedom given by

$$df = \frac{(V_1 + V_2)^2}{\frac{V_1^2}{n_1 - 1} + \frac{V_2^2}{n_2 - 1}},$$

where $V_1 = \frac{s_1^2}{n_1}$ and $V_2 = \frac{s_2^2}{n_2}$, and *truncate (i.e. round down)* the result.

Step V. **Look up the p-Value on Table A3**
For this we use the degrees of freedom indicated in step 5.

Step VI. **Make a Conclusion**

Conclusions are made as before. When the test statistic lies in the rejection region, we reject the null hypothesis in favor of the alternative. Otherwise we fail to reject the null hypothesis. For observational studies, the conclusion refers to the difference between the population means. For experimental studies, the conclusion refers to the treatment, and when groups are divided by random selection, a rejection of the null hypothesis allows us to infer that the treatment is the cause for the observed difference.

Example 8.3.2 For students who complete the Summer Bridge mathematics program at Mt. San Antonio College without drawing, a sample of 28 students was randomly selected, and the mean grade was 3.04 on a 0 to 4 point scale, with a standard deviation of 1.10. For a control group of 27 mathematics students, the mean grade was 2.07, with a standard deviation of 1.33. Assume that the populations represented by these samples are normal. Using a 1% significance level, test the claim that the mean grade for the Summer Bridge students is greater than the mean for the control group of ordinary math students. (Sample data for this problem were collected by Imelda Plascencia, an honors student at Mt. San Antonio College).

Step I. **Gather Sample Data and Verify that Requirements Are Met**

We define sample1/population 1 to be the Summer Bridge math students, and sample 2/population 2 to be the control group of ordinary math students. Thus, $\bar{x}_1 = 3.04$, $s_1 = 1.10$, and $n_1 = 28$. Also, $\bar{x}_2 = 2.07$, $s_2 = 1.33$, and $n_2 = 27$. For this example, $\alpha = 0.01$. The requirements for the test are met under the assumptions that the populations are normal.

Step II. **Identify the Claims**

The null hypothesis is $H_0 : \mu_1 = \mu_2$, and the alternative hypothesis is $H_1 : \mu_1 > \mu_2$. This is a right-tail test.

Step III. **Compute the Test Statistic (T.S.)**

Here, our test statistic is

$$T.S. : t = \frac{(\bar{x}_1 - \bar{x}_2) - (\mu_1 - \mu_2)}{\sqrt{\frac{s_1^2}{n_1} + \frac{s_2^2}{n_2}}}$$

$$= \frac{(3.04 - 2.07) - 0}{\sqrt{\frac{1.10^2}{28} + \frac{1.33^2}{27}}}$$

$$= 2.942,$$

where the zero in the numerator is due to the null hypothesis that $\mu_1 = \mu_2$.

Step IV. **Look Up the Critical Value(s) (C.V.) on Table A2.**

The tricky part now is the computation of the degrees of freedom. First, we compute

$$V_1 = \frac{s_1^2}{n_1} = \frac{1.10^2}{28} = 0.0432,$$

and

$$V_2 = \frac{s_2^2}{n_2} = \frac{1.33^2}{27} = 0.0655.$$

These give the degrees of freedom below.

$$df = \frac{(V_1 + V_2)^2}{\frac{V_1^2}{n_1 - 1} + \frac{V_2^2}{n_2 - 1}}$$

$$= \frac{(0.0432 + 0.0655)^2}{\frac{0.0432^2}{28 - 1} + \frac{0.0655^2}{27 - 1}}$$

$$= 50.47,$$

which we must *truncate* (*round down*) to $df = 50$. Now we may find our critical value in Table A2.

	One Tail Applications		
df	$\alpha = .005$	$\alpha = .01$	$\alpha = .02$
49	2.680	2.405	2.110
50	2.678	**2.403**	2.109
75	2.643	2.377	2.090

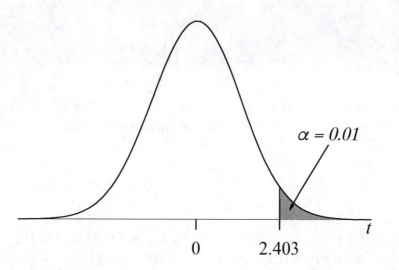

Figure 8.5: The critical value for Example 8.3.2.

For this right-tail test, we have a critical value $C.V. : t = 2.403$. The test statistic lies in the rejection region, so once again, we will reject the null hypothesis.

Step V. **Look up the *p*-Value on Table A3**

In using Table A3, recall that we round our test statistic to one place after the decimal, so we look for $t = 2.9$ with 50 degrees of freedom.

	Degrees of Freedom		
T.S.	40	**50**	60
2.8	.996	.996	.997
2.9	.997	**.997**	.997
3.0	.998	.998	.998

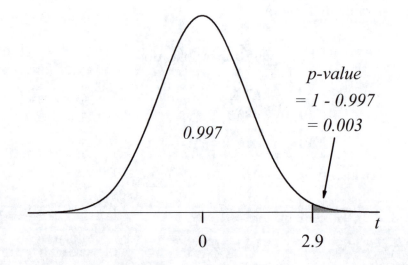

Figure 8.6: The *p-value* for Example 8.3.2.

The area to the left of the test statistic, found in the table, is 0.997, but since this is a right-tail test, we use the rule of complements to get *p-value* = $1 - 0.997 = 0.003$. This is smaller than α, indicating a rejection of the null hypothesis, and a highly significant difference between the average grade for the Summer Bridge math students, and the control group.

Step VI. **Make a Conclusion**

The null hypothesis stated that there was no difference in the mean grade between the Summer Bridge math students and the control group, but we have rejected this in favor of the alternative, which stated that the mean grade for the Summer Bridge students was higher than the mean for the control group. This was the original claim, so we conclude as follows.

The data support the claim that the mean grade for math students in the Summer Bridge program at Mt. San Antonio College is higher than the mean grade for math students in the control group. The difference between the sample means was highly significant as indicated by a p-value of 0.003.

Example 8.3.3 Here we revisit the cloud seeding *experiment* from Chapter 7, but now we treat the two means properly using sample data to represent each population. The first sample is drawn from the population of rainfall totals for clouds seeded with chemicals to increase rainfall amounts. For 540 future storms, 270 were randomly selected in advance for seeding. From the storms where clouds were seeded, the mean rainfall was 1.53 inches, with a standard deviation of 1.34 inches. From 270 storms whose clouds were not seeded, the mean rainfall was 1.74 inches, with a standard deviation of 1.45 inches. At 6% significance, we will test the claim that the mean rainfall was the same for seeded and for unseeded clouds, against the claim that there was a difference between these means. (Data are taken from "Analyzing the results of a cloud-seeding experiment in Tasmania", by Miller, A.J., Shaw, D.E., Veitch, L.G. & Smith, E.J. (1979), *Communications in Statistics - Theory & Methods*, vol.A8(10), 1017-1047).

Step I. **Gather Sample Data and Verify that Requirements Are Met**
We define sample 1, drawn from population 1, to be the rainfall totals for seeded clouds, and sample 2, drawn from population 2, to be the control group of rainfall totals for unseeded clouds. Thus, $\bar{x}_1 = 1.53$, $s_1 = 1.34$, and $n_1 = 270$. Also, $\bar{x}_2 = 1.74$, $s_2 = 1.45$, and $n_2 = 270$. For this example we are using $\alpha = 0.06$. The requirements for the test are met since both sample sizes are greater than 30.

Step II. **Identify the Claims**
The null hypothesis is that the means are the same for seeded and unseeded clouds, $H_0 : \mu_1 = \mu_2$, and the alternative hypothesis is that these means are different, $H_1 : \mu_1 \neq \mu_2$. This is a two-tail test.

Step III. **Compute the Test Statistic (T.S.)**
Here, our test statistic is,

$$T.S. : t = \frac{(\bar{x}_1 - \bar{x}_2) - (\mu_1 - \mu_2)}{\sqrt{\frac{s_1^2}{n_1} + \frac{s_2^2}{n_2}}}$$
$$= \frac{(1.53 - 1.74) - 0}{\sqrt{\frac{1.34^2}{270} + \frac{1.45^2}{270}}}$$
$$= -1.748,$$

where the zero in the numerator is due to the null hypothesis that $\mu_1 = \mu_2$.

Step IV. **Look Up the Critical Value(s) (C.V.) on Table A2.**
Once again, we must compute of the degrees of freedom.

$$V_1 = \frac{s_1^2}{n_1} = \frac{1.34^2}{270} = 0.00665,$$

and

$$V_2 = \frac{s_2^2}{n_2} = \frac{1.45^2}{270} = 0.00779.$$

These yield the degrees of freedom.

$$df = \frac{(V_1 + V_2)^2}{\frac{V_1^2}{n_1-1} + \frac{V_2^2}{n_2-1}}$$

$$= \frac{(0.00665 + 0.00779)^2}{\frac{0.00665^2}{270-1} + \frac{0.00779^2}{270-1}}$$

$$= 534.69,$$

which we must *truncate* (*round down*) to $df = 534$. Unfortunately, the table does not include this value for the degrees of freedom, so we pick the closest value in the table, namely $df = 500$. Note that the values above and below are very similar, indicating that, as degrees of freedom become large, the critical values do not change much. From the table, we find the value $t = 1.885$.

Two Tail Applications			
df	$\alpha = .05$	$\alpha = .06$	$\alpha = .08$
250	1.969	1.889	1.758
500	1.965	**1.885**	1.754
1000	1.962	1.883	1.752

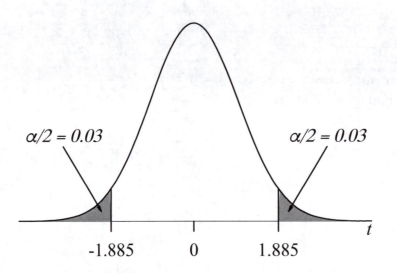

Figure 8.7: The critical values for Example 8.3.3.

This is a two-tail test, so we have critical values $C.V. : t = \pm 1.885$. The test statistic does not lie within the rejection region, so this time we will fail to reject the null hypothesis.

Step V. **Look up the *p*-Value on Table A3**

Using Table A3, recall that we round our test statistic to one place after the decimal, and since this is a two-tail test we take the absolute value, so we look for $|T.S.| : |t| = 1.7$, with 534 degrees of freedom. Once again, this value for the degrees of freedom is not in the table, so we must choose between 100 or *infinity* (∞). Because 534 is so much bigger than 100, we will use ∞, noting that these choices correspond to areas that are nearly equal. From the table we see an area, under ∞ degrees of freedom, of 0.955.

	Degrees of Freedom	
T.S.	100	∞
1.6	.944	.945
1.7	.954	**.955**
1.8	.963	.964

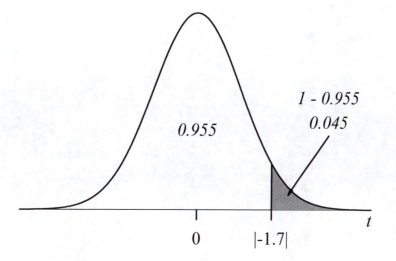

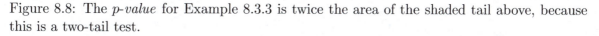

Figure 8.8: The *p-value* for Example 8.3.3 is twice the area of the shaded tail above, because this is a two-tail test.

For the two-tail test, we use

$$\begin{aligned}
p\text{-}value &= 2 \cdot (1 - area) \\
&= 2 \cdot (1 - 0.955) \\
&= 2 \cdot 0.045 \\
&= 0.090.
\end{aligned}$$

This is larger than α, so again we know that we are failing to reject the null hypothesis.

Step VI. **Make a Conclusion**

The null hypothesis said that there was no difference between the mean rainfall amounts for seeded storm clouds and unseeded storm clouds. This was also the original claim that we are testing. Because we fail to reject the null hypothesis, we are failing to reject the original claim. Keep in mind that failing to reject the claim does not mean that it is true, and the wording of the conclusion reflects this. Finally, note that this study was an experiment, because the seeding treatment was applied to clouds selected by random assignment. This tells us that, as a treatment, the cloud seeding was ineffective.

The sample data do not give sufficient evidence to warrant rejection of the claim that the mean rainfall amounts are the same for seeded and unseeded storm clouds. The p-value of 0.090 indicated a difference between the sample means that was only moderately significant – and this did not meet the level required by $\alpha = 0.06$. *The cloud seeding treatment appears to be ineffective.*

These storms were all monitored in Tanzania, where the methods used to seed these clouds do not appear to be effective in increasing rainfall amounts.

8.3.2   Homework

Test the claims below, including each step outlined.

1. Claim: $\mu_1 < \mu_2$; $\bar{x}_1 = 3.1$, $s_1 = 1.8$, $n_1 = 31$, $\bar{x}_2 = 2.4$, $s_2 = 0.9$, $n_2 = 37$, $\alpha = 0.05$.

2. Claim: $\mu_1 > \mu_2$; $\bar{x}_1 = 96$, $s_1 = 11$, $n_1 = 82$, $\bar{x}_2 = 92$, $s_2 = 9$, $n_2 = 71$, $\alpha = 0.01$.

3. Claim: $\mu_1 \neq \mu_2$; $\bar{x}_1 = 0.096$, $s_1 = 0.0019$, $n_1 = 28$, $\bar{x}_2 = 0.095$, $s_2 = 0.0014$, $n_2 = 25$, $\alpha = 0.02$. Data appear to come from normal populations.

4. Test the claim at 4% significance that the mean from Population 1 is less than the mean from Population 2. In support of this, the mean from Sample 1 is 85.6 with a standard deviation of 15.9 and a sample size of 162, while the mean from Sample 2 is 93.9 with a standard deviation of 21.7 and a sample size of 148.

5. Test the claim at 2% significance that the mean from Population 1 is not equal to the mean from Population 2. In support of this, Sample 1 has a mean, standard deviation, and size of 0.294, 0.012 and 35 respectively, while Sample 2 has mean, standard deviation, and size of 0.301, 0.018, and 43 respectively.

Answer the following questions.

6. An algebra professor taught two sections of intermediate algebra. In the first section, she had students work through carefully designed activities in groups. In the second section, she taught the same material through traditional lectures. The mean final exam score for the workgroup section was significantly higher, as demonstrated by the p-value of 0.035. Is it appropriate to conclude that group work is associated with a higher mean final exam score? Explain your answer.

7. The research study in Problem 6 was observational. Redesign the study so that a *causal* conclusion can be made.

Test the following claims. Conclusions should refer to populations in observational studies, and to treatments in experimental studies.

8. As mentioned previously, at Mt. San Antonio College, there are two ways to complete Intermediate Algebra – this can be done by taking Math 71, a single semester approach, or by taking Math 71A and Math 71B, a two semester approach, and student is not considered successful until completing Math 71B. In a random sample of 932 Math 71 students, the mean grade is 1.92 on the 4 point scale, with a standard deviation of 1.33. In an random sample of 54 Math 71B students, the mean grade is 1.51, with a sample standard deviation of 1.26. Test the claim at 1% significance that the mean grade in Math 71 is higher than the mean grade in Math 71B. What would your conclusion be at 2% significance (referring to the *p-value* only).

9. In the 2000 U.S. presidential elections, the inaccuracy of voting technology became the subject of worldwide interest. In Florida, the two front-runner candidates, George W. Bush and Al Gore, were so close in the tally that votes had to be recounted by hand to verify the totals. In the process of recounting, it was found that the most unreliable voting machine was the *Votomatic*, with the most under-counted (votes placed but not counted) and over-counted (votes counted that were never placed) votes of any other technology. The under-counted votes for 15 different Florida counties are listed below.

$$\{4946, 2070, 5090, 466, 5431, 1044, 1975, 2410, 10570, 634, 10134, 1763, 4240, 1846, 596\}.$$

The over-counted votes for 15 *Votomatic* machines are listed below.

$$\{7826, 1134, 21855, 520, 3640, 790, 2531, 890, 17833, 1039, 19218, 2124, 4258, 994, 170\}.$$

It has been argued that these discrepancies are not relevant, because while there tend to be many under- and over-counted votes, that these errors tend to cancel one another out, so that the end result is a relatively accurate count. Using these two samples, test the claim at 1% significance that the mean number of under-counted votes is equal to the mean number of over-counted votes (with the Votomatic machines), and thus the overall errors truly do cancel.

10. In Fall 2005, the Mathematics Department at Mt. San Antonio College gave grades to 3647 students in developmental math, with an average grade of 1.93 on the 4 point scale, with a sample standard deviation of 1.34. In Fall of 2001, the department gave 3638 students an average grade of 1.75, with a standard deviation of 1.35. Test the claim that the mean grade was higher in Fall 2005, compared to the mean from Fall 2001. Would it be reasonable to claim that grades have risen significantly? Use 1% significance for this test.

11. Consider the data in the table below, which shows the number of years' remission from symptoms in two groups of AIDS patients: group A (the treatment group) who received a new experimental drug, and group B (the control group) who received a placebo.

$$\text{Treatment: } \{3, 5, 4, 2, 3, 2, 0, 4, 1, 4, 3\}$$
$$\text{Placebo: } \{1, 0, 2, 1, 2, 1, 1, 0, 1, 0\}$$

Test the claim at 5% significance that the mean number of years in remission is higher for the treatment group. Assume both samples are drawn from normal populations.

12. For 1532 schizophrenic patients, the mean eye gain ratio is 0.840, with a sample standard deviation of 0.118. For 1374 healthy control group members, the mean was 0.807 with a sample standard deviation was 0.134. Test the claim at 1% significance that the mean gain ratio is higher for schizophrenic patients than for the control group.

13. Do statistics students spend as much time studying as calculus students? The following data were gathered by Mt. San Antonio College honors student Jessica Yang and give the mean time spent studying per week for randomly selected statistics and calculus students at Mt. San Antonio College. Test the claim that Statistics students spend less time per week studying, on average, than calculus students. Use 4% significance.

Statistics	Calculus
$n = 32$	$n = 32$
$\overline{x} = 2.519$	$\overline{x} = 3.625$
$s = 1.359$	$s = 1.431$

14. This problem is based on the Abecedarian (A-B-C-Darian) Project, where a group of children born in the 1970s were randomly divided into two groups. A treatment group received quality preschool care for 6 to 8 hours per week. A second group received no preschool care, but both groups received nutritional supplements, health care, and social services to help control confounding variables. Among the treatment group of 53 children, the mean reading achievement as young adults was 11.1 (academic grade equivalent) with a standard deviation of 4.2. Among the control group of 51 children, the mean reading achievement as 9.3 with a standard deviation of 3.1. Test the claim at 1% significance that quality preschool care increases the young adult reading achievement level. (Source: *Poverty and Early Childhood Education Intervention*, by Pungello and Campell.)

15. Refer again to the Abecedarian Project from Problem 14. Among the treatment group of 53 children, the mean math achievement as young adults was 9.2 (academic grade equivalent) with a standard deviation of 3.3. Among the control group of 51 children, the mean math achievement as 7.9 with a standard deviation of 3.0. Test the claim at 2% significance that quality preschool care increases the young adult math achievement level. (Source: *Poverty and Early Childhood Education Intervention*, by Pungello and Campell.)

16. Do male and female college students have the same mean heights? The following table contains random heights measured in centimeters. Under the assumption that these samples were drawn from normal populations, test the claim that the mean height for male college students is greater than that for female college students at 1% significance.

Male Heights(cm)	Female Heights(cm)
167.7	148.0
164.1	152.2
159.8	155.2
165.4	152.7
166.6	155.1
171.2	155.5
170.6	156.8
168.6	158.4
168.5	158.3
172.4	156.2
168.0	158.6
172.7	159.9
169.7	164.2
176.3	157.9
174.1	159.8
172.8	160.9
177.6	162.3
178.9	158.0
177.0	159.8
175.5	156.9
179.4	162.4
174.8	160.2
175.5	166.9
182.2	167.2
178.3	167.6
179.1	164.0
176.3	170.4
178.4	175.9
177.1	176.4
189.1	184.1

17. A shoe company plans on selling a certain type of sandal as a unisex sandal suitable for young men and women. They also plan to make the sandal a "one size fits all" size. This would make sense if men and women have the same mean foot length. Under the assumption that these samples were drawn from normal populations, test the claim at 1% significance that the mean foot length is the same for young men as for young women. The following values were measured in centimeters.

Male	Female
24.43	18.16
24.25	20.43
25.40	21.53
25.78	20.35
26.32	21.10
24.73	19.13
25.07	20.61
24.60	20.43
26.09	21.72
26.75	21.64
25.08	22.95
25.76	20.90
26.40	20.04
25.69	21.81
25.63	21.01
26.23	22.25
26.18	22.68
26.78	22.81
26.40	21.78
26.77	23.74
26.32	21.32
26.38	22.66
24.53	22.11
25.64	22.60
25.71	22.04
24.94	21.58
26.99	22.75
27.76	22.06
27.59	24.18
27.57	23.67

8.4   Intervals for Differences of Two Parameters

Now that we have covered all of our two sample hypothesis tests, it is time to give an overview for each of the corresponding two sample confidence intervals. The development will be quick, as all of the theoretical sampling distribution work is done.

Each procedure here assumes that we are using simple random samples. This should ensure that they will represent their populations well.

The first confidence interval regards the difference between two population means where the samples are matched in pairs. This confidence interval is exactly the same as the t-distribution interval covered in Chapter 6, but now we update the mean and standard deviation to the mean difference, μ_d, and the standard deviation of differences, s_d, respectively.

8.4.1   The Difference Between Means with Paired Samples

Summary 8.4.1 The confidence interval for the difference between two population means with paired samples is

$$\bar{d} - E < \mu_d < \bar{d} + E,$$

where $E = t_{\alpha/2} \frac{s_d}{\sqrt{n}}$, and $df = n - 1$. This procedure requires that the population of differences, d, is normal.

Example 8.4.2 Prozac is a drug which is meant to elevate the moods of people who suffer from depression. In the following sample (provided by the WebStat website of the University of New England, School of Psychology), mood levels were recorded before and after taking Prozac. In these values, a higher score means a less depressed psychological state. We will label the *Before Prozac* values as x, and the *After Prozac* values as y. We will compute the 95% confidence interval for the mean difference.

$x = Before$	$y = After$
3	5
0	1
6	5
7	7
4	10
3	9
2	7
1	11
4	8

In Section 7.5, we computed the differences as below.

x	y	$d = x - y$
3	5	-2
0	1	-1
6	5	1
7	7	0
4	10	-6
3	9	-6
2	7	-5
1	11	-10
4	8	-4

In computing the mean and standard deviation of the sample d values, we have $\bar{d} = -3.667$, $s_d = 3.5$, and we note that the sample size (the number of differences) is $n = 9$. We will assume that the population of *differences* is normal.

With these, we may proceed directly to the confidence interval. The maximum error is

$$E = t_{\alpha/2} \frac{s_d}{\sqrt{n}} = 2.306 \cdot \frac{3.5}{\sqrt{9}} = 2.6903,$$

and using this, we compute the confidence interval.

$$\bar{d} - E < \mu_d < \bar{d} + E$$

$$-3.667 - 2.6903 < \mu_d < -3.667 + 2.6903$$

$$-6.3573 < \mu_d < -0.9767$$

Our confidence interval does not allow for a possible zero mean difference, so it is likely that our two samples are drawn from populations with different means.

Next, we move into the sampling distributions of Chapter 8. These were for the differences of two proportions and of two means, and we will recall these as we proceed. First, a reminder of the general form for confidence intervals from Chapter 6.

Summary 8.4.3 The Generalized Confidence Interval
Assuming x is normally distributed, and that x is an unbiased estimate of μ, and that the standard deviation of all x values is σ, the maximum error in the estimate is $E = z_{\alpha/2}\sigma$, and the corresponding confidence interval is $x - E < \mu < x + E$. To correctly interpret the confidence interval, it must be stated that the proportion of such intervals that actually contain the mean is equal to the level of confidence.

Next, we recall the sampling distribution of the difference between two sample proportions.

8.4.2   The Sampling Distribution of Differences, $\hat{p}_1 - \hat{p}_2$

The mean of this population is

$$\mu_{\hat{p}_1 - \hat{p}_2} = p_1 - p_2$$

Noting that the difference of sample proportions, $\hat{p}_1 - \hat{p}_2$, is an estimate of the difference of population proportions $p_1 - p_2$, since

$$E\left(\hat{p}_1 - \hat{p}_2\right) = p_1 - p_2,$$

we may say that $\hat{p}_1 - \hat{p}_2$ is an unbiased estimator of $p_1 - p_2$.

The variance of this population is:

$$Var\left(\hat{p}_1 - \hat{p}_2\right) = \frac{p_1 q_1}{n_1} + \frac{p_2 q_2}{n_2}$$

The standard deviation is the square root of the variance.

$$\sigma_{\hat{p}_1 - \hat{p}_2} = \sqrt{\frac{p_1 q_1}{n_1} + \frac{p_2 q_2}{n_2}}$$

We know that the differences, $\hat{p}_1 - \hat{p}_2$, are approximately normal in their distribution as long as $\hat{p}_1$ and $\hat{p}_2$ are individually approximately normal in their respective distributions, and this is true if $n_1 p_1 \geq 10$, $n_1 q_1 \geq 10$, $n_2 p_2 \geq 10$ and $n_2 q_2 \geq 10$. Applying this sampling distribution to the general form of confidence intervals we have the following:

$$\hat{p}_1 - \hat{p}_2 - E < \mu_{\hat{p}_1 - \hat{p}_2} < \hat{p}_1 - \hat{p}_2 + E.$$

Substituting, this becomes

$$\hat{p}_1 - \hat{p}_2 - E < p_1 - p_2 < \hat{p}_1 - \hat{p}_2 + E.$$

The error formula becomes $E = z_{\alpha/2} \cdot \sigma_{\hat{p}_1 - \hat{p}_2} = z_{\alpha/2} \cdot \sqrt{\frac{p_1 q_1}{n_1} + \frac{p_2 q_2}{n_2}}$, which we will have to approximate by using $E = z_{\alpha/2} \cdot \sqrt{\frac{\hat{p}_1 \hat{q}_1}{n_1} + \frac{\hat{p}_2 \hat{q}_2}{n_2}}$.

We summarize these findings below.

Summary 8.4.4 The large sample confidence interval for a difference between two population proportions

$$\hat{p}_1 - \hat{p}_2 - E < p_1 - p_2 < \hat{p}_1 - \hat{p}_2 + E,$$

where

$$E = z_{\alpha/2} \cdot \sqrt{\frac{\hat{p}_1 \hat{q}_1}{n_1} + \frac{\hat{p}_2 \hat{q}_2}{n_2}}.$$

This analysis assumes that $n_1 p_1 \geq 10$, $n_1 q_1 \geq 10$, $n_2 p_2 \geq 10$ and $n_2 q_2 \geq 10$. There is some sense that these assumptions may be satisfied if each sample contains at least 10 successes and 10 failures.

Example 8.4.5 If you were a landlord, would you rent an apartment to a person with AIDS? The authors of "Accommodating Persons with AIDS: Acceptance and Rejection in Rental Situations (J. of Applied Social Psychology (1999): 261 – 270), chose from newspapers two randomly selected samples of advertisement for apartments. In the first sample, 80 calls were made to landlords, asking to rent a room, where it was mentioned that the renter had AIDS, and, after being released from the hospital, would need a place to live. Out of those 80 calls, 32 landlords were willing to rent the room in question. For a control, 80 additional calls were made where no reference to AIDS was mentioned. Of those, 61 landlords were willing to rent the room in question. We will construct a 99% confidence interval for the difference between these two proportions.

To verify the assumptions for this test, it is easy to note that there are more than 10 successes and failures in each sample.

The calls with an AIDS reference were mentioned first, so they will be sample 1/population 1. The control group will be sample 2/population 2. $\hat{p}_1 = x_1/n_1 = 32/80 = 0.400$, $n_1 = 80$, $\hat{p}_2 = x_2/n_2 = 61/80 = 0.763$, and $n_2 = 80$.

The maximum error is

$$E = z_{\alpha/2} \cdot \sqrt{\frac{\hat{p}_1 \hat{q}_1}{n_1} + \frac{\hat{p}_2 \hat{q}_2}{n_2}}$$

$$= 2.576 \cdot \sqrt{\frac{0.400 \cdot 0.600}{80} + \frac{0.763 \cdot 0.237}{80}}$$

$$= 0.1868,$$

and the corresponding confidence interval is computed below.

$$\hat{p}_1 - \hat{p}_2 - E < p_1 - p_2 < \hat{p}_1 - \hat{p}_2 + E$$

$$0.400 - 0.763 - 0.1868 < p_1 - p_2 < 0.400 - 0.763 + 0.1868$$

$$-0.5498 < p_1 - p_2 < -0.1762$$

Our confidence interval does not include a 0 difference as a possibility, so it is likely that our two population proportions are not equal.

Our last application constructs a confidence interval for the difference between two sample means. The information for the sampling distribution of such differences is given below.

8.4.3 The Sampling Distribution of Differences, $\bar{x}_1 - \bar{x}_2$

The mean of this population is

$$\mu_{\bar{x}_1 - \bar{x}_2} = E(\bar{x}_1 - \bar{x}_2)$$

$$= E(\bar{x}_1) - E(\bar{x}_2)$$

$$= \mu_{\bar{x}_1} - \mu_{\bar{x}_2}$$

$$= \mu_1 - \mu_2$$

Noting that the difference of sample means, $\bar{x}_1 - \bar{x}_2$, is an estimate of the difference of population means $\mu_1 - \mu_2$, since $E(\bar{x}_1 - \bar{x}_2) = \mu_1 - \mu_2$, we may say that $\bar{x}_1 - \bar{x}_2$ is an unbiased estimator of $\mu_1 - \mu_2$.

The variance of this population is:

$$\begin{aligned}
\sigma^2_{\bar{x}_1 - \bar{x}_2} &= Var(\bar{x}_1 - \bar{x}_2) \\
&= Var(\bar{x}_1) + Var(\bar{x}_2) \\
&= \frac{\sigma^2_1}{n_1} + \frac{\sigma^2_2}{n_2}
\end{aligned}$$

The standard deviation is the square root of the variance:

$$\begin{aligned}
\sigma_{\bar{x}_1 - \bar{x}_2} &= \sqrt{Var(\bar{x}_1 - \bar{x}_2)} \\
&= \sqrt{\frac{\sigma^2_1}{n_1} + \frac{\sigma^2_2}{n_2}}.
\end{aligned}$$

We know that the differences $\bar{x}_1 - \bar{x}_2$ are normally distributed as long as $\bar{x}_1$ and $\bar{x}_2$ are normally distributed, and this is approximately true if either $n_1 > 30$, or Population 1 is normal, and either $n_2 > 30$ or Population 2 is normal.

Applying this sampling distribution to the general form for confidence intervals we have the following:

$$\bar{x}_1 - \bar{x}_2 - E < \mu_{\bar{x}_1 - \bar{x}_2} < \bar{x}_1 - \bar{x}_2 + E.$$

Substituting, this becomes

$$\bar{x}_1 - \bar{x}_2 - E < \mu_1 - \mu_2 < \bar{x}_1 - \bar{x}_2 + E.$$

The error formula becomes $E = z_{\alpha/2} \cdot \sigma_{\bar{x}_1 - \bar{x}_2} = z_{\alpha/2} \cdot \sqrt{\frac{\sigma^2_1}{n_1} + \frac{\sigma^2_2}{n_2}}$, which we will have to approximate by using $E = t_{\alpha/2} \cdot \sqrt{\frac{s^2_1}{n_1} + \frac{s^2_2}{n_2}}$, where the degrees of freedom are given by

$$df = \frac{(V_1 + V_2)^2}{\frac{V_1^2}{n_1 - 1} + \frac{V_2^2}{n_2 - 1}},$$

where $V_1 = \frac{s^2_1}{n_1}$ and $V_2 = \frac{s^2_2}{n_2}$.

We summarize these findings below.

Summary 8.4.6 The confidence interval for the difference between two population means with independent samples is

$$\bar{x}_1 - \bar{x}_2 - E < \mu_1 - \mu_2 < \bar{x}_1 - \bar{x}_2 + E,$$

where

$$E = t_{\alpha/2} \cdot \sqrt{\frac{s^2_1}{n_1} + \frac{s^2_2}{n_2}},$$

using degrees of freedom are given by

$$df = \frac{(V_1 + V_2)^2}{\frac{V_1^2}{n_1-1} + \frac{V_2^2}{n_2-1}},$$

where $V_1 = \frac{s_1^2}{n_1}$ and $V_2 = \frac{s_2^2}{n_2}$.

This test requires that both sample sizes be greater than 30, or that their respective populations are normal.

Example 8.4.7 Here we revisit the cloud seeding experiment which we have examined a few times previously. The first sample is drawn from the population of rainfall totals for clouds seeded with chemicals to increase rainfall amounts. From 270 storms where clouds are seeded, the mean rainfall was 1.53 inches, with a standard deviation of 1.34 inches. From 270 storms whose clouds were not seeded, the mean rainfall was 1.74 inches, with a standard deviation of 1.45 inches. We will construct the 96% confidence interval for the difference between the two mean rainfall amounts. (Data are taken from "Analyzing the results of a cloud-seeding experiment in Tasmania", by Miller, A.J., Shaw, D.E., Veitch, L.G. & Smith, E.J. (1979), *Communications in Statistics - Theory & Methods*, vol.A8(10), 1017-1047).

We define Sample 1, drawn from Population 1, to be the rainfall totals for seeded clouds, and Sample 2, drawn from Population 2, to be the control group of rainfall totals for unseeded clouds. Thus, $\bar{x}_1 = 1.53$, $s_1 = 1.34$, and $n_1 = 270$. Also, $\bar{x}_2 = 1.74$, $s_2 = 1.45$, and $n_2 = 270$. The requirements for this procedure are met since both sample sizes are greater than 30.

First, we must compute the degrees of freedom for our critical value. We begin by computing $V_1 = \frac{s_1^2}{n_1} = \frac{1.34^2}{270} = 0.00665$, and $V_2 = \frac{s_2^2}{n_2} = \frac{1.45^2}{270} = 0.00779$. This gives the degrees of freedom:

$$\begin{aligned}
df &= \frac{(V_1 + V_2)^2}{\frac{V_1^2}{n_1-1} + \frac{V_2^2}{n_2-1}} \\
&= \frac{(0.00665 + 0.00779)^2}{\frac{0.00665^2}{270-1} + \frac{0.00779^2}{270-1}} \\
&= 534.69,
\end{aligned}$$

which we must *truncate* (*round down*) to $df = 534$. Unfortunately, the table does not include this value for the degrees of freedom, so we pick the closest value in the table, namely $df = 500$.

Next we compute the maximum error,

$$E = t_{\alpha/2} \cdot \sqrt{\frac{s_1^2}{n_1} + \frac{s_2^2}{n_2}}$$

$$= 2.059 \cdot \sqrt{\frac{1.34^2}{270} + \frac{1.45^2}{270}}$$

$$= 0.2474.$$

Finally, we compute our confidence interval.

$$\bar{x}_1 - \bar{x}_2 - E < \mu_1 - \mu_2 < \bar{x}_1 - \bar{x}_2 + E$$

$$1.53 - 1.74 - 0.2474 < \mu_1 - \mu_2 < 1.53 - 1.74 + 0.2474$$

$$-0.4574 < \mu_1 - \mu_2 < 0.0374.$$

This time our confidence interval does contain zero, so it is plausible that the difference between our population means is zero, and in such a case there would be no difference between them.

8.4.4 Homework

Perform the following hypothesis tests.

1. A study was completed to investigate whether oat bran lowers serum cholesterol levels in people (Pagano & Gauvreau, 1993, p. 252). Fourteen individuals were randomly assigned a diet that included either oat bran or corn flakes. After two weeks on the initial diet, serum low-density lipoprotein levels (mg/dl) were measured. Each subject was then "crossed-over" to the alternate diet. After two weeks on this second diet, low-density lipoprotein levels were once again recorded. The data are recorded below.

Corn Flake Lipoprotein (x)	Oat Bran Lipoprotein (y)
4.61	3.84
6.42	5.57
5.40	5.85
4.54	4.80
3.98	3.68
3.82	2.96
5.01	4.41
4.34	3.72
3.80	3.49
4.56	3.84
5.35	5.26
3.89	3.73
2.25	1.84
4.24	4.14

Assuming the population of differences is normal, construct a 95% confidence interval for the mean difference of lipoprotein levels.

2. To determine the effect that exercise has on lactic acid levels in the blood, the authors of "A Descriptive Analysis of Elite-Level Racquetball" (Research Quarterly for Exercise and Sport, 1991, pp. 109 – 114), listed the blood levels of 8 different players, before and after playing three games of racquetball.

Before	13	20	17	13	13	16	15	16
After	18	37	40	35	30	20	33	19

Compute the 98% confidence interval for the mean difference of lactic acid levels. Assume that the population of differences is normal.

3. Worldwide there appears to be a discrepancy between the proportion of girls who are allowed to attend elementary school, and the like proportion for boys. The following data are based on a study published by Unicef in 2006. In a sample of 752 girls (of school age or older) randomly sampled from the Earth's population, 632 have

been allowed to attend elementary school. In a sample of 783 boys (of school age or older) randomly sampled from the Earth's population, 681 were allowed to attend elementary school. Construct a 96% confidence interval for the difference between these proportions.

4. The values in this problem are, again, based on a study published by Unicef in 2006. A commonly used measure of poverty is the proportion of people in a given area who live on less than $1 a day. In a random sample of 879 people from Eastern and South Africa, 334 of these live on less than $1 a day. In sampling 1032 people from South Asia, 340 of these live on less than $1 a day. Construct a 94% confidence interval for the difference between these proportions.

5. In Fall 2005, the Mathematics Department at Mt. San Antonio College gave grades to 3647 students in developmental math, with an average grade of 1.93 on the 4 point scale, with a sample standard deviation of 1.34. In Fall of 2001, the department gave 3638 students an average grade of 1.75, with a standard deviation of 1.35. Compute a 95% confidence interval for the difference between these means.

6. For 1532 schizophrenic patients, the mean eye gain ratio is 0.840, with a sample standard deviation of 0.118. For 1374 healthy control group members, the mean was 0.807 with a sample standard deviation was 0.134. Construct a 96% confidence interval for the difference between these means.

Chapter 9

Correlation and Regression

In Chapter 2 we summarized single variable data sets where a single observation is made for each member of sample. Here, in Chapter 9 we will look closely at bivariate quantitative data where *two* observations are gathered from each sample member. In such cases we generally make efforts to determine if the outcomes of such paired (x, y) observations are *dependent* – if they influence one another. When they do, we say that a *relationship* exists between the variables. The discussion begins with *linear correlation*, where we look for a *linear* relationship between two quantitative variables. When such relationships exist, we are able to determine the equation of the *line of best fit*, also known as the *regression line*, which may be used to estimate the value of one variable which corresponds to a chosen value of the other variable.

We begin now with our discussion of linear correlation and regression.

9.1   Linear Correlation and Regression

In this section we introduce a few of the methods of multivariate statistics, where we will analyze bivariate data – data that are measured in (x, y) pairs. We will no longer consider our x and y values to be independent. You may recall that when variables are dependent, the outcome of one will affect the likelihoods of outcomes for the other. Here, our outcomes will be the possible values of two random variables, which *may* be related in some way. In some cases, this relationship can lead to a method for using an outcome for one variable to predict an outcome for the other variable. When such a relationship exists between two random variables, we will say that they *correlate*.

In a bivariate observation, (x, y), we refer to x as the *explanatory* variable, and y as the *response* variable. Be careful though – the terminology can be misleading. We say that x *explains* y, but this does not mean that the variation in x is the *cause* for the variation observed in y. Cause and effect can only be inferred in a well conducted controlled experiment. In a controlled experiment, the values of x must be randomly *assigned* to (not *randomly observed* among) sample members, and then the response variable, y, is measured to observe an effect. In such cases, when x and y correlate, a causal inference can be made regarding the variables.

When the values of x and y are random (not controlled) observations of members in a sample, then the study is observational. In an observational study, inferences merely describe the relationship between the variables, but do not suggest cause.

In Section 9.1, we will use a graph called a *scatterplot* to attempt to determine the geometry of possible relationships between variables. We will focus our efforts on identifying linear correlation (when the geometry is linear), and develop methods for determining the strength of this relationship. Once we determine that two variables exhibit linear correlation, we will learn a method for finding a linear equation that best describes the linear relationship between the two variables.

9.1.1   Relationships Between Quantitative Variables

The Food and Agriculture Organization (the FAO) of the United Nations leads in many ways internationally with the goal to end hunger all over the world. In many parts of the world, fish is an important food source, but *overfishing* is a huge problem. It is important that fish are gathered in ways that allow fish *and* humans to thrive. The FAO helps societies manage food resources like fish in sustainable ways. The data in the following example are based on information provided by the FAO.

When fish are studied, there are many variables that are of interest to scientists. One is the age of a fish. The age of a fish captured in the wild cannot be observed directly, so scientists have to predict the age of a fish using other variables that *can* be observed directly. Variables that might *predict* or *explain* the age of a fish are sometimes obvious (like the length or weight of a fish), and sometimes not obvious (by examining growth lines on their scales).

In fresh water, juvenile Atlantic salmon look different from adults and are easy to identify. Juveniles are smaller, brown, spotted, and camouflaged, while adults are silvery.

Four juvenile Atlantic salmon were raised in captivity, so their ages (y) are known. Their ages are in weeks as follows: 10.5, 16.5, 23.5, 21.5. The mean age is $\overline{y} = 18$ weeks. Two ages are above average, and two are below. Figure 9.1.1 includes a dotplot of these data.

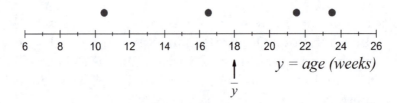

Dotplot of ages of Atlantic salmon.

There is no information here that would allow us to predict the age of a random juvenile Atlantic salmon, other than the mean. If we needed to predict the age of a randomly selected juvenile salmon based on the data gathered, we would use the mean age, $\overline{y} = 18$ weeks. Such a prediction would be unreliable, but better than a random guess.

To *better* predict ages of juvenile Atlantic salmon in the wild, we need to know something more about these fish – an explanatory variable that will help us predict (explain) the age of other fish. The table below contains *weights* (in grams) for those same juvenile salmon, along with their respective ages. We want to determine whether weight can be used to explain (predict) their age.

Weight (g)	4	8	12	16
Age (weeks)	10.5	16.5	23.5	21.5

When the value of one variable can be used to predict (or explain) another with some reliability that is better than a random guess, we say there is a *relationship* between the two variables. The easiest way to determine if there is a relationship between quantitative variables is to plot them in a scatterplot and look for a pattern. The data are plotted in Figure 9.1 as (x, y) pairs where $x = weight$, and $y = age$.

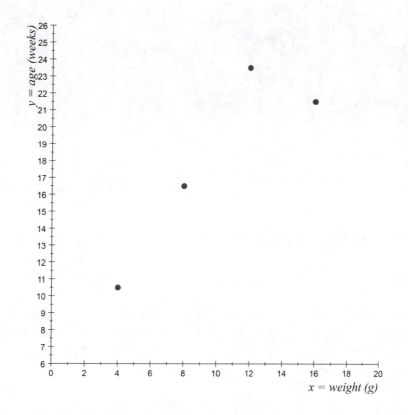

Figure 9.1: Scatterplot of weight vs. age for juvenile Atlantic salmon.

We know that juvenile fish get heavier as they get older, so we expect to see a relationship between *weight* and *age* that is positive – as one variable increases (*weight*), so does the other (*age*). The scatterplot gives some support of this by exhibiting a pattern that goes upward in *age* as *weight* increases. This pattern is evidence that there is a relationship between *weight* and *age* that is, admittedly, rather insufficient since we have only four data points.

As stated, if there were no relationship between weight and age, we would be forced to use the mean age ($\overline{y} = 18$ weeks) for making predictions on age. The data points all deviate from the mean, and we would have no way to explain these deviations. But there does appear to be such a relationship, so we can use the explanatory variable to try to explain the deviations.

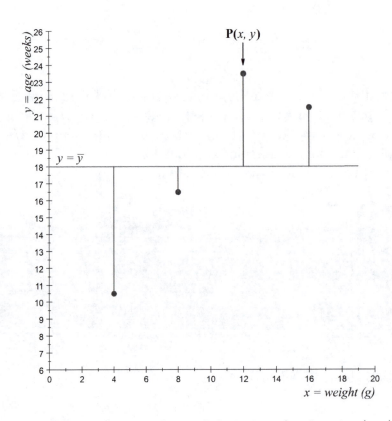

Figure 9.2: The vertical lines represent the total deviation of each point (x, y) from the mean of y values ($\overline{y}$).

The horizontal line in Figure 9.2 represents the mean of y-values, $\overline{y} = 18$. The vertical lines that extend from the mean to each point represent the deviations from the mean of y-values. We will use the explanatory variable, x, to explain these deviations using the *line of best fit*, plotted with the scatterplot in Figure 9.3.

We see that the line of best fit has a positive slope (upward while moving right) that matches the positive relationship between the variables, and it essentially represents the

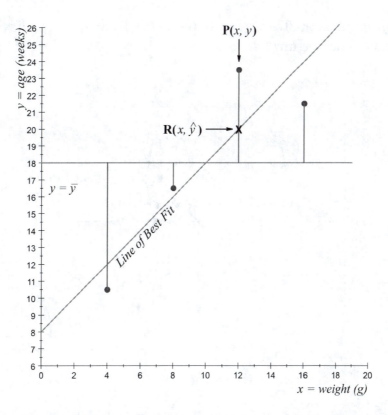

Figure 9.3: The line of best fit, an observed point $P(x, y)$ and a predicted point, $R(x, \widehat{y})$.

center of the upward trend exhibited by the data points. If we want to predict a value for y corresponding to a given x, we will find the point on the line with that x-coordinate, and determine its y-value.

The equation of any non-vertical line can be written as $\widehat{y} = a + bx$, where b, is called the *slope* of the line, and a is called the *y-intercept* (since this is the value of $\widehat{y}$ when $x = 0$ on the y-axis). We will determine a and b so that the line becomes one of *best fit*. We will denote the y-coordinates of points on the line of best fit as $\widehat{y}$ instead of y because we are not observing these values from a sample – we are predicting them.

For any given value of the explanatory variable, x, the predicted value of y is $\widehat{y} = a + bx$. Point R on Figure 9.3 with coordinates $(x, \widehat{y})$ is one of many such points that make up the line.

Using the same x-coordinate as in point R, consider the point $P(x, y)$ on the scatterplot (Fig. 9.3). It is *not* on the line of best fit. Points P and R have different y-coordinates, $\widehat{y} \neq y$, so there is error when we use x to predict y (there is always *some* error when we make predictions). As we develop the line of best fit, our goal will to be to somehow minimize total error among all data points.

For our point $P(x, y)$, we see that y deviates *above* the mean, $\overline{y}$. When considering, x, the line of best fit explains (predicts) a y-coordinate that is above average (at point R), but not as high as we actually observe at point P. To analyze the deviation from the mean carefully, consider Figure 9.4.

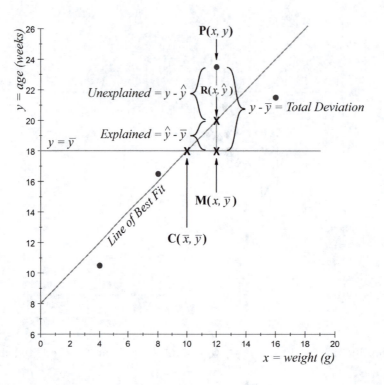

Figure 9.4: The line of best fit, and the total deviation of one point, $P(x, y)$, from the mean broken into explained and unexplained components.

Without an explanatory variable, we have said that the predicted value of y should be $\overline{y}$. This is the y-coordinate of point M in Figure 9.4. The *total deviation* between points M and P is the difference between these y-coordinates,

$$Total\ Deviation = y - \overline{y}.$$

But we *do* have an explanatory variable, x, which leads us to predict (explain) the y-value of $\widehat{y}$ in point R on the line of best fit. The part of total deviation that is *explained* by x (using the line of best fit) is the difference between the y coordinates of points M and R,

$$Explained\ Deviation = \widehat{y} - \overline{y}.$$

Of course, if part (not all) of the total deviation in y is explained by x, then part of it is *unexplained* as well. The deviation in y that is unexplained by x is that part which extends beyond the predicted value ($\widehat{y}$) to the actual value of y between points R and P,

$$Unexplained\ Deviation = y - \widehat{y}.$$

The *unexplained deviation* for each point is also referred to the *residual*, or sometimes the *error*, $E = y - \widehat{y}$, in the estimate, $\widehat{y}$.

For a single point (x, y), Figure 9.4 makes it clear that

$$Total\ Deviation = Explained\ Deviation + Unexplained\ Deviation.$$

For points to be close to the line of best fit is for the unexplained deviations (i.e. residuals or errors, $E = y - \widehat{y}$) to be as small as possible. To get some sense of total error for all data points, it seems logical to sum these residuals. This proves useless, however, because it is always the case that the sum of the unexplained deviations, $\sum E$, is always zero. We will prove this in due course.

Thus, we cannot measure total unexplained deviation by summing in this way. A similar problem occurred for us when averaging deviations from the mean in the development of standard deviation. The sum of deviations from the mean is always zero, so we squared the deviations to make them positive. We solve our current problem in the same way. Instead of summing the unexplained deviations (i.e. residuals or errors), we will sum their squares. This sum will be referred to as the *Sum of Squared Errors*, denoted *SSE*.

Definition 9.1.1 The sum of squared errors, *SSE*, sums the squares of unexplained deviations of all points from the line of best fit.

$$SSE = \sum (y - \widehat{y})^2$$

For a line to fit data points in a way that is *best*, we need to minimize total error of all points away from the line. It is very important to remember that the sum of squared errors, *SSE*, is the total we want to minimize. We will create our line of best fit so that *SSE* is as small as possible.

9.1.2   Finding the Line of Best Fit

The development we are about to begin determines a and b values that will define the equation for our line of best fit, $\widehat{y} = a + bx$. There is a short list of facts that we must establish to simplify things as we proceed. The first idea is that when we add all z-scores from a sample of values, x, we always get zero.

Theorem 9.1.2 For any sample of values, x, the sum of corresponding z-scores is 0.

Proof. This is straightforward: $\sum z = \sum \left(\frac{x - \bar{x}}{s}\right) = \frac{1}{s}\left(\sum x - \sum \bar{x}\right) = \frac{1}{s}\left(\sum x - n\,\bar{x}\right) = \frac{1}{s}\left(\sum x - n\frac{\sum x}{n}\right) = \frac{1}{s}\left(\sum x - \sum x\right) = 0.$ ∎

From this, we can quickly establish the fact that the mean of all z-scores from a sample is always zero.

Theorem 9.1.3 For any sample of values, x, the mean of corresponding z-scores is $\overline{z} = 0$.

Next, there is a simple identity that tells us that the sum of all squared z-scores is always equal to the sample size minus 1.

Theorem 9.1.4 For any sample of n values of x, with corresponding z-scores, $\sum z^2 = n - 1$.

Proof.

$$
\begin{aligned}
\sum z^2 &= \sum \left(\frac{x - \overline{x}}{s} \right)^2 \\
&= \frac{n-1}{s^2} \frac{\sum (x - \overline{x})^2}{n - 1} \\
&= \frac{n-1}{s^2} \cdot s^2 \\
&= n - 1.
\end{aligned}
$$

∎

The last of our preliminary facts allows us to work with z-scores instead of y-values as we simplify $SSE = \sum (y - \widehat{y})^2$.

Theorem 9.1.5 For any pair of values, y_1 and y_2, chosen from a sample with mean $\overline{y}$ and standard deviation s_y, it is true that $(y_1 - y_2)^2 = s_y^2 (z_1 - z_2)^2$, where z_1 and z_2 are the z-scores of y_1 and y_2.

Proof.

$$
\begin{aligned}
s_y^2 (z_1 - z_2)^2 &= s_y^2 \left(\frac{y_1 - \overline{y}}{s_y} - \frac{y_2 - \overline{y}}{s_y} \right)^2 \\
&= s_y^2 \left(\frac{y_1 - y_2}{s_y} \right)^2 \\
&= \frac{s_y^2}{s_y^2} (y_1 - y_2)^2 \\
&= (y_1 - y_2)^2
\end{aligned}
$$

∎

Our goal is to somehow minimize the unexplained variation for *all* points, which is quantified by $SSE = \sum (y - \widehat{y})^2$. As mentioned, we will focus on z-scores to simplify the algebra involved.

The statistics of our sample data of weights and ages of juvenile Atlantic salmon are as follows.

Variable	Mean	Std. Dev.
x	$\overline{x} = 10$	$s_x = 5.1640$
y	$\overline{y} = 18$	$s_y = 5.8023$

If we take z-scores of each value of x $\left(z_x = \frac{x-\overline{x}}{s_x}\right)$ and also of each y $\left(z_y = \frac{y-\overline{y}}{s_y}\right)$, the results are as follows.

x	y	z_x	z_y
4	10.5	-1.1619	-1.2926
8	16.5	-0.3873	-0.2585
12	23.5	0.3873	0.9479
16	21.5	1.1679	0.6032

These z-scores don't mean much until we graph them and compare to the original xy-scatterplot. These are plotted below.

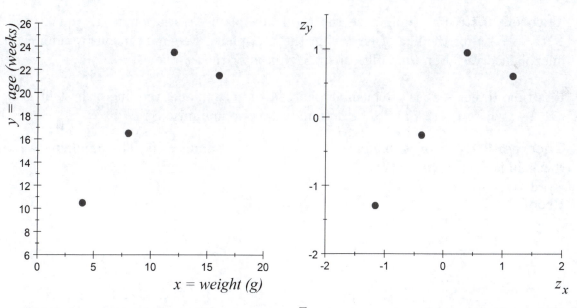

Weight vs. age for Atlantic Salmon. Z-scores, z_x vs. z_y, of Atlantic salmon data.

The main point to make here is that the data distributions are the same whether we look at xy-values or $z_x z_y$-values. Because of this, we would expect the lines of best fit to be arranged similarly in either plane. Our goal is to find the line of best fit in the $z_x z_y$-plane, then transform it into the xy-plane.

As stated, the line of best fit will minimize SSE, which is the total unexplained deviation of all points from this line. In the $z_x z_y$-plane, the equation of the line of best fit must be in the form $\widehat{z}_y = r\, z_x + k$, where r is the slope and k is the intercept of the vertical z_y-axis. In the xy-plane we want the z-score of $\widehat{y}$ to be $\widehat{z}_y$

$$\widehat{z}_y = \frac{\widehat{y} - \overline{y}}{s_y}.$$

If we solve for $\widehat{y}$, we arrive at the equation we will use to define it.

$$\widehat{y} = s_y \widehat{z}_y + \overline{y}$$

Now we can use Theorem 9.1.5, to rewrite SSE as follows.

$$
\begin{aligned}
SSE &= \sum (y - \widehat{y})^2 \\
&= s_y^2 \sum (z_y - \widehat{z}_y)^2 .
\end{aligned}
$$

As mentioned, it is much easier to determine the line of best fit in the $z_x z_y$-plane. As we do this, we minimize the unexplained variation of points away from the corresponding line of best fit, quantified by $\sum (z_y - \widehat{z}_y)^2$. But since SSE is so closely related to this quantity, $SSE = s_y^2 \sum (z_y - \widehat{z}_y)^2$, minimizing the first expression in z_y naturally minimizes the second expression in y. This is made official in the following theorem.

Theorem 9.1.6 If we minimize the total unexplained variation in z_y (quantified by $\sum (z_y - \widehat{z}_y)^2$) for the line of best fit in the $z_x z_y$-plane, we simultaneously minimize the unexplained variation (quantified by SSE) in y for the xy-plane.

It is time to get serious. Remember that, in the $z_x z_y$-plane, the line of best fit is $\widehat{z}_y = r\, z_x + k$, where r and k are to be determined as we minimize SSE.

Theorem 9.1.7 SSE is minimized when $r = \frac{\sum z_x z_y}{n-1}$ and $k = 0$. The minimum value of this sum is $SSE = (n-1)\left(1 - r^2\right) s_y^2$.

Proof.

$$
\begin{aligned}
SSE &= \sum (y - \widehat{y})^2 \\
&= s_y^2 \sum (z_y - z_{\widehat{y}})^2 \\
&= s_y^2 \sum (z_y - r z_x - k)^2 \\
&= s_y^2 \sum \left(z_y^2 - 2r z_x z_y - 2k z_y + r^2 z_x^2 + 2rk z_x + k^2\right) \\
&= s_y^2 \left(\sum z_y^2 - 2r \sum z_x z_y - 2k \sum z_y + r^2 \sum z_x^2 + 2rk \sum z_x + \sum k^2\right) \\
&= s_y^2 \left((n-1) - 2r \sum z_x z_y - 2k \cdot 0 + r^2 (n-1) + 2rk \cdot 0 + nk^2\right) \\
&= s_y^2 \left((n-1) - 2r \sum z_x z_y + r^2 (n-1) + nk^2\right)
\end{aligned}
$$

Notice the only mention of k is in the nk^2 term. If we want to minimize SSE, this term is smallest when $k = 0$ (since nk^2 is always positive). So our smaller SSE continues below.

$$SSE = s_y^2 \left((n-1) - 2r \sum z_x z_y + r^2 (n-1) + n \cdot 0^2 \right)$$
$$= s_y^2 \left((n-1) r^2 - 2r \sum z_x z_y + (n-1) \right)$$

This is a quadratic in r. You may recall from algebra that quadratics of the form $y = Ax^2 + Bx + C$, where $A > 0$, are minimized when $x = -\frac{B}{2A}$. In the quadratic for SSE, $A = (n-1)$, and $B = -2 \sum z_x z_y$. Thus, to minimize SSE, we need

$$r = -\frac{B}{2A}$$
$$= -\frac{-2 \sum z_x z_y}{2(n-1)}$$
$$= \frac{\sum z_x z_y}{n-1}.$$

With this value of r, note that $\sum z_x z_y = r(n-1)$. Substituting into our expression for SSE yields the following.

$$SSE = s_y^2 \left((n-1) r^2 - 2r \cdot r(n-1) + (n-1) \right)$$
$$= s_y^2 \left((n-1) - (n-1) r^2 \right)$$
$$= s_y^2 (n-1) \left(1 - r^2 \right)$$

■

We now know the equation of the line of best fit in the $z_x z_y$-plane is $\widehat{z}_y = r z_x$ where the slope of this line is $r = \frac{\sum z_x z_y}{n-1}$. This must be transformed to the xy-plane using the definition for $\widehat{y}$ given above.

$$\widehat{y} = s_y \widehat{z}_y + \overline{y}$$
$$= s_y \cdot r z_x + \overline{y}$$
$$= r s_y \frac{x - \overline{x}}{s_x} + \overline{y}$$
$$= \frac{r s_y}{s_x} (x - \overline{x}) + \overline{y}$$
$$= \left(\overline{y} - \frac{r s_y}{s_x} \overline{x} \right) + \frac{r s_y}{s_x} x$$

The resulting expression is in the form $\widehat{y} = a + bx$. The slope is the x-coefficient: $b = \frac{r s_y}{s_x}$. The y-intercept is the constant, $a = \overline{y} - \frac{r s_y}{s_x} \overline{x} = \overline{y} - b\overline{x}$.

The line of best fit is often called the *regression line*, a term whose meaning has very little to do with the mathematics behind it, but will be discussed in Section 9.6.

An important fact is that the centroid, $(\overline{x}, \overline{y})$, is always on the line of best fit. This is true because if $x = \overline{x}$, then $\widehat{y} = \overline{y}$.

$$\widehat{y} = \left(\overline{y} - \frac{r\, s_y}{s_x}\overline{x}\right) + \frac{r\, s_y}{s_x}\overline{x}$$
$$= \overline{y}$$

Theorem 9.1.8 The equation of the line of best fit (i.e. the *regression line*) is $\widehat{y} = a + bx$, where the slope is

$$b = \frac{r\, s_y}{s_x}$$

and the y-intercept is

$$a = \overline{y} - b\overline{x}.$$

The slope and y-intercept are computed using r, whose value minimizes SSE, the unexplained deviation of all points from this line,

$$r = \frac{\sum z_x z_y}{n - 1}.$$

With this value of r, the minimum value of SSE is $s_y^2\,(n - 1)\,(1 - r^2)$.
Finally, the centroid, $(\overline{x}, \overline{y})$, is *always* on the line of best fit.

As a bit of housekeeping, we mentioned that the sum of unexplained deviations, $\sum E = \sum (y - \widehat{y})$, among y-values is always zero. This is easily proved with what we have learned as follows: $\sum (y - \widehat{y}) = s_y \sum (z_y - r\, z_x) = s_y \left(\sum z_y - r \sum z_x\right)$, which must be zero since any sum of sample z-scores is always zero. This is why we must square deviations before adding if we want a measure of true variability.

Returning to our example data involving juvenile Atlantic salmon, recall that we have computed the following statistics.

Variable	Mean	Std. Dev.
$x = weight$	$\overline{x} = 10$	$s_x = 5.1640$
$y = age$	$\overline{y} = 18$	$s_y = 5.8023$

To compute r, we multiply each z_x by z_y, then add.

x	y	z_x	z_y	$z_x \cdot z_y$
4	10.5	-1.1619	-1.2926	1.5019
8	16.5	-0.3873	-0.2585	0.1001
12	23.5	0.3873	0.9479	0.3671
16	21.5	1.1679	0.6032	0.7009
			Total:	2.6700

Thus we have $\sum z_x z_y = 2.6700$. This yields the following value for r.

$$r = \frac{\sum z_x z_y}{n-1}$$
$$= \frac{2.6700}{4-1}$$
$$= 0.8900$$

With this and the statistics given above, we can compute the slope and y-intercept of the line of best fit. First we compute the slope, b.

$$b = \frac{r\, s_y}{s_x}$$
$$= \frac{0.8900 \cdot 5.8023}{5.1640}$$
$$= 1.0000$$

Next, the y-intercept, a.

$$a = \overline{y} - b\overline{x}$$
$$= 12 - 1.0000 \cdot 10$$
$$= 2.0000$$

In spite of the rounding involved, the rounding error surprisingly turns out to be zero, and these values are actually exactly $b = 1$, and $a = 8$. The equation of the line of best fit is $\widehat{y} = a + bx$,

$$\widehat{y} = 8 + x.$$

The line of best fit is plotted with the scatterplot on Figure 9.5. The centroid $(\overline{x}, \overline{y}) = (10, 18)$ (seen as point **C** in the figure) is always on the line, as is the y-intercept (point **I** in the figure). These two points are sufficient to sketch the line of best fit, provided that the y-axis can be fit into the picture.

There is more to learn about r, especially when we quantify *explained* variation from the mean, but first we focus on *total* variation. Recall that the total variation of y-values from their mean is represented by $SST = \sum (y - \overline{y})^2$. The following theorem gives a simplified formula for SST.

Theorem 9.1.9 $SST = (n - 1)\, s_y^2$.

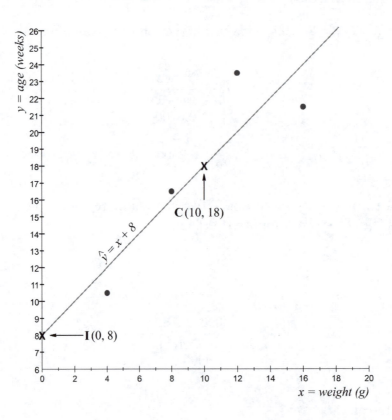

Figure 9.5: The line of best fit $y = 8 + x$ always passes through the y-intercept **I**, and the centroid **C**.

Proof. This is relatively straightforward since $s_y = \sqrt{\frac{\sum(y-\bar{y})^2}{n-1}}$, and the numerator in the square root is SST. So $s_y^2 = \frac{SST}{n-1}$, and thus $SST = (n-1)\, s_y^2$. ∎

We have stated that total deviation from the mean is composed of explained deviation and unexplained deviation. For a given data point, (x, y), the explained deviation from $\bar{y}$ is that which is explained by the explanatory variable x by the prediction, $\hat{y} = a + bx$, so this is $\hat{y} - \bar{y}$. The sum of all such variation from the mean is called SSR, *Sum of Squares due to Regression*, since it measures deviation from the mean to the regression line (i.e. the line of best fit).

Theorem 9.1.10 The Sum of Squares due to Regression (SSR) measures the explained variation of all points from the line of best fit: $SSR = (n-1)\, r^2 s_y^2$.

Proof. Recall that, in the $z_x z_y$-plane, the line of best fit is $z_{\widehat{y}} = r z_x + k$. Recall, however that we minimized SSE by choosing $k = 0$. Therefore, $z_{\widehat{y}} = r z_x$.

$$SSR = \sum \left(\widehat{y} - \overline{y}\right)^2$$
$$= s_y^2 \sum \left(z_{\widehat{y}} - z_{\overline{y}}\right)^2$$
$$= s_y^2 \sum \left(r z_x - 0\right)^2$$
$$= s_y^2 r^2 \sum z_x^2$$
$$= s_y^2 r^2 \left(n - 1\right)$$

∎

Next we want to show that total variability in y (SST) can be expressed as the sum of explained variability (SSR) and unexplained variability (SSE).

Theorem 9.1.11 The total variability, in y, from the mean, $\overline{y}$, is equal to the sum of explained and unexplained variability.

$$SST = SSR + SSE$$

Proof.

$$SSR + SSE = (n - 1) r^2 s_y^2 + (n - 1) \left(1 - r^2\right) s_y^2$$
$$= (n - 1) s_y^2 \left(r^2 + \left(1 - r^2\right)\right)$$
$$= (n - 1) s_y^2$$
$$= SST$$

∎

The total explained (SSR) and unexplained (SSE) variations in y are two parts of a whole (SST). Since this is the case, it makes sense to express them in terms of proportions. The expression $SST = SSR + SSE$, can be divided by SST to arrive at the following.

$$\frac{SSR}{SST} + \frac{SSE}{SST} = 1$$

The expression $\frac{SSR}{SST}$ represents the *proportion* of total variation in y that is *explained* by x, while $\frac{SSE}{SST}$ represents the proportion of total variation in y that is *unexplained* by x, using a line of best fit. The facts gathered to this point lead us to two interesting facts related to r.

Theorem 9.1.12 The proportion of total variation in y that is *explained* by x using the line of best fit is r^2. This quantity, r^2, is called the *coefficient of determination*.

Proof. The straightforward proof draws upon Theorems 9.1.10 and 9.1.9, given above.

$$\frac{SSR}{SST} = \frac{(n - 1) r^2 s_y^2}{(n - 1) s_y^2}$$
$$= r^2$$

∎

Theorem 9.1.13 The proportion of total variation in y that is *unexplained* by x using the line of best fit is $1 - r^2$.

Proof. This is trivial since the unexplained variation is $\frac{SSE}{SST} = 1 - \frac{SSR}{SST} = 1 - r^2$. ∎
Because the definitions of SSR, SSE, and SST are all sums of squared values, it is clear that each of these values is positive. This, plus the fact that $\frac{SSR}{SST} + \frac{SSE}{SST} = 1$ tells us that $0 \le \frac{SSR}{SST} \le 1$. Since $r^2 = \frac{SSR}{SST}$, we know that $0 \le r^2 \le 1$. This fact implies that $-1 \le r \le 1$.

Suppose that all points in our data set lie on the line of best fit. This tells us that the total unexplained deviation from this line is zero, thus $\frac{SSE}{SST} = 0$. Since $\frac{SSR}{SST} + \frac{SSE}{SST} = 1$, we would have $\frac{SSR}{SST} = r^2 = 1$. This tells us that $r = \pm 1$. Note that r is the slope of the line of best fit in the $z_x z_y$-plane, and this mirrors the line of best fit in the xy-plane. With this we can now say that if $r = 1$, all points in our data set lie on a line with positive slope (so there is a positive relationship between x and y). Likewise, when $r = -1$, all points lie on a line with negative slope (and there is a negative relationship between x and y).

A similar argument holds when points are *close* to the line of best fit, where $\frac{SSE}{SST} \approx 0$, implying that $r^2 = \frac{SSR}{SST} \approx 1$. This tells us that $r \approx \pm 1$, and when $r \approx 1$ the line has a positive slope (and the variables x and y have a positive relationship) and when $r \approx -1$ the slope is negative (and x and y have a negative relationship).

If $r^2 = \frac{SSR}{SST} \approx 0$ (and thus $r \approx 0$), it must be true that most of the variation in y is unexplained by x using a line of best fit, and that there is very little linear relationship (or none) between x and y.

Much information about the relationship between x and y is revealed by r, which is called the *linear correlation coefficient* (or sometimes *Pearson's Linear Correlation Coefficient*). All of this is summarized in the following theorem.

Theorem 9.1.14 The linear correlation coefficient, r, has the following properties.

- When $r = \pm 1$, then all (x, y) points lie on the line of best fit.

- When $r \approx 1$, then points lie near a line with positive slope, and we say that the relationship between x and y is strong and positive.

- When $r \approx -1$, then points lie near a line with negative slope, and we say that the relationship between x and y is strong and negative.

- When $r \approx 0$, then the linear relationship between x and y is weak because the proportion of variation in y that is explained by x through a line of best fit is nearly zero. Either there is no relationship, or the relationship between x and y is very nonlinear. In either case, a line of best fit is not an appropriate model for predicting a y-value that corresponds to a given value of x.

In the theorem above, we describe relationships as either *strong* or *weak*. When we say that a relationship is strong, we are saying that there is a somewhat reliably predictive relationship between x and y, where values of x may used to predict, or explain, values of y, and these predictions are better, on average, than using the mean of y values ($\bar{y}$) to predict y.

Figure 9.6 gives the range of values for r, and suggestions describing the strength of the relationship between x and y.

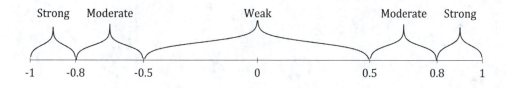

Figure 9.6: It is always true that $-1 \leq r \leq 1$, but the location of r in this range tells us about the strength and direction of a possible linear relationship between x and y.

Recall that for the relationship between the weight (x) and age (y) of juvenile Atlantic salmon, we computed $r = 0.8900$. In examining Figure 9.6, we see that the relationship among the four data points is *strong*. If we were working with a reasonably large random sample (and this is not the case *here*), we could use weight of a juvenile Atlantic salmon to predict its age using the line of best fit.

When describing the relationship between two quantitative variables, we refer to the *form*, *strength* and (when the relationship is linear) *direction* of the relationship.

A scatterplot should be used to assess the *form* (linear, nonlinear, or none) of any relationship that may exist between (x, y) data pairs. When the relationship is determined to be somewhat linear, the correlation coefficient, r, should be used to determine the strength of that relationship. The stronger the linear relationship, the more the explanatory variable, x, can be used to make somewhat reliable predictions of y using the line of best fit, $\hat{y} = a + bx$. When the correlation coefficient, r, of the line of best fit is positive (as will be the slope), the *direction* of the relationship is also called *positive*. Likewise, when r is negative (as will be the slope) the direction of the relationship is *negative*.

Table 9.1 relates different strengths and directions of linear relationships to varying values of r.

Note, in Table 9.1 that the strongest relationships have r very close to -1 or 1. We will refer to a linear relationship as being *perfect* when all points lie on a line and $r = \pm 1$.

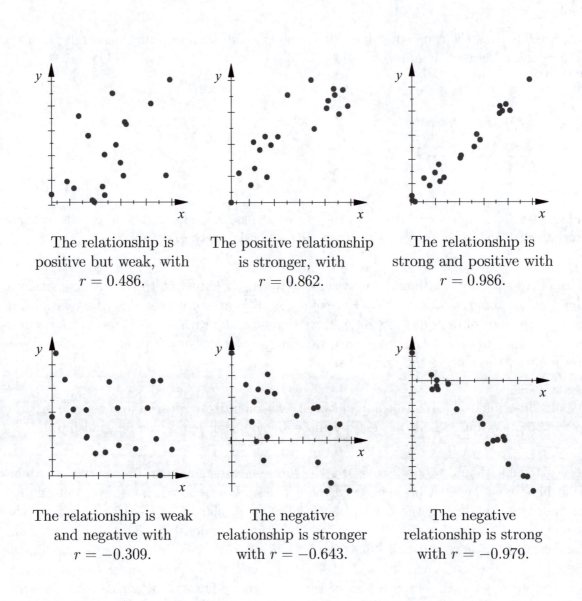

The relationship is positive but weak, with $r = 0.486$.

The positive relationship is stronger, with $r = 0.862$.

The relationship is strong and positive with $r = 0.986$.

The relationship is weak and negative with $r = -0.309$.

The negative relationship is stronger with $r = -0.643$.

The negative relationship is strong with $r = -0.979$.

Table 9.1: The various graphs here demonstrate the way that r may be used to discover the strength of a linear relationship between x and y.

9.1.3 Nonlinear Forms

As mentioned, we should not use the correlation coefficient, r, to determine the *form* (linear, nonlinear, or none) of a relationship between two variables. Scatterplots or other displays should be used for that.

Consider the graphs in Figures 9.7 and 9.8. Both data sets have correlation coefficients that are nearly zero, but in Figure 9.7 we see a strong but very nonlinear relationship between x and y. In Figure 9.8 there appears to be no relationship between the variables. Thus, the fact that $r \approx 0$, does not imply that the variables are unrelated – it only implies that there is no *linear* relationship.

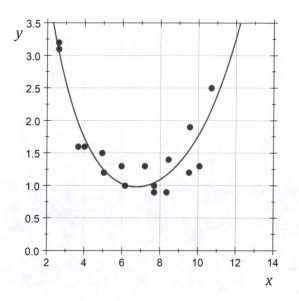

Figure 9.7: Here the correlation coefficient, r, is nearly zero with an apparent nonlinear relationship between x (pH of battery electrolyte) and y (battery electric current).

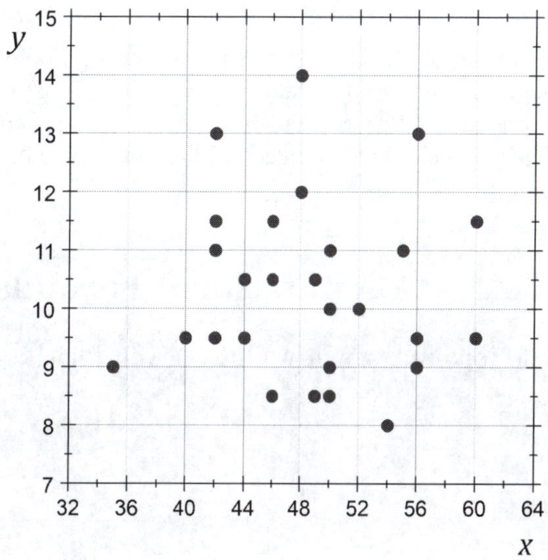

Figure 9.8: Again, the correlation coefficient, r, is nearly zero but here there appears to be no relationship between x (individual's average sleep hours) and y (individual's shoe size).

Next, consider the graphs in Figures 9.9 and 9.10. In each case the correlation coefficient is nearly one, but in neither case is it clear that a linear model is appropriate to represent the data pairs. In Figure 9.9 we are reminded that some patterns are nearly linear, but would be better represented as curves. In Figure 9.10 we see how a single influential outlier can take a pair of variables that would otherwise demonstrate no relationship and yield a linear correlation coefficient that is nearly 1. In either case, using a line to represent how the variables are related (or not) is inappropriate.

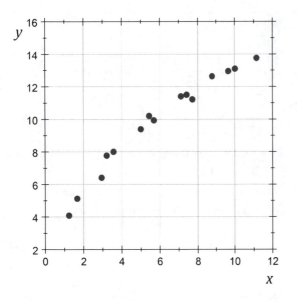

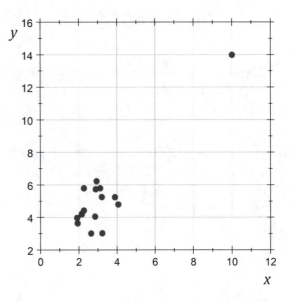

Figure 9.9: In this case, the correlation coefficient, r, is nearly one, and clearly the relationship is positive. The points are close to a line, but clearly a curved or nonlinear model would be more appropriate than a line.

Figure 9.10: With an influential outlier at (10,14) the data have a correlation coefficient of $r = 0.906$. Ignoring the outlier yields a correlation coefficient of $r = 0.296$, and there is little evidence of a linear relationship.

9.1.4   Computational Formulas

The formula which we have given for the linear correlation coefficient, r,

$$r = \frac{\sum z_x z_y}{n - 1}$$

is useful theoretical discussions. It does not take long, however, to realize that it is a difficult computation to perform, and prone to larger than necessary rounding error. The following computational formula is algebraically equivalent to the formula above, and yields the same result, but with less rounding error.

Theorem 9.1.15 Pearson's Linear Correlation Coefficient can be computed with the formula below. This formula is less prone to rounding error.

$$r = \frac{\sum xy - \frac{1}{n}\left(\sum x\right)\left(\sum y\right)}{\sqrt{\sum x^2 - \frac{1}{n}\left(\sum x\right)^2}\sqrt{\sum y^2 - \frac{1}{n}\left(\sum y\right)^2}}$$

If we apply this formula to the juvenile Atlantic salmon data, we first need to tabulate x^2, y^2, and xy values. Totals of these are given in the last row of the table below.

x	y	x^2	y^2	xy
4	10.5	16	110.25	42
8	16.5	64	272.25	132
12	23.5	144	552.25	282
16	21.5	256	462.25	344
40	72	480	1397.00	800

With these, computing r is relatively straightforward.

$$r = \frac{\sum xy - \frac{1}{n}\left(\sum x\right)\left(\sum y\right)}{\sqrt{\sum x^2 - \frac{1}{n}\left(\sum x\right)^2}\sqrt{\sum y^2 - \frac{1}{n}\left(\sum y\right)^2}}$$

$$= \frac{800 - \frac{1}{4}(40)(72)}{\sqrt{480 - \frac{1}{4}(40)^2}\sqrt{1397 - \frac{1}{4}(72)^2}}$$

$$= 0.890$$

This yields the same value as before, but note that the only rounding that occurs is in the last step. Previously, rounding occurred at *every step*. Regarding a *rule* for rounding, we will use the same rule for r and r^2 as we will use for proportions – round to 3 significant figures (where leading zeros are not considered significant). The slope, y-intercept, and $\widehat{y}$ values should be rounded using the same rule for means and standard deviations, as explained below.

Procedure 9.1.16 Rounding Rule for Correlation Statistics

- Round r and r^2 to no fewer than 3 significant digits.

- Round the slope, y-intercept and $\widehat{y}$ values to one place more than the observed y-data values.

The slope statistic, b, can be rewritten as well by substiting the above formula.

$$b = \frac{r\, s_y}{s_x}$$

$$= \frac{\frac{\sum xy - \frac{1}{n}(\sum x)(\sum y)}{\sqrt{\sum x^2 - \frac{1}{n}(\sum x)^2}\sqrt{\sum y^2 - \frac{1}{n}(\sum y)^2}} \cdot \sqrt{\frac{\sum y^2 - \frac{1}{n}(\sum y)^2}{n-1}}}{\sqrt{\frac{\sum x^2 - \frac{1}{n}(\sum x)^2}{n-1}}}$$

$$= \frac{\sum xy - \frac{1}{n}\left(\sum x\right)\left(\sum y\right)}{\sum x^2 - \frac{1}{n}\left(\sum x\right)^2}$$

Theorem 9.1.17 A computational formula for the slope statistic is as follows.

$$b = \frac{r\, s_y}{s_x}$$

$$= \frac{\sum xy - \frac{1}{n}\left(\sum x\right)\left(\sum y\right)}{\sum x^2 - \frac{1}{n}\left(\sum x\right)^2}$$

Applying this formula to the juvenile Atlantic salmon data yields the following.

$$b = \frac{\sum xy - \frac{1}{n}\left(\sum x\right)\left(\sum y\right)}{\sum x^2 - \frac{1}{n}\left(\sum x\right)^2}$$

$$= \frac{800 - \frac{1}{4}\cdot 40 \cdot 72}{480 - \frac{1}{4}\left(40\right)^2}$$

$$= 1$$

This is the same value obtained originally with the original formula for the sample slope.

9.1.5 More Data Yield Better Predictions

The table below contains weights (in grams) and known ages (in weeks) of 15 random juvenile Atlantic salmon.

Weight (x)	18.76	32.06	7.76	11.87	25.15	15.87	25.44	24.97	32.44	17.92	13.82	15.28	21.05	15.42	18.08
Age (y)	24.45	40.84	17.87	18.95	34.57	25.33	38.48	35.82	44.26	27.78	26.10	25.22	29.02	21.63	28.96

The scatterplot for these data is provided in Figure 9.11. Now that we have much more data, it is reasonable to start making some informal inferences regarding the relationship between weight and age for juvenile Atlantic salmon. The first observation is that the relationship between these variables appears somewhat linear and positive (for the range of weights provided).

Here, with more data points that are recorded with more precision (with two places after the decimal), computations are messier, and it is reasonable to use technology to compute the statistics that we need. Most scientific calculators will compute statistics for bivariate quantitative data, and it is advised that technology be employed in such cases.

For these data, the relevant statistics are as follows.

$\overline{x}$	s_x	$\overline{y}$	s_y	r
19.7260	7.0961	29.2853	7.9138	0.965

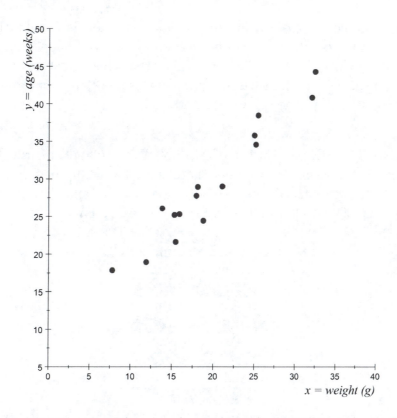

Figure 9.11: Scatterplot for weight (in grams) versus age (in weeks) for juvenile Atlantic salmon.

We have visually classified the relationship between x and y as positive and linear. Referring again to Figure 9.6 we see that, with $r = 0.965$ in the larger data set, the relationship is quite strong. Thus, it makes sense to use the line of best fit to describe the relationship.

The slope is computed using these statistics as follows.

$$
\begin{aligned}
b &= \frac{r\, s_y}{s_x} \\
&= \frac{0.965 \cdot 7.9138}{7.0961} \\
&= 1.076
\end{aligned}
$$

The y-intercept is computed next.

$$
\begin{aligned}
a &= \overline{y} - b\,\overline{x} \\
&= 29.2853 - 1.076 \cdot 19.7260 \\
&= 8.060
\end{aligned}
$$

The line of best fit is $\widehat{y} = 8.060 + 1.076x$ This is similar to the equation found with the smaller data set, but should give better predictions because it is founded on more evidence. For example, suppose a juvenile Atlantic salmon is caught that weighs 23 grams. Based on this, our best estimate for its age is $\widehat{y} = 8.060 + 1.076 \cdot 23 \approx 32.8$, approximately 33 weeks old. We should remember that this is just an estimate, not the exact age of the fish, but one that can be considered somewhat accurate because of the strength of the relationship between weight and age.

Example 9.1.18 The FBI and Bureau of Alcohol, Tobacco and Firearms have compiled data with the goal of determining whether there is a relationship between the number of registered automatic weapons (in thousands) in a given city and that city's murder rate (in murders per 100,000 people). These data include the following for 8 different cities.

Automatic Weapons	Murder Rate
11.6	13.1
8.3	10.6
3.6	10.1
0.6	4.4
6.9	11.5
2.5	6.6
2.4	3.6
2.6	5.3

We begin our analysis with the scatterplot in Figure 9.12, including the line of best fit.

The relationship between the variables x and y appears to be positive, roughly linear, with a positive slope. To compute the correlation coefficient, we label $x = Automatic$ $Weapons$ (in thousands), and $y = Murder\ Rate$ (murders per 100,000 people), and add x^2, y^2, and xy columns to the table. We will compute these for each (x, y) pair and total every column in the table.

$x = Automatic$ $Weapons$	$y = Murder$ $Rate$	x^2	y^2	$x \cdot y$
11.6	13.1	134.56	171.61	151.96
8.3	10.6	68.89	112.36	87.98
3.6	10.1	12.96	102.01	36.36
0.6	4.4	0.36	19.36	2.64
6.9	11.5	47.61	132.25	79.35
2.5	6.6	6.25	43.56	16.5
2.4	3.6	5.76	12.96	8.64
2.6	5.3	6.76	28.09	13.78
38.5	65.2	283.15	622.2	397.21

We have now what we call the *summary statistics* for these values. They are: $\sum x = 38.5$, $\sum y = 65.2$, $\sum x^2 = 283.15$, $\sum y^2 = 622.2$, and $\sum x \cdot y = 397.21$. With these, we should

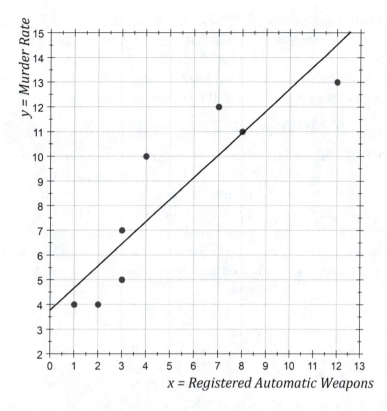

Figure 9.12: Quantity of registered weapons vs. murder rates (murders per 100,000 people) in randomly selected American cities.

use the new *computational formulas* (Theorems 9.1.15 and 9.1.17) to compute r and b.

$$r = \frac{\sum xy - \frac{1}{n}\left(\sum x\right)\left(\sum y\right)}{\sqrt{\sum x^2 - \frac{1}{n}\left(\sum x\right)^2}\sqrt{\sum y^2 - \frac{1}{n}\left(\sum y\right)^2}}$$

$$= \frac{397.21 - \frac{1}{8} \cdot 38.5 \cdot 65.2}{\sqrt{283.15 - \frac{1}{8}\left(38.5\right)^2}\sqrt{622.2 - \frac{1}{8}\left(65.2\right)^2}}$$

$$= \frac{83.435}{\sqrt{97.86875}\sqrt{90.82}}$$

$$= 0.885$$

After rounding, $r = 0.885$, and we can conclude (referring to Figure 9.6) that the data exhibit a strong positive linear relationship. With the form being linear and the relationship being strong, it makes sense to give the equation of the regression line. First the slope.

$$b = \frac{\sum xy - \frac{1}{n}\left(\sum x\right)\left(\sum y\right)}{\sum x^2 - \frac{1}{n}\left(\sum x\right)^2}$$

$$= \frac{397.21 - \frac{1}{8} \cdot 38.5 \cdot 65.2}{283.15 - \frac{1}{8}\left(38.5\right)^2}$$

$$\approx 0.8525$$

The y-intercept requires $\overline{x} = \frac{\sum x}{n} = \frac{38.5}{8} = 4.8125$ and $\overline{y} = \frac{\sum y}{n} = \frac{65.2}{8} = 8.15$. Thus the y-intercept is $a = \overline{y} - b\overline{x} = 8.15 - 0.8525 \cdot 4.8125 = 4.047$. Thus the line of best fit is $\widehat{y} = a + bx = 4.05 + 0.85\,x$. If we wanted to predict the murder rate in a city with 6000 registered automatic weapons ($x = 6$) we would compute $\widehat{y} = 4.05 + 0.85 \cdot 6 = 9.15$. This means that we expect 9.15 murders per 100,000 people in such a city.

Example 9.1.19 Do you believe that there is a relationship between the literacy rate, x (the percent of adults who can read) in a given country and the infant mortality rate, y (the number of deaths per 1000 children in their first year) for that country? It seems reasonable that countries where people tend to be more educated also tend to be wealthy, with better health care for children. One might guess that such countries would be likely to have lower infant mortality rates. Data for 50 countries are plotted in Figure 9.13. We will use our computational formula to compute r. The necessary totals are computed in Table 9.2.

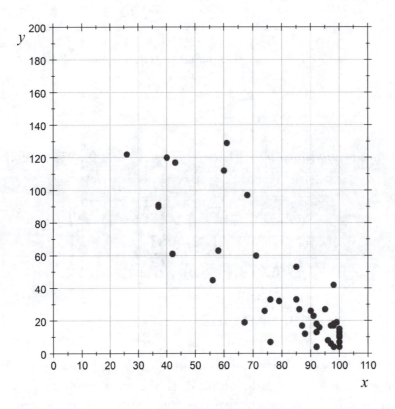

Figure 9.13: Adult literacy (x) versus infant mortality (y) for 50 countries.

The summary statistics follow.

$\sum x$	$\sum y$	$\sum x^2$	$\sum y^2$	$\sum xy$
3968	1952	338402	148974	119175

x	y	x^2	y^2	xy
74	26	5476	676	1924
92	18	8464	324	1656
85	53	7225	2809	4505
86	27	7396	729	2322
40	120	1600	14400	4800
43	117	1849	13689	5031
100	11	10000	121	1100
98	42	9604	1764	4116
68	97	4624	9409	6596
100	15	10000	225	1500
58	63	3364	3969	3654
42	61	1764	3721	2562
87	17	7569	289	1479
92	13	8464	169	1196
26	122	676	14884	3172
100	4	10000	16	400
67	19	4489	361	1273
98	18	9604	324	1764
76	7	5776	49	532
61	129	3721	16641	7869
71	60	5041	3600	4260
91	23	8281	529	2093
96	8	9216	64	768
97	6	9409	36	582
100	13	10000	169	1300
100	10	10000	100	1000
98	4	9604	16	392
96	8	9216	64	768
99	19	9801	361	1881
37	90	1369	8100	3330
85	33	7225	1089	2805
92	4	8464	16	368
37	91	1369	8281	3367
100	7	10000	49	700
93	16	8649	256	1488
98	4	9604	16	392
79	32	6241	1024	2528
97	17	9409	289	1649
76	33	5776	1089	2508
90	26	8100	676	2340
60	112	3600	12544	6720
98	17	9604	289	1666
95	27	9025	729	2565
88	12	7744	144	1056
56	45	3136	2025	2520
100	8	10000	64	800
38	126	1444	15876	4788
57	63	3249	3969	3591
56	54	3136	2916	3024
95	5	9025	25	475
3968	1952	338402	148974	119175

Table 9.2: The summary statistics for Example 9.1.19.

With these, we compute r.

$$
\begin{aligned}
r &= \frac{\sum xy - \frac{1}{n}\left(\sum x\right)\left(\sum y\right)}{\sqrt{\sum x^2 - \frac{1}{n}\left(\sum x\right)^2}\sqrt{\sum y^2 - \frac{1}{n}\left(\sum y\right)^2}} \\
&= \frac{119175 - \frac{1}{50}\cdot 3968 \cdot 1952}{\sqrt{338402 - \frac{1}{50}\left(3968\right)^2}\sqrt{148974 - \frac{1}{50}\left(1952\right)^2}} \\
&= \frac{-35735.72}{\sqrt{23501.52}\sqrt{72767.92}} \\
&= -0.864
\end{aligned}
$$

In this case, r is close to -1. This fact, along with the scatterplot, lead us to conclude that the data exhibit a strong negative linear relationship. This is sufficient evidence to motivate finding the line of best fit. The slope of the line is

$$
\begin{aligned}
b &= \frac{\sum xy - \frac{1}{n}\left(\sum x\right)\left(\sum y\right)}{\sum x^2 - \frac{1}{n}\left(\sum x\right)^2} \\
&= \frac{119175 - \frac{1}{50}\cdot 3968 \cdot 1952}{338402 - \frac{1}{50}\left(3968\right)^2} \\
&\approx -1.5206
\end{aligned}
$$

To compute the y-intercept, we need $\overline{y} = \frac{\sum y}{n} = \frac{1952}{50} = 39.04$ and $\overline{x} = \frac{\sum x}{n} = \frac{3968}{50} = 79.36$. Thus, the y-intercept is $a = \overline{y} - b\overline{x} = 39.04 - (-1.5206)\,79.36 = 159.7$. The line of best fit is $\widehat{y} = a + bx = 159.7 - 1.5\,x$. If we wanted to predict the infant mortality rate for a country with 50% literacy ($x = 50$), we would compute $\widehat{y} = 159.7 - 1.5 \cdot 50 = 84.7$. We predict that 84.7 deaths per thousand infants will die in their first year in such a country.

Example 9.1.20 For our last example, we wish to examine the relationship between x, the number of deaths (in thousands) for children under 5 in a given country, and y, the corresponding gross national income (GNI) per capita for that country. The table of values for 18 randomly selected countries, and necessary totals are totaled in Table 9.3.

The summary statistics follow.

$\sum x$	$\sum y$	$\sum x^2$	$\sum y^2$	$\sum xy$
124	24030	1828	44168100	119940

x	y	x^2	y^2	xy
0	1490	0	2220100	0
17	890	289	792100	15130
22	430	484	184900	9460
8	970	64	940900	7760
1	2760	1	7617600	2760
5	930	25	864900	4650
0	2090	0	4368100	0
5	2310	25	5336100	11550
1	1170	1	1368900	1170
1	1590	1	2528100	1590
0	1600	0	2560000	0
6	310	36	96100	1860
0	2360	0	5569600	0
6	2200	36	4840000	13200
21	2150	441	4622500	45150
14	190	196	36100	2660
15	140	225	19600	2100
2	450	4	202500	900
124	24030	1828	44168100	119940

Table 9.3: The summary statistics for Example 9.1.20.

With these we are able to compute the sample correlation coefficient.

$$
\begin{aligned}
r &= \frac{\sum xy - \frac{1}{n}\left(\sum x\right)\left(\sum y\right)}{\sqrt{\sum x^2 - \frac{1}{n}\left(\sum x\right)^2}\sqrt{\sum y^2 - \frac{1}{n}\left(\sum y\right)^2}} \\
&= \frac{119940 - \frac{1}{18}\cdot 124 \cdot 24030}{\sqrt{1828 - \frac{1}{18}\left(124\right)^2}\sqrt{44168100 - \frac{1}{18}\left(24030\right)^2}} \\
&= \frac{-45600}{\sqrt{973.7778}\sqrt{12088050}} \\
&= -0.420
\end{aligned}
$$

This time, our correlation coefficient is negative, but it is not so close to -1. Figure 9.6 suggests that we classify the any linear relationship between x and y as *weak*. In this case, we are more reluctant to say that a negative relationship exists between these variables. Referring the scatterplot in Figure 9.14, we see visually what we suspect – that there is very little evidence of a linear relationship between these variables.

In cases like this, it is may be best to not use a line of best fit to imply a relationship between the variables.

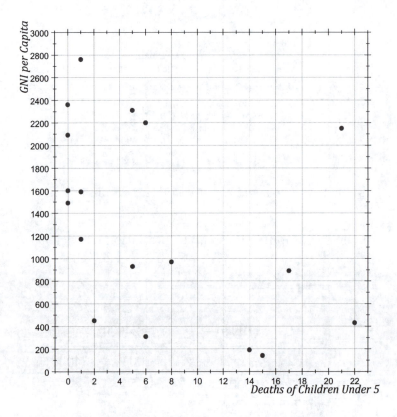

Figure 9.14: There is a very weak relationship, if any, between a country's number of deaths of children under 5 (in thousands) and the GNI per capita.

9.1.6 Homework

For the following problems, use a calculator to answer the given questions a through h. All hand computations should be carefully rounded to four places *after* the decimal.

a. Plot the points by hand on a scatterplot.
b. Describe the form (linear, nonlinear, or none) of the relationship between x and y.
c. Do the data pairs have a positive or negative relationship?
d. How strong does the relationship appear visually?
e. What are the sample mean and standard deviation of the x values?
f. What are the sample mean and standard deviation of the y values?
g. What is r? If the form between x and y is linear, use r to describe the strength of the relationship.
h. If the relationship between x and y is linear in form and moderate or strong, give the equation of the line of best fit.

1. _____

x	y
3.6	23.3
5.8	29.9
1.1	15.8
3.8	23.9
4.7	26.6
9.9	42.2
4.5	26.0
6.6	32.3

2. _____

x	y
4.1	8.3
2.6	11.3
5.7	5.1
7.5	1.5
5.1	6.3

3. _____

x	y
5.6	11.7
5.2	12.9
1.1	25.2
2.4	21.3
4.7	14.4
9.3	0.6
4.5	15.0
7.2	6.9

4. _____

x	y
1.1	5.8
3.6	10.8
7.2	18.0
3.3	10.2
9.8	23.2

For the following problems, use a calculator to answer the given questions a through j. All hand computations should be carefully rounded to four places *after* the decimal.

a. Plot the points by hand on a scatterplot.
b. Describe the form (linear, nonlinear, or none) of the relationship between x and y.
c. Do the points have a positive or negative relationship?
d. How strong does the relationship appear visually?
e. What are the mean and standard deviation of the x values?
f. What are the mean and standard deviation of the y values?
g. What is the value of r? If the form between x and y is linear, use r to describe the strength of the relationship.
h. If the relationship between x and y is linear in form and moderate or strong, give the equation of the line of best fit.
i. Give and interpret the value of r^2.
j. Use the line of best fit to find the predicted value of y corresponding to the given value of x.

5. _____

x	y
5.3	15.8
4.9	12.7
8.5	18.7
6.5	15.9
4.8	12.6

Value:

$x = 5$

6. _____

x	y
3.6	18.1
5.5	21.7
1.6	10.8
3.9	19.0
4.6	20.6
9.5	35.1
4.4	20.3
6.9	26.1

Value:

$x = 8$

7. _____

x	y
14.5	1.3
3.8	22.3
9.5	10.3
6.5	17.5
5.7	18.7

Value:

$x = 11$

8. _____

x	y
2.8	13.0
3.9	12.5
1.1	16.2
3.7	11.5
5.2	7.6
7.9	2.3
6.3	5.5
9.1	1.2

Value:

$x = 6$

For each data set, use technology (such as *Statcato* at www.statcato.org) to plot a scatterplot, and compute the linear correlation coefficient, r. Give the equation of the line of best fit ($\widehat{y} = a + bx$), using it to predict y for the given value of x. Graph the line of best fit on the scatterplot. Finally, compute the coefficient of determination (r^2) and interpret its value in context.

9. The following data are assessments of tuna quality. We compare the Hunter L measure of lightness (x) to the averages of consumer panel scores (y) (recoded as integer values from 1 to 6 and averaged over 80 such values) in 9 lots of canned tuna. (*Data are from Hollander & Wolfe, 1973, p. 187f.*). Predict the score if a given lot of tuna has a lightness score of 55.

x	y
44.4	2.6
45.9	3.1
41.9	2.5
53.3	5.0
44.7	3.6
44.1	4.0
50.7	5.2
45.2	2.8
60.1	3.8

10. Below are yearly precipitation totals (x), in inches, paired with annual lowest lake levels (in feet down from a full lake) (y) at Big Bear Lake, in California, for collection of randomly selected years. (*Data are provided by the Big Bear Municipal Water District*). *Predict the lowest lake level for a year with 20 inches of rain.*

x	y
15.0	14.0
30.6	9.9
24.8	7.4
50.4	1.2
27.0	5.7
41.0	4.6
49.0	2.5
31.8	3.8
22.2	14.5
17.3	12.2
24.2	9.5
27.5	7.0
35.2	5.6
22.4	6.5

11. The following are mean GPA and success rates in developmental math courses offered at Mt. San Antonio College for a random selection of semesters.

GPA	Success
1.7536	45.9
1.7090	43.8
1.8416	49.7
1.7615	46.8
1.7467	47.3
1.7915	49.9
1.8266	48.9
1.8587	50.3
1.8386	50.0
1.8892	51.9
1.9109	53.0
1.9365	54.1
1.9346	54.2

12. In an anthropological study of fossils, researchers wanted to predict a human's age at death (y) by examining one of his or her teeth. From a given tooth, the percentage (x) of the root with transparent dentine was chosen as the explanatory variable. *(American Journal of Physical Anthropology, 1991, pp. 25 – 30)*. Predict the age of such a fossil if 35% of its tooth has transparent dentine.

x	15	19	31	39	41	44	47	48	55	65
y	23	52	65	55	32	60	78	59	61	60

13. The table below summarizes measurements taken for cancer patients prior to surgery. The x values represent CAT scan measurements, while the y values are tumor mass. Give the best predicted value for tumor mass (y) corresponding to a CAT scan measurement (x) of 22.*(Statistics for Research 2nd ed., by Dowdy & Weardon, 1985, pp. 269 – 270, John Wiley & Sons)*.

x	18	28	20	24	16	15	19	24	23	13
y	20	25	16	21	22	10	23	27	18	11

14. The x values below represent carbon monoxide concentration measured on random occasions in New York City, while the y values are corresponding benzopyrene concentrations *(Carcinogenic Air Pollutants in Relation to Automobile Traffic in New York City, Environmental Science and Technology, 1971, pp. 145 – 150)*.

x	y
2.8	0.5
15.5	0.1
19.0	0.8
6.8	0.9
5.5	1.0
5.6	1.1
9.6	3.9
13.3	4.0
5.5	1.3
12.0	5.7
5.6	1.5
19.5	6.0
11.0	7.3
12.8	8.1
5.5	2.2
10.5	9.5

15. Is there a relationship between homework scores and overall course percentages in a statistics class? The author randomly selected 35 of his own students and recorded their overall homework scores (x), each paired with the corresponding overall course percentage (y). For your convenience, the values are programmed into *Statcato* (*www.statcato.org*) or in spreadsheet form at *http://math.mtsac.edu/statistics/data*.

x	y
33	42.6
38	34.2
35	67.2
78	52.2
85	80.2
110	64.9
90	86.4
83	63.8
78	77.7
78	59.0
113	97.9
113	85.1
85	80.0
65	64.7
103	86.0
53	77.2
118	65.4
75	72.7
108	98.4
108	98.4
25	43.6
115	95.4
50	58.6
105	94.8
70	82.7
90	84.9
70	78.0
103	53.1
85	68.2
80	62.2
115	85.4
105	83.5
113	96.4
115	98.8
100	83.9

16. The following table gives arm spans and corresponding heights of 32 randomly selected adults. Does arm span correlate to height? *(These values were measured in
centimeters by Mt. San Antonio College honors student Alphonso Tran).*

Arm Span	Height
170.9	172.0
175.0	177.0
177.0	173.0
177.0	176.0
178.1	178.1
183.9	180.1
188.0	188.0
188.2	186.9
185.9	183.9
188.0	182.1
188.2	181.1
188.5	192.0
194.1	193.0
196.1	183.9
199.9	185.9
150.1	156.0
152.9	159.0
156.0	162.1
157.0	159.8
159.0	162.1
159.8	154.9
161.0	159.8
161.0	162.1
162.8	167.1
162.1	170.2
165.1	166.1
170.2	170.2
170.4	167.1
173.0	184.9
199.9	176.0
167.9	168.9

17. Is there a correlation between foot length and height for young women? *(The following random data pairs are measured in centimeters).*

Height (cm)	Foot (cm)
148.0	18.16
152.2	20.43
155.2	21.53
152.7	20.35
155.1	21.10
155.5	19.13
156.8	20.61
158.4	20.43
158.3	21.72
156.2	21.64
158.6	22.95
159.9	20.90
164.2	20.04
157.9	21.81
159.8	21.01
160.9	22.25
162.3	22.68
158.0	22.81
159.8	21.78
156.9	23.74
162.4	21.32
160.2	22.66
166.9	22.11
167.2	22.60
167.6	22.04
164.0	21.58
170.4	22.75
175.9	22.06
176.4	24.18
184.1	23.67

18. In this example, we examine success rates in developmental mathematics courses at Mt. San Antonio College, over time. Is student success improving during this time period? No $\hat{y}$ computation is necessary for this problem.

Year	Percent
1999.5	45.9
2000.0	43.8
2000.5	49.7
2001.0	46.8
2001.5	47.3
2002.0	49.9
2002.5	48.9
2003.0	50.3
2003.5	50.0
2004.0	51.9
2004.5	53.0
2005.0	54.1
2005.5	54.2

19. Here, we examine overall developmental mathematics GPA at Mt. San Antonio College, over time. Are average grades improving during this time period? No $\hat{y}$ computation is necessary for this problem.

Year	GPA
1999.5	1.75360
2000.0	1.70896
2000.5	1.84157
2001.0	1.76150
2001.5	1.74670
2002.0	1.79150
2002.5	1.82661
2003.0	1.85865
2003.5	1.83863
2004.0	1.88923
2004.5	1.91088
2005.0	1.93647
2005.5	1.93466

9.2 Assessing Linear Fit with Residuals

We have used the linear correlation coefficient, r, to determine whether a linear model is appropriate for a given collection of data points. Unfortunately, the correlation coefficient is unable to clearly distinguish cases where points deviate randomly from a line from other cases where points follow a trend that is slightly nonlinear.

Example 9.2.1 For example, consider the scatterplot in Figure 9.15. The sample of (x, y) pairs have regression line equation $\widehat{y} = -21.05 + 10.90x$, and linear correlation coefficient $r = 0.969$. From this r value, we might conclude that the data pairs exhibit a strong positive linear trend. This has some truth, but it may be possible that there are other algebraic nonlinear curves that would better fit the trend that is evident in the scatterplot.

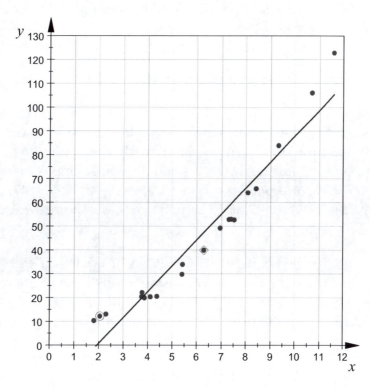

Figure 9.15: The sample of (x, y) pairs plotted here from Example 9.2.1 appear to have a slightly non-linear trend. In spite of this fact, the linear correlation coefficient is $r = 0.969$. We could conclude that a strong positive linear trend exists, but are we missing anything?

Because of this, r alone cannot determine when a collection of (x, y) points exhibit a trend that is truly linear. We need other tools for making a more complete assessment.

One tool which may be used for such an assessment is the scatterplot, as in Figure 9.15. A scatterplot can help us see when the points are close to a line or when there is nonlinearity in their trend.

Another tool which can complement the correlation coefficient in the analysis of a trend is called a *residual plot*. The residual plot is simply a scatterplot which uses residuals, which we review briefly below.

9.2.1   Residual Plots

For any pair, (x, y) the *residual* from a linear regression model, $\widehat{y} = a + bx$, is $E = (y - \widehat{y})$. Solving for y, we see that, for any pair (x, y), $y = \widehat{y} + E$. This tells us that each observed value of y is composed of two parts, the part that is *explainable* by the regression model $(\widehat{y})$ and the residual, E. The residual is the part of y that is *unexplainable* by the regression equation.

Now that we know what residuals are, we are able to define the residual plot.

Definition 9.2.2 Given a sample of (x, y) pairs, the *residual plot* is simply a scatterplot of (x, E) pairs, where $E = (y - \widehat{y})$.

The scatterplot in Figure 9.15 was generated by the (x, y) pairs in Table 9.4.

In the collection of (x, y) pairs in this Table 9.4, note that the residuals are negative when observed y values are less than predicted $\widehat{y}$ values. In this table, notice two rows highlighted in grey. The first row is light grey, containing the data point $(x, y) = (6.29, 39.95)$. This point is plotted in Figure 9.15 inside a small diamond symbol, below the regression line. The value of $\widehat{y}$ that corresponds is $\widehat{y} = -21.05 + 10.90 \cdot 6.29 = 47.51$, with residual $E = (y - \widehat{y}) = 39.95 - 47.51 = -7.56$. The point is below the regression line, and that is why its residual is negative.

Also in Table 9.4, notice the point $(x, y) = (2.06, 12.16)$, whose row is shaded in dark grey. This point is plotted in Figure 9.15 inside a small circle above the regression line, and corresponds to the estimate $\widehat{y} = -21.05 + 10.90 \cdot 2.06 = 1.40$, with residual $E = (y - \widehat{y}) = 12.16 - 1.40 = 10.76$. The point lies above the regression line, so its residual is positive.

The entire collection of (x, E) pairs are plotted on the residual plot in Figure 9.16. Note that the point $(x, y) = (6.29, 39.95)$ which lies below the regression line (plotted in a small diamond in Figure 9.15) corresponds to the point in the residual plot $(x, E) = (6.29, -7.56)$, plotted below the x-axis in a small diamond. Generally speaking, any (x, y) point that is below the regression line has negative residual, and its corresponding (x, E) point lies below the x-axis on the residual plot.

Note also that the point $(x, y) = (2.06, 12.16)$ lies above the regression line (plotted in a small circle in Figure 9.15). This corresponds to the point, $(x, E) = (2.06, 10.76)$, which lies above the x-axis on the residual plot, also plotted in a small circle. In general, any (x, y) point that is above the regression line has positive residual, so its corresponding (x, E) point lies above the x-axis on the residual plot.

x	y	$\widehat{y}$	E
1.81	10.29	−1.32	11.61
5.40	29.74	37.81	−8.07
7.40	52.91	59.61	−6.70
3.87	19.85	21.13	−1.28
8.08	64.01	67.02	−3.01
4.11	20.30	23.75	−3.45
9.33	83.83	80.65	3.18
7.53	52.63	61.03	−8.40
11.60	122.78	105.39	17.39
7.30	52.79	58.52	−5.73
3.77	22.05	20.04	2.01
8.41	65.75	70.62	−4.87
3.76	20.35	19.93	0.42
2.31	13.02	4.13	8.89
5.42	33.88	38.03	−4.15
4.38	20.50	26.69	−6.19
6.29	39.95	47.51	−7.56
2.06	12.16	1.40	10.76
10.68	105.98	95.36	10.62
6.95	49.19	54.71	−5.52

Table 9.4: Above are residuals $E = (y - \widehat{y})$ for Example 9.2.1. Each residual, E, requires $\widehat{y}$, computed using the equation of the regression line $\widehat{y} = a + bx$ for each x.

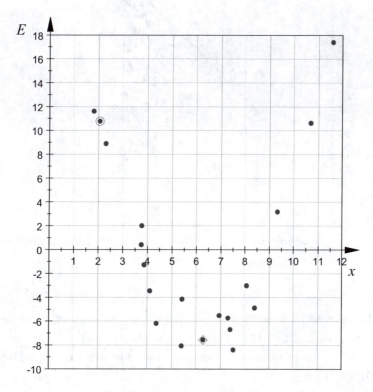

Figure 9.16: This is the residual plot of all (x, E) pairs for the residuals of points in Example 9.2.1. Note the non-linear trend in the residuals.

Note the pattern in the residual plot. The points follow a clear path which begins above the x-axis, moves below the axis then later rises above it. This corresponds to the more subtle way that the points in the scatterplot (Figure 9.15) begin above the regression line, move below it, and later rise above it. The residual plot enhances and isolates the nonlinearity. A curved pattern in a residual plot warns us of a nonlinear pattern in an associated scatterplot. When a nonlinear pattern exists, it is possible that a linear model is not the *best* model.

Example 9.2.3 The sample (x, y) pairs for this example are listed in Table 9.5.

The scatterplot for these values is pictured in Figure 9.17, revealing a collection of data points that seem randomly scattered on either side of the regression line.

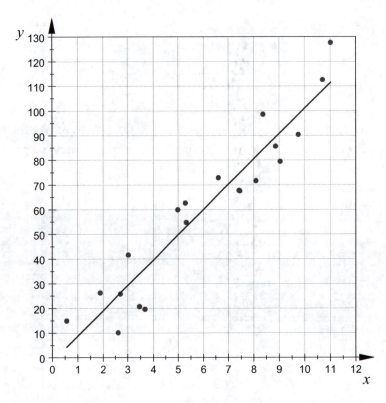

Figure 9.17: The sample (x, y) pairs from Example 9.2.3 are plotted above.

The correlation coefficient for these data points is $r = 0.949$. This indicates a strong linear relationship between x and y, but the value of r is smaller than in Example 9.2.1. While the value of r is smaller this time, the scatterplot does not seem to portray a nonlinear pattern in the data points. To emphasize this, we examine the residual plot of (x, E) pairs in Figure 9.18.

In the residual plot, there is no obvious pattern that the points seem to follow. Because of this, we conclude that a linear model is a good model for these data points.

x	y	$\widehat{y}$	E
11.02	127.61	111.55	16.06
5.29	62.64	52.76	9.88
3.03	41.59	29.57	12.02
9.76	90.38	98.62	−8.24
7.42	67.78	74.61	−6.83
6.61	72.86	66.3	6.56
4.99	59.94	49.68	10.26
7.45	67.55	74.92	−7.37
8.87	85.64	89.49	−3.85
8.38	98.6	84.46	14.14
2.72	25.76	26.39	−0.63
1.88	26.18	17.77	8.41
3.47	20.65	34.08	−13.43
9.05	79.46	91.33	−11.87
0.57	14.86	4.33	10.53
5.33	54.73	53.17	1.56
10.71	112.59	108.36	4.23
8.09	71.65	81.48	−9.83
3.68	19.57	36.24	−16.67
2.63	10.11	25.46	−15.35

Table 9.5: This table lists all residuals $E = (y - \widehat{y})$ for Example 9.2.3. Each residual, E, requires $\widehat{y}$, which is computed using the equation of the regression line $\widehat{y} = a + bx$ for each x.

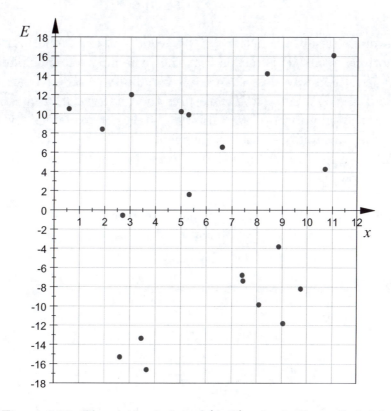

Figure 9.18: The residual plot of (x, E) pairs in Example 9.2.3.

In conclusion, remember that when the points in a residual plot have a curved pattern, a linear model may not be the best model for the sample data pairs. When the points in the residual plot seem to be randomly distributed above and below the horizontal axis, a linear model is a good choice.

9.2.2   Sums of Squares as a Measure of Variation.

We have discussed the sum of squared residuals, SSE, whose name implies its formula. SSE is a measure of fit. When the SSE is small, then the sample data points are close to their regression line. When the SSE is large, then there is much deviation from the line.

$$SSE = \sum (y - \widehat{y})^2$$

The SSE measures the total *unexplained* variation of points away from the regression line, where *variation* is measured as a squared deviation.

Another type of variation among y values is deviation from the sample mean, $\overline{y}$ (as with the *standard* deviation). For any given y, the deviation from the sample mean is $y - \overline{y}$, and the variation from $\overline{y}$ is the square of the deviation, $(y - \overline{y})^2$. The total variation of points away from $\overline{y}$ is the *sum of squares total (SST)*,

$$SST = \sum (y - \overline{y})^2 .$$

A least squares model is always better (or no worse) at explaining variability among y-values than nothing at all. This means that the variability left unexplained by a least squares model (SSE) is always less than total variability (SST). That is, it is always the case that $SSE \leq SST$. Since SST measures the total variation in sample y values, the ratio, $\frac{SSE}{SST}$, is the proportion of variability that is not explained by the regression line. The remaining variability is the complement, $1 - \frac{SSE}{SST}$. This is the proportion of total variation of y values away from $\overline{y}$ that *is explained by* the explanatory variable, x, through the regression line.

In Theorem 9.1.12 on page 415, we proved that r^2 is exactly equal to this proportion of total variation in y that explained by x using the line of best fit.

Definition 9.2.4 The coefficient of determination,

$$\begin{aligned} r^2 &= 1 - \frac{SSE}{SST} \\ &= \frac{SSR}{SST}, \end{aligned}$$

is the proportion of total variation of y values that is explained by the explanatory variable, x, through the line of best fit.

When we wish to compute r^2, we may use the formula given above, but it may be easier to compute r with technology or using the formula in Theorem 9.1.15 on page 420, then square the result to get r^2.

Example 9.2.1 Continued: Returning to our first collection of (x, y), where $r = 0.969$, we reproduce the data set with each type of variation, $(y - \widehat{y})^2$ and $(y - \overline{y})^2$ in Table 9.6.

The unexplained variation of y values from the regression line is $SSE = \sum (y - \widehat{y})^2 = 1162$, and the total variation of y values from their mean is $SST = \sum (y - \overline{y})^2 = 18929$. The coefficient of determination is:

$$r^2 = 1 - \frac{SSE}{SST}$$
$$= 1 - \frac{1162}{18929}$$
$$= 0.939.$$

Using the value of $r = 0.969$, given above, we could also square this to get the same value,

$$r^2 = 0.969^2$$
$$= 0.939.$$

From this we may conclude that 93.9% of total variation of sample y coordinates can be explained by x through the regression line. The unexplained variation is the complement, $1 - 0.939 = 0.061 = 6.1\%$. Thus, 6.1% of total variation among sample y values is unexplained by the variation along the regression line.

9.2.3   The Standard Deviation about the Regression Line

Given that we have the sum of squared deviations from the regression line, $SSE = \sum (y - \widehat{y})^2$, it makes sense to compute the average deviation in the same way that we compute standard deviation from a mean. The sample variance of y values about the regression line is denoted by s_e^2, where we divide SSE by its degrees of freedom, df, for an unbiased estimate of the population variance about the population regression line.

$$s_e^2 = \frac{SSE}{df}$$

The degrees of freedom, df, in $SSE = \sum (y - \widehat{y})^2$ is $n - 2$. Remember that for $\widehat{y}$ we are using $\widehat{y} = a + bx$, where b is a sample slope and a is a sample y-intercept. The regression line, $\widehat{y} = a + bx$, minimizes SSE for the y values in the sample data (that is why it is a *least squares* line). This is the point of the regression line, but in actuality, the value of SSE is smaller than we want.

Computing SSE using any other slope or y-intercept would give a larger result. The SSE computed using the population slope and y-intercept the resulting SSE would be

x	y	$\widehat{y}$	$(y-\widehat{y})^2$	$(y-\overline{y})^2$
1.81	10.29	−1.32	134.79	1177.04
5.40	29.74	37.81	65.12	220.76
7.40	52.91	59.61	44.89	69.09
3.87	19.85	21.13	1.64	612.46
8.08	64.01	67.02	9.06	376.83
4.11	20.30	23.75	11.90	590.39
9.33	83.83	80.65	10.11	1539.15
7.53	52.63	61.03	70.56	64.51
11.60	122.78	105.39	302.41	6112.43
7.30	52.79	58.52	32.83	67.11
3.77	22.05	20.04	4.04	508.41
8.41	65.75	70.62	23.72	447.41
3.76	20.35	19.93	0.18	587.97
2.31	13.02	4.13	79.03	997.17
5.42	33.88	38.03	17.22	114.88
4.38	20.50	26.69	38.32	580.71
6.29	39.95	47.51	57.15	21.60
2.06	12.16	1.40	115.78	1052.22
10.68	105.98	95.36	112.78	3767.75
6.95	49.19	54.71	30.47	21.09
$\sum$	891.96	–	1162	18929

Table 9.6: This table lists all squared residuals $E^2 = (y-\hat{y})^2$ for Example 9.2.3. Additionally, the squares of the deviations, $(y-\overline{y})^2$. The totals of the respective columns are SSE and SST respectively.

larger, but this would be the true value of SSE. Unfortunately, the population slope and y-intercept are unknown. We are forced to use estimates from our sample instead.

Starting from the population regression line, we substitute first the sample slope for the unknown population slope. The sample slope is from the least squares line for the given sample, so the corresponding SSE is made smaller. Subtracting 1 from the number of deviations in the sum removes the resulting bias in the variance. The degrees of freedom are reduced by one. If we next substitute the sample y-intercept, a, for the unknown population intercept, the resulting SSE is smaller still, and again we subtract 1 from the remaining number of deviations to remove the additional bias introduced by using another statistic in place of a parameter. The unbiased variance about the regression line therefore has $n - 2$ degrees of freedom.

Definition 9.2.5 The sample variance of points away from their regression line is s_e^2, which is computed using the formula

$$s_e^2 = \frac{SSE}{n - 2}$$
$$= \frac{\sum (y - \widehat{y})^2}{n - 2}.$$

The sample standard deviation of points about the regression line is the square root of the variance.

$$s_e = \sqrt{\frac{SSE}{n - 2}}$$
$$= \sqrt{\frac{\sum (y - \widehat{y})^2}{n - 2}}.$$

The standard deviation of points away from the regression line is an average of the vertical distances of sample y values from the regression line. Obviously, when this is small, the points are close to the line. Because of this, s_e, is another measure of the quality of the fit of regression lines to sample data.

Example 9.2.1 Continued: Referring again to Table 9.5, where $SSE = \sum (y - \widehat{y})^2 = 1162$ in a sample with $n = 20$ data points, the sample standard deviation of points about the regression line is

$$s_e = \sqrt{\frac{SSE}{n - 2}}$$
$$= \sqrt{\frac{1162}{20 - 2}}$$
$$= 8.03$$

The average deviation from the regression line is $s_e = 8.03$.

9.2.4   A Computational Formula for SSE

As is typical for formulas involving sums of squares, SSE, has a computational formula, given below.

Theorem 9.2.6 A computational formula for SSE is

$$SSE = \sum y^2 - a \sum y - b \sum xy$$

Example 9.2.1 Continued: Referring again to Example 9.2.1 with regression line

$$\widehat{y} = -21.05 + 10.90x,$$

the y-intercept and slope are, more accurately, $a = -21.0469$, and $b = 10.8990$. The data points are in Table 9.4, and summary statistics are as follows:

$$\sum y^2 = 58708.6076,$$
$$\sum y = 891.96,$$

and

$$\sum xy = 7002.4308.$$

The computational formula for SSE gives the following.

$$
\begin{aligned}
SSE &= \sum y^2 - a \sum y - b \sum xy \\
&= 58708.6076 - (-21.0469) \cdot 891.96 - 10.8990 \cdot 7002.4308 \\
&\approx 1162.1
\end{aligned}
$$

This answer is only slightly different from the answer obtained earlier in this section, and this difference is due to rounding.

Standard deviations and variances are required in most methods of inference which we will cover in Part IV of this text. The standard deviation of residuals, s_e, is used for inferences in linear correlation and regression.

9.2.5   Homework

Compute the standard deviation of residuals, s_e, and the coefficient of determination using the formula $r^2 = 1 - \frac{SSE}{SST}$. Compare your answer to r^2, where r was computed in the indicated problem.

1. Compare to Problem 5 in Section 9.1.

x	y
5.3	15.8
4.9	12.7
8.5	18.7
6.5	15.9
4.8	12.6

2. Compare to Problem 6 in Section 9.1.

x	y
3.6	18.1
5.5	21.7
1.6	10.8
3.9	19.0
4.6	20.6
9.5	35.1
4.4	20.3
6.9	26.1

3. Compare to Problem 7 in Section 9.1.

x	y
14.5	1.3
3.8	22.3
9.5	10.3
6.5	17.5
5.7	18.7

4. Compare to Problem 8 in Section 9.1.

x	y
2.8	13.0
3.9	12.5
1.1	16.2
3.7	11.5
5.2	7.6
7.9	2.3
6.3	5.5
9.1	1.2

For each data set, use technology (such as *Statcato* at www.statcato.org) to plot a residual plot, compute the standard deviation of residuals, s_e, and the coefficient of determination

5. The following data are assessments of tuna quality. We compare the Hunter L measure of lightness (x) to the averages of consumer panel scores (y) (recoded as integer values from 1 to 6 and averaged over 80 such values) in 9 lots of canned tuna. (*Data are from Hollander & Wolfe, 1973, p. 187f*).

x	y
44.4	2.6
45.9	3.1
41.9	2.5
53.3	5.0
44.7	3.6
44.1	4.0
50.7	5.2
45.2	2.8
60.1	3.8

6. Below are yearly precipitation totals (x), in inches, paired with annual lowest lake levels (in feet down from a full lake) (y) at Big Bear Lake, in California, for collection of randomly selected years. (*Data are provided by the Big Bear Municipal Water District*).

x	y
15.02	13.98
30.62	9.89
24.82	7.38
50.40	1.22
27.00	5.67
41.04	4.57
49.00	2.48
31.78	3.78
22.20	14.45
17.32	12.17
24.18	9.48
27.49	7.00
35.16	5.63
22.40	6.48

Compute the standard deviation of residuals using $s_e = \sqrt{\frac{SSE}{n-2}}$, where SSE is computed with the computational formula

$$SSE = \sum y^2 - a \sum y - b \sum xy.$$

7. The following are GPA and success rates in developmental math courses offered at Mt. San Antonio College for a random selection of semesters. Use the values of a and b computed in Problem 11 in Section 9.1.

GPA	Success
1.7536	45.9
1.7090	43.8
1.8416	49.7
1.7615	46.8
1.7467	47.3
1.7915	49.9
1.8266	48.9
1.8587	50.3
1.8386	50.0
1.8892	51.9
1.9109	53.0
1.9365	54.1
1.9346	54.2

8. In an anthropological study of fossils, researchers' objective was to predict age (y) from the percentage (x) of a tooth's root with transparent dentine. *(American Journal of Physical Anthropology, 1991, pp. 25 – 30)*. Use the values of a and b computed in Problem 12 in Section 9.1

x	y
15	23
19	52
31	65
39	55
41	32
44	60
47	78
48	59
55	61
65	60

9. The table below summarizes measurements taken for cancer patients prior to surgery. The x values represent CAT scan measurements, while the y values are tumor mass. *(Statistics for Research 2nd ed., by Dowdy & Weardon, 1985, pp. 269 – 270, John Wiley & Sons).* Use the values of a and b computed in Problem 13 in Section 9.1.

x	y
18	20
28	25
20	16
24	21
16	22
15	10
19	23
24	27
23	18
13	11

10. The x values below represent carbon monoxide concentration measured on random occasions in New York City, while the y values are corresponding benzopyrene concentrations *(Carcinogenic Air Pollutants in Relation to Automobile Traffic in New York City, Environmental Science and Technology, 1971, pp. 145 – 150).* Use the values of a and b computed in Problem 14 in Section 9.1.

x	y
2.8	0.5
15.5	0.1
19.0	0.8
6.8	0.9
5.5	1.0
5.6	1.1
9.6	3.9
13.3	4.0
5.5	1.3
12.0	5.7
5.6	1.5
19.5	6.0
11.0	7.3
12.8	8.1
5.5	2.2
10.5	9.5

9.3   Inference in Correlation and Regression

We have previously learned about the correlation coefficient, r, which may be used to determine the direction and strength of a possible linear relationship between random paired (x, y) observations.

A quick review of what we know about r is in order. The formula for r is

$$r = \frac{\sum z_x z_y}{n - 1}.$$

With some work, we can show that r may also be computed using the formula,

$$r = \frac{\sum xy - \frac{1}{n}\left(\sum x\right)\left(\sum y\right)}{\sqrt{\sum x^2 - \frac{1}{n}\left(\sum x\right)^2}\sqrt{\sum y^2 - \frac{1}{n}\left(\sum y\right)^2}}.$$

We also know that in every instance, $-1 \leq r \leq 1$. When $r \approx 1$, the sample (x, y) pairs exhibit a strong positive relationship. When $r \approx -1$, the sample (x, y) pairs exhibit a strong negative relationship. When $r \approx 0$, the sample (x, y) pairs do not exhibit a linear relationship.

When a linear relationship exists between x and y, it is appropriate to use the line of best fit, the *regression line*, $\widehat{y} = a + bx$, to estimate a value for y that corresponds to a given x. Of course, there is typically a difference between what is predicted by the regression line, $\widehat{y}$, and what is actually observed, y. This difference is called a *residual*, and is denoted by $e = y - \widehat{y}$. The slope, b, of the line of best fit is computed using

$$b = \frac{r\, s_y}{s_x},$$

while the y-intercept, a, may be computed using

$$a = \overline{y} - b\overline{x}.$$

The *sample* linear correlation coefficient, r, is an estimate of ρ (pronounced *rho*), the *population* correlation coefficient. The sample slope, b, is an estimate of the population slope, β (pronounced *beta*). We make inferences regarding the parameters ρ and β in this section.

The inferences in this section make the following assumptions.

Assumption 9.3.1 The inference methods of linear correlation and regression assume the following.

1. The distribution of residuals, $e = y - \hat{y}$, corresponding to any value of x, is normal.

2. The standard deviation, σ, of all residuals $e = y - \hat{y}$ is the same for all x.

3. The expected value of residuals is zero: $E(e) = 0$.

4. The residuals, e, associated to any given value of x, are independent.

9.3.1   Inferences Regarding the Linear Correlation Parameter

We begin with a hypothesis test for the population correlation coefficient, ρ. Our null hypothesis for this test will be that ρ is equal to zero. This is equivalent to stating that, in the population of (x, y) pairs, there exists no linear relationship between x and y.

It was shown by Ronald A. Fisher that, when a population of (x, y) pairs has zero correlation, and under the assumptions of this section, the *studentized* test statistic

$$t = \frac{r}{\sqrt{\frac{1-r^2}{n-2}}}$$

is distributed according to Student's t-distribution, using $n - 2$ degrees of freedom. This will be the test statistic for the hypothesis test. We use $n - 2$ degrees of freedom because the linear relationship is described using *estimates* of two parameters – the population slope, and the population y-intercept.

This is a test for linear correlation between the explanatory variable, x, and the response variable, y. Remember though, that correlation does not imply cause. If we wish to infer a causal relationship between x and y, then varying values of the explanatory variable (x) must be assigned to experimental groups which are created by random assignment. Thus, if x represents one of many treatment (or *dosage*) levels, randomly assigned to sample members, and y represents a response variable, a causal connection can be inferred if x and y correlate. Such conclusions are only possible in controlled experiments.

More often, x and y are uncontrolled observations from randomly selected population members, and when this is the case, the study is observational and we cannot make causal conclusions.

Procedure 9.3.2 Testing For Linear Correlation

    Step I. **Verify Assumptions, and Gather Data**

           For sample data, we need r, n, and α, the level of significance. The test requires that the residuals are normally distributed with equal standard deviation for each value of x. Visually, this appears in the scatterplot as a swath of points which is roughly uniform in width, scattered about the regression line. The regression line is in the center of the swath, with higher point density near the line.

    Step II. **Identify the Claims**

           The null and alternative hypotheses are always the same for this test: H_0 : $\rho = 0$, and $H_1 : \rho \neq 0$. The null hypothesis claims that the population of (x, y) pairs does not exhibit linear correlation. The alternative says that the data do exhibit linear correlation.

    Step III. **Compute the Test Statistic**

$$T.S. : t = \frac{r}{\sqrt{\frac{1-r^2}{n-2}}}$$

Step IV. **Look up Critical Values in Table A2**

This test is always a two-tail test. Remember to use $n - 2$ degrees of freedom.

Step V. **Look up the *p-value* in Table A3**

This is done as always, but again, we are using $n - 2$ degrees of freedom.

Step VI. **Make a conclusion**

When we reject the null hypothesis, we are rejecting the claim that the data do not exhibit linear correlation. To reject this, the correlation coefficient needs to be rather close to 1 or -1, so when we reject the null hypothesis, it is because the data exhibit linear correlation. In this case, we that the population of (x, y) pairs exhibits linear correlation. Otherwise we conclude that we are unable to reject the claim that the data do not exhibit linear correlation.

When values of the explanatory variable, x, are randomly assigned to sample members (as a treatment level), a rejection of the null hypothesis allows us to infer a causal relationship between x and y. Otherwise, a rejection of the null hypothesis only allows an inference that describes the population of (x, y) pairs.

Example 9.3.3 In Section 9.1, Example 9.1.18, we computed the linear correlation coefficient for the relationship between x, the number of registered automatic weapons in a city, and y, that city's murder rate. The sample data follow.

Automatic Weapons	Murder Rate
11.6	13.1
8.3	10.6
3.6	10.1
0.6	4.4
6.9	11.5
2.5	6.6
2.4	3.6
2.6	5.3

Note that it is impossible to randomly assign the number of automatic weapons, x, to a city. This is *not* a controlled experiment. The scatterplot of these points is given in Figure 9.19. From these data, we computed $r = 0.885$. Under the assumptions of this section, we now test the claim, at 1% significance, that linear correlation exists between these variables.

Step I. **Verify Assumptions, and Gather Data**

We have already computed $r = 0.885$, and $n = 8$. In the current example, we have $\alpha = 0.01$. We have assumed that the requirements for this test are satisfied.

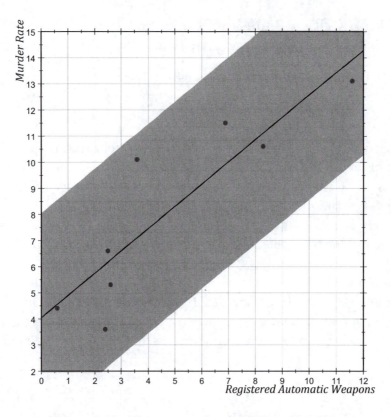

Figure 9.19: The scatterplot for Example 9.3.3.

Step II. Identify the Claims
We will always use $H_0 : \rho = 0$, and $H_1 : \rho \neq 0$ for this test.

Step III. Compute the Test Statistic

$$T.S. : t = \frac{r}{\sqrt{\frac{1-r^2}{n-2}}}$$

$$= \frac{0.885}{\sqrt{\frac{1-0.885^2}{8-2}}}$$

$$= 4.656$$

Step IV. Look up Critical Values in Table A2
We have a two-tail test with $df = 8 - 2 = 6$, and $\alpha = 0.01$.

Two Tail Applications		
df	$\alpha = .01$	$\alpha = .02$
5	4.032	3.365
6	**3.707**	3.143
7	3.499	2.998

In this case, our critical values are $CV : t = \pm 3.707$. The test statistic is in the right tail, and thus we will reject the null hypothesis.

Step V. **Look up the** *p-value* **in Table A3**

	Degrees of Freedom		
T.S.	5	**6**	7
4.6	.997	.998	.999
4.7	.997	**.998**	.999
4.8	.998	.999	.999

Since this is a two-tail test, we use $p\text{-}value = 2 \cdot (1 - 0.998) = 2 \cdot 0.002 = 0.004$. This is smaller than α, emphasizing that we are rejecting the null hypothesis.

Step VI. **Make a conclusion**

We have rejected the null hypothesis in favor of the alternative. So the data support the claim that the population exhibits linear correlation. Since r is positive, the data exhibit positive linear correlation. Since this study is not an experiment, we make an inference about the population, and do not imply a causal relationship.

The data support the claim that a positive linear correlation exists between the number of registered automatic weapons in cities, and the respective murder rates. This correlation was highly significantly different from zero, as emphasized by the p-value of 0.004.

Example 9.3.4 Now we look more closely at the literacy rates (x), and infant mortality rates (y) from 50 countries. The (x, y) pairs are provided in Table 9.7. Note that it is again impossible to randomly assign a literacy rate to a country, so this study is *descriptive*, not *experimental*. In Example 9.1.19 on page 426 we computed the value of r which corresponds to these values. The result was $r = -0.86414$. The value is recorded with much precision because we want an accurate test statistic. In this case, r appears to be close to -1, indicating that the population of (x, y) pairs may exhibit negative linear correlation. These points are plotted in Figure 9.22. We now test for correlation objectively with our hypothesis test, at 2% significance.

As before, our null hypothesis is that the population exhibits no linear correlation ($\rho = 0$), and we are testing this against the claim that linear correlation exists.

Step I. **Verify Assumptions, and Gather Data**

We have already computed $r = -0.86414$, and $n = 50$. Here, we are using $\alpha = 0.02$. If we look at the scatterplot above, we see a swath of points which seems to be consistently wide, with more points near the regression line (not plotted in the graph). This helps to verify that the requirements for this section are satisfied.

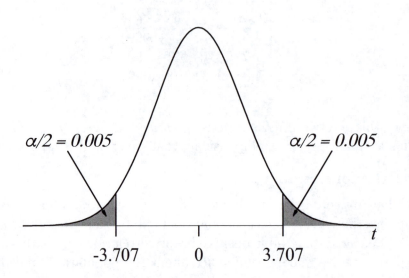

Figure 9.20: The critical values for Example 9.3.3.

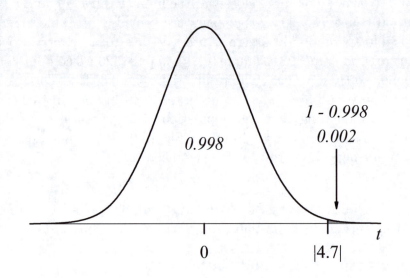

Figure 9.21: The *p-value* for Example 9.3.3 is twice the area of the shaded region above.

x	y	x	y
74	26	100	10
92	18	98	4
85	53	96	8
86	27	99	19
40	120	37	90
43	117	85	33
100	11	92	4
98	42	37	91
68	97	100	7
100	15	93	16
58	63	98	4
42	61	79	32
87	17	97	17
92	13	76	33
26	122	90	26
100	4	60	112
67	19	98	17
98	18	95	27
76	7	88	12
61	129	56	45
71	60	100	8
91	23	38	126
96	8	57	63
97	6	56	54
100	13	95	5

Table 9.7: The summary statistics, computed for Example 9.3.4. The paired values are divided amongst the two tables above.

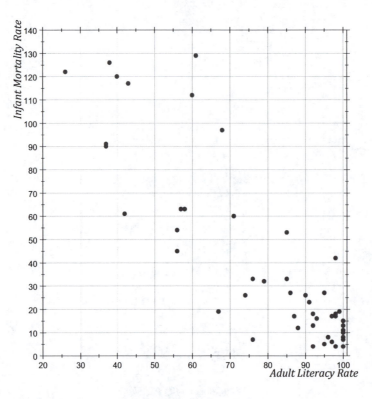

Figure 9.22: The scatterplot for Example 9.3.4.

Step II. Identify the Claims

Again, we will always use $H_0 : \rho = 0$, and $H_1 : \rho \neq 0$ for this test.

Step III. Compute the Test Statistic

$$T.S. : t = \frac{r}{\sqrt{\frac{1-r^2}{n-2}}}$$

$$= \frac{-0.86414}{\sqrt{\frac{1-(-0.86414)^2}{50-2}}}$$

$$= -11.896$$

Step IV. Look up Critical Values in Table A2

We have a two-tail test with $df = 50 - 2 = 48$, and $\alpha = 0.02$.

Two Tail Applications			
df	$\alpha = 0.01$	$\boldsymbol{\alpha = 0.02}$	$\alpha = 0.04$
47	2.685	2.408	2.112
48	2.682	**2.407**	2.111
49	2.680	2.405	2.110

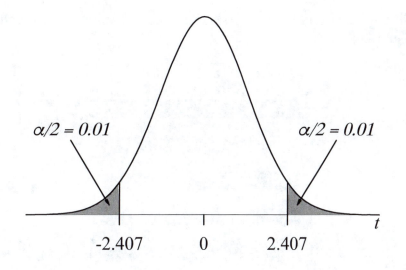

Figure 9.23: The critical values for Example 9.3.4.

In this case, our critical values are $CV : t = \pm 2.407$. The test statistic is deep in the left tail; we will again reject the null hypothesis.

Step V. **Look up the p-Value in Table A3**

Once again, we are faced with a situation where our p-value is not in Table A3, because of the large magnitude of our negative test statistic $T.S. : t = -11.889$. Using computer software, we are able to find the area to the left of the test statistic to be 0.000000000000000316, but we must double this for the two-tail test, so our p-value is 0.000000000000000632. Of course, this is extraordinarily small, indicating a correlation coefficient which is highly significantly different from zero. This means, again, that we must reject the null hypothesis.

Step VI. **Make a conclusion**

In examining both the test statistic and the p-value, we see that they are both highly extreme, both indicating that we should reject the null hypothesis, in favor of the alternative that the population data do exhibit linear correlation. With r negative we say that the correlation is negative. Since the study is descriptive and not experimental, our conclusion describes the population of (x, y) pairs, but does not infer a causal relationship.

The data support the claim that there is a negative linear correlation between literacy rates and infant mortality. Countries with higher literacy appear to have lower infant mortality, and likewise, countries with lower literacy have higher infant mortality. This correlation was strong with the sample data, as emphasized by an extraordinarily small p-value.

Example 9.3.5 For our last example, we wish to examine the relationship between the gross national income (GNI) per capita for a country, and that country's number of deaths for children under 5 in the year 2003. The table of values for 18 randomly selected countries is listed below.

Annual deaths (x) under 5 (1000s)	GNI (y) Per Capita
0	1490
17	890
22	430
8	970
1	2760
5	930
0	2090
5	2310
1	1170
1	1590
0	1600
6	310
0	2360
6	2200
21	2150
14	190
15	140
2	450

The explanatory variable, x, represents a death rate which cannot be assigned randomly to a country, so this study is *descriptive*. In Example 9.1.20 on page 428 we computed the correlation coefficient for these values. The result follows.

$$r = -0.420$$

The correlation coefficient is negative, but it is not so close to -1. The scatterplot of these points is provided in Figure 9.24.

The test statistic is as follows.

$$T.S. : t = \frac{r}{\sqrt{\frac{1-r^2}{n-2}}}$$

$$= \frac{-0.420}{\sqrt{\frac{1-(-0.420)^2}{18-2}}}$$

$$= -1.851$$

With critical values are $C.V. : t = \pm 2.120$, with 16 degrees of freedom and 5% significance. The *p-value* is .076, which indicates that the negative correlation in the sample data was only moderately different from zero, so we fail to reject the null hypothesis that the data do not exhibit linear correlation.

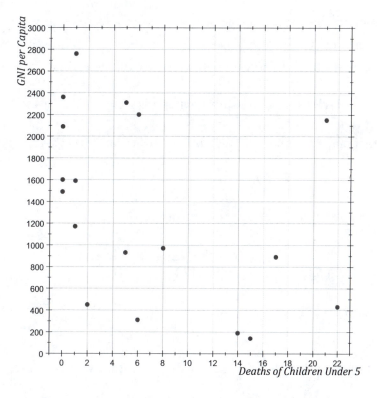

Figure 9.24: The scatterplot for Example 9.3.5.

It should be mentioned that the scatterplot above gives less evidence of satisfying the requirements for this test, because the data points are closer to a regression line at the upper-left than they are in the center. This seems to violate the assumption that the standard deviation of residuals is the same for all values of x.

From these sample data, it does not appear that a country's gross national income per capita linearly correlates with the number of deaths for children under 5 years of age.

9.3.2 Inferences Regarding the Slope Parameter

Under the conditions that the population of (x, y) pairs exhibit bivariate normality and the standard deviation, σ, of all residuals $e = y - \hat{y}$ is the same for any given x, estimates, b, of slope parameter, β, may be studentized for statistical inferences.

Theorem 9.3.6 Under the assumptions of this section the following are true.

1. The sample slope, b, is an unbiased estimate of the population slope, β. That is $\mu_b = E(b) = \beta$. This is proven in Theorem 9.5.1 on page 497.

2. The variance of residuals, e, assumed to be the same for all x, is σ^2, has the unbiased estimator, $s_e^2 = \frac{SSE}{n-2}$. This is proven in Theorem 9.5.6 on page 500.

3. The variance of the sample slope statistic, b, is $Var\,(b) = \sigma_b^2 = \frac{\sigma^2}{\sum(x-\bar{x})^2}$, where σ^2 is the variance of residuals e (also the variance of, y). This is assumed to be the same for each value of x. The unbiased estimator of σ_b^2 is $s_b^2 = \frac{s_e^2}{\sum(x-\bar{x})^2}$.

4. The standard deviation of the sample slope is $\sigma_b = \sqrt{Var\,(b)} = \frac{\sigma}{\sqrt{\sum(x-\bar{x})^2}}$.

5. A computational Formula for SSE is

$$SSE = \sum y^2 - a\sum y - b\sum xy.$$

From this we are able to give a computational formula for s_b^2.

$$s_b^2 = \frac{\sum y^2 - a\sum y - b\sum xy}{(n-2)\left(\sum x^2 - \frac{1}{n}\left(\sum x\right)^2\right)}$$

6. The *studentized* statistic

$$t = \frac{b - \beta}{s_b}$$

is distributed according to Student's t-distribution, with $n - 2$ degrees of freedom.

9.3.3 Why $n - 2$ Degrees of Freedom?

Since the variance of the sample slope, $s_b^2 = \frac{s_e^2}{S_{xx}}$, is directly related to s_e^2, they both have the same degrees of freedom. We therefore direct this explanation toward the degrees of freedom for s_e^2.

We have stated that s_b contains $n - 2$ degrees of freedom. There have been hints as to why this is true. The sample variance of residuals, $s_e^2 = \frac{SSE}{n-2}$, is an unbiased estimate of σ^2 *only* if this mean of squares has $n - 2$ in its denominator. This is proven in Theorem 9.5.6 on page 500. In general, mean square estimates are only unbiased when their denominator uses the appropriate degrees of freedom.

Another clue as to why s_e^2 has $n-2$ degrees of freedom lies in its formula. The statistic s_e^2 is an estimate of σ^2, which is the expected value of all $(y - \hat{y})^2$ values. Since $\hat{y} = \alpha + \beta x$, $(y - \hat{y})^2 = (y - \alpha - \beta x)^2$. In using s_e^2 to estimate σ^2, we recognize that the slope and y-intercept parameters (β and α) are unknown. Thus, we must use the *sample* slope and y-intercept (b and a) in their places. As a consequence, the values of $(y - a - bx)^2$ tend to be smaller, on average, than $(y - \alpha - \beta x)^2$ if β and α were known. This occurs because b and a are computed from the same sample in which each (x, y) pair belongs. The *sample* data points are closer to the *sample* regression line $\hat{y} = a + bx$ than they are to the *population* regression line $\hat{y} = \alpha + \beta x$.

Each sample produces its own b and a, perfectly designed to give the smallest possible value of $SSE = \sum (y - \hat{y})^2$. This is done *too* well with the sample slope and y-intercept, and the sample SSE is generally smaller than it would be if β and α were used. This yields a *biased* estimate of the population SSE.

The sum, $\sum (y - \alpha - \beta x)^2$ would give the best, *unbiased* value for SSE, but we are forced to substitute b for β and a for α. Consider these substitutions, one at a time, to observe their effect. Since, β is unknown, substituting b for β (but not a for α yet) would tend to make the result a bit smaller, introducing bias. This could be corrected by dividing the sum of squares $\sum (y - \alpha - bx)^2$ by $n - 1$. This would remove all bias in this estimate. Next, substitute a for α. As mentioned, this tends to make the resulting $SSE = \sum (y - a - bx)^2$ smaller still. We have shown in Theorem 9.5.6 that this bias is removed by subtracting 1, once more, from the denominator. This is why s_e^2 contains $n - 2$ degrees of freedom. Since s_b^2 is closely related to s_e^2, they both contain $n - 2$ degrees of freedom.

The sample variance of residuals, $s_e^2 = \frac{SSE}{n-2}$, uses $n - 2$ degrees of freedom in the denominator, contains no bias in its estimate of σ^2.

9.3.4 Hypothesis Tests Regarding the Slope Parameter

We know that the sample slope, b, may be *studentized* under the assumptions of this section, we are able to introduce the corresponding hypothesis test.

Procedure 9.3.7 Testing Claims Regarding the Slope Parameter

Step I. **Verify Assumptions, and Gather Data**
The assumptions are listed in Assumption 9.3.1 on page 453. The normality of residuals can be verified visually, or by using techniques from Section 9.4. Computationally, we need the sample slope and y-intercept (a and b), the summary statistics n, $\sum x$, $\sum x^2$, $\sum y$, $\sum y^2$, and $\sum xy$. The standard deviation of sample slope values should be computed, using the formula

$$s_b = \sqrt{\frac{\sum y^2 - a \sum y - b \sum xy}{(n - 2)\left(\sum x^2 - \frac{1}{n}\left(\sum x\right)^2\right)}}.$$

We also require a level of significance, α.

Step II. **Identify the Claims**
The null hypothesis is $H_0 : \beta = value$, while H_1 is one of following: $\beta > value$, $\beta < value$, or $\beta \neq value$.

Step III. **Compute the Test Statistic**

$$T.S. : t = \frac{b - \beta}{s_b}$$

Step IV. **Look up Critical Values in Table A2**
Remember to use $n - 2$ degrees of freedom.

Step V. **Look up the** *p-value* **in Table A3**
This is done as always, but again, use $n - 2$ degrees of freedom.

Step VI. **Make a conclusion**
When we reject the null hypothesis, we may say that the data support the alternative hypothesis. Otherwise, the best we can say that we fail to reject the null hypothesis.

Example 9.3.8 In Example 9.1.18 on page 424 we computed the regression line for sample (x, y) pairs, where $x = registered\ automatic\ weapons$ (in thousands) and $y = murder\ rate$ (murders per 100,000 people). We also computed $r = 0.885$, concluding that a positive linear relationship exists. The summary statistics for these data values were $\sum x = 38.5$, $\sum y = 65.2$, $\sum x^2 = 283.15$, $\sum y^2 = 622.2$, $\sum xy = 397.21$, with $n = 8$. The slope for the regression line was $b = 0.8525$, and the y-intercept was $a = 4.0473$. The equation of the regression line was therefore $\hat{y} = 4.0473 + 0.8525x$. The scatterplot is given again in Figure 9.25.

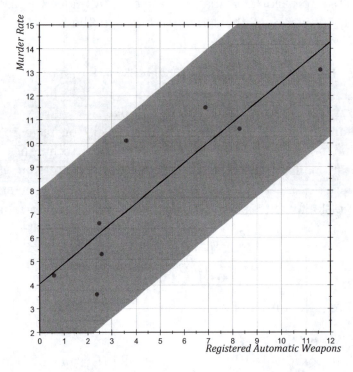

Figure 9.25: Registered automatic weapons are in thousands, where the murder rate is in murders per hundred thousand people.

We would like to test the claim that the population slope is equal to one, at 5% significance. We assume that the assumptions of this section hold true.

Step I. **Verify Assumptions, and Gather Data**

The assumptions are listed in Assumption 9.3.1 on page 453. Visually, we see in the scatterplot a swath of points which is somewhat uniform in width, and where more points are concentrated near the regression line. Computationally, we need the sample slope and y-intercept (a and b), and the summary statistics $n = 8$, $\sum x = 38.5$, $\sum x^2 = 283.15$, $\sum y = 65.2$, $\sum y^2 = 622.2$, and $\sum xy = 397.21$. The slope and y-intercept are $b = 0.8525$, and $a = 4.0473$. The standard deviation of sample slope values is computed next

$$s_b = \sqrt{\frac{\sum y^2 - a \sum y - b \sum xy}{(n-2)\left(\sum x^2 - \frac{1}{n}\left(\sum x\right)^2\right)}}$$

$$= \sqrt{\frac{622.2 - 4.0473 \cdot 65.2 - 0.8525 \cdot 397.21}{(8-2)\left(283.15 - \frac{1}{8} \cdot 38.5^2\right)}}$$

$$= 0.1831$$

For this hypothesis test, we have chosen $\alpha = 0.05$.

Step II. **Identify the Claims**

We are claiming that the population slope is equal to 1. The null hypothesis is therefore $H_0 : \beta = 1$, and we will counter this with $H_1 : \beta \neq 1$.

Step III. **Compute the Test Statistic**

$$T.S. : t = \frac{b - \beta}{s_b} = \frac{0.8525 - 1}{0.1831} = -0.806.$$

This is not far from zero, so it will be difficult to reject the null hypothesis.

Step IV. **Look up Critical Values in Table A2**

Using $df = n - 2 = 8 - 2 = 6$, on a two-tail test with $\alpha = 0.05$ gives $CV : t = \pm 2.447$. The test statistic lies between these, so we will not reject the null hypothesis.

Step V. **Look up the *p-value* in Table A3**

With $df = n - 2 = 6$, and the absolute value of the test statistic rounded to the nearest tenth, $|T.S.| = 0.8$, we see the value 0.773 in the table, but the *p-value* is

$$p\text{-}value = 2\left(1 - 0.773\right) = 0.454.$$

This is quite large and does not indicate a significant departure from the assumption of the null hypothesis.

Step VI. **Make a conclusion**

We have failed to reject the null hypothesis. There is, therefore, insufficient data to warrant rejection of the claim that the population slope is equal to one.

9.3.5   Confidence Intervals for the Slope Parameter

Given that we know the expected value and variance of the sample slope, b, we may use the general form of confidence intervals to create a confidence interval for the population slope.

Theorem 9.3.9 Under the assumptions of this section, the maximum error in the sample slope is

$$E = t_{\alpha/2}\, s_b,$$

and the confidence interval is

$$b - E < \beta < b + E.$$

Example 9.3.10 In Example 9.3.8, we computed the standard deviation of the sample slope, $s_b = 0.1831$. The sample slope was $b = 0.8525$, and the sample size $n = 8$. If we wish to compute a 99% confidence interval we look up $t_{\alpha/2}$ on Table A2 with $n - 2 = 6$ degrees of freedom. The table yields $t_{\alpha/2} = 3.707$, and thus,

$$\begin{aligned}
E &= t_{\alpha/2} \cdot s_b \\
&= 3.707 \cdot 0.1831 \\
&= 0.6788
\end{aligned}$$

The confidence interval follows directly.

$$\begin{aligned}
b - E &< \beta < b + E \\
0.8525 - 0.6788 &< \beta < 0.8525 + 0.6788 \\
0.1737 &< \beta < 1.5313
\end{aligned}$$

Clearly this interval supports the hypothesis that the population slope is $\beta = 1$, as was tested in Example 9.3.8.

Example 9.3.11 We now refer again to Example 9.3.4 which gives literacy rates (x), and infant mortality rates (y) for 50 randomly selected countries. In Example 9.1.19 on page 426 we computed the summary statistics for these data pairs. They are as follows.

$$\sum x = 3968$$
$$\sum y = 1952$$
$$\sum x^2 = 338402$$
$$\sum y^2 = 148974$$
$$\sum xy = 119175$$

The following values will be computed with much precision because we want accurate results. From the summary statistics we first compute the sample slope.

$$b = \frac{\sum xy - \frac{1}{n}\left(\sum x\right)\left(\sum y\right)}{\sum x^2 - \frac{1}{n}\left(\sum x\right)^2}$$

$$= \frac{119175 - \frac{1}{50} \cdot 3968 \cdot 1952}{338402 - \frac{1}{50} \cdot 3968^2}$$

$$= -1.52057$$

Next, we compute the sample y-intercept.

$$a = \overline{y} - b\overline{x}$$

$$= \frac{1952}{50} - (-1.52057) \cdot \frac{3968}{50}$$

$$= 159.71248$$

The standard deviation of sample slopes follows.

$$s_b = \sqrt{\frac{\sum y^2 - a\sum y - b\sum xy}{(n-2)\left(\sum x^2 - \frac{1}{n}\left(\sum x\right)^2\right)}}$$

$$= \sqrt{\frac{148974 - 159.71248 \cdot 1952 - (-1.52057) \cdot 119175}{(50-2)\left(338402 - \frac{1}{50} \cdot 3968^2\right)}}$$

$$= 0.12782$$

We would like to test the claim that the slope parameter is zero, $H_0 : \beta = 0$, against the claim that the slope is greater than zero, $H_1 : \beta < 0$. Since the slope and correlation coefficient formulas have the same numerator, the null hypothesis is equivalent to claiming that the correlation coefficient is zero, $H_0 : \rho = 0$. Note the value of the test statistic.

$$T.S. : t = \frac{b - \beta}{s_b}$$

$$= \frac{-1.52057 - 0}{0.12782}$$

$$= -11.896$$

This is the same value obtained in the correlation test in Example 9.3.4 on page 457. Not only are the null hypotheses equivalent for these two tests, the test statistics are identical as well. Theorem 9.5.7 proves that this is always the case. Claiming that $H_0 : \beta = 0$ is always equivalent to claiming that $H_0 : \rho = 0$. Note that $\rho = 0$ is the only null hypothesis allowed by the correlation test, but the slope test allows any null hypothesis at all, and is, in that sense, more useful.

The critical value for this example will be different, because we are conducting a left-tail test. Choosing $\alpha = 0.01$ with $n - 2 = 6$ degrees of freedom gives the critical value $C.V. : t = -3.143$. We easily reject the null hypothesis and conclude that the claim that the slope is less than zero is strongly supported by the sample data.

9.3.6 Homework

Use the r value computed in Sections 9.1 to answer the following questions. State whether the study is descriptive or experimental, then test the claim that the population (x, y) pairs exhibit linear correlation ($H_1 : \rho \neq 0$). Assume that the requirements for the test are satisfied.

1. Test the claim that the following paired data exhibit linear correlation at 2% significance. The values of x were randomly assigned levels of an experimental treatment.

x	4	2	8	3	1
y	5	4	15	8	2

2. Test the claim that the following paired data exhibit linear correlation at 4% significance. The values of x were randomly assigned levels of an experimental treatment.

x	11	17	21	9	14
y	19	14	9	18	17

3. The following data are assessments of tuna quality. We compare the Hunter L measure of lightness (x) to the averages of consumer panel scores (y) (recoded as integer values from 1 to 6 and averaged over 80 such values) in 9 lots of canned tuna. (Hollander & Wolfe, 1973, p. 187f.). Test the claim at 8% significance that the population of (x, y) pairs exhibits linear correlation.

x	y
44.4	2.6
45.9	3.1
41.9	2.5
53.3	5.0
44.7	3.6
44.1	4.0
50.7	5.2
45.2	2.8
60.1	3.8

4. Below are yearly precipitation totals (x), in inches, paired with annual lowest lake levels (in feet down from a full lake) (y) at Big Bear Lake, in California, for collection of randomly selected years. Test the claim at 6% significance that the population of (x, y) pairs exhibits linear correlation. (Data provided by the Big Bear Municipal Water District).

x	y
15.02	13.98
30.62	9.89
24.82	7.38
50.40	1.22
27.00	5.67
41.04	4.57
49.00	2.48
31.78	3.78
22.20	14.45
17.32	12.17
24.18	9.48
27.49	7.00
35.16	5.63
22.40	6.48

5. The following are GPA and success rates in developmental math courses offered at Mt. San Antonio College for a random selection of semesters. Test the claim at 5% significance that the population of (x, y) pairs exhibits linear correlation.

GPA	Success
1.7536	45.9
1.7090	43.8
1.8416	49.7
1.7615	46.8
1.7467	47.3
1.7915	49.9
1.8266	48.9
1.8587	50.3
1.8386	50.0
1.8892	51.9
1.9109	53.0
1.9365	54.1
1.9346	54.2

6. In an anthropological study of fossils, researchers' objective was to predict age (y) from the percentage (x) of a tooth's root with transparent dentine. The following data are from anterior teeth. Test the claim at 5% significance that the population of (x, y) pairs exhibits linear correlation. (The American Journal of Physical Anthropology, 1991, pp. 25 – 30).

x	y
15	23
19	52
31	65
39	55
41	32
44	60
47	78
48	59
55	61
65	60

7. The table below summarizes measurements taken for cancer patients prior to surgery. The x values represent CAT scan measurements, while the y values are tumor mass. (*Statistics for Research* 2nd ed., by Dowdy & Weardon, 1985, pp. 269 – 270, John Wiley & Sons). At 6% significance, test the claim that the data exhibit linear correlation.

x	y
18	20
28	25
20	16
24	21
16	22
15	10
19	23
24	27
23	18
13	11

8. The data below are from *Carcinogenic Air Pollutants in Relation to Automobile Traffic in New York City*, Environmental Science and Technology (1971), pp. 145 – 150. The x values represent carbon monoxide concentration, while the y values are corresponding benzopyrene concentrations. Test the claim at 5% significance that the population of (x, y) pairs exhibits linear correlation.

x	y
2.8	0.5
15.5	0.1
19.0	0.8
6.8	0.9
5.5	1.0
5.6	1.1
9.6	3.9
13.3	4.0
5.5	1.3
12.0	5.7
5.6	1.5
19.5	6.0
11.0	7.3
12.8	8.1
5.5	2.2
10.5	9.5

9. Is there a correlation between homework scores and overall course percentages in a statistics class? The author randomly selected 35 of his own students and recorded their overall homework scores (x), each of paired with the corresponding overall course percentage (y). The values are recorded below. They are also stored in *Statcato* (www.statcato.org) or in spreadsheet form at http://math.mtsac.edu/statistics/data. Test the claim at 4% significance that these data come from populations which exhibit linear correlation.

x	y
33	42.6
38	34.2
35	67.2
78	52.2
85	80.2
110	64.9
90	86.4
83	63.8
78	77.7
78	59.0
113	97.9
113	85.1
85	80.0
65	64.7
103	86.0
53	77.2
118	65.4
75	72.7
108	98.4
108	98.4
25	43.6
115	95.4
50	58.6
105	94.8
70	82.7
90	84.9
70	78.0
103	53.1
85	68.2
80	62.2
115	85.4
105	83.5
113	96.4
115	98.8
100	83.9

10. The following table gives arm spans and corresponding heights of 32 randomly selected adults. Does a person's arm span correlate to his or her height? Test this claim at 5% significance. (Measured in centimeter by Mt. San Antonio College honors student Alphonso Tran).

Arm Span	Height
170.9	172.0
175.0	177.0
177.0	173.0
177.0	176.0
178.1	178.1
183.9	180.1
188.0	188.0
188.2	186.9
185.9	183.9
188.0	182.1
188.2	181.1
188.5	192.0
194.1	193.0
196.1	183.9
199.9	185.9
150.1	156.0
152.9	159.0
156.0	162.1
157.0	159.8
159.0	162.1
159.8	154.9
161.0	159.8
161.0	162.1
162.8	167.1
162.1	170.2
165.1	166.1
170.2	170.2
170.4	167.1
173.0	184.9
199.9	176.0
167.9	168.9

11. Is there a correlation between foot length and height for young women? (Random data pairs are given in centimeters). Test the claim at 5% significance that linear correlation exists between these variables.

Height (cm)	Foot (cm)
148.0	18.16
152.2	20.43
155.2	21.53
152.7	20.35
155.1	21.10
155.5	19.13
156.8	20.61
158.4	20.43
158.3	21.72
156.2	21.64
158.6	22.95
159.9	20.90
164.2	20.04
157.9	21.81
159.8	21.01
160.9	22.25
162.3	22.68
158.0	22.81
159.8	21.78
156.9	23.74
162.4	21.32
160.2	22.66
166.9	22.11
167.2	22.60
167.6	22.04
164.0	21.58
170.4	22.75
175.9	22.06
176.4	24.18
184.1	23.67

Perform the following inferences regarding the population slope parameter. Assume that the requirements for each inference are satisfied.

12. The following data are assessments of tuna quality. We compare the Hunter L measure of lightness (x) to the averages of consumer panel scores (y) (recoded as integer values from 1 to 6 and averaged over 80 such values) in 9 lots of canned tuna. (*Hollander & Wolfe (1973), p. 187f.*). Test the claim that the population slope is greater than zero at 5% significance.

x	44.4	45.9	41.9	53.3	44.7	44.1	50.7	45.2	60.1
y	2.6	3.1	2.5	5.0	3.6	4.0	5.2	2.8	3.8

13. Below are yearly precipitation totals (x), in inches, paired with annual lowest lake levels (in feet down from a full lake) (y) at Big Bear Lake, in California, for collection of randomly selected years. Test the claim at 1% significance that the population slope is negative. (*Data are provided by the Big Bear Municipal Water District*).

x	y
15.0	14.0
30.6	9.9
24.8	7.4
50.4	1.2
27.0	5.7
41.0	4.6
49.0	2.5
31.8	3.8
22.2	14.5
17.3	12.2
24.2	9.5
27.5	7.0
35.2	5.6
22.4	6.5

14. The following are GPA and success rates in developmental math courses offered at Mt. San Antonio College for a random selection of semesters. Test the claim at 6% significance that the population slope is greater than 30.

GPA	Success
1.7536	45.9
1.7090	43.8
1.8416	49.7
1.7615	46.8
1.7467	47.3
1.7915	49.9
1.8266	48.9
1.8587	50.3
1.8386	50.0
1.8892	51.9
1.9109	53.0
1.9365	54.1
1.9346	54.2

15. In an anthropological study of fossils, researchers' objective was to predict age (y) from the percentage (x) of a tooth's root with transparent dentine. Test the claim at 4% significance that the population slope is greater than 0.15. *(American Journal of Physical Anthropology, 1991, pp. 25 – 30).*

x	15	19	31	39	41	44	47	48	55	65
y	23	52	65	55	32	60	78	59	61	60

16. Construct a 99% confidence interval for the slope parameter using the sample data in Problem 12.

17. Construct a 95% confidence interval for the slope parameter using the sample data in Problem 13.

18. Construct a 98% confidence interval for the slope parameter using the sample data in Problem 14.

19. Construct a 96% confidence interval for the slope parameter using the sample data in Problem 15.

9.4   The Ryan-Joiner Test for *Non*-Normality

We have learned many procedures in this book which have had requirements of normality. All of our tests for means require normality when sample sizes go below 30, and the correlation and regression procedures required bivariate normality. The analysis of variance procedures in Chapter 11 require normality as well. It is high time that we offer a tool for determining when normality is plausible.

Like most tests in statistics, results can be rather unsatisfying at times. In this test, our null hypothesis will assume that the population is normal, and then we will attempt to disprove this. When the data give evidence that they are not from a normal population, we will reject our null hypothesis, and make a strong conclusion. In such a case we will say that the data do not appear to be from a normal population. Unfortunately, if we fail to reject the null hypothesis, this does not make it true. We will be in the unfortunate position of saying we cannot disprove the claim that the population is normal, but that does not mean that it actually *is* normal. In this sense, the test we are about to learn is more a test for *non*-normality than a test for normality. The best that we can say is that it is *plausible* that the data are taken from a normal population.

Normal quantile plots are one way of testing for non-normality. Our approach will be to begin with a sorted sample, and compare it to another sorted sample which is *perfectly normal*. Just what that means will be defined soon. The way that we will compare these samples will be to pair the sorted sample, as *y* values, with the perfectly normal sample of the same size (as *x* values) and create a scatterplot. This scatterplot is called a *normal quantile plot*.

If the sample is distributed similarly to the perfectly normal sample, then the scatterplot will nearly resemble points on a straight line. If the sample is very non-normal, then the scatterplot will resemble points on a highly curved path. Our tool for measuring the linearity of this path will be as before – the sample linear correlation coefficient, r (technically, r is not being used as a true *correlation coefficient* here, since the *perfectly normal* sample values are not random). If the null hypothesis is true, then r will be close to 1. The more non-normal the sample appears, the closer r will be to zero.

We should note, that since both the sample and the perfectly normal sample are sorted from low to high, that they both increase together, so the slope of the regression line is positive, and therefore r can never be negative.

9.4.1   Constructing the Perfectly Normal Sample

We begin the construction of the perfectly normal sample by gathering z-scores from Table A1. Finding the z-scores in a perfectly normal sample is similar to finding z-score percentiles. The kth percentile is the z-score whose corresponding left area is k. There are 99 such percentiles, and they represent a perfect normal sample, because they are separated by 100 equal proportions of area $1/100 = 0.01$ under the normal density function. For a sample of size 99, the perfect normal sample we would compare these to consists of these 99 percentiles. Similarly, to construct a perfect normal sample of size n, we need to find the n values, the *quantiles*, on the normal distribution, which are separated by equal proportions of $1/(n+1)$ under the standard normal density function. These z-scores could be converted to x values using our old formula $x = \mu + z\sigma$ (we would have to use $\bar{x}$ and s), but this is not necessary as we have learned that the value of r does not depend on the units used for the sample observations.

Example 9.4.1 Consider the sample of 15 values: $\{1.0, 2.0, 2.5, 2.6, 2.7, 3.2, 4.2, 5.0,$ $5.3, 5.5, 5.6, 5.7, 5.9, 6.2, 7.0\}$. These values are represented by the histogram and boxplot in Figure 9.26. The histogram shows a bimodal distribution of values – very non-normal in appearance.

We compare the given values to a perfectly normal sample. To construct a perfectly normal sample of size 15, we recognize that the 15 values should be separated by 16 regions of equal area, $\frac{1}{16}$, under the standard normal density function. We will find the smallest value first, then move on to finding all consecutive values. As we move from value to consecutive value the area to the left will accumulate. Thus, we will begin with an area of $\frac{1}{16} = 0.0625$, then repeatedly add 0.0625 to the accumulated area to find the next corresponding z-score. The z-scores will initially be negative.

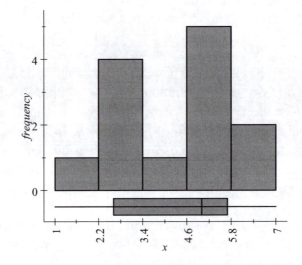

Figure 9.26: A histogram and boxplot for the sample given in Example 9.4.1.

The closest area to 0.0625 in the Table A1 is 0.0630, associated with a z-score of $z = -1.53$.

Negative z-Scores			
z	.02	**.03**	.04
-1.6	.0526	.0516	.0505
-1.5	.0643	**.0630**	.0618
-1.4	.0778	.0764	.0749

Next, we add 0.0625 to the previous area to get 0.1250. The closest area to 0.1250 in the table is 0.1251, associated with a z-score of $z = -1.15$.

Negative z-Scores			
z	.04	**.05**	.06
-1.2	.1075	.1056	.1038
-1.1	.1271	**.1251**	.1230
-1.0	.1492	.1469	.1446

Again, we add 0.0625 to get 0.1875. The closest area to this in the table is 0.1867, associated with a z-score of $z = -0.89$.

Negative z-Scores			
z	.07	.08	**.09**
-0.9	.1660	.1635	.1611
-0.8	.1922	.1894	**.1867**
-0.7	.2206	.2177	.2148

Adding 0.0625 to the previous area gives us 0.2500. The closest area to this is 0.2514, with an associated z-score of $z = -0.67$.

Negative z-Scores			
z	.06	**.07**	.08
-0.7	.2236	.2206	.2177
-0.6	.2546	**.2514**	.2483
-0.5	.2877	.2843	.2810

Next, we add 0.0625 to get an area of 0.3125. The closest area to this in the table is 0.3121, with the corresponding z-score of $z = -0.49$.

Negative z-Scores			
z	.07	.08	**.09**
-0.5	.2843	.2810	.2776
-0.4	.3192	.3156	**.3121**
-0.3	.3557	.3520	.3483

Again, adding 0.0625 to the last area gives us 0.3750. The closest area to this is 0.3745, which is associated to a z-score of $z = -0.32$.

Negative z-Scores			
z	.01	**.02**	.03
-0.4	.3409	.3372	.3336
-0.3	.3783	**.3745**	.3707
-0.2	.4168	.4129	.4090

Next, we add 0.0625 to get 0.4375. The closest area to this in the table is 0.4364, with the corresponding z-score, $z = -0.16$.

Negative z-Scores			
z	.05	**.06**	.07
-0.2	.4013	.3974	.3936
-0.1	.4404	**.4364**	.4325
-0.0	.4801	.4761	.4721

When we add 0.0625 once more to the last area, we get exactly 0.5000, which is associated to $z = 0$, the 50th percentile of the z-distribution. Now that we have arrived at $z = 0$, the remaining z-scores are positive, and due to the symmetry of the z-distribution, the positive z-scores are the absolute values of their negative counterparts. With these, we have created the perfectly normal sample of z-scores: $\{-1.53, -1.15, -0.89, -0.67, -0.49, -0.32, -0.16, 0.00, 0.16, 0.32, 0.49, 0.67, 0.89, 1.15, 1.53\}$.

These values are plotted on the bell curve in Figure 9.27, with emphasis made on the equal areas separating them.

Note the way that the points tend to be closer together in the center, and further apart when farther from the center. This is the nature of normally distributed values.

Of course, no population is perfectly normal, but still we find close approximations, and we are coming close to finding a way to determine when such a population's normality is plausible. We will do this by constructing the scatterplot of our (z, x) pairs.

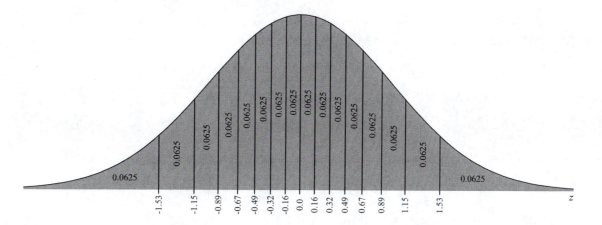

Figure 9.27: The values of a perfectly normal sample are separated by regions of equal probability.

z	x
−1.53	1.0
−1.15	2.0
−0.89	2.5
−0.67	2.6
−0.49	2.7
−0.32	3.2
−0.16	4.2
0.00	5.0
0.16	5.3
0.32	5.5
0.49	5.6
0.67	5.7
0.89	5.9
1.15	6.2
1.53	7.0

In the table above, we have our perfectly distributed z-scores, paired with our x values. In Figure 9.28 we see the normal quantile plot. The curvy nature of the plot is a visual demonstration of the non-normality of our x values. The points should be closely packed in the middle, but instead they are densely packed on either side of this. The correlation coefficient of these pairs is 0.9723, indicating that a positive linear relationship exists, but it is an imperfect one.

Example 9.4.2 Next, we have another sample of size 15: {1.52, 2.21, 2.68, 3.08, 3.40, 3.71, 4.00, 4.29, 4.58, 4.87, 5.18, 5.51, 5.91, 6.38, 7.07}. These values have mean and standard deviation similar the sample in Example 9.4.1, but they are distributed differently. Notice the symmetry, and bell-shaped nature of the histogram.

This sample appears to have a bell-shaped distribution. Note the crudeness of this statement in comparison to the refined approach of the normal quantile plot. In examining

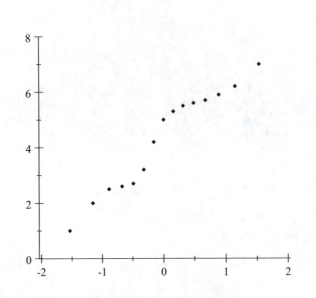

Figure 9.28: The normal quantile plot for the data in Example 9.4.1.

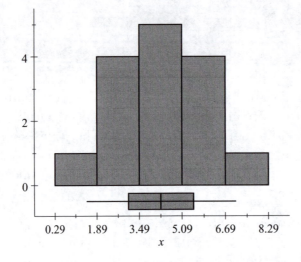

Figure 9.29: The histogram and boxplot of data values from Example 9.4.2.

the histogram, we are only considering the frequencies corresponding to arbitrarily chosen classes. In this assessment, not a single value was taken into account; instead we considered only a picture that can vary considerably depending on the choices for the class limit.

z	x
-1.53	1.52
-1.15	2.21
-0.89	2.68
-0.67	3.08
-0.49	3.40
-0.32	3.71
-0.16	4.00
0.00	4.29
0.16	4.58
0.32	4.87
0.49	5.18
0.67	5.51
0.89	5.91
1.15	6.38
1.53	7.07

Thus, we turn again to the normal quantile plot of our new sample, given in Figure 9.30. This time, when pairing the sorted x values with the perfectly normal z-scores, we see in the scatterplot with points appearing to lie on a straight line.

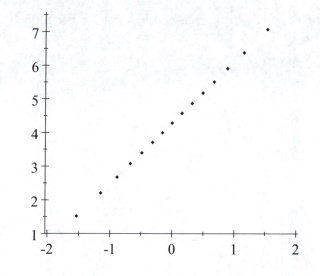

Figure 9.30: The normal quantile plot for Example 9.4.2.

With this example, the correlation coefficient is equal to 0.999993; the relationship is nearly perfect. Notice that r is now being used to measure the normality of the sample, and the closer r is to one, the more normal the sample appears, and the more plausible it is that the population is normal.

Of course, the difficulty of this process was in that we had to look up so many z-scores to construct the perfectly normal sample. This makes the normal quantile plot nearly impossible for a human to construct when sample sizes get very large. Generally speaking, this is a procedure which a computer should perform for all but the smallest of samples.

Example 9.4.3 Next, we consider the values, $\{1.2, 1.6, 2.0, 2.4, 2.8, 3.2, 3.6, 4.0, 4.4,$ $4.8, 5.2, 5.6, 6.0, 6.4, 6.8\}$. The histogram and normal quantile plots are given in Figures 9.31 and 9.32. The values are uniformly distributed, and the normal quantile plot has an S shaped appearance.

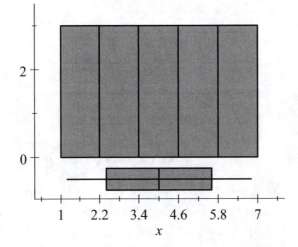

Figure 9.31: The histogram and boxplot for Example 9.4.3.

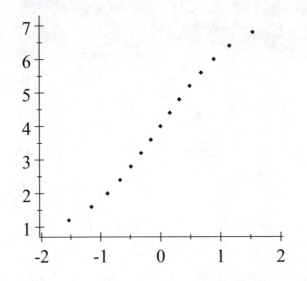

Figure 9.32: The normal quantile plot for Example 9.4.3.

Example 9.4.4 Our next sample is skewed to the right as demonstrated in Figure 9.33: {1.2, 1.6, 1.8, 1.9, 1.9, 2.0, 2.2, 2.6, 3.0, 3.5, 4.0, 4.8, 5.6, 6.6, 7.6}. The normal quantile plot in Figure 9.34 has a small rate of increase in areas where the values are more densely packed than the normal sample is, and is more steeply inclined in places where the data are less dense than a normal data set would be.

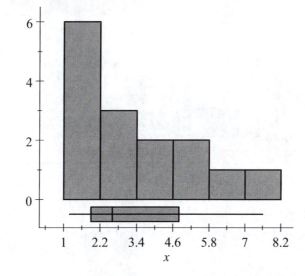

Figure 9.33: The histogram and boxplot for Example 9.4.4.

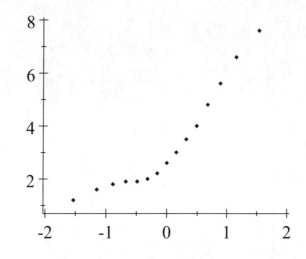

Figure 9.34: The normal quantile plot for Example 9.4.4.

Example 9.4.5 Finally, we have a sample that is skewed to the left as pictured in Figure 9.35. Again we see steep slopes on the normal quantile plot in Figure 9.36 where points are less dense than they should be, and shallow slopes where the data points are more dense than the normal sample.

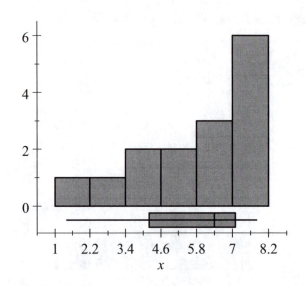

Figure 9.35: Above are the histogram and boxplot for Example 9.4.5.

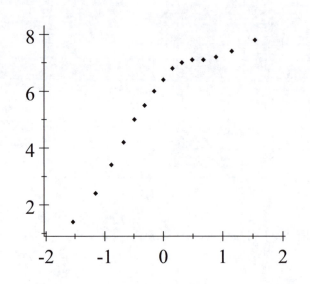

Figure 9.36: Above is the normal quantile plot for Example 9.4.5.

9.4.2   Testing for *Non*-Normality

The visual analysis of the normal quantile plot is quite useful, and much more information can be determined from this than is possible with a histogram, due to the arbitrary nature of the choices for classes. With all of its refinement, the analysis is still subjective, and of course in a statistics course, that cannot be tolerated! We need a critical value and a test statistic!

What we are about to do begins to approach the fringes of research on this subject. To this point, all of our hypothesis tests have used known distributions for statistics that follow them faithfully (with the one exception of the two independent sample t-test). The distribution of the r correlation coefficient for normal quantile plots has not been described analytically, but researchers have studied the problem closely and have developed approximate critical value formulas with good properties. These tables are reproduced at the end of this section.

The problem with this test is that it is, like all hypothesis tests, only strongly decisive in the rejection of the null hypothesis, but weak in non-rejection of the null hypothesis. Our null hypothesis will be that the population is normal. If we reject this, it will be because our data give strong evidence against this claim. If we fail to reject the null hypothesis, we will not be able to say much. This will not necessarily mean that the population is normal. It will only mean that we are unable to reject the claim that the population is normal, but that is a rather unsatisfying conclusion. Still, this is the best we can do.

The reason which we cannot use the test for correlation covered in Section 9.3 is that the claim is different. In that test, we assumed that the population correlation coefficient was equal to zero. We provided critical values that get closer to zero as n increases, making it easier to reject that assumption.

With the normal quantile plot, we are claiming that the population is normal, and if that is true, then the (so-called) null hypothesis is stating that the *population* correlation coefficient is equal to one, and in such a case, the test statistic covered in Section 9.3 does not follow Student's t-distribution. That statistic is only valid when we are claiming that the population correlation coefficient is zero (and when both variables are random, remembering that our perfectly normal z-scores are not random). We want to claim that the (so-called) population correlation coefficient is equal to one, then have critical values that are between zero and one, which approach one as n increases, making it easier to reject the null hypothesis (of normality).

Recently, several statisticians have attempted to solve this problem using computer simulations, rather than with a purely theoretical proof. The first to give critical values for r in this application were Thomas A. Ryan, Jr. and Brian L. Joiner of the Statistics Department, at Pennsylvania State University in 1976. Using computers to generate thousands of random samples from normal populations, make normal quantile plots, and calculate r from these, critical r values were found by finding the 1st, 5th, and 10th percentiles of r values corresponding to $\alpha = 0.01$, 0.05, and 0.10. These values are tabulated in Table A6, which is reproduced below.

n	α=.01	α=.05	α=.10
\multicolumn{4}{c}{Table A6 – Ryan Joiner Test}

n	α=.01	α=.05	α=.10
5	0.832	0.880	0.903
6	0.841	0.889	0.911
7	0.852	0.898	0.919
8	0.862	0.905	0.925
9	0.872	0.912	0.930
10	0.880	0.918	0.935
11	0.888	0.923	0.939
12	0.895	0.928	0.942
13	0.901	0.932	0.945
14	0.906	0.935	0.948
15	0.911	0.938	0.951
16	0.915	0.941	0.953
17	0.919	0.944	0.955
18	0.923	0.946	0.957
19	0.926	0.948	0.958
20	0.929	0.950	0.960
21	0.932	0.952	0.961
22	0.934	0.954	0.963
23	0.937	0.955	0.964
24	0.939	0.957	0.965
25	0.941	0.958	0.966
26	0.943	0.960	0.967
27	0.944	0.961	0.968
28	0.946	0.962	0.969
29	0.948	0.963	0.970
30	0.949	0.964	0.971
31	0.950	0.965	0.971
32	0.952	0.966	0.972
33	0.953	0.967	0.973
34	0.954	0.967	0.973
35	0.955	0.968	0.974
36	0.956	0.969	0.975
37	0.957	0.970	0.975
38	0.958	0.970	0.976
39	0.959	0.971	0.976
40	0.960	0.971	0.977
41	0.960	0.972	0.977
42	0.961	0.973	0.978
43	0.962	0.973	0.978
44	0.963	0.974	0.978
45	0.963	0.974	0.979
46	0.964	0.975	0.979
47	0.965	0.975	0.980
48	0.965	0.976	0.980
49	0.966	0.976	0.980
50	0.966	0.976	0.981
60	0.971	0.980	0.983
70	0.974	0.982	0.986
80	0.977	0.984	0.987
90	0.979	0.986	0.989
100	0.980	0.987	0.990
150	0.985	0.992	0.993
200	0.988	0.994	0.996
250	0.989	0.996	0.997
400	0.997	0.996	0.997
600	0.998	0.998	0.997
1000	0.999	0.998	0.998

The null hypothesis is that the sample is drawn from a normal population. The test statistic is the "correlation" coefficient r, and the critical values are from Table A6. If $r < C.V.$ then the hypothesis of normality is unlikely to be true. If $r > C.V.$ then it is plausible that the sample is taken from a normal population.

Note that the critical values get closer to one as n gets larger. This is because the distributions of smaller samples tend to be more erratic, and thus r values tend to vary more. Also note that the critical values are smaller with smaller values of α, because this is, effectively, a left tailed test.

Example 9.4.6 This "real world" dataset is the result of a study by H. S. Lew of the Structures Division of the Center for Building Technology at the National Institute of Standards & Technology (NIST). The purpose of the study was to characterize the physical behavior of steel-concrete beams under periodic load. The response variable is deflection (from a rest point) of the steel-concrete beam. The 200 observations were collected equally-spaced in time.

				Beam Deflections					
-579	-579	-578	-578	-577	-577	-577	-576	-574	-572
-572	-571	-570	-568	-568	-568	-566	-564	-564	-559
-554	-553	-552	-550	-543	-541	-540	-538	-536	-535
-534	-533	-531	-515	-513	-507	-506	-504	-504	-501
-495	-492	-477	-474	-472	-465	-463	-462	-460	-457
-449	-442	-437	-431	-430	-424	-422	-420	-420	-414
-408	-406	-391	-385	-384	-381	-374	-372	-361	-360
-355	-351	-346	-344	-338	-338	-321	-311	-300	-287
-281	-280	-266	-262	-254	-244	-243	-236	-235	-224
-220	-220	-218	-213	-211	-198	-190	-186	-175	-164
-160	-147	-135	-125	-125	-125	-124	-118	-112	-107
-106	-90	-83	-81	-80	-74	-64	-52	-49	-46
-45	-44	-35	-32	-28	-21	-15	-13	-13	-8
-6	0	11	15	17	25	27	32	48	48
66	68	72	74	83	83	83	83	92	92
96	99	103	110	115	131	131	134	136	137
138	139	139	141	142	143	154	156	156	161
162	162	164	169	172	172	172	174	180	182
187	188	190	193	194	194	194	194	198	199
200	201	201	201	202	203	204	205	205	206

A frequency histogram and normal quantile plot for these values are plotted in Figures 9.37 and 9.38. Obviously these values are highly non-normal in appearance. Correspondingly, the normal quantile plot appears curved.

In the normal quantile plot, $r = 0.954$, and consulting Table A6 we find a critical value of $C.V. : r = 0.994.$ (using 5% significance), and since our r value is less than the critical value, we conclude that it is unlikely that these values were taken from a normal

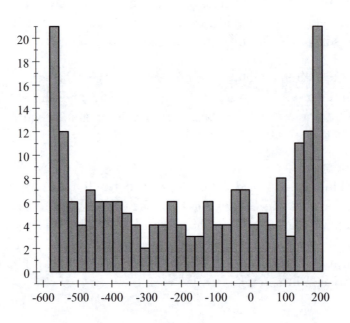

Figure 9.37: The histogram of data values in Example 9.4.6.

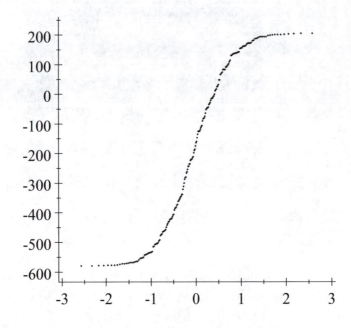

Figure 9.38: The normal quantile plot for Example 9.4.6.

population. Pearson's kurtosis for this sample is 1.5, much less than the standard value of 3, indicating how non-normal the sample is.

Example 9.4.7 Returning to the speed of light measurements from Chapter 2, we list these values again below, sorted from low to high.

Speed of Light Estimates (1000 km/s)									
299.62	299.65	299.72	299.72	299.72	299.74	299.74	299.74	299.75	299.76
299.76	299.76	299.76	299.76	299.77	299.78	299.78	299.79	299.79	299.79
299.80	299.80	299.80	299.80	299.80	299.81	299.81	299.81	299.81	299.81
299.81	299.81	299.81	299.81	299.81	299.82	299.82	299.83	299.83	299.84
299.84	299.84	299.84	299.84	299.84	299.84	299.84	299.85	299.85	299.85
299.85	299.85	299.85	299.85	299.85	299.86	299.86	299.86	299.87	299.87
299.87	299.87	299.88	299.88	299.88	299.88	299.88	299.88	299.88	299.88
299.88	299.88	299.89	299.89	299.89	299.90	299.90	299.91	299.91	299.92
299.93	299.93	299.94	299.94	299.94	299.95	299.95	299.95	299.96	299.96
299.96	299.96	299.97	299.98	299.98	299.98	300.00	300.00	300.00	300.07

The histogram and normal quantile plot are plotted in Figures 9.39 and 9.40.

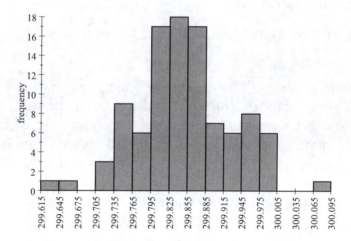

Figure 9.39: The histogram for the speed of light estimates in Example 9.4.7.

In this normal quantile plot, $r = 0.992$, with a critical value from Table A6 of $C.V. : r = 0.987$ (using 5% significance). This time our r value is greater than the critical value, so we fail to reject the null hypothesis that the population is normal. It is plausible that the speed of light data are taken from a normal population.

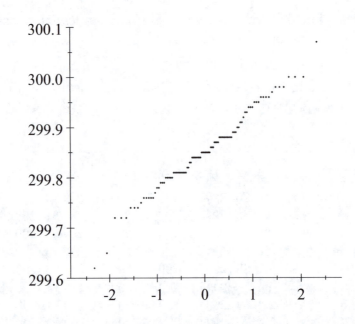

Figure 9.40: The normal quantile plot of values in Example 9.4.7.

9.4.3   Homework

Answer the following questions.

1. Construct a perfectly normal sample of size 9, using Table A1 – without using a computer.

2. Pair the sample of values x: $\{1.3, 1.4, 1.6, 1.7, 5.0, 8.3, 8.4, 8.6, 8.7\}$ with the perfectly normal sample constructed in Problem 1 above. Sketch a scatterplot of these paired values – that is, sketch the normal quantile plot. Compute the correlation coefficient, r, and compare this to the critical value in Table A6 with 5% significance to determine whether it is plausible that this sample was drawn from a normal population.

Test to determine if it is plausible that the following samples are taken from normal populations. For each of these problems, use *Statcato* (available at www.statcato.org) to make this analysis. Alternately, the data may be sorted, and pasted into the normal quantile plot spreadsheet under the Chapter 9 section of the website http://math.mtsac.edu/statistics/excel. In each case, use a 1% significance level.

3. The data below are from *Carcinogenic Air Pollutants in Relation to Automobile Traffic in New York City*, Environmental Science and Technology (1971), pp. 145 – 150. The values represent carbon monoxide concentration on randomly selected days. $\{2.8, 15.5, 19.0, 6.8, 5.5, 5.6, 9.6, 13.3, 5.5, 12.0, 5.6, 19.5, 11.0, 12.8, 5.5, 10.5\}$.

4. The following values are hours worked per week for randomly selected students at Mt. San Antonio College in Walnut, California. $\{15, 20, 40, 16, 20, 25, 19, 40, 40, 20, 20, 40, 0, 40, 35, 30, 35, 25, 29, 0, 0, 40, 16, 0, 40, 30, 25, 16, 35, 0\}$.

5. The following are grade point averages for the same students whose work hours are listed above. { 2.7, 1.7, 2.0, 3.7, 3.1, 4.0, 3.4, 3.7, 2.5, 2.8, 3.0, 3.6, 3.3, 3.2, 3.1, 3.2, 2.6, 2.8, 2.0, 4.0, 2.7, 3.9, 3.1, 2.5, 3.7, 2.9, 3.5, 3.7, 2.2, 3.8 }.

6. The following data are overall grades for your instructor's statistics students gathered from Fall, 2005. { 59.3, 60.2, 68.4, 69.1, 70.8, 71.7, 72.3, 73.9, 75.1, 76.1, 76.1, 80.1, 84.9, 85.3, 86, 88.1, 88.3, 93, 94, 98.6 }.

7. The over-counted votes for 15 *Votomatic* machines are listed below. { 7826, 1134, 21855, 520, 3640, 790, 2531, 890, 17833, 1039, 19218, 2124, 4258, 994, 170 }.

8. The following are rainfall totals (in inches) for clouds chemically seeded to increase rainfall amounts. (Data are taken from Miller, A.J., Shaw, D.E., Veitch, L.G. & Smith, E.J. (1979). "Analyzing the results of a cloud-seeding experiment in Tasmania", Communications in Statistics - Theory & Methods, vol.A8(10), pp. 1017-1047). These values are built-into *Statcato*, or downloadable at http://math.mtsac.edu/statistics/data.

{0.81, 0.86, 2.39, 0.36, 2.06, 2.48, 4.61, 4.16, 2.16, 6.00, 0.37, 0.84, 0.26, 0.47, 0.90, 0.42, 1.23, 0.13, 0.59, 0.91, 0.88, 1.32, 1.87, 0.58, 2.97, 1.25, 1.00, 2.04, 0.71, 2.22, 1.11, 0.80, 1.46, 1.48, 0.40, 1.09, 3.56, 0.07, 2.26, 2.08, 0.79, 1.43, 1.62, 1.16, 2.87, 0.76, 1.83, 1.50, 0.41, 2.56, 0.24, 0.61, 0.05, 0.38, 0.90, 2.35, 4.29, 4.24, 1.67, 5.48, 1.63, 3.31, 2.21, 2.36, 3.25, 1.08, 3.17, 0.80, 2.25, 2.79}

9. For the 1532 schizophrenic patients embedded as a data set in *Statcato*, or downloadable as a spreadsheet at *http://math.mtsac.edu/statistics/data*, test the eye gain ratios for normality.

10. For the 1374 healthy control group members embedded as a data set in *Statcato*, or downloadable as a spreadsheet at *http://math.mtsac.edu/statistics/data*, test the eye gain ratios for normality.

9.5 Proofs for Correlation and Regression

In this section, our goal is to prove a few ideas in the context of the least squares regression line. For inferences regarding the slope parameter β, we have given formulas for the expected value and variance of its estimator, b. These will be proven in this section, along with a theorem that gives the expected value for the standard error of residuals.

The first measure of variation is one we have already discussed. It totals the squared deviations of sample y values from predicted $\hat{y}$ values. These are the squared residuals, $e^2 = (y - \hat{y})^2$. The sum of the squared residuals we have denoted $SSE = \sum (y - \hat{y})^2$. This total variation is considered *unexplained*, because it measures the variation *away* from the regression line, the line that allows us to use x as an *explanatory* variable for y.

9.5.1 Correlation and Regression Theorems

Previously, we have given a computational formula for the slope statistic, b, as follows.

$$b = \frac{\sum (x - \overline{x})(y - \overline{y})}{\sum (x - \overline{x})^2}$$

We have defined and simplifed parts of this expression as

$$S_{xy} = \sum (x - \overline{x})(y - \overline{y})$$
$$= \sum xy - \frac{1}{n}\left(\sum x\right)\left(\sum y\right),$$

and

$$S_{xx} = \sum (x - \overline{x})^2$$
$$= \sum x^2 - \frac{1}{n}\left(\sum x\right)^2.$$

With these, we can also express b as follows.

$$b = \frac{S_{xy}}{S_{xx}}$$

When we say that a population of (x, y) data pairs is linear in form, we mean that there is a line that represents the center of the relationship between the variables. The central value of y for a given value of x is the expected value of y, given x, denoted $E(y)$, and this mean should be determined by a linear function:

$$E(y) = \alpha + \beta x$$

where α and β are the population slope and y-intercept for the population regression line.

To perform inferences on the population slope parameter, β, we need information about the statistical estimate, b. The argument below determines the expected value of b.

$$E(b) = E\left(\frac{\sum (x - \bar{x})(y - \bar{y})}{\sum (x - \bar{x})^2}\right) \qquad \text{This is our reworked formula for } b.$$

$$= \frac{\sum (x - \bar{x}) E(y - \bar{y})}{\sum (x - \bar{x})^2} \qquad E(ky) = kE(y).$$

$$= \frac{\sum (x - \bar{x})(\alpha + \beta x - \alpha - \beta \bar{x})}{\sum (x - \bar{x})^2} \qquad \text{For each given } x, E(y) = \alpha + \beta x.$$

$$= \frac{\sum (x - \bar{x}) \beta (x - \bar{x})}{\sum (x - \bar{x})^2} \qquad \text{Simplify}$$

$$= \beta \qquad \text{Rearrange and reduce.}$$

Not only does this tell us that $E(b) = \beta$, but also that b is an unbiased estimator of β. This argument is summarized in the theorem below.

Theorem 9.5.1 The sample slope,

$$b = \frac{r\, s_y}{s_x},$$

is an unbiased estimate of the population slope parameter, β. This means that the expected value of the sampe slope is the population slope: $E(b) = \beta$.

Next, we need the variance of the sample slope which leads us to a formula for its standard error.

Theorem 9.5.2 The variance of the sample slope statistic, b, is $Var(b) = \frac{\sigma^2}{\Sigma(x-\bar{x})^2}$, where σ^2 is the variance of y, assumed to be the same for each value of x.

Proof. In the discussion below, we assume x is not random, but y is. So statistics regarding x are regarded as constants.

$$Var\left(b\right) = Var\left(\frac{\sum\left(x - \overline{x}\right)\left(y - \overline{y}\right)}{\sum\left(x - \overline{x}\right)^2}\right) \qquad \text{We substite } s_x = \frac{\sum\left(x - \overline{x}\right)^2}{\left(n - 1\right)}.$$

$$= \frac{\sum\left(x - \overline{x}\right)^2\left(Var\left(y - \overline{y}\right)\right)}{\left(\sum\left(x - \overline{x}\right)^2\right)^2} \qquad \text{Distribute, noting } Var\left(y\right) = k^2 Var\left(y\right).$$

$$= \frac{\sum\left(x - \overline{x}\right)^2\left(Var\left(y\right) + Var\left(\overline{y}\right)\right)}{\left(\sum\left(x - \overline{x}\right)^2\right)^2} \qquad Var\left(x + y\right) = Var\left(x\right) + Var\left(y\right)$$

$$= \frac{\sum\left(x - \overline{x}\right)^2\left(\sigma^2 + 0\right)}{\left(\sum\left(x - \overline{x}\right)^2\right)^2} \qquad \text{The variance of a constant is 0.}$$

$$= \frac{\sigma^2 \sum\left(x - \overline{x}\right)^2}{\left(\sum\left(x - \overline{x}\right)^2\right)^2} \qquad \text{Here we factor out } \sigma^2.$$

$$= \frac{\sigma^2}{\sum\left(x - \overline{x}\right)^2} \qquad \text{We reduce to finish.}$$

■

The next two theorems give intermediate algebraic simplifications.

Theorem 9.5.3 The sum $S_{xy} = \sum\left(x - \overline{x}\right)\left(y - \overline{y}\right)$ can be simplified as $S_{xy} = b\sum\left(x - \overline{x}\right)^2 = bS_{xx}$.

Proof. This is straightforward since $bS_{xx} = \frac{S_{xy}}{S_{xx}} \cdot S_{xx} = S_{xy}$. ■

Theorem 9.5.4 The sum $S_{yy} = \sum\left(y - \overline{y}\right)^2$ can be simplified as $S_{yy} = \sum y^2 - n\overline{y}^2$

Proof. This is relatively straight forward.

$$\begin{aligned} S_{yy} &= \sum\left(y - \overline{y}\right)^2 \\ &= \sum y^2 - \frac{1}{n}\left(\sum y\right)^2 \\ &= \sum y^2 - \frac{n^2}{n}\left(\frac{\sum y}{n}\right)^2 \\ &= \sum y^2 - n\overline{y}^2 \end{aligned}$$

■

The following theorem gives the expected value of SSE – a result with important consequences.

Theorem 9.5.5 The expected value of SSE is $(n-2)\sigma^2$.

Proof. This is more involved. This uses the fact that $\overline{y} = a + b\overline{x}$, so $a = \overline{y} - b\overline{x}$, $S_{xy} = \sum (x - \overline{x})(y - \overline{y})$, $S_{xx} = \sum (x - \overline{x})^2$ and $S_{yy} = \sum (y - \overline{y})^2$.

$$E(SSE) = E\left(\sum (y - \widehat{y})^2\right) \qquad \text{Def. of } SSE$$

$$= E\left(\sum (y - a - bx)^2\right) \qquad \text{Def. of } \widehat{y}.$$

$$= E\left(\sum (y - \overline{y} + b\overline{x} - bx)^2\right) \qquad a = \overline{y} - b\overline{x}$$

$$= E\left(\sum ((y - \overline{y}) - b(x - \overline{x}))^2\right) \qquad \text{Factoring}$$

$$= E\left(\sum (y - \overline{y})^2 + b^2 \sum (x - \overline{x})^2 - 2bS_{xy}\right) \qquad \text{Expand}$$

$$= E\left(\sum (y - \overline{y})^2 + b^2 \sum (x - \overline{x})^2 - 2b^2 \sum (x - \overline{x})^2\right) \qquad \text{Thm. 9.5.3.}$$

$$= E\left(\sum (y - \overline{y})^2 - b^2 \sum (x - \overline{x})^2\right) \qquad \text{Combine}$$

$$= E\left(\sum y^2 - n\overline{y}^2 - b^2 S_{xx}\right) \qquad \text{Thm. 9.5.4.}$$

$$= \sum E(y^2) - nE(\overline{y}^2) - S_{xx}E(b^2) \qquad \text{Prop. of } E.$$

Here we have the expected values of two squared variables, $E(\overline{y}^2)$ and $E(a^2)$. Simplifying these requires the property that $E(x^2) = Var(x) + E(x)^2$.

$$E(SSE) = \sum E(y^2) - nE(\overline{y}^2) - S_{xx}E(b^2) \qquad \text{Previous result.}$$

$$= \sum \left(Var(y) + E(y)^2\right) - n\left(Var(\overline{y}) + E(\overline{y})^2\right) \qquad E(x^2)$$

$$\qquad - S_{xx}\left(Var(b) + E(b)^2\right) \qquad = Var(x) + E(x)^2$$

$$= n\sigma^2 + \sum (\alpha + \beta x)^2 - n\left(\frac{\sigma^2}{n} + (\alpha + \beta\overline{x})^2\right) \qquad Var(\overline{y}) = \frac{\sigma^2}{n}$$

$$\qquad - S_{xx}\left(\frac{\sigma^2}{S_{xx}} + \beta^2\right) \qquad Var(a) = \frac{\sigma^2}{S_{xx}}$$

$$= n\sigma^2 - \sigma^2 - \sigma^2 + \sum (\alpha + \beta x)^2 - n(\alpha + \beta\overline{x})^2 - S_{xx}\beta^2 \qquad \text{Simplify}$$

$$= n\sigma^2 - 2\sigma^2 + \sum \left((\alpha + \beta x)^2 - (\alpha + \beta\overline{x})^2 - \beta^2(x - \overline{x})^2\right) \qquad \text{Simplify}$$

The terms in the sum on the right may be simplified as follows.

$$(\alpha + \beta x)^2 - (\alpha + \beta\overline{x})^2 - \beta^2(x - \overline{x})^2$$
$$= (\alpha^2 + 2\alpha\beta x + \beta^2 x^2) - (\alpha^2 + 2\alpha\beta\overline{x} + \beta^2\overline{x}^2) - (\beta^2 x^2 - 2\beta^2 x\overline{x} + \beta^2\overline{x}^2)$$

With further simplification we have the following.

$$
\begin{aligned}
(\alpha + \beta x)^2 - (\alpha + \beta \bar{x})^2 - \beta^2 (x - \bar{x})^2 &= -2\bar{x}^2 \beta^2 + 2x\bar{x}\beta^2 - 2\alpha\bar{x}\beta + 2x\alpha\beta \\
&= 2\beta^2 \bar{x}(x - \bar{x}) + 2\alpha\beta(x - \bar{x}) \\
&= 2\beta(\alpha + \beta\bar{x})(x - \bar{x}) \\
&= 2\beta\bar{y}(x - \bar{x})
\end{aligned}
$$

The sum of these terms is therefore

$$
\sum \left((\alpha + \beta x)^2 - (\alpha + \beta \bar{x})^2 - \beta^2 (x - \bar{x})^2 \right)
$$
$$
= 2\beta\bar{y} \sum (x - \bar{x}) = 0,
$$

since the sum of deviations from $\bar{x}$ is always zero. The expected value of SSE becomes the following.

$$
\begin{aligned}
E(SSE) &= E\left(\sum (y - \hat{y}^2) \right) \\
&= n\sigma^2 - 2\sigma^2 + \sum \left((\alpha + \beta x)^2 - (\alpha + \beta\bar{x})^2 - \alpha^2 (x - \bar{x})^2 \right) \\
&= n\sigma^2 - 2\sigma^2 + 0 \\
&= (n - 2)\sigma^2
\end{aligned}
$$

Thus concludes a rather difficult proof. ∎

The last result, $E(SSE) = (n - 2)\sigma^2$, tells us that $E\left(\frac{SSE}{n-2}\right) = \sigma^2$, so if we need an unbiased estimate of σ^2, we will use

$$
s_e^2 = \frac{SSE}{n - 2} = \frac{\sum (y - \hat{y})^2}{n - 2}.
$$

This is stated formally in the following theorem.

Theorem 9.5.6 The sample variance of residuals, $s_e^2 = \frac{SSE}{n-2} = \frac{\sum (y - \hat{y})^2}{n-2}$, is an unbiased estimate of σ^2.

Proof. This follows immediately from Theorem 9.5.5. ∎

The estimator, s_e^2, of σ^2 is only unbiased if we divide by $n - 2$. This is why we use $n - 2$ in the denominator of s_e^2 instead of $n - 1$, and that *studentized* statistics which use $s_e = \sqrt{s_e^2}$ have $n - 2$ degrees of freedom.

Theorem 9.5.7 When the population correlation coefficient is zero, $\rho = 0$, the population slope is zero as well, $\beta = 0$. In this case, the test statistics $t = \frac{r}{\sqrt{\frac{1-r^2}{n-2}}}$, for the correlation test, and $t = \frac{b - \beta}{s_b}$, for the slope test, are equivalent.

Proof. This theorem uses the fact that $SST = SSR + SSE$, so $SSE = SST - SSR$, and that $r^2 = \frac{SSR}{SST}$.

$$t = \frac{b - \beta}{s_b}$$
This is the $T.S.$ for the slope test.

$$= \frac{b - 0}{s_b}$$
The null hypothesis states that $\beta = 0$.

$$= \frac{S_{xy}/S_{xx}}{s_e/\sqrt{S_{xx}}}$$
$b = S_{xy}/S_{xx}$ and $s_b = \frac{s_e}{\sqrt{S_{xx}}}$.

$$= \frac{\frac{S_{xy}}{\sqrt{S_{xx}}}}{\sqrt{\frac{SSE}{n-2}}}$$
$s_e = \sqrt{\frac{SSE}{n-2}}$ & simplification.

$$= \frac{\frac{S_{xy}}{\sqrt{S_{xx}}}}{\sqrt{\frac{SST-SSR}{n-2}}}$$
$SSE = SST - SSR$

$$= \frac{\frac{S_{xy}}{\sqrt{S_{xx}}\sqrt{SST}}}{\sqrt{\frac{1-\frac{SSR}{SST}}{n-2}}}$$
Rearranging factors.

$$= \frac{\frac{S_{xy}}{\sqrt{S_{xx}}\sqrt{S_{yy}}}}{\sqrt{\frac{1-r^2}{n-2}}}$$
$S_{yy} = \sum (y - \overline{y})^2 = SST$, $r^2 = \frac{SSR}{SST}$

$$= \frac{r}{\sqrt{\frac{1-r^2}{n-2}}}$$
$r = \frac{S_{xy}}{\sqrt{S_{xx}}\sqrt{S_{yy}}}$

■

9.6   The History of Correlation and Regression

If you have glanced at the proofs at the end of Section 9.5, you may have gathered that the development of the theory of correlation and regression must be quite complicated. What is interesting is that the methods of *least squares* for finding the slope and y-intercept of the regression line existed for nearly one-hundred years before a clear formula for the correlation coefficient existed. This is surprising because we have proven all that we have claimed about the correlation coefficient, but have avoided proving the formulas for the slope and y-intercept because the proofs require calculus of several variables.

Regardless, the method of least squares which is used to find the slope and y-intercept of the regression line was developed first. This is probably because there was an obvious need for it. Scientists saw the relationship between variables in sample data and wanted to find the best equation to approximate this relationship.

Legendre is credited as being the first to describe the method of least squares in 1805. Carl Friedrich Gauss claimed to have known the method years before this, but there is no proof of this, other than his own word. The method can be used to fit lines to data, but also to fit other curves to data that exhibit nonlinear relationships. The need that drove scientists to develop least squares methods was the same need that drove the invention of much of the material that is now known as calculus – the need to describe the motions of the planets. Astronomers had so much empirical data which had been gathered over years of planetary observation, but since planets obviously followed smooth curves, scientists desired to find equations of these curves. Knowing these equations would allow them to predict the position of planets at any moment in time. When Legendre presented his least squares methods, they became an instant success.

Possibly the reason that methods for finding the coefficients of the equation of least squares were discovered before the correlation coefficient was that there was no clear idea of why it was needed. It is interesting that the mathematics of the correlation coefficient were present in the works of Auguste Bravais (1811-1863), professor of astronomy and physics, but Bravais never recognized them for what they were. He was not looking for a way of quantifying correlation, and thus never recognized it in his own work.

The first person to recognize correlation as we currently understand it was Francis Galton, pictured in Figure 9.41. Galton was a cousin of Charles Darwin, the father of evolutionary theory. Galton, upon reading the work of his cousin, was so inspired that he began a lifelong career of looking for hereditary characteristics in humans and other life forms. Galton was not a mathematician – most of his work involved the gathering and comparing of data, while he often left the mathematical development to others at his request.

The term *regression* was coined by Galton, and is seen in a paper which he wrote,

REGRESSION *towards* MEDIOCRITY *in* HEREDITARY STATURE.
By FRANCIS GALTON, F.R.S., &c.

entitled *Regression Towards Mediocrity in Hereditary Stature*, where he compares heights of parents to heights of their adult children.

In examining Galton's from this paper, there is a clear relationship between these data pairs. The taller children tend to come from taller parents, while shorter children tend to come from shorter parents. The data are replicated in Figure 9.42, providing Galton's original data for this example.

The *mid-parent* heights (representing the average of the parents' heights) are in the left column of the table in Figure 9.42, while the children's heights are in the top row. The values in the midst of the table are frequencies. For example, for a mid-parent height of 68.5 in there were 22 cases in which the respective adult child height was 66.2 in. The correlation coefficient for these (x, y) pairs was $r = 0.4056$. There were 892 pairs recorded. With a sample as large as this, the test for linear correlation results in a rejection of the claim that $\rho = 0$. So there appears to be a linear relationship between these variables. The data do show strong evidence of bivariate normality, where the highest frequencies are near the center of the table, with the frequencies getting smaller as we move toward the sides of the table in any direction.

One way to visualize this type of data set is with an enhanced scatterplot called a *bubble plot*, depicted in Figure 9.43. A bubble plot is a scatterplot where the points are enlarged to circles, whose radii are proportional to the frequencies of the (x, y) pairs.

The linear regression line is plotted on the bubble plot as well. Upon examining the data, the equation is $\hat{y} = 29.0395 + 0.5772x$. The reason that Galton coined the term *regression* becomes evident when we consider what happens when we enter values above average or below average. The mean parent height in the sample data is 68.0 inches, while the mean child height is 68.3 inches. Suppose we wish to estimate the height of a child whose parent is above average at 71 inches. We use the regression line,

$$\hat{y} = 29.0395 + 0.5772 \cdot 71$$
$$= 70.02$$

inches. From a parent who is above average in height, we estimate the child will be shorter. Now consider a parent whose height is below average at 64 inches. Again, we use the regression line,

$$\hat{y} = 29.0395 + 0.5772 \cdot 64$$
$$= 65.98.$$

From a parent who is below average in height, we estimate a taller child. From a parent whose height is 68.0, exactly average for the parents, we estimate the child's height to be

$$\hat{y} = 29.0395 + 0.5772 \cdot 68.0$$
$$= 68.3,$$

the average height for children.

Figure 9.41: Sir Francis Galton, probably taken in the 1850s or early 1860s. From Karl Pearson's *The Life, Letters, and Labors of Francis Galton*, now in the public domain.

TABLE I.

NUMBER OF ADULT CHILDREN OF VARIOUS STATURES BORN OF 205 MID-PARENTS OF VARIOUS STATURES.

(All Female heights have been multiplied by 1·08).

Heights of the Mid-parents in inches.	Heights of the Adult Children.														Total Number of		Medians.
	Below	62·2	63·2	64·2	65·2	66·2	67·2	68·2	69·2	70·2	71·2	72·2	73·2	Above	Adult Children.	Mid-parents.	
Above ..	..	..	..	..	..	..	..	..	..	..	1	3	..	..	4	5	..
72·5	..	..	..	..	..	..	..	1	2	1	2	7	2	4	19	6	72·2
71·5	..	..	..	..	1	3	4	3	5	10	4	9	2	2	43	11	69·9
70·5	1	..	1	..	1	1	3	12	18	14	7	4	3	3	68	22	69·5
69·5	..	..	1	16	4	17	27	20	33	25	20	11	4	5	183	41	68·9
68·5	1	..	7	11	16	25	31	34	48	21	18	4	3	..	219	49	68·2
67·5	..	3	5	14	15	36	38	28	38	19	11	4	..	..	211	33	67·6
66·5	..	3	3	5	2	17	17	14	13	4	..	..	..	..	78	20	67·2
65·5	1	..	9	5	7	11	11	7	7	5	2	1	..	..	66	12	66·7
64·5	1	1	4	4	1	5	5	..	2	..	..	..	..	..	23	5	65·8
Below ..	1	..	2	4	1	2	2	1	1	..	..	..	..	..	14	1	..
Totals ..	5	7	32	59	48	117	138	120	167	99	64	41	17	14	928	205	..
Medians ..	..	..	66·3	67·8	67·9	67·7	67·9	68·3	68·5	69·0	69·0	70·0	..	..	..	..	..

NOTE.—In calculating the Medians, the entries have been taken as referring to the middle of the squares in which they stand. The reason why the headings run 62·2, 63·2, &c., instead of 62·5, 63·5, &c., is that the observations are unequally distributed between 62 and 63, 63 and 64, &c., there being a strong bias in favour of integral inches. After careful consideration, I concluded that the headings, as adopted, best satisfied the conditions. This inequality was not apparent in the case of the Mid-parents.

Figure 9.42: The data gathered by Galton tabulate frequencies for (x, y) data pairs where x represents mid-parent height (the average of both parents' heights), and y represents height of their respective adult children.

As the parents are different from average, the linear relationship indicates that the children will be closer to average. This is happening because the slope of the regression line is less than 1. A deviation of x from $\bar{x}$ is always accompanied by a smaller deviation of y from $\bar{y}$, so when the input is far from average, the output is closer to average.

Galton called this pulling toward the mean *regression*, because he viewed this tendency of children to be less exceptional than their parents as a trend toward mediocrity, and the term stuck. In any situation where the regression line has slope larger than 1 or less than -1, the term does not make sense, and yet it is there as a kind of memorial to Galton and his work.

Galton was the first to attempt to compute the correlation coefficient, r, even though the regression equation had been used for decades before this time. His ideas were carried on by Karl Pearson, after whom the Pearson Correlation Coefficient, r, is named. As is so often true in mathematics, the name of the thing is often attributed to someone who was not the originator of the idea. Certainly, Pearson did a great deal of work to simplify and clarify Galton's ideas, but it all began with Galton.

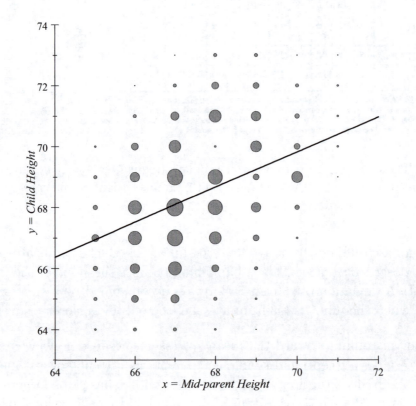

Figure 9.43: A bubble-plot of Galton's data comparing mid-parent heights to corresponding adult child heights. The size of each bubble is proportional to its respective frequency.

Chapter 10

Chi-Square Tests

Until now, all of our testing procedures have used either the normal or Student's t distribution. In this chapter, we will learn to use an entirely new probability distribution called the *chi-square* distribution, as applied to *goodness of fit tests*. The goodness of fit tests were developed primarily by a man who represents one of greatest names in statistics – Karl Pearson, pictured in Figure 10.1. Pearson was a mathematics professor at University College in London, and for the first 33 years of his life showed no particular interest in statistics, a subject which was in its early stages of development. In 1889, Pearson read a book, *Natural Inheritance*, by Francis Galton. Galton was the cousin of Charles Darwin, whose ideas are responsible for the development of *Pearson's* correlation coefficient. Pearson was fascinated by Galton's book, and it was then that Pearson began his professional work in statistics.

10.0.1 Karl Pearson

Between 1893 and 1912, Pearson wrote 18 papers which contained his most important work as a statistician. *Pearson's correlation coefficient*, r, initially developed by Galton, was named after Karl Pearson because of his work in helping to develop a theory for computing and understanding it.

As previously mentioned in this text, the correlation coefficient took some time to take shape, and while the process was begun by Galton, Pearson had a great deal to do with the development as well, so the statistic bears Pearson's name.

Pearson published the *chi-square* goodness of fit tests in 1900 in order to find a way to study data that were not distributed according to the normal or t distributions. The *chi-square* tests are used for determining *goodness of fit* in what is called a *multinomial experiment*. The goodness of fit tests are used for testing hypotheses about categorical, or nominal data. While Pearson was the first to introduce the goodness of fit tests, he was not the first to discover the *chi-square* distribution. The first person to discover the *chi-square* distribution was Friedrich Helmert in 1875, when he discovered that inferences concerning sample variances can be made using the *chi-square* distribution.

Figure 10.1: Karl Pearson, pencil drawing by F.A. de Biden Footner, 1924. Used by permission.

Pearson's use of goodness of fit tests was a new approach to using the *chi-square* distribution, and to this day they remain among the most important tools of statistical research. It is interesting to note that it was discovered later by Ronald A. Fisher that Pearson's use of the test was incorrect in some cases, because Pearson had been applying degrees of freedom incorrectly. We will learn about the great statistician Ronald A. Fisher in the next chapter of this text, entitled *Analysis of Variance*.

10.1 The Goodness of Fit Test

Knowing the definition of a binomial experiment may be enough to guess intelligently the definition of a multinomial experiment. A multinomial experiment consists of a sample of n independent trials. Each trial will have k possible outcomes, and the probabilities, or *proportions*, remain constant from trial to trial. We will not concern ourselves with computing multinomial probabilities directly as we did with the binomial probabilities, but will proceed directly to the hypothesis test.

The hypothesis test is relatively straight forward. We will make an assumption about the values of each hypothetical proportion. From each proportion we will compute an expected frequency, denoted by E, which is the number of times a given outcome is expected to occur. The value for E will be the same as the mean in the binomial experiment: np, where n is the sample size and p is the proportion or probability of a given outcome. Next, we will look at our sample data to determine the actual observed frequencies, denoted by O, which are the number of times each outcome actually occurred in the sample. We will use Karl Pearson's *chi-square* test statistic: $\chi^2 = \sum (O - E)^2 / E$, with corresponding degrees of freedom equal to $k - 1$. The symbol, χ, is a Greek lower case letter, pronounced *chi*.

We will provide some motivation for the origins of this test statistic at the end of this section. Suffice it to say that this test statistic is the *square* of the z test statistic which we used for the binomial experiment in Chapter 7 (the test for a single population proportion), with one degree of freedom. The critical value that we will use will come from the *chi-square* table (Table A4), and while this is a completely new table we will see that, with one degree of freedom, that the critical values there are simply the squares of the z-critical values from Tables A1 and A2. Table A4 is in the appendix, and is also reproduced at the end of this chapter.

Before we begin with a procedure, let us discuss the properties of the *chi-square* distribution. First, it should be noted that the test statistic uses only positive values in the summation, so *chi-square* critical values are always positive. Because of this, the *chi-square* distribution is not symmetric – it begins at zero and is skewed to the right. Also, degrees of freedom will again be an issue since the *chi-square* distribution changes with degrees of freedom. In fact, *chi-square* critical values get larger and larger as the degrees of freedom increase. Degrees of freedom for a *chi-square* test statistic will represent the number of independent observations in the test statistic.

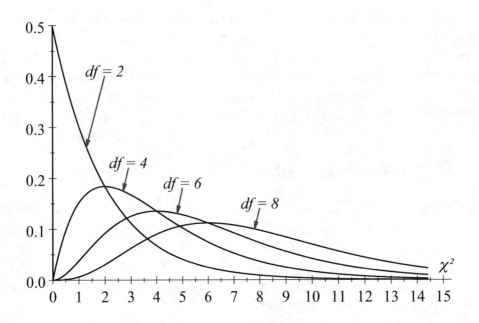

Figure 10.2: The χ^2 (*chi-square*) distribution, with varying degrees of freedom (*df*).

It does not take much imagination to see that the test statistic,

$$\chi^2 = \sum \frac{(O-E)^2}{E},$$

can be used as a measure of *fit*. This statistic compares the fit between what is observed, O, and what is expected, E, by subtracting these values, and as usual, squaring the deviations so that the sum is a non-negative number. Remember that the expected values are based on whatever the null hypothesis says they should be.

If the observed values fit well with the expected values the differences $O - E$ will be close to zero, so the test statistic will be small, and we should not reject the null hypothesis. When the test statistic is large, the O values do not fit well with the expected E values (as predicted by the null hypothesis), and so we should reject the null hypothesis. For this reason, the *goodness of fit tests are all right-tail tests*. You will notice that the table for *chi-square* critical values also has left-tail critical values, but these are not used for goodness of fit tests.

To use Table A4, we simply find the column with the appropriate α value under the right-tail section of the table, then we find the row heading with the appropriate degrees of freedom. The intersection of this column and row will contain the right-tail *chi-square* critical value.

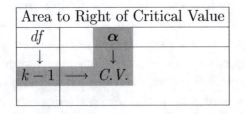

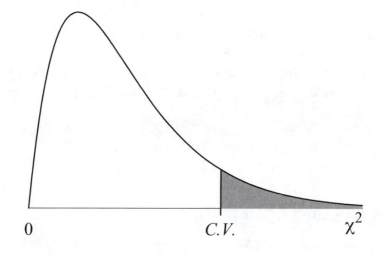

Figure 10.3: A right-tail critical value on the χ^2 distribution.

For example, with $\alpha = 0.05$ and 5 degrees of freedom, the associated critical value is $C.V. : \chi^2 = 11.0705$.

	Area to Right of Critical Value		
df	0.10	**0.05**	0.01
4	7.7794	9.4877	13.2767
5	9.2364	**11.0705**	15.0863
6	10.6446	12.5916	16.8119

The distribution is pictured with this critical value in Figure 10.4.

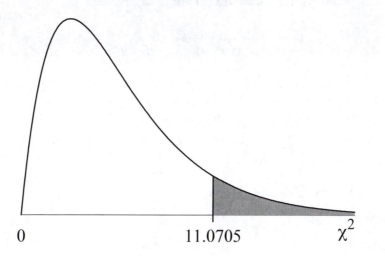

Figure 10.4: The χ^2 right-tail critical value corresponding to $\alpha = 0.05$ with 5 degrees of freedom.

The requirements for using the *chi-square* distribution to approximate probabilities in a multinomial distribution are similar to those for using the normal distribution to approximate the binomial distribution. With the binomial experiment we required np and nq, the *expected* number of successes and failures, respectively, to be at least 10. Ideally, we would use the same requirement here, but with the added categories (more than two) in the multinomial experiment, this requirement begins to become difficult to satisfy. We, therefore, ease that requirement here, and assume that the test statistic is approximately distributed according to the *chi-square* distribution as long as each expected frequency, E, is *at least five*.

The last comment that needs to be made is that we will not have a table for looking up *p-values* for the *chi-square* distribution. Chi-square *p*-values are best left to computers.

10.1.1   The Goodness of Fit Test

Now that we are able to use Table A4, we may outline the procedure for the goodness of fit test.

Procedure 10.1.1 Goodness of Fit Tests

  Step I. **Gather Sample Data**
    In this test, we will have k categories, or outcomes, in which each member of the sample may reside. For each category, we count the total observed frequencies, O. As usual, we will need α, the level of significance.

  Step II. **Identify Claims, Compute Expected Frequencies, and Verify that Requirements are Met**
    The null hypothesis may take various forms, the simplest being a list of hypothetical proportions for each category, as follows:

    $H_0 : p_1 = h_1, p_2 = h_2, \ldots, p_k = h_k$;
    H_1: *At least one of these proportions differs from its claimed value.*

    The k proportions must add up to one: $h_1 + h_2 + \ldots + h_k = 1$. To compute the expected frequencies, E, multiply each hypothetical proportion by the total sample size $n = \sum O$. The requirement for this test is that each expected frequency must be at least 5.

  Step III. **Compute the Test Statistic**

$$T.S. : \chi^2 = \sum \frac{(O - E)^2}{E}.$$

  Step IV. **Look Up the Right Tail Critical Value (C.V.) on Table A4**
    Because the k proportions add up to one, the first $k - 1$ are free, the last proportion must be a value that allows this sum to equal one. For this reason, there are $k - 1$ degrees of freedom for this test. Remember that the goodness of fit tests are always right-tail tests.

Step V. **Make a Conclusion**

As always, we reject the null hypothesis if the test statistic lies in the rejection region. This will always happen when the test statistic is greater than the critical value.

Example 10.1.2 Below are sample data which show observed numbers of births for each day of the week from a random sample of 59,247 births.

Day	Sunday	Monday	Tuesday	Wednesday	Thursday	Friday	Saturday	Total
Births	7,701	7,527	8,825	8,859	9,043	9,208	8,084	59,247

Test the claim at $\alpha = 0.01$ that the proportions are the same for each day of the week. (Thanks to Jeffrey S. Simonoff, Professor of Statistics, Leonard N. Stern School of Business, New York University, for providing these values on the StatLib Archive, and permitting their use here. These values come from his book *Analyzing Categorical Data*, published by Springer-Verlag, July 2003.)

Step I. **Gather Sample Data**

The observed frequencies are in the table above. For this example, we have chosen $\alpha = 0.01$.

Step II. **Identify Claims, Compute Expected Frequencies, and Verify that Requirements are Met**

The claim is that the proportion of babies born on a given day of the week is the same as every other day of the week. So for 7 days, with equal proportions, the claims are:

H_0: $p_1 = 1/7$, $p_2 = 1/7$, $p_3 = 1/7$, $p_4 = 1/7$, $p_5 = 1/7$, $p_6 = 1/7$, $p_7 = 1/7$;
H_1: *At least one of these proportions differs from what is claimed.*

From these we compute expected frequencies. We do this by multiplying the proportions by the sample size. In each case, this is $E = 59,247 \cdot (1/7) = 8463.86$. Clearly the requirement that each expected frequency is at least 5 is met.

Step III. **Compute the Test Statistic**

The test statistic is

$$T.S. : \chi^2 = \sum \frac{(O - E)^2}{E}.$$

We compute the terms $\frac{(O-E)^2}{E}$ in the table below, and add them.

Day	O	E	$\frac{(O-E)^2}{E}$
Sunday	7701	8463.86	68.7572
Monday	7527	8463.86	103.6999
Tuesday	8825	8463.86	15.4095
Wednesday	8859	8463.86	18.4476
Thursday	9043	8463.86	39.6281
Friday	9208	8463.86	65.4251
Saturday	8084	8463.86	17.0480
Totals:	59247	59247.00	328.4154

Our test statistic is

$$T.S. : \chi^2 = \sum \frac{(O-E)^2}{E}$$
$$= 328.4154.$$

Step IV. **Look Up the Right Tail Critical Value (C.V.) on Table A4**
In this example, the number of categories is $k = 7$. We look up our test statistic with $\alpha = 0.01$ and $df = k - 1 = 7 - 1 = 6$.

	Area to Right of Critical Value		
df	0.10	0.05	**0.01**
5	9.2364	11.0705	15.0863
6	10.6446	12.5916	**16.8119**
7	12.0170	14.0671	18.4753

The critical value is $C.V. : \chi^2 = 16.8119$, and is pictured in Figure 10.5.

Clearly the test statistic lies deep within the rejection region, so we will be rejecting the null hypothesis.

As far as *p-values* are concerned, we have no table for looking up *chi-square* *p*-values. What we can say is that with 6 degrees of freedom, and $\alpha = 0.01$, we rejected the null hypothesis, so we know that the *p*-value is less than 0.01. For exact values, computer software is required. In this case, the *p-value* is extraordinarily small and can only be easily written in scientific notation: *p-value* $= 6.62 \times 10^{-68}$.

Step V. **Make a Conclusion**
We have rejected the null hypothesis in favor of the alternative. The original claim was the null hypothesis, and so we have rejected the original claim.

The data do not support the claim that births occur in equal proportions on the seven days of the week.

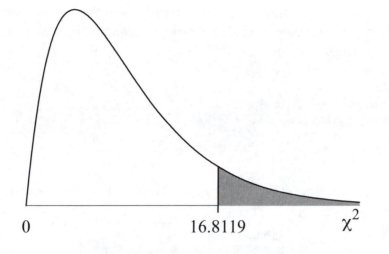

Figure 10.5: The critical value for Example 10.1.2.

It is interesting that, now that we know that the proportions are statistically significantly different, that the days with the lowest proportions are Saturday, Sunday, and Monday. Why this is true is beyond the ability of statistics to answer, but we could wonder if the more relaxing nature of the weekend could be the reason.

Example 10.1.3 The multinomial experiment is a generalization of the binomial experiment. In Chapter 7 we learned a method for testing a claim about a single population proportion, using ideas learned from binomial experiments. In this example, we will repeat a problem from Chapter 7 to show that our goodness of fit method gives an equivalent result.

We have stated on several occasions that the proportion of people in the world who are female is 49.8%. Suppose you survey 1000 randomly selected individuals, and find 528 of these are females. We conclude from this that the sample contained 472 males. Test the claim that the proportion of people who are female in the world is equal 49.8%, and that the proportion for men is 50.2%, using a 1% level of significance.

Step I. **Gather Sample Data**
The observed frequencies are 528 for the females, and 472 for the males. For this latest example we have chosen $\alpha = 0.01$.

Step II. **Identify Claims, Compute Expected Frequencies, and Verify that Requirements are Met**
The claim is that the proportion of people who are female is 0.498, and for males it is 0.502.

H_0: $p_1 = 0.498$, $p_2 = 0.502$;
H_1: *At least one of these proportions differs from what is claimed.*

From these we compute expected frequencies. We do this by multiplying the proportions by the sample size. For the females, this is $E = 1000 \cdot 0.498 = 498$. For the males, this is $E = 1000 \cdot 0.502 = 502$. Both expected frequencies are at least 5, so we may proceed.

Step III. Compute the Test Statistic

The test statistic is

$$T.S. : \chi^2 = \sum \frac{(O - E)^2}{E}.$$

We compute the terms $\frac{(O-E)^2}{E}$ in the table below, and sum them.

Gender	O	E	$\frac{(O-E)^2}{E}$
Female	528	498	1.80723
Male	472	502	1.79283
Totals:	1000	1000	3.60006

Our test statistic is therefore

$$T.S. : \chi^2 = \sum \frac{(O - E)^2}{E}$$
$$= 3.6001.$$

Step IV. Look Up the Right Tail Critical Value (C.V.) on Table A4

In this example, the number of categories is $k = 2$. We look up our test statistic with $\alpha = 0.01$ and

$$df = k - 1 = 2 - 1 = 1.$$

Area to Right of Critical Value			
df	0.10	0.05	**0.01**
1	2.7055	3.8415	**6.6349**

The critical value is $C.V. : \chi^2 = 6.6349$, as depicted in Figure 10.6.

The test statistic is $\chi^2 = 3.6001$, which does not lie in the rejection region, we fail to reject the null hypothesis.

For *p-values*, in looking at the table above, it is evident that with $\alpha = 0.10$, the critical value would have been $\chi^2 = 2.7055$, and in this case we would have rejected the null hypothesis, but with $\alpha = 0.05$, we would not have rejected the null hypothesis. With a reject with $\alpha = 0.10$, and a fail to reject with $\alpha = 0.05$, we must conclude that $0.05 < p\text{-}value < 0.10$. Using computer software we are able to determine $p\text{-}value = 0.0578$.

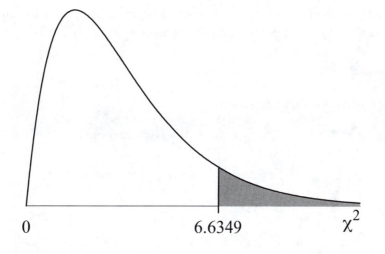

Figure 10.6: The critical value for Example 10.1.3.

Step V. **Make a Conclusion**

We have failed to reject the null hypothesis that 49.8% of the world's population is female and 50.2% is male. This was our original claim, so we are failing to reject that original claim.

The data do not give sufficient evidence to warrant rejection of the claim that 49.8% of the world's population is female and 50.2% is male.

10.1.2   The Goodness of Fit with One Degree of Freedom

The best motivation for the test statistic

$$T.S. : \chi^2 = \sum \frac{(O - E)^2}{E}$$

is to note that this statistic, when there is one degree of freedom, is equivalent to the test statistic for testing a claim regarding one population proportion, covered in Chapter 7,

$$T.S. : z = \frac{\hat{p} - p}{\sqrt{\frac{pq}{n}}}.$$

If we repeat Example 10.1.2 using the null hypothesis that the proportion of women in the world's population is equal to 49.8%, $H_0 : p = 0.498$ (with $q = 1 - p = 0.502$), and compute $\hat{p} = 528/1000 = 0.528$, recalling that the observed number of females in the sample was 528, we get the test statistic,

$$T.S. \;\; : \;\; z = \frac{\hat{p} - p}{\sqrt{\frac{pq}{n}}}$$

$$= \frac{0.528 - 0.498}{\sqrt{\frac{0.498 \cdot 0.502}{1000}}}$$

$$= 1.897.$$

The alternative hypothesis is $H_1 : p \neq 0.498$. The critical values for this two-tail test, using $\alpha = 0.01$, are $C.V. : z = \pm 2.576$, with *p-value* $= 2(1 - 0.9713) = 0.0574$.

Note the following facts.

- The square of the z test statistic is

$$TS^2 = z^2 = 1.897^2 = 3.60,$$

  which is the test statistic from the *chi-square* version of this test.

- The square of the z critical value is

$$CV^2 = z^2 = 2.576^2 = 6.635,$$

  which is the critical value from the *chi-square* version of this test.

- The *p*-values for the z and *chi-square* versions of this test are the same.

These demonstrate the point that with one degree of freedom the *chi-square* test statistic is equal to the square of the test statistic for a single population proportion. That is:

$$\chi^2 = \sum \frac{(O - E)^2}{E} = \left(\frac{\hat{p} - p}{\sqrt{pq/n}} \right)^2 = z^2.$$

This fact applies to all *chi-square* tests with one degree of freedom – the squares of the z test statistic and critical value are equal to the corresponding *chi-square* values, and their *p*-values are identical.

10.1.3 Proof that $\chi^2 = z^2$ for One Degree of Freedom

We now *prove* that the test statistics, χ^2 and z^2, are equal when the statistic has one degree of freedom. Consider the test for a single population proportion p, with $q = 1 - p$. Note that p is the probability of success, $E = \mu = np$ is the expected number of successes, and $O = x$ is the observed number of success. For failures, q is the probability of failure, and the expected number of failures is $E = nq$. The observed number of failures would be $n - x$.

$$\chi^2 = \sum \frac{(O-E)^2}{E} \qquad\qquad \text{This is the } \chi^2 \text{ T.S.}$$

$$= \frac{(O-E)^2}{E}\Big|_{successes} + \frac{(O-E)^2}{E}\Big|_{failures} \qquad\qquad \text{The sum has 2 terms.}$$

$$= \frac{(x-np)^2}{np} + \frac{(n-x-nq)^2}{nq} \qquad\qquad \text{Substitute}$$

$$= \frac{q(x-np)^2}{npq} + \frac{p(n-nq-x)^2}{npq} \qquad\qquad \text{Common denominator}$$

$$= \frac{q(x-np)^2 + p(n(1-q)-x)^2}{npq} \qquad\qquad \text{Add}$$

$$= \frac{q(x-np)^2 + p(np-x)^2}{npq} \qquad\qquad \text{Substitute } p = 1 - q.$$

$$= \frac{(q+p)(x-np)^2}{npq} \qquad\qquad \text{Factor}$$

$$= \frac{1(x-np)^2}{npq} \qquad\qquad q + p = 1$$

$$= \left(\frac{x-np}{\sqrt{npq}}\right)^2 \qquad\qquad \text{Simplify}$$

$$= \left(\frac{(x-np)\frac{1}{n}}{(\sqrt{npq})\frac{1}{n}}\right)^2 \qquad\qquad \text{Multiply by } \frac{1}{n}.$$

$$= \left(\frac{\frac{x}{n}-p}{\frac{\sqrt{npq}}{n}}\right)^2 \qquad\qquad \text{Simplify}$$

$$= \left(\frac{\widehat{p}-p}{\sqrt{\frac{pq}{n}}}\right)^2 \qquad\qquad \text{Simplify}$$

$$= z^2 \qquad\qquad \text{This is the } z^2 \text{ T.S.}$$

10.1.4 Homework

Test the following claims.

1. Assume that the observations in a given population of categorical data fall into five possible categories, with sample data giving observed frequencies of 13, 25, 11, 18, and 19, respectively. Test the claim that the proportions of frequencies for the entire population are the same for each category at 5% significance.

2. Assume that the observations in a given population of categorical data fall into three possible categories, with sample data giving observed frequencies of 140, 109, and 102, respectively. Test the claim that the proportions of frequencies for the entire population are the same for each category at 5% significance.

3. Between Spring 2001 and Spring 2004, the Mathematics Department at Mt. San Antonio College offered Math 72, which was an Intermediate Algebra class with less time spent on review. Over this period of time, the enrollments fluctuated somewhat, and when in Spring 2004 the enrollment was only 41 people, the department decided to stop offering the course. Test the claim, at 5% significance, that the differences seen in the enrollment counts are only due to sampling variability, and that the overall proportions were not significantly different.

Math 72 Enrollment by Semester	
Spring 2004	41
Fall 2003	75
Spring 2003	159
Fall 2002	67
Spring 2002	112
Fall 2001	37
Spring 2001	39

4. In March 2006, the author had reason to believe that his two children, Ryan and Gary, possibly aided by his wife, Alicia, were eating his Skittles candy. Skittles come in red, orange, yellow, green, and purple. Alicia's favorite color is purple, Gary's is orange, and Ryan's is red. The author has no color preference. In order to attempt to prove the guilt of his children and wife, the author counted the skittles colors.

Color	Frequency
Red	25
Orange	51
Yellow	40
Green	50
Purple	30

Upon investigation, the author learned that there is a Skittles hotline (this is actually true), and according to this hotline, Skittles colors are produced in equal proportions

(20% for each color) prior to being mixed together and then bagged. Being a statistics professor, the author understands that certain variability in the frequencies is to be expected, and that it is highly unlikely for each color to be equally represented in his bowl of Skittles. The real question is whether the deviations present are large enough to determine guilt.

Test the claim that the proportions of Skittles in the author's Skittles bowl are significantly different from the advertised proportions of 20% each. Use a 1% level of significance.

5. For the collection of 1532 gain ratio measurements for schizophrenic patients (*http://math.mtsac.edu/statistics/data*) encountered previously, we wish to use a goodness of fit test to determine whether it is plausible that these data are taken from a normal population. The way which we will do this is to recall the empirical rule, which describes the proportions of a normal population which lie within given ranges, determined by multiples of the standard deviation from the mean. The diagram in Figure 10.7 illustrates these proportions. Keeping these proportions in mind,

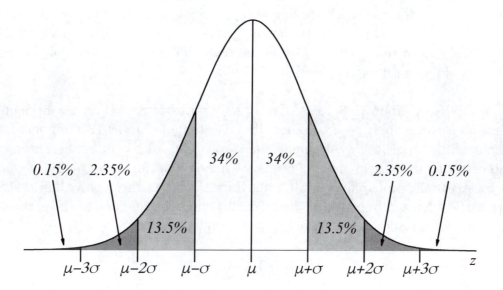

Figure 10.7: The distribution of values as indicated by the Empirical Rule.

and knowing that the sample has 52 values, we use the proportions to construct the expected frequencies in the table below.

Range in Standard Deviations	Lower Limit	Upper Limit	O	E	Proportion
More than 2 std. dev's below μ	$-\infty$	0.604	29	38.30	0.025
From 2 to 1 std. dev's below μ	0.605	0.722	198	206.82	0.135
From 1 to 0 std. dev's below μ	0.723	0.840	543	520.88	0.340
From 0 to 1 std. dev's above μ	0.841	0.957	564	520.88	0.340
From 1 to 2 std. dev's above μ	0.958	1.075	153	206.82	0.135
More than 2 std dev's above μ	1.076	∞	45	38.30	0.025

Use a 0.5% significance level (0.005) to test the claim, using the observed and expected frequencies in the table above, that these values are sampled from a normal population.

6. In the article "Is There a Season for Homicide?" (Criminology, 1988, pp. 287 – 296), the author gave the seasons in which 1361 homicides were committed. These values are summarized below.

Season	Frequency
Winter	328
Spring	334
Summer	372
Fall	327

Test the claim at $\alpha = 0.05$ that the proportions of homicides in the 4 seasons are all equal (to 25%).

7. Mathematician John Kerrich tossed a coin 10,000 times while interned in a prison camp in Denmark during World War I. At various stages of the experiment, the relative frequency would climb or fall below the theoretical probability of 0.5, but as the number of tosses increased, the relative frequency tended to vary less and stay near 0.5, or 50 percent. (Only in a prison camp would a mathematician toss a coin 10,000 times and record the results!) Overall he observed 5067 heads out of these 10,000 tosses, yielding a sample proportion of 50.67%.

If we randomly group these tosses into 1250 groups of 8, and count the number of heads in each group, we can consider this collection of data as 1250 binomial experiments, with 8 trials each. Assuming that probability of a coin landing heads is 0.5, and with x equal to the number of heads, we may use the formula $P(x) =_n C_x \cdot p^x \cdot q^{n-x}$ to construct the probability distribution below. Next to these probabilities, we construct the expected frequencies, E, by multiplying the number of times the experiment was conducted by the probability of each value of x.

x	$P(x)$	E	O
0	0.004	5.0	4
1	0.031	38.8	34
2	0.109	136.3	124
3	0.219	273.8	289
4	0.273	341.3	327
5	0.219	273.8	287
6	0.109	136.3	127
7	0.031	38.8	51
8	0.004	5.0	7

The observed frequencies, O, in the table are the number of times that exactly x (respectively) heads occurred in the 8 coin-toss experiment. The only assumption made in computing these expected frequencies was that the coin was unbiased – that the probability of heads is 0.5. Test this assumption at 5% significance.

10.2 Contingency Tables: Testing For Independence

We turn to a slightly different, and rather interesting goodness of fit test. We will, again, use the test statistic

$$\chi^2 = \sum \frac{(O - E)^2}{E}$$

for this test, but the degrees of freedom will be different. Our goal is to use a two-way or *contingency table* to test whether two variables (measuring nominal level data) are dependent or independent.

As an example, let us return to the example of drinking levels for male and female college students, which we examined in Chapter 3. (Data are from "Relationship of Health Behaviors to Alcohol and Cigarette Use by College Students," in the *Journal of College Student Development*, 1992).

Drinking Level	Male	Female	Totals
None	140	186	326
Low	478	661	1139
Moderate	300	173	473
High	63	16	79
Totals	981	1036	2017

The data in the table above are our observed frequencies, O. The expected frequencies are yet to be determined.

We would like to determine whether, for college students, drinking levels are independent of gender. Our null hypothesis will be that they are independent, and our expected frequencies will be derived from this assumption. Our approach will be as follows.

Suppose that gender and drinking level really are independent. If this is true, the old multiplication rule for computing probabilities should hold true:

$$P(A \ \& \ B) = P(A) \cdot P(B).$$

We will assume all of the row categories are independent of the column categories. For example, being *male* should be independent of having a *low* drinking level. If this is true then

$$\begin{aligned}
P(Male \ \& \ Low) &= P(Male) \cdot P(Low) \\
&= \frac{981}{2017} \cdot \frac{1139}{2017} \\
&= 0.274651 \\
&= 27.47\%.
\end{aligned}$$

Since the probability of *Male & Low* is 27.47%, we would expect 27.47% of the sample to fall in this category. That is, our expected frequency is $E = 0.274651 \cdot 2017 = 553.97$

people. What we actually observed were $O = 478$ people. If drinking levels are truly independent of gender then we should have seen about 554 males in the low drinking category, but we got quite a bit fewer – only 478. Maybe these events are *dependent* after all! We will not know until we complete the test. For now, let us look a little closer at how we got our expected frequency.

To get the expected number of people in the *Male & Low* category, we multiplied the probability of *Male & Low* times the sample size. That is,

$$
\begin{aligned}
E &= 553.97 \\
&= 0.274651 \cdot 2017 \\
&= \frac{981}{2017} \cdot \frac{1139}{2017} \cdot 2017 \\
&= \frac{1139 \cdot 981}{2017} \\
&= \frac{(row\ total) \cdot (column\ total)}{(grand\ total)}.
\end{aligned}
$$

This approach will allow us to compute an expected frequency corresponding to any observed frequency in the table.

The *expected frequency*, E, corresponding to an observed frequency, O, is equal to

$$
E = \frac{(row\ total) \cdot (column\ total)}{(grand\ total)},
$$

where we total the row and column corresponding to the position of the observed frequency, O.

As was mentioned, when this test was originally invented by Karl Pearson, there was some confusion about the degrees of freedom for the test. It was Ronald A. Fisher who pointed this out, and was ignored about this for years until it was discovered that he was correct. There are fewer degrees of freedom than you might guess. Consider the first row of the table where the drinking level is *None*. Given that the total is 326, once we know the number of males is 140 in this category, the number of females is not free – this number must be $326 - 140 = 186$. The same could be said of each value in the second column. The entire second column is dependent on the values in the first. There are two columns, but only $2 - 1$ of these are free. Now consider the first column. Given that the first column totals to 981, and that the first three values are 140, 478, and 300, the last value is predetermined to be $981 - 140 - 478 - 300 = 63$. This is true of every value in the last row. The last row values are all determined by the rows above, so even though there are 4 rows, only $4 - 1$ of these are free. To count the total number of free observations, the degrees of freedom, we multiply the number of free rows times the number of free columns. In this case,

$$df = (4 - 1)(2 - 1) = 3 \cdot 1 = 3.$$

In general, the contingency table test statistic has

$$df = (rows - 1) \cdot (columns - 1)$$

degrees of freedom. Be careful not to count the *totals* row or column when computing the degrees of freedom! They do not contain categorical data – only summaries.

10.2.1 Contingency Tables

Now we have all of the information we need to develop a procedure for this test. It is as follows.

Procedure 10.2.1 Contingency Tables – A Chi-Square Test for Independence

Step I. **Gather Sample Data**
In this test, the data will be summarized in a table. We will need a level of significance, α, as well.

Step II. **Identify Claims, Compute E Values, & Verify Requirements**

H_0: *The row and column events are independent.*
H_1: *The row and column events are dependent.*

The expected frequencies are

$$E = \frac{(row\ total) \cdot (column\ total)}{(grand\ total)},$$

and each of these must be at least 5.

Step III. **Compute the Test Statistic**
The test statistic is:

$$T.S. : \chi^2 = \sum \frac{(O - E)^2}{E}.$$

Step IV. **Look Up the Critical Value on Table A4**
The degrees of freedom are

$$df = (rows - 1) \cdot (columns - 1).$$

Remember that the goodness of fit tests are always right-tail tests.

Step V. **Make a Conclusion**
As always, we reject the null hypothesis if the test statistic lies in the rejection region. This will always happen when the test statistic is greater than the critical value. If we reject the null hypothesis we will say that the data do not support the claim that the row and column events are independent, and they do support the claim that they are dependent. If we fail to reject the null hypothesis, we will say that we are unable to reject the claim that the row and column events are independent, and any appearance of dependence is due to sampling error.

Example 10.2.2 At last we are able to test whether drinking levels and gender are independent for college students. We will use a 1% significance level.

Step I. **Gather Sample Data**

The observed frequencies are below, with totals.

Drinking Level	Male	Female	Totals
None	140	186	326
Low	478	661	1139
Moderate	300	173	473
High	63	16	79
Totals	981	1036	2017

For this example, we have chosen $\alpha = 0.01$.

Step II. **Identify Claims, Compute E Values, & Verify Requirements**

H_0: Drinking levels for college students are independent of gender.
H_1: Drinking levels for college students are dependent on gender.

The expected frequencies are

$$E = \frac{(row\ total) \cdot (column\ total)}{(grand\ total)}.$$

Let us show the work for the *Males* in the *None* drinking category. In this case the row total is 326, and the column total is 981. The grand total is 2017, so the expected frequency is:

$$E = \frac{(row\ total) \cdot (column\ total)}{(grand\ total)} = \frac{326 \cdot 981}{2017} = 158.56$$

The rest of the expected frequencies are in the table below.

Drinking Level	Male	Female
None	158.56	167.44
Low	553.97	585.03
Moderate	230.05	242.95
High	38.42	40.58

All of the expected frequencies are at least 5, as required.

Step III. **Compute the Test Statistic**

The observed and expected frequencies are brought together so we can total the $(O - E)^2 / E$ values.

O	E	$\frac{(O-E)^2}{E}$
140	158.56	2.17251
478	553.97	10.41833
300	230.05	21.26930
63	38.42	15.72557
186	167.44	2.05730
661	585.03	9.86521
173	242.95	20.13996
16	40.58	14.88853
	Total:	96.53669

The test statistic is

$$T.S. : \chi^2 = \sum \frac{(O-E)^2}{E} = 96.54.$$

Step IV. **Look Up the Critical Value on Table A4**
The degrees of freedom are

$$df = (rows - 1) \cdot (columns - 1) = (4-1)(2-1) = 3.$$

This is a right-tail test with $\alpha = 0.01$.

	Area Right of Critical Value	
df	0.05	**0.01**
2	5.9915	9.2103
3	7.8147	**11.3449**

In this case we find the critical value $C.V. : \chi^2 = 11.3449$. This critical value is pictured in Figure 10.8.

The test statistic, $T.S. : \chi^2 = 96.54$, is deep within the rejection region, so we will reject the null hypothesis. Using computer software, we learn that *p-value* $= 8.62 \times 10^{-21}$. Once again, this is extraordinarily small, emphasizing that the observed frequencies are significantly different from what was expected under the assumption that drinking habits for college students are independent of gender. Our data support the opposite claim.

Step V. **Make a Conclusion**
We have rejected the null hypothesis which claimed that drinking habits are independent of gender. This was our original claim, thus we are rejecting the original claim.

The data do not support the claim that drinking habits are independent of gender.

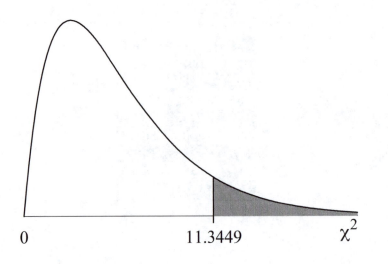

Figure 10.8: The critical value for Example 10.2.2.

Example 10.2.3 For a fictional business school, a lawsuit has been filed for gender discrimination. The claim is that women are being rejected for admission at a higher rate than men. In fact, for a sample of 511 admissions for a particular year, nearly 25% of qualified women candidates are denied admission, where only about 15% of qualified men are denied admission. The data are summarized in the table below. Is this discrepancy high enough to be considered statistically significant, or could it be due to nothing more than sampling error? Test the claim at 1% significance that admission into the college is independent of gender.

Gender	Admit	Deny
Male	310	55
Female	110	36

Step I. Gather Sample Data

The observed frequencies are below, with totals.

Gender	Admit	Deny	Totals
Male	310	55	365
Female	110	36	146
Totals	420	91	511

For this example, we have chosen $\alpha = 0.01$.

Step II. Identify Claims, Compute E Values, & Verify Requirements

H_0: *College admission is independent of gender;*
H_1: *College admission is dependent on gender.*

Let us show the work for the *Males* in the *Admit* category. In this case the row total is 365, and the column total is 420. The grand total is 511, so the expected frequency is:

$$E = \frac{(row\ total) \cdot (column\ total)}{(grand\ total)} = \frac{365 \cdot 420}{511} = 300.00$$

The rest of the expected frequencies are given in the table below.

Gender	Admit	Deny
Male	300.00	65.00
Female	120.00	26.00

Step III. **Compute the Test Statistic**

The observed and expected frequencies are brought together so we can total the $(O - E)^2/E$ values.

O	E	$\frac{(O-E)^2}{E}$
310	300	0.3333
110	120	0.8333
55	65	1.5385
36	26	3.8463
Total:		6.5514

The test statistic is

$$T.S.: \chi^2 = \sum \frac{(O - E)^2}{E} = 6.55.$$

Step IV. **Look Up the Critical Value on Table A4**

The degrees of freedom are

$$\begin{aligned}
df &= (rows - 1) \cdot (columns - 1) \\
&= (2 - 1)(2 - 1) \\
&= 1.
\end{aligned}$$

This is a right-tail test with $\alpha = 0.01$.

	Area Right of Critical Value		
df	0.10	0.05	**0.01**
1	2.7055	3.8415	**6.6349**

This time we find the critical value $C.V. : \chi^2 = 6.6349$, as pictured in Figure 10.9.

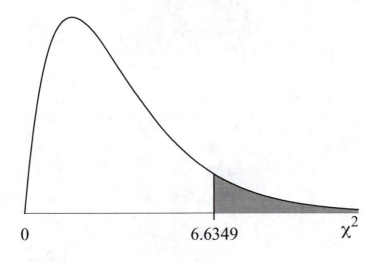

Figure 10.9: The critical value for Example 10.2.3.

The test statistic, $T.S. : \chi^2 = 6.5514$, is not in the rejection region, so we fail to reject the null hypothesis. Using computer software, we learn that *p-value* = 0.0105. This is small, emphasizing that the observed frequencies are significantly different from what was expected under the null hypothesis but not significant enough at the required level of $\alpha = 0.01$.

Step V. **Make a Conclusion**

We have failed to reject the null hypothesis that admission into this college is independent of gender. This was the original claim, so we are failing to reject the original claim.

The data do not give sufficient evidence to warrant rejection of the claim that admission into this college is independent of gender. Still the differences between the observed admission and denial frequencies and those that were expected under the assumption of independence were significantly different, as indicated by the p-value of 0.0105.

This concludes our discussion on the goodness of fit tests. We will conclude this chapter with Section 10.3, which gives a brief overview of the history of the *chi-square* distribution, and a discussion on the nature of the distribution of sample variances.

10.2.2   Homework

Perform the following hypothesis tests.

1. In the article, "Attitudes About Marijuana and Political Views" in Psychological Reports, 1973, pp. 1051 – 1054, the following frequencies were reported.

Political Views	Never Smoke	Rarely Smoke	Frequently Smoke	Totals
Liberal	479	173	119	771
Conservative	214	47	15	276
Other	172	45	85	302
Totals	865	265	219	1349

Altogether, 1,349 people were surveyed. Test the claim that marijuana use is independent of political views at the 1% significance level.

2. The table below is based on "Ignoring a covariate: An example of Simpson's Paradox" by Appleton, D.R. French, J.M. and Vanderpump, M.P (1996, American Statistician, 50, 340-341). In 1972-1994 a one-in-six survey of the electoral roll, largely concerned with thyroid disease and heart disease was carried out in Wichkham, a mixed urban and rural district near Newcastle upon Tyne, in the UK. Twenty years later, a follow-up study was conducted to see which study members were still alive.

Here are the results for females aged 65 to 74, assuming that 7425 women were selected. Test the claim at 5% significance that being alive in this age group is independent of smoking.

Status	Dead	Alive	Totals
Smokers	1305	315	1620
Non-smokers	4545	1260	5805
Totals	5850	1575	7425

3. A random sample of 291 male and female college student grades for a math class were gathered, and these are summarized in the table below. The letter grade "D" represents any substandard grade (D, F, or W). Test the claim at 1% significance that letter grades in this math class are independent of student gender.

Grade	Female	Male
A	9	9
B	57	54
C	57	63
D	12	30

4. The following data describe the state of grief of 66 mothers who had suffered a neonatal death. The table relates this to the level of support given to these women. Test the claim at 5% significance that the mothers' grief state is independent of the support they receive.

Level	Good	Fair	Poor
I	34	18	16
II	12	10	2
III	8	14	18

5. The following data represent the numbers of men and women who died or survived in the infamous Donner Party which was stranded in the California Sierras near Lake Tahoe in the winter of 1847. Test the claim that gender is independent of survival at 1% significance.

Gender	Survived	Died
Male	23	29
Female	25	9

6. The following data represent the numbers of people who died and survived in the infamous Donner Party which was stranded in the California Sierras near Lake Tahoe in the winter of 1847 for various age groups. Test the claim at 5% significance that survival is independent of age category.

Age	Survived	Died
0 – 9	18	11
10 – 19	16	3
20 – 29	5	9
30 and older	9	15

7. The following data were gathered by Mt. San Antonio College honors student Helentina Pang, regarding genders of sample members and whether respective members have ever been in a car accident.

Gender	Accident	No Accident
Female	15	15
Male	20	10

Using the given sample data, test the claim at 5% significance that car accidents are independent of gender.

8. The following data give game rating preferences by gender for randomly selected college students. These data were gathered by Sean Meshkin, honors student at Mt. San Antonio College.

Gender	Rated-E	Rated-T	Rated-M
Male	7	12	15
Female	5	17	5

Test the claim at 10% significance that game rating preferences are independent of gender.

9. Is the grade assigned for a college student in a mathematics course at all dependent on the number of units the student is attempting? Test the claim at 5% significance that they are dependent, using the following data which were gathered by Mt. San Antonio College honors student Lily Bai.

Units Attempted	A	B	C or Lower
0 – 12 Units	9	21	9
12 – 13 Units	45	63	27
> 13 Units	43	22	9

10.3   A Chi-Square Test for Population Variance

As previously mentioned, the first person to discover the *chi-square* distribution was Friedrich Helmert in 1875, when he discovered that inferences concerning sample variances can be made using the *chi-square* distribution.

So far, very little has been said about the distribution of sample variances or standard deviations. To rectify this, we will summarize the facts relevant to this here.

First of all, we must point out that the sample variance, s^2, is an unbiased estimator of the population variance, σ^2. This means that the mean, or expected value, of all sample variances is exactly equal to the population variance. This unbiased nature is only assured when the sample variance has an $n-1$ (not n) in the denominator. The proof of this is included at the end of this section. In addition, we add that the sample variance is the best point estimate of population variance.

Unfortunately, we cannot say the same about standard deviation. Taking the square root of sample variance to get sample standard deviation makes the sample standard deviation a biased estimator of the population standard deviation.

When x is taken from a normal population the statistic

$$\chi^2 = \frac{(n-1)\ s^2}{\sigma^2}$$

is distributed according to the *chi-square* distribution with $n-1$ degrees of freedom. Using this, we can perform left, right, and two-tailed tests and confidence intervals regarding the population variance or standard deviation.

The difficulty with this *chi-square* application is that it is very sensitive to the normality requirement. Unlike most of the tests covered in this text, this procedure is not robust – that is, it is unreliable when populations are not normal. The ramification of this is that rejection of a null hypothesis may happen when it is not true, but it may also happen because the condition of normality was not met. Such non-robustness has made this procedure rather unpopular.

Table A4 – Chi-Square Distribution Critical Values

df			Area to the Right of Critical Value			
	0.99	0.95	0.90	0.10	0.05	0.01
1	0.0002	0.0039	0.0158	2.7055	3.8415	6.6349
2	0.0201	0.1026	0.2107	4.6052	5.9915	9.2103
3	0.1148	0.3518	0.5844	6.2514	7.8147	11.3449
4	0.2971	0.7107	1.0636	7.7794	9.4877	13.2767
5	0.5543	1.1455	1.6103	9.2364	11.0705	15.0863
6	0.8721	1.6354	2.2041	10.6446	12.5916	16.8119
7	1.2390	2.1673	2.8331	12.0170	14.0671	18.4753
8	1.6465	2.7326	3.4895	13.3616	15.5073	20.0902
9	2.0879	3.3251	4.1682	14.6837	16.9190	21.6660
10	2.5582	3.9403	4.8652	15.9872	18.3070	23.2093
11	3.0535	4.5748	5.5778	17.2750	19.6751	24.7250
12	3.5706	5.2260	6.3038	18.5493	21.0261	26.2170
13	4.1069	5.8919	7.0415	19.8119	22.3620	27.6882
14	4.6604	6.5706	7.7895	21.0641	23.6848	29.1412
15	5.2293	7.2609	8.5468	22.3071	24.9958	30.5779
16	5.8122	7.9616	9.3122	23.5418	26.2962	31.9999
17	6.4078	8.6718	10.0852	24.7690	27.5871	33.4087
18	7.0149	9.3905	10.8649	25.9894	28.8693	34.8053
19	7.6327	10.1170	11.6509	27.2036	30.1435	36.1909
20	8.2604	10.8508	12.4426	28.4120	31.4104	37.5662
21	8.8972	11.5913	13.2396	29.6151	32.6706	38.9322
22	9.5425	12.3380	14.0415	30.8133	33.9244	40.2894
23	10.1957	13.0905	14.8480	32.0069	35.1725	41.6384
24	10.8564	13.8484	15.6587	33.1962	36.4150	42.9798
25	11.5240	14.6114	16.4734	34.3816	37.6525	44.3141
26	12.1981	15.3792	17.2919	35.5632	38.8851	45.6417
27	12.8785	16.1514	18.1139	36.7412	40.1133	46.9629
28	13.5647	16.9279	18.9392	37.9159	41.3371	48.2782
29	14.2565	17.7084	19.7677	39.0875	42.5570	49.5879
30	14.9535	18.4927	20.5992	40.2560	43.7730	50.8922
31	15.6555	19.2806	21.4336	41.4217	44.9853	52.1914
32	16.3622	20.0719	22.2706	42.5847	46.1943	53.4858
33	17.0735	20.8665	23.1102	43.7452	47.3999	54.7755
34	17.7891	21.6643	23.9523	44.9032	48.6024	56.0609
35	18.5089	22.4650	24.7967	46.0588	49.8018	57.3421
36	19.2327	23.2686	25.6433	47.2122	50.9985	58.6192
37	19.9602	24.0749	26.4921	48.3634	52.1923	59.8925
38	20.6914	24.8839	27.3430	49.5126	53.3835	61.1621
39	21.4262	25.6954	28.1958	50.6598	54.5722	62.4281
40	22.1643	26.5093	29.0505	51.8051	55.7585	63.6907
41	22.9056	27.3256	29.9071	52.9485	56.9424	64.9501
42	23.6501	28.1440	30.7654	54.0902	58.1240	66.2062
43	24.3976	28.9647	31.6255	55.2302	59.3035	67.4593
44	25.1480	29.7875	32.4871	56.3685	60.4809	68.7095
45	25.9013	30.6123	33.3504	57.5053	61.6562	69.9568
46	26.6572	31.4390	34.2152	58.6405	62.8296	71.2014
47	27.4158	32.2676	35.0814	59.7743	64.0011	72.4433
48	28.1770	33.0981	35.9491	60.9066	65.1708	73.6826
49	28.9406	33.9303	36.8182	62.0375	66.3386	74.9195
50	29.7067	34.7643	37.6886	63.1671	67.5048	76.1539
55	33.5705	38.9580	42.0596	68.7962	73.3115	82.2921
60	37.4849	43.1880	46.4589	74.3970	79.0819	88.3794
65	41.4436	47.4496	50.8829	79.9730	84.8206	94.4221
70	45.4417	51.7393	55.3289	85.5270	90.5312	100.4252
75	49.4750	56.0541	59.7946	91.0615	96.2167	106.3929
80	53.5401	60.3915	64.2778	96.5782	101.8795	112.3288
85	57.6339	64.7494	68.7772	102.0789	107.5217	118.2357
90	61.7541	69.1260	73.2911	107.5650	113.1453	124.1163
95	65.8984	73.5198	77.8184	113.0377	118.7516	129.9727
100	70.0649	77.9295	82.3581	118.4980	124.3421	135.8067
150	112.6676	122.6918	128.2751	172.5812	179.5806	193.2077
200	156.4320	168.2786	174.8353	226.0210	233.9943	249.4451
250	200.9386	214.3916	221.8059	279.0504	287.8815	304.9396
300	245.9725	260.8781	269.0679	331.7885	341.3951	359.9064

Chapter 11

An Introduction to Analysis of Variance

Probably the most prominent statistician of the twentieth century was Ronald A. Fisher. Fisher's book *Statistical Methods for Research Workers* is perhaps the most influential statistics book of all time. This book, for the first time, brought together most of the inference methods discussed in this text. It streamlined and corrected procedures of inference relating to z and t distribution tests, settled the issues with degrees of freedom for t and *chi-square* tests, introduced inferences regarding Pearson's correlation coefficient, and introduced Fisher's own methods for Analysis of Variance, which we introduce in this chapter. Analysis of variance uses the F test statistic and its distribution, where the F is used in honor of Fisher.

Fisher was a very controversial man. He challenged the most prominent statistician of the early 20th century, Karl Pearson, in areas that Pearson pioneered, stating that Pearson was mistaken in his approach to degrees of freedom in his *chi-square* tests (which we covered in chapter ten). Pearson did not respond well to Fisher's criticisms, and the two were mutually unfriendly for years.

Eventually, Fisher was shown to be correct in this matter, but Pearson apparently ignored this.

William *Student* Gosset (the developer of Student's t-distribution) was a friend of both Pearson and Fisher – quite a feat considering how confrontational and egotistical they both reputedly were. Fisher helped Gosset in his development of the Student's t-distribution, and was able to provide a theory to generalize the distribution with formulas (most of the values in Gosset's original tables were developed empirically through carefully examining distributions of data).

While Gosset developed Student's t critical values, it was Fisher who determined that the statistic

$$t = \frac{\bar{x} - \mu}{s/\sqrt{n}}$$

was distributed according to Student's distribution, and Fisher who settled the matter of degrees of freedom with regard to this statistic as well.

Figure 11.1: Sir Ronald Fisher, on his desk calculator at Whittinghome Lodge in 1952. *Copyright photograph by A. Barrington Brown; reproduced with the permission of the Fisher Memorial Trust.*

Fisher was also the man who effectively accused the great genetics scientist, Gregor Mendel, of data tampering in 1936 when he challenged the authenticity of Mendel's data in his famous pea experiments. This mystery was never resolved, but again underscores the tenacity of Fisher and his talent for confrontation and controversy.

Fisher is responsible for developing much important statistical theory that is beyond the scope of this text. He was truly a statistical genius – solving many problems that are vital to modern research as we know it today.

11.1   One Way Analysis of Variance

In its simplest form, the F test is a comparison of two variances or standard deviations. The comparison is made by dividing the variances of two samples.

$$T.S. : F = \frac{s_1^2}{s_2^2}$$

If this proportion is much greater than one, then the variance represented by the statistic in the numerator is significantly greater than the variance represented by the statistic in the denominator. If the statistic is less than one, and close to zero then the variance represented by the numerator is significantly smaller than the variance represented by the denominator. Generally speaking, the F test can be right, left, or two tailed, but our main application, analysis of variance, is always a right-tail test.

Of course, with the two variances in this test statistic, we have two degree of freedom measurements, which we will refer to as df_1, the *numerator* degrees of freedom (referring to the sample variance in the numerator of the test statistic) and df_2, the denominator degrees of freedom (referring to the denominator of the test statistic).

In appearance, the distribution of the statistic $F = s_1^2/s_2^2$ looks like the *chi-square* distribution, with the exception that the critical values get smaller, not larger, as the degrees of freedom increase. The critical values are always larger than one, but they approach one as the degrees of freedom become large. The mode of the distribution is always less than one, and the mean is always greater than one, but both of these approach one as degrees of freedom increase. The distribution starts at zero and includes positive F values only, since the test statistic, being the ratio of two positive numbers, can only be positive. Critical F distribution values are tabulated in Table A5. The density function's graph is plotted in Figure 11.2, with a right-tail critical value.

Analysis of Variance, abbreviated ANOVA, uses a more complicated test statistic than the one given above, but it always uses two measures of variation in a ratio, and is *always a right tail test*. The simplest form of ANOVA is called *one-way* or *single factor* ANOVA. In one-way ANOVA, we will develop a way for comparing means from 3 or more populations, or treatments. In ANOVA, the characteristic by which the populations or treatments differ is called the *factor* which we are studying, and the different populations

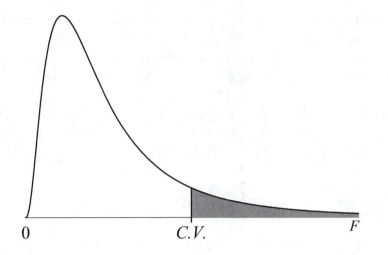

Figure 11.2: The F distribution with a right-tail critical value.

or treatments are the *levels* of the factor. In one-way ANOVA, we only consider a single factor, but with multiple levels.

The assumptions of one-way ANOVA are, first, that the populations which are being studied are all normal, and that they have equal variances. Also, it is assumed that samples are chosen independently of one another, and that observations within any sample are independent. This procedure has been shown to be *robust*, meaning that results tend to be reliable, even when the assumptions for the test are slightly violated.

Suppose that we have k samples, drawn randomly from k populations. Each of these samples have sample means, $\bar{x}_1, \bar{x}_2, \ldots , \bar{x}_k$, and variances, $s_1^2, s_2^2, \ldots ,s_k^2$. Initially, we will assume that the samples sizes are all the same. We will use n to denote this common sample size. As mentioned, we are assuming that these populations all have a common variance, which we will denote by σ^2. Our F test will compare two different estimates of this common variance, one for the numerator and one for the denominator of the F test statistic.

For the denominator estimate of the common variance σ^2, we will simply pool (using a weighted mean) the sample variances, s_i^2 for all i from 1 to k. We will denote this pooled variance by s_p^2, and it is called the *variance within the samples* because it is the variance of the samples combined. Under the assumption that the sample sizes are all equal, it is unnecessary to use weights in this mean, so the pooled variance will simply be the sum of the sample variances, divided by k. When the samples are of a different size, the weights will be the degrees of freedom for each variance.

For the numerator estimate of the common variance, σ^2, we take a more advanced approach. Recalling that the standard deviation of all sample means taken from samples of size n is equal to $\sigma_{\bar{x}} = \sigma/\sqrt{n}$, we can solve for σ to get $\sigma = \sqrt{n} \cdot \sigma_{\bar{x}}$, so $\sigma^2 = n \cdot \sigma_{\bar{x}}^2$. Now, we do not know the value of $\sigma_{\bar{x}}^2$, but we could estimate it by computing the sample variance of the given sample means, which we will denote by $s_{\bar{x}}^2$. So, $\sigma^2 = n \cdot \sigma_{\bar{x}}^2 \approx n \cdot s_{\bar{x}}^2$.

This estimate of σ^2 is called the *variance between the samples*. It simply measures total variation using the variance of the sample means.

We can now compute our test statistic, dividing our two estimates of σ^2, as

$$F = \frac{n \cdot s_{\bar{x}}^2}{s_p^2} = \frac{variance\, between\, samples}{variance\, within\, samples}.$$

This is the simplest form of the test statistic. The degrees of freedom for the numerator are the number of deviations in the numerator minus one. Since there are k deviations in the numerator, the degrees of freedom for the numerator are $k - 1$. The denominator is slightly more complicated, with the mean of k variances, each from samples of size n, with $n - 1$ degrees of freedom each. In total, this is $k \cdot (n - 1)$ degrees of freedom for the denominator.

The null hypothesis for this test will be H_0: $\mu_1 = \mu_2 = \ldots = \mu_k$. When the null hypothesis is true, the sample means will usually be similar, so the variance between them (the numerator of the test statistic) will be close to zero, and thus the test statistic will be close to zero. Because of this, we should not reject the null hypothesis when the test statistic is close to zero. Conversely, when the null hypothesis is false, the sample means should be quite different, so the variance between them (in the numerator of the test statistic) will be large, and thus the test statistic will be large. Thus, we will reject the null hypothesis when the test statistic is large. This is why this test is always right-tailed.

The use of Table A5 is quite simple. The first two pages have 0.01 in the right tail, and the last two pages have 0.05 in the right tail. Look up the appropriate numerator and denominator degrees of freedom for your test, and the intersection of the corresponding row and column contains the right tail critical value.

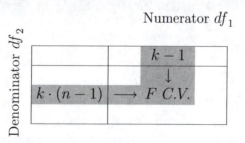

Once again, the only reasonable way to find a *p-value* for an F test is to use computer software.

We are now ready to define our procedure for one-way ANOVA with equal sample sizes.

Procedure 11.1.1 One-Way ANOVA with Equal Sample Sizes

Step I.   **Gather Sample Data and Verify Requirements**
This test requires that each population is normal, with a common population variance. The robustness of this test allows these conditions to be violated somewhat. Normal quantile plots may be used to test the plausibility of the normality assumption. For each sample, we will need a sample mean, $\bar{x}$, and sample variance, s^2. From these, we compute $s_{\bar{x}}^2$, the variance of the $\bar{x}$ values (the square of the standard deviation of the $\bar{x}$ values), and s_p^2, the mean of the sample variances. It is important to remember that s_p^2 is the mean of the *variances*, so if we are given standard deviations, we must *square* these values before averaging them. Identify the common sample size, n, and number of samples, k. A level of significance, α, must be chosen as well.

Step II.   **Identify Claims**
The null hypothesis is always the same for this test. H_0: $\mu_1 = \mu_2 = \ldots = \mu_k$
The alternative hypothesis simply says that at least one of these differs from the others.

Step III.   **Compute the Test Statistic**
The test statistic is T.S.: $F = \frac{n \cdot s_{\bar{x}}^2}{s_p^2}$. In this statistic, $s_{\bar{x}}^2$ is the variance of the sample means, and s_p^2 is the mean of the sample variances.

Step IV.   **Look Up the Critical Value on Table A5**
Use numerator degrees of freedom $df1 = k - 1$, and denominator degrees of freedom $df2 = k \cdot (n - 1)$. Here, k is the number of samples, and n is the common sample size.

Step V.   **Make a Conclusion**
Conclusions are to be made as before, always referring to the data as having led us to our conclusion.

Example 11.1.2 Do different types of mutual funds have better returns on investments? In the table below, five different types of mutual funds are summarized by their 5 year average investment performance. Assuming that these samples we drawn from normal populations with equal variances, test the claim that there is no difference in the mean 5 year investment performance for these five types of mutual funds, using a 1% level of significance.

Statistic	Balanced	Equity Income	Growth Income	Growth	Aggressive Growth
n	21	21	21	21	21
$\bar{x}$	14402.59	14160.24	15446.24	16411.18	18023
s^2	509898.76	1406514.69	3904565.69	4235215.78	6555676.38

Source: Data are based on an article which was © 1993 by Consumers Union of U.S., Inc.

Yonkers, NY 10703-1057. Used by Permission

Step I. **Gather Sample Data and Verify Requirements**

We have assumed that the samples are drawn from normal populations with equal variances, so we proceed to compute $s_{\bar{x}}^2$. The sample standard deviation of the sample means is $s_{\bar{x}} = 1{,}582.718356$, which we square to get $s_{\bar{x}}^2 = 2{,}504{,}997.396$. Next, the mean of the variances is $s_p^2 = 3{,}322{,}374.26$. The common sample size is $n = 21$, and the number of samples is $k = 5$. We have chosen $\alpha = 0.01$ for this test.

Step II. **Identify Claims**

The null hypothesis is: H_0: $\mu_1 = \mu_2 = \mu_3 = \mu_4 = \mu_5$.
The alternative hypothesis simply says that at least one of these differs from the others.

Step III. **Compute the Test Statistic**

The test statistic is

$$T.S. : F = \frac{n \cdot s_{\bar{x}}^2}{s_p^2} = \frac{21 \cdot 2{,}504{,}997.396}{3{,}322{,}374.26} = 15.834.$$

Step IV. **Look Up the Critical Value on Table A5**

For the numerator degrees of freedom, we have $df_1 = k - 1 = 5 - 1 = 4$, and denominator degrees of freedom $df_2 = k \cdot (n - 1) = 5 \cdot (21 - 1) = 100$. Remember that we have chosen $\alpha = 0.01$ for this test, so we will look at the first page of the table.

Numerator df_1

Denominator df_2	3	4	5
50	4.1993	3.7195	3.4077
100	3.9837	**3.5127**	3.2059
200	3.8810	3.4143	3.1100

We see that the critical value is $C.V. : F = 3.5217$, as depicted in Figure 11.3.

Our test statistic is T.S.: $F = 15.834$, and this lies in the rejection region, so we reject the null hypothesis. Using computer software for this test, we see that the *p-value* is $4.52 \cdot 10^{-10} = 0.000000000452$. Obviously this is very small, so the variation among the sample means is highly significant, indicating that the null hypothesis should be rejected.

Step V. **Make a Conclusion**

We have rejected the null hypothesis H_0: $\mu_1 = \mu_2 = \mu_3 = \mu_4 = \mu_5$, which means that the data support the alternative that at least one of these population means differs from the others.

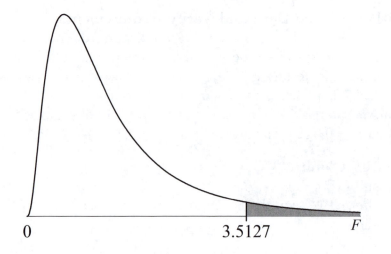

Figure 11.3: The critical value for Example 11.1.2.

The data do not support the claim that there is no difference in the mean 5 year investment performance for these five types of mutual funds. There is a significant amount of variation between the sample means as indicated by the p-value of 0.000000000452.

Example 11.1.3 Next, we have an example in computer engineering. Computer chips, or semiconductors, are made from silicon wafers. *Resistance* is a measurement of what amounts to friction in an electrical circuit. Measurements of bulk resistivity of silicon wafers were made at the National Institute for Standards in Technology (NIST) with 5 probing instruments on each of 5 days. The wafers were doped with phosphorous by neutron transmutation doping in order to have nominal resistivities of 200 ohm/cm. Each data point is the average of 6 measurements at the center of each wafer. Measurements were carried out with four-point DC probes according to ASTM Standard F84-93, "*Standard Method for Measuring Resistivity of Silicon Wafers with an In-Line Four-Point Probe.*" These data were provided by NIST. Assume the corresponding populations are normal with equal variances.

Day 1	Day 2	Day 3	Day 4	Day 5
196.3052	196.3042	196.1303	196.2795	196.2119
196.1240	196.3825	196.2005	196.1748	196.1051
196.1890	196.1669	196.2889	196.1494	196.1850
196.2569	196.3257	196.0343	196.1485	196.0052
196.3403	196.0422	196.1811	195.9885	196.2090

Test the claim at 5% significance that there is no difference in the mean resistivity for each of the five days.

Step I. **Gather Sample Data and Verify Requirements**
We have more work to do on this problem, since none of the statistics are provided for us. Thus, we must compute the mean and variance for each of the five days listed. Remember that sample variance is the square of sample standard deviation.

Statistic	Day 1	Day 2	Day 3	Day 4	Day 5
$\bar{x}$	196.2431	196.2443	196.1670	196.1481	196.1432
s	0.08747	0.13797	0.09372	0.10423	0.08845
s^2	0.00765	0.01904	0.00878	0.01086	0.00782

Next, we need the variance of the sample means. The standard deviation is $s_{\bar{x}} = 0.0506$, so the variance is $s_{\bar{x}}^2 = 0.0506^2 = 0.002560$. Also, we need the mean of the variances, s_p^2. This is just the average of the s^2 values,

$$s_p^2 = \frac{0.00765 + 0.01904 + 0.00878 + 0.01086 + 0.00782}{5} = 0.01083.$$

These numbers are quite small, so it is important that we are careful to not round too much.

On this problem, the number of samples is $k = 5$, and the common sample size is $n = 5$. We have assumed that the populations are normal, with equal variances. We have chosen $\alpha = 0.05$ for this test.

Step II. **Identify Claims**
The null hypothesis is H_0: $\mu_1 = \mu_2 = \mu_3 = \mu_4 = \mu_5$.
The alternative hypothesis simply says that at least one of these differs from the others.

Step III. **Compute the Test Statistic**
The test statistic is

$$T.S. : F = \frac{n \cdot s_{\bar{x}}^2}{s_p^2} = \frac{5 \cdot 0.002560}{0.01083} = 1.1819.$$

Step IV. **Look Up the Critical Value on Table A5**
For the numerator degrees of freedom, we have $df_1 = k - 1 = 5 - 1 = 4$, and denominator degrees of freedom $df_2 = k \cdot (n - 1) = 5 \cdot (5 - 1) = 20$. Remember that we have chosen $\alpha = 0.05$ for this test, so we will look at the third page of the table.

Numerator df_1

Denominator df_2	3	4	5
19	3.1274	2.8951	2.7401
20	3.0984	**2.8661**	2.7109
21	3.0725	2.8401	2.6848

We see that the critical value is $C.V. : F = 2.8661$. This critical value is pictured in Figure 11.4.

Our test statistic is $T.S. : F = 1.1819$, and this does not lie within the rejection region, so we fail to reject the null hypothesis. Using computer software, we are able to determine that the *p-value* is 0.3495. This is rather large, and clearly larger than α, so the variation among the sample means is not significant, indicating that the null hypothesis should not be rejected. It is possible that any differences between the sample means may be nothing more than sampling error.

Step V. Make a Conclusion

We have failed to reject the null hypothesis that the mean resistance is the same for each of the five days, and so our final conclusion follows.

Based on our sample data, we are unable to reject the claim that the mean resistivity is the same for each of the five days in which the silicon wafers we doped with phosphorous by neutron transmutation doping. This is indicated by a large p-value of 0.3495.

11.1.1   One-Way Analysis of Variance with Unequal Sample Sizes

Our test statistic $F = \frac{n \cdot s_{\bar{x}}^2}{s_p^2}$ looks rather simple, but with the common sample size, n, in the numerator, it is clear that this will not apply with the sample sizes are different. In order to rectify this problem, we must look closely at the computations required to compute $s_{\bar{x}}^2$ and s_p^2.

Before doing so, we must recall the formula for the weighted mean from chapter two:

$$Weighted\ mean = \frac{\sum w \cdot x}{\sum w}.$$

Originally, s_p^2 was used as an estimate for the common population variance σ^2. We still need this estimate, and will continue to average the s^2 values to estimate it, but now we will use the weighted mean in this average because s^2 values from the larger samples are more reliable. We will, therefore, use the degrees of freedom for each variance as the weight.

For notation, we will now use N as the sum of the different sample sizes. To distinguish the different sample sizes, we will use subscripts: $n_1, n_2, \ldots, n_k$.

Also, to distinguish the variances, we use $s_1^2, s_2^2, \ldots, s_k^2$. With $N = \sum n_i$, and the degrees of freedom are $n_i - 1$ for each variance. The pooled variance becomes:

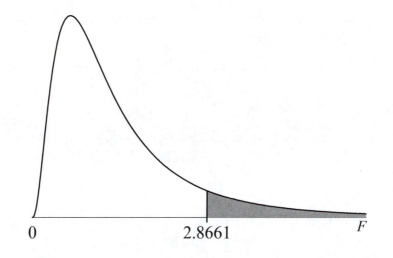

Figure 11.4: The critical value for Example 11.1.3.

$$s_p^2 = \textit{variance within the samples}$$
$$= \frac{\sum (n_i - 1) s_i^2}{\sum (n_i - 1)}$$
$$= \frac{\sum (n_i - 1) s_i^2}{\sum n_i - \sum 1}$$
$$= \frac{\sum (n_i - 1) s_i^2}{N - k}.$$

That settles the denominator for the test statistic.

Next, we must alter the numerator of the test statistic. To distinguish the sample means, we will use subscripts again, $\bar{x}_1, \bar{x}_2, \ldots, \bar{x}_k$. In computing $s_{\bar{x}}^2$, we will refer to the formula for sample variance

$$s^2 = \frac{\sum (x - \bar{x})^2}{n - 1},$$

and alter this to create a formula for the variance of the k sample means. We need the mean of the sample means, and we will call this $\bar{\bar{x}}$, the *grand mean*, but again we will use a weighted mean to computed it. This time the sample sizes will be the weights.

$$\bar{\bar{x}} = \frac{\sum n_i \cdot \bar{x}_i}{\sum n_i} = \frac{\sum n_i \cdot \bar{x}_i}{N}$$

Thus, the sample variance of the sample means would be

$$s_{\bar{x}}^2 = \frac{\sum (\bar{x}_i - \bar{\bar{x}})^2}{k - 1}.$$

All that remains to reconstruct the original numerator, $n \cdot s_{\bar{x}}^2$ (the variance between the samples), is to multiply by sample size. Since the sample size varies now, we distribute into the sum in the sample variance, so that the numerator, $n \cdot s_{\bar{x}}^2$ becomes:

$$Variance\ between\ the\ samples = \frac{\sum n_i\,(\bar{x}_i - \bar{\bar{x}})^2}{k-1}.$$

We now can change our test statistic

$$F = \frac{variance\ between\ samples}{variance\ within\ samples}$$

to its new form:

$$F = \frac{\frac{\sum n_i(\bar{x}_i - \bar{\bar{x}})^2}{k-1}}{\frac{\sum (n_i-1)s_i^2}{N-k}},$$

where $N = \sum n_i$, and $\bar{\bar{x}} = \frac{\sum n_i \cdot \bar{x}_i}{N}$.

The degrees of freedom are visible in the test statistic. Since the numerator contains the total variance of k sample means, it contains $k-1$ degrees of freedom. The denominator contains k sample variances, each with $n_i - 1$ degrees of freedom. Altogether the denominator contains $\sum_{i=1}^{k}(n_i - 1) = \sum_{i=1}^{k} n_i - \sum_{i=1}^{k} 1 = N - k$ degrees of freedom. We summarize all of this below.

Summary 11.1.4 One-Way ANOVA with Unequal Sample Sizes
The test statistic is
$$T.S.: F = \frac{\frac{\sum n_i(\bar{x}_i - \bar{\bar{x}})^2}{k-1}}{\frac{\sum (n_i-1)s_i^2}{N-k}},$$

where $N = \sum n_i$, and $\bar{\bar{x}} = \frac{\sum n_i \cdot \bar{x}_i}{N}$. This is a right-tail F test with numerator degrees of freedom $df_1 = k - 1$, and denominator degrees of freedom $df_2 = N - k$.

The procedure and requirements for this test are identical to those for the equal sample size version of this test. The level of difficulty of this has risen to the point where it is best to use computer software to make necessary computations. Still, using the above formulas, we can do the procedure by hand if necessary.

ANOVA uses special notation universally, and it would be a disservice to not cover this notation here. The letters *SS* always stand for *Sum of Squares*, where *MS* stands for *Mean of Squares*. To compute a mean of squares, the sum of squares must be divided by the degrees of freedom in that sum. This is how sample variances are computed. The squared deviations from the mean are summed, then we divide by the degrees of freedom to get variance. The notations for the expressions in the test statistic are:

$$SS\,(treatment) = \sum n_i\,(\bar{x}_i - \bar{\bar{x}})^2,$$

and

$$MS\left(treatment\right) = \frac{SS\left(treatment\right)}{k-1}$$
$$= \frac{\sum n_i \left(\bar{x}_i - \bar{\bar{x}}\right)^2}{k-1}.$$

Additionally,

$$SS\left(error\right) = \sum \left(n_i - 1\right) s_i^2,$$

which gives

$$MS\left(error\right) = \frac{SS\left(error\right)}{N-k} = \frac{\sum \left(n_i - 1\right) s_i^2}{N-k}.$$

With these, the test statistic is often expressed as:

$$F = \frac{MS\left(treatment\right)}{MS\left(error\right)},$$

with numerator degrees of freedom $df_1 = k - 1$, and denominator degrees of freedom $df_2 = N - k$.

Example 11.1.5 As an example of one-way ANOVA with unequal sample sizes, consider the following data which were collected by Lorne Schweitzer and David Stein, winners of the 2005 Canadian Virtual Science Fair. The data represent voltages collected by solar panels on differing weather conditions.

Rainy	Overcast	Partly Cloudy	Sunny
15.43	16.89	19.30	21.09
13.81	14.97	17.12	20.43
8.72	12.84		20.88
12.19			20.55

We will test the claim that the mean voltages collected by these panels in differing weather conditions are not all the same, using 1% significance. For each sample, we compute the mean, variance and sample size, tabulated below.

Statistic	Rainy	Overcast	Cloudy	Sunny
$\bar{x}$	12.5375	14.9	18.21	20.7375
s^2	8.226625	4.1043	2.3762	0.091425
n	4	3	2	4

We need to compute the grand mean,

$$\bar{\bar{x}} = \frac{\sum n_i \cdot \bar{x}_i}{N}.$$

First we compute $N = \sum n_i = 4 + 3 + 2 + 4 = 13$. Next, we tabulate and sum the values $n_i \cdot \bar{x}_i$ below.

Weather	n	$\bar{x}$	$n_i \cdot \bar{x}_i$
Rainy	4	12.5375	50.15
Overcast	3	14.9	44.70
Cloudy	2	18.21	36.42
Sunny	4	20.7375	82.95
Totals:	13	—	214.22

The grand mean is computed next.

$$\bar{\bar{x}} = \frac{\sum n_i \cdot \bar{x}_i}{N} = \frac{214.22}{13} = 16.4785.$$

Next, we tabulate and total the values $n_i \left(\bar{x}_i - \bar{\bar{x}}\right)^2$ and $(n_i - 1) s_i^2$ below.

Weather	n	$\bar{x}$	s^2	$n_i(\bar{x}_i - \bar{\bar{x}})^2$	$(n_i - 1)s_i^2$
Rainy	4	12.5375	8.226625	62.12471139	24.679875
Overcast	3	14.9	4.1043	7.474622485	8.2086
Cloudy	2	18.21	2.3762	5.996450888	2.3762
Sunny	4	20.7375	0.091425	72.55763447	0.274275
Totals:	13	—	—	148.1534192	35.53895

Now we have most of what we need. Remembering that $k = 4$,

$$SS\,(treatment) = \sum n_i \left(\bar{x}_i - \bar{\bar{x}}\right)^2 = 148.1534,$$

and $MS(treatment) = \frac{SS(treatment)}{k-1} = \frac{148.1534}{4-1} = 49.3844.$

Also,

$$SS\,(error) = \sum (n_i - 1)\, s_i^2 = 35.53895,$$

so $MS(error) = \frac{SS(error)}{N-k} = \frac{35.53895}{13-4} = 3.9488.$

Finally, we can compute our test statistic,

$$T.S. \ : \ F = \frac{MS\,(treatment)}{MS\,(error)}$$
$$= \frac{49.3844}{3.9488}$$
$$= 12.5062.$$

To look up the critical value, we note that $df_1 = k - 1 = 4 - 1 = 3$, and $df_2 = N - k = 13 - 4 = 9$.

Using $\alpha = 0.01$, we look to Table A5.

Numerator df_1

	2	3	4
8	8.6491	7.5910	7.0061
9	8.0215	**6.9919**	6.4221
10	7.5594	6.5523	5.9943

Denominator df_2

Our critical value is $C.V. : F = 6.9919$, pictured in Figure 11.5. The test statistic of

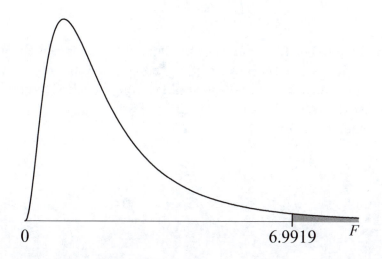

Figure 11.5: The critical value for Example 11.1.5.

$T.S. : F = 12.5062$ is well within the rejection region, so we reject the null hypothesis. Using computer software, we find that the *p-value* is equal to 0.00146. This is much less than α, emphasizing that the differences observed were highly significant.

In conclusion, we may say that our data indicate that the mean voltages collected by these panels under different conditions are not all the same. The observed differences were highly significant as emphasized by the p-value of 0.00146.

As mentioned, computer software is usually employed to conduct ANOVA. Below is a typical printout for one-way ANOVA.

	Groups	Count	Sum	Average	Variance		
Summary	Precipitation	4	50.15	12.5375	8.22663		
	Overcast	3	44.7	14.9	4.1043		
	Partly Cloudy	2	36.42	18.21	2.3762		
	Sunny	4	82.95	20.7375	0.09143		
ANOVA	Source of Variation	SS	df	MS	F	P-value	F crit
	Between Groups	148.15342	3	49.3845	12.5063	0.001462	6.9919
	Within Groups	35.53895	9	3.9488			
	Total	183.69237	12				

Note that, under the *ANOVA* section, the sums under *SS*, between the groups (*SS(Treatment)*), and within the groups (*SS(Error)*), the agreement with our totals, along with the degrees of freedom and the mean of squares (*MS*) values, again in agreement with ours. Next to the *MS* column, is the test statistic F, the *p-value*, and critical value, all in agreement with ours.

11.1.2   Homework

Test the following claims.

1. In the influential book, *Statistical Methods for Research Workers*, the great statistics genius Ronald Fisher provides the following example. In an experiment on the accuracy of counting soil bacteria, a soil sample was divided into four parallel samples, and from each of these after dilution, seven plates were inoculated. The number of bacteria colonies on each plate is shown below. Test the claim at 5% significance that the mean number of colonies is the same for each of the four samples. Assume these samples are drawn from normal populations with equal variances.

\multicolumn{4}{c}{Sample}			
I	II	III	IV
72	74	78	69
69	72	74	67
63	70	70	66
59	69	58	64
59	66	58	62
53	58	56	58
51	52	56	54

2. The following data represent tea leaf pluckings from sixteen different plots of tea bushes intended for experimental use in Ceylon. The pluckings are divided into four different treatment groups, and we wish to determine if the means under the four different treatments are significantly different. Test this using 1% significance, assuming that these samples are drawn from normal populations with equal variances. These values are also from *Statistical Methods for Research Workers* by Ronald Fisher, but the data were gathered by T. Eden.

Treatment 1	Treatment 2	Treatment 3	Treatment 4
88	102	91	88
94	110	109	118
109	105	115	94
88	102	91	96

3. There are many tools in statistics which are useful for classifying and separating sample data into separate populations. The table below includes lengths of cuckoo's eggs (in millimeters) found in nests of bird foster parents of differing species. Do the data indicate that there may be significant differences amongst the means of these seven populations? Test this using 1% significance, assuming that these samples are drawn from normal populations with equal variances. These historical figures are extracted from *Latter* & abbreviated for simplicity by M.G. Bulmer.

Meadow-Pipit	Tree-Pipit	Hedge-Sparrow	Robin	Reed-Warbler	Pied-Wagtail	Wren
21.7	22.7	22.0	21.8	23.2	23.0	19.8
22.6	23.3	23.9	23.0	22.0	23.4	22.1
20.9	24.0	20.9	23.3	22.2	24.0	21.5
21.6	23.6	23.8	22.4	21.2	23.3	20.9
22.2	22.1	25.0	22.4	21.6	23.1	22.0
22.5	21.8	24.0	23.0	21.6	22.4	21.0
22.2	21.1	21.7	23.0	21.9	21.8	22.3
24.3	23.4	23.8	23.0	22.0	21.8	21.0
22.3	23.8	22.8	23.9	22.9	24.9	20.3
22.6	23.3	23.1	22.3	22.8	24.0	20.9

4. The following data were provided by NIST, the National Institute of Standards and Technology. In this example we have measurements from five different machines making the same part and we take five random samples from each machine to obtain the following diameter data. Test that the parts have different mean diameters using 5% significance.

Machine				
#1	#2	#3	#4	#5
.125	.118	.123	.126	.118
.127	.122	.125	.128	.129
.125	.120	.125	.126	.127
.126	.124	.124	.127	.120
.128	.119	.126	.129	.121

5. The following are temperatures of patients after taking different treatments for reducing fevers. Test the claim that there are no differences between the mean temperatures for these 5 different treatments. Test this using 1% significance, assuming that these samples are drawn from normal populations with equal variances.

Placebo	Aspirin	Anacin	Tylenol	Bufferin
98.3	96	97.2	95.1	96.1
98.7	96.6	97.4	95.7	96.5
98.1	95.3	96.9	94.4	95.4
99.9	98.1	97.8	97.2	98.2
96.2	94.6	94.7	93.7	94.5
98.6	96.6	97	95.7	96.5
98.3	96.2	96.83	95.3	96.2

6. This example is from the SPSS 7.5 "Applications Guide." The mean and standard deviation of ages are given for seven different politically inclined groups. Test the claim at 5% significance that the mean age is independent of the group. This example uses unequal sample sizes. Assume that the populations involved are normal with equal variances.

Political Views	n	$\bar{x}$	s
Extremely liberal	30	39.07	15.94
Liberal	162	45.27	15.90
Slightly liberal	192	42.06	16.38
Moderate	527	45.53	17.38
Slightly conservative	247	44.43	16.63
Conservative	240	50.76	18.23
Extremely conservative	40	54.55	17.50

Chapter 12

Distribution Free Tests

In several of the tests covered in this book, certain assumptions are made, often with respect to the distribution of the populations which are being studied. In each t-test with small sample sizes, normality is required. The analysis of variance procedures also required normality. In chapter nine we introduced a test for normality which may be used to determine when normality is plausible, and when the assumption of normality should be rejected.

When normality is not plausible, there are still times when claims must be tested, and thus statisticians have developed *distribution free tests*, also known as *non-parametric tests*. Distribution free tests do not make assumptions about the shape of the distributions in question, and thus may be applied in cases where other tests may not be appropriate. Distribution free tests are not quite as powerful as their counterparts when their own distribution requirements are met (so the *power* of these tests is not as close to 1), but they perform *almost* as well.

12.1   The Signs Tests

Our first distribution free test is called The Signs Test. The Signs Test is an excellent introduction to distribution free tests because it is easy to conduct, and it has no requirements at all. It can be conducted on any sample of quantitative values, or paired values. The simplest version of the Signs Test is used to test a claim about a single population median. Another version of the test is used to compare the means of two symmetric populations using matched-pair samples as defined in chapter seven. We focus our attention first to the median of a single population.

Using the Signs Test for a Single Population Median
The Signs Test is a clever application that arises from a simple fact – the median of a population is the population's 50th percentile. Thus, 50 percent of all values lie below the median, and 50 percent of all values lie above the median. With this in mind, any sample of size n drawn from such a population could be thought of as a binomial experiment with n trials and two outcomes – either a given value is greater than the median or it is not.

The median of a sample will be different from the population median, but still we should expect sample values to be distributed about the population median, with about half of the values above the population median. Of course, we would not expect to see *exactly* half of sample values to be above the median – there would always be some randomness in this regard, but when the distribution becomes very imbalanced we might question any assumptions made about the median. Our testing procedure will involve a null hypothesis which assumes a value for the population median. This assumption will be rejected whenever the distribution is so unbalanced on either side of the assumed median that it is quite unlikely to have occurred by chance. In such a case we will conclude that our assumed value for the population median is incorrect, and reject the null hypothesis.

Example 12.1.1 For our example from the journal *Environmental Concentration and Toxicology* published the article *"Trace Metals in Sea Scallops"* (vol. 19, pp. 326 - 1334), giving the cadmium amounts (in mg) from sea scallops observed at a number of different stations in North Atlantic waters, suppose we wish to claim that the median cadmium amounts in these scallops is less than 14 mg. The sample values are: {5.1, 14.4, 14.7, 10.8, 6.5, 5.7, 7.7, 14.1, 9.5, 3.7, 8.9, 7.9, 7.9, 4.5, 10.1, 5.0, 9.6, 5.5, 5.1, 11.4, 8.0, 12.1, 7.5, 8.5, 13.1, 6.4, 18.0, 27.0, 18.9, 10.8, 13.1, 8.4, 16.9, 2.7, 9.6, 4.5, 12.4, 5.5, 12.7, 17.1}.

Next to each value we denote values that are greater than 14 with a positive (+) sign, and values that are less than 14 with a negative (–) sign. The collection of (–) and (+) symbols are the *signs* in our *Signs Test*.

Sample Values with Signs				
5.1 (–)	14.4 (+)	14.7 (+)	10.8 (–)	6.5 (–)
5.7 (–)	7.7 (–)	14.1 (+)	9.5 (–)	3.7 (–)
8.9 (–)	7.9 (–)	7.9 (–)	4.5 (–)	10.1 (–)
5.0 (–)	9.6 (–)	5.5 (–)	5.1 (–)	11.4 (–)
8.0 (–)	12.1 (–)	7.5 (–)	8.5 (–)	13.1 (–)
6.4 (–)	18.0 (+)	27.0 (+)	18.9 (+)	10.8 (–)
13.1 (–)	8.4 (–)	16.9 (+)	2.7 (–)	9.6 (–)
4.5 (–)	12.4 (–)	5.5 (–)	12.7 (–)	17.1 (+)

Our claim is that the population from which the above sample was taken has a median less than 14. The symbol which we will use for the population median will be $\tilde{\mu}$, and with this, our alternative hypothesis is $H_1 : \tilde{\mu} < 14$, paired with the null hypothesis $H_0 : \tilde{\mu} = 14$. If the null hypothesis is true, we might expect to see around half of the sample values greater than 14. Our sample size is 40, so we expect to see around 20 values greater than 14. In fact, we only observe 8 values greater than 14. If the null hypothesis is true, the question we must answer is whether observing only 8 values out of 40 greater than 14 is unusual enough to warrant rejection of the null hypothesis?

12.1.1   Determining p-Values

To answer this question, we will compute the probability of getting 8 or fewer values greater than 14, $P(x_s \leqslant 8)$, where x_s represents the possible numbers of sample values which are greater than 14. This is, in fact, the *p-value* for our hypothesis test. But how shall we compute this?

As always, we shall assume the null hypothesis is true – that the population median is equal to 14. If this is true, then with a continuous random variable, half of the population values are greater than 14. If we randomly sample any value from the population, there are two possible outcomes: either $S =$ *"value is greater than 14"*, or $F =$ *"value is not greater than 14"*. The probability for each of these is $p = P(S) = 0.5$, and $q = P(F) = 1 - p = 0.5$. If we sample $n = 40$ values, we have 40 trials, and we are conducting a binomial experiment with $p = 0.5$ and $n = 40$.

To compute the *p-value* for our test, we use the binomial probability formula,

$$P(x_s) =_n C_x \cdot p^{x_s} \cdot q^{n-x_s}.$$

For the *p-value*, we compute

$$p\text{-}value = P(x_s \leqslant 8) = P(0 \text{ or } 1 \text{ or } 2 \text{ or } 3 \text{ or } 4 \text{ or } 5 \text{ or } 6 \text{ or } 7 \text{ or } 8)$$
$$= P(0) + P(1) + P(2) + P(3) + P(4) + P(5) + P(6) + P(7) + P(8).$$

And of course we need to apply our binomial probability formula several times. As a refresher, we compute $P(x_s = 8)$ below.

$$\begin{aligned} P(x_s = 8) &= {}_nC_{x_s} \cdot p^{x_s} \cdot q^{n-x_s} \\ &= {}_{40}C_8 \cdot p^8 \cdot q^{40-8} \\ &= {}_{40}C_8 \cdot p^8 \cdot q^{32} \\ &= 76,904,685 \cdot 0.5^8 \cdot 0.5^{32} \\ &= 0.00006994 \end{aligned}$$

The rest of the probabilities and their total are summarized in the following table.

x_s	$P(x_s)$
0	0.000000000000909
1	0.000000000036380
2	0.000000000709406
3	0.000000008985808
4	0.000000083118721
5	0.000000598454790
6	0.000003490986273
7	0.000016956219042
8	0.000069944403549
$\sum$	0.000091082914878

The *p-value* is the sum of the computed probabilities.

$$p\text{-}value = \sum_{x_s=0}^{8} P(x_s)$$
$$= 0.00009108$$
$$\approx 0.0001.$$

This is very small, and indicates that if the median of this population is truly 14, that to get 8 or fewer values greater than 14 is nearly impossible to do, and yet we have done it. Nearly impossible things happen quite rarely, and because of this we must question our assumption that the population median is 14. We will, in fact, reject this assumption and favor the alternative that the population median is less than 14.

On the previous example, we conducted a *left tail test*, and to do this we were counting the number of values greater than 14, or the number of (+) symbols in our collection of signs, and this number should be substantially less than half of the sample if we are to reject the null hypothesis. In the case where we wish to conduct a *right tail test* with alternate hypothesis H_1: $\tilde{\mu} > a$, we must recognize that, in attempting to prove this, the number of values in the sample less than a, corresponding to the number of (−) signs, should be substantially smaller than half of the overall sample. In this case, to compute the *p-value* we will count the number of (−) signs in our collection and compute the probability of getting less than or equal to this amount.

For a two tail test, we will use the minimum of the number of negative (−) signs and the number of positive (+) signs, compute the probability of getting less than or equal to this number of signs, and multiply by two to obtain our *p-value*.

Table A8 gives *p*-values for the Signs Test for sample sizes ranging from 1 to 10. The table is reproduced below.

Table A8: *p*-Values for the Sign Test										
					$x_s=s$					
n	0	1	2	3	4	5	6	7	8	9	10
1	0.5000	1.0000									
2	0.2500	0.7500	1.0000								
3	0.1250	0.5000	0.8750	1.0000							
4	0.0625	0.3125	0.6875	0.9375	1.0000						
5	0.0313	0.1875	0.5000	0.8125	0.9688	1.0000					
6	0.0156	0.1094	0.3438	0.6563	0.8906	0.9844	1.0000				
7	0.0078	0.0625	0.2266	0.5000	0.7734	0.9375	0.9922	1.0000			
8	0.0039	0.0352	0.1445	0.3633	0.6367	0.8555	0.9648	0.9961	1.0000		
9	0.0020	0.0195	0.0898	0.2539	0.5000	0.7461	0.9102	0.9805	0.9981	1.0000	
10	0.0010	0.0107	0.0547	0.1719	0.3770	0.6231	0.8281	0.9453	0.9893	0.9990	1.0000

Values above are p-values for one-tail tests. Multiply by 2 for a two-tail test p-value.

Right tail test: s = number of (−) signs. Left tail test: s = number of (+) signs

Two tail test: s = minimum number of (+) or (−) signs.

Computing the *p-value* can be difficult with large sample sizes. Recalling from the discussion above that the number of signs is distributed according to the binomial distribution with $p = \frac{1}{2}$, it is possible to approximately compute the probability corresponding to the *p-value* using the *normal distribution* as an approximation to the binomial. This approximation was introduced in Chapter 6, and was valid whenever np and nq are at least 10. To ease the level of complexity introduced by this requirement, we will modify the "*at least* 10" rule to a older version: "*at least* 5". With $p = \frac{1}{2} = 0.5$, and $q = 1 - p = 0.5$, the "*at least* 5" requirements both become $n \cdot 0.5 \geqslant 5$, or simply that $n \geqslant \frac{5}{0.5} = 10$. Thus, we use the normal approximation whenever $n > 10$, but require the exact binomial distribution whenever $n \leqslant 10$. You may also remember that the normal approximation to the binomial required what we referred to as a *continuity correction*, where we add or subtract 0.5 to or from the endpoints of the interval in question.

We will use s to denote number of signs counted, and with the continuity correction the interval $x_s \leqslant s$ becomes $x_s < (s + 0.5)$. Taking z-scores requires the mean and standard deviation for the binomial distribution: $\mu = np = 0.5 \cdot n$, and $\sigma = \sqrt{npq} = \sqrt{n \cdot 0.5 \cdot 0.5} = 0.5 \cdot \sqrt{n}$. In computing $P(x_s \leqslant s)$, we proceed as follows:

$$
\begin{aligned}
P(x_s \leqslant s) &= P(x_s < s + 0.5) \\
&= P\left(z < \frac{(s + 0.5) - \mu}{\sigma}\right) \\
&= P\left(z < \frac{(s + 0.5) - 0.5 \cdot n}{0.5 \cdot \sqrt{n}}\right).
\end{aligned}
$$

The test statistic will therefore be

$$
T.S. : z = \frac{(s + 0.5) - 0.5 \cdot n}{0.5 \cdot \sqrt{n}}.
$$

Using the normal approximation of the binomial distribution, we can now compute a *p-value* corresponding to any sample size. Whenever this *p-value* is less than our chosen level of significance, α, we will reject the null hypothesis in favor of the alternative.

Determining Critical Values

Finally, because the binomial distribution is a *discrete* probability distribution, finding exact critical values according to a particular value for α becomes impossible. It *is* possible to give critical values which *nearly* correspond to a chosen α value, as is demonstrated below.

Suppose in our example we wish to determine a critical value which corresponds as nearly as possible to a given level of significance, α. To do this, we sum probabilities as we did for the *p-value* but continue to add probabilities until we come to a value c where $P(x_s \leqslant c) \approx \alpha$. The sum of probabilities from $x_s = 0$ to $x_s = c$ is called a *cumulative probability*. When we determine a value c where the cumulative probability $P(x_s \leqslant c) \approx \alpha$, we have found our critical value. Suppose we wish to use $\alpha = 0.01$ in our example above. To find the closest corresponding critical value, we search for the cumulative probability that is closest to $\alpha = 0.01$. The probabilities and cumulative probabilities for several x_s values starting at zero, with $n = 40$, are given below.

$x_s = c$	$P(x_s = c)$	$P(x_s \leqslant c)$
0	0.000000000000909	0.000000000000909
1	0.000000000036380	0.000000000037289
2	0.000000000709406	0.000000000746695
3	0.000000008985808	0.000000009732503
4	0.000000083118721	0.000000092851224
5	0.000000598454790	0.000000691306013
6	0.000003490986273	0.000004182292287
7	0.000016956219042	0.000021138511329
8	0.000069944403549	0.000091082914878
9	0.000248691212619	0.000339774127497
10	0.000770942759118	0.001110716886615
11	0.002102571161231	0.003213288047846
12	0.005081213639642	0.008294501687487
13	0.010944152454613	0.019238654142100

Notice that there are two cumulative probabilities which are near to our $\alpha = 0.01$. We see that one is approximately 0.00829, while the other is approximately 0.0192. The closest of these to $\alpha = 0.01$ is the first, which corresponds to a critical value of $x_s = 12$. In our right tail test, we should reject the null hypothesis if the number of negative $(-)$ signs is less than or equal to $x_s = 12$.

It is important to note, however, that our critical value does not correspond *exactly* to our $\alpha = 0.01$ level of significance. If our decision is made by using the critical value, then the true level of significance for this test is the cumulative probability. In using $x_s = 12$ as a critical value in our example, the true value for α is 0.00829.

For our example, we observed $x_s = 8$ negative $(-)$ signs, and this is less than our critical value, so we reaffirm that the null hypothesis should be rejected in favor of the alternative.

Table A7 gives critical values for the Signs Test for values of $n \leqslant 10$. This table is constructed by applying the process used above for various levels of significance applied to one- and two-tailed tests.

We are in need of a procedure to summarize the steps needed for this test, and this is provided below.

Procedure 12.1.2 The Signs Test for a Single Population Median

    Step I. **Gather Data.** All we need for this test is the original sample data, sample size, and α, the level of significance.

Step II. **Identify Claims**

The null hypothesis is H_0: $\tilde{\mu} = a$. The possible alternative hypotheses are $H_1 : \tilde{\mu} > a$, $H_1 : \tilde{\mu} < a$, or $H_1 : \tilde{\mu} \neq a$.

Step III. **Compute the Test Statistic and p-Value**

To do this, we first assign our signs for the Signs Test. For each value in the sample which is less than the assumed median, a, we assign a negative ($-$) sign. For values greater than a, we assign a positive ($+$) sign. No sign is associated to values equal to a.

Next, we determine the value for s.

- For a right tail test, s is the number of negative ($-$) signs.

- For a left tail test, s is the number of positive ($+$) signs.

- For a two tail test, s is the minimum of the numbers of ($+$) and ($-$) signs, respectively.

After finding s, we determine the test statistic and, as an option, the p-value. The p-value on a one-tail test is $P(x_s \leqslant s)$, where x_s is distributed according to the binomial distribution. The p-value on a two-tail test is p-value $= 2 \cdot P(x_s \leqslant s)$

- If $n \leqslant 10$, the *test statistic is simply* s. If the p-value is desired, we may look up $P(x_s \leqslant s)$ on Table A8. For one tail tests, this probability is our p-value. *Multiply this by two* to get the p-value on a two tail test.

- If $n > 10$, we use the test statistic

$$T.S. : z = \frac{(s + 0.5) - 0.5 \cdot n}{0.5 \cdot \sqrt{n}}$$

and the normal approximation to the binomial. To find the optional p-value, find this z-score in Table A1 to get $P(x_s \leqslant s)$. This is the p-value for a one-tail test. For a two tail test, *multiply this by two* to obtain the p-value.

Step IV. **Look Up the Critical Value**

- If $n \leqslant 10$, find s in Table A7 to obtain the critical value. It should be noted that the true level of significance will be slightly different from the chosen level as discussed previously.

- If $n > 10$, look in the bottom row of Table A2 to find the appropriate critical value. The z critical value *should be made negative* for this test.

Step V. **Make a Decision**

If the test statistic is less than or equal to the critical value, reject the null hypothesis in favor of the alternative. Whenever the null hypothesis is rejected, the p-value will be less than α.

Step VI. **Make a Conclusion**

As usual, conclusions should be written in the context of the problem, without the use of symbolic notation.

Example 12.1.1 Continued: Let us repeat our first example using the formal process outlined above. Our data values and associated signs are reproduced below.

		Sample Values with Signs		
5.1 (−)	14.4 (+)	14.7 (+)	10.8 (−)	6.5 (−)
5.7 (−)	7.7 (−)	14.1 (+)	9.5 (−)	3.7 (−)
8.9 (−)	7.9 (−)	7.9 (−)	4.5 (−)	10.1 (−)
5.0 (−)	9.6 (−)	5.5 (−)	5.1 (−)	11.4 (−)
8.0 (−)	12.1 (−)	7.5 (−)	8.5 (−)	13.1 (−)
6.4 (−)	18.0 (+)	27.0 (+)	18.9 (+)	10.8 (−)
13.1 (−)	8.4 (−)	16.9 (+)	2.7 (−)	9.6 (−)
4.5 (−)	12.4 (−)	5.5 (−)	12.7 (−)	17.1 (+)

Step I. **Gather Data.** Our original sample is above with a sample size of $n = 40$. For this test, we will pick $\alpha = 0.01$.

Step II. **Identify Claims.** Our hypotheses are H_0: $\tilde{\mu} = 14$, and H_1: $\tilde{\mu} < 14$.

Step III. **Compute the Test Statistic and p-Value**

Compute the Test Statistic and p-*value*. Since our sample size is greater than 10, we will use the normal approximation to the binomial in computing the p-*value* as indicated by our procedure. First we count s, the number of positive (+) signs in our sample. As before we see that $s = 8$. Next we compute the test statistic,

$$T.S. : z = \frac{(s + 0.5) - 0.5 \cdot n}{0.5 \cdot \sqrt{n}}$$
$$= \frac{(8 + 0.5) - 0.5 \cdot 40}{0.5 \cdot \sqrt{40}}$$
$$= -3.64.$$

The associated p-*value* is $P(x_s \leqslant s) = P(z < -3.64) = 0.0001$. This is found by looking up $z = -3.64$ in Table A1 as depicted below.

z	.03	.04	.05
− 3.7	.0001	.0001	.0001
−3.6	.0001	**.0001**	.0001
− 3.5	.0002	.0002	.0002

This p-*value* is a four digit approximation of the more accurate p-*value* found previously using the binomial distribution formula.

Step IV. **Look Up the Critical Value**

With $\alpha = 0.01$ and using the normal approximation, we see in the bottom row of Table A2 that $z = 2.326$. We will always make this negative regardless of the type of test we are conducting (left- right- or two-tailed), so we use the critical value $C.V. : z = -2.326$.

Step V. **Make a Decision**

Our test statistic is less than the critical value in this test, so we reject null hypothesis in favor of the alternative. Our *p-value* of 0.0001 affirms this decision, and indicates a highly significant departure from the null hypothesis.

Step VI. **Make a Conclusion**

We conclude by stating that our sample data support the claim that the median cadmium amounts for sea scallops is less than 14 mg. Our sample data gave highly significant support for this conclusion as indicated by the *p-value* of 0.0001.

Example 12.1.3 In chapter two, we gave the miles per gallon measurements for ten round-trip commutes in a Toyota Prius Hybrid vehicle from Lake Arrowhead to Walnut, California. The miles per gallon measurements are: {46.2, 44.9, 47.0, 46.3, 45.6, 46.0, 44.6, 45.9, 46.3, 45.6}. We would like to test the claim at 5% significance that the median miles per gallon on this round-trip commute is equal to 45 miles per gallon against the claim that the median is not 45 miles per gallon.

Step I. **Gather Data**

Our original sample is above with a sample size of $n = 10$. For this test, we have chosen $\alpha = 0.05$.

Step II. **Identify Claims**

The null hypothesis is $H_0 : \tilde{\mu} = 45$, and the alternative hypothesis is $H_1 : \tilde{\mu} \neq 45$.

Step III. **Compute the Test Statistic and *p*-Value**

Here our sample size is $n = 10$, so our test statistic is s, and on a two-tail test s is the minimum of the numbers of positive (+) and negative (−) signs, respectively, and we optionally look up the *p-value* using the cumulative binomial distribution with $p = 0.5$ on Table A8.

Our sample data are reproduced below with positive (+) signs assigned to values which are above the assumed median of 45, and negative (−) signs to values which are less than 45.

46.2	44.9	47.0	46.3	45.6	46.0	44.6	45.9	46.3	45.6
(+)	(−)	(+)	(+)	(+)	(+)	(−)	(+)	(+)	(+)

Here we have two (−) signs, and eight (+) signs, and picking the smaller, we have our test statistic, $s = 2$. To find the *p-value*, we look up $s = 2$ with $n = 10$ on Table A8 to get $P(x_s \leqslant 2)$.

In the table we see that $P(x_s \leqslant 2) = 0.054688$, but since this is a two-tail test, we multiply by two to get *p-value* $= 2 \cdot P(x_s \leqslant 2) = 2 \cdot 0.0547 = 0.1094$.

Step IV. **Look up the Critical Value**

This time, since $n \leqslant 10$, we look up our critical value on Table A7. Part of this table is reproduced below. Our critical value is $C.V. : x_s = 1$.

Table A7 – Sign Test Critical Values
Two Tail Applications

n	$\alpha \approx 0.01$	$\alpha \approx 0.02$	$\alpha \approx 0.05$	$\alpha \approx 0.10$
5	−	−	−	0
6	−	−	0	0
7	−	0	0	0
8	0	0	0	1
9	0	0	1	1
10	0	0	1	1
11	0	1	1	2

Step V. **Make a Decision**

Since our test statistic, $s = 2$, is not less than or equal to our critical value, $x_s = 1$, we fail to reject the null hypothesis. Our *p-value* $= 0.109376$ affirms this, since it is not less than $\alpha = 0.05$.

Step VI. **Make a Conclusion**

The sample data do not give sufficient evidence to warrant rejection of the claim that the median miles per gallon for the Toyota Prius on the commute mentioned above is equal to 45 miles per gallon. The *p-value* indicated insignificant evidence to warrant rejection of this claim.

12.1.2   The Signs Test for a Mean Difference

The Signs Test can be used to compare matched pair samples as we did in chapter seven. When we last studied matched pair samples, we made inferences regarding these samples by subtracting each y value from its corresponding x value, and our inferences were made with regard to the mean difference, which is the difference of the two means. The Signs Test is primarily a test regarding a population median, but when we use differences from matched pair samples it is not clear about which parameter we are making claims. In fact, the claim regards the median of all differences, but this is not the same as the difference of the individual population medians. We may make our claim with regard to the median of all differences, but this may be easier to understand if we make an assumption about the population of all differences, $x - y$, which we are studying. We will assume that the population of differences is continuous and symmetric (with *skewness* = 0) because the median of a symmetric population is equal its mean. If our population of differences is symmetric, then the median difference will be the mean difference, and this is the difference of the two population means. This is the approach we shall use in our examples here.

For our two continuous populations of values x, and y, which have a natural one-to-one pairing, (x, y), it can be proven that if these population differ only with respect to their means (so they have the same shape and variation), then the population of differences $x - y$ is symmetric even if x and y are not individually symmetric.

Example 12.1.4 A study was conducted to investigate whether oat bran lowers serum cholesterol levels in people (Pagano & Gauvreau, 1993, p. 252). Fourteen individuals were randomly assigned a diet that included either oat bran or corn flakes. After two weeks on the initial diet, serum low-density lipoprotein levels (mg/dl) were measured. Each subject was then 'crossed-over" to the alternate diet. After two-weeks on this second diet, low-density lipoprotein levels were once again recorded. The data are recorded in the following table.

Corn Flake Lipoprotein (x)	Oat Bran Lipoprotein (y)
4.61	3.84
6.42	5.57
5.40	5.85
4.54	4.8
3.98	3.68
3.82	2.96
5.01	4.41
4.34	3.72
3.80	3.49
4.56	3.84
5.35	5.26
3.89	3.73
2.25	1.84
4.24	4.14

This application uses matched pair samples to compare the means from two populations by considering the mean of the differences between the paired sample values. We assume that the population of differences is symmetric and continuous, so the median difference will be equal to the mean difference. Signs are computed by subtracting sample y values from corresponding sample x values. Whenever the difference is negative, a negative sign $(-)$ is applied. When the difference is positive, a positive sign $(+)$ is applied. No sign is associated when the difference is zero.

Below, the signs are applied to each (x, y) pair in our example. The matched-pair test procedure is essentially the same as before, but we give the slightly modified version as follows.

Corn Flake Lipoprotein (x)	Oat Bran Lipoprotein (y)	Sign of $(x-y)$ difference
4.61	3.84	(+)
6.42	5.57	(+)
5.40	5.85	(−)
4.54	4.80	(−)
3.98	3.68	(+)
3.82	2.96	(+)
5.01	4.41	(+)
4.34	3.72	(+)
3.80	3.49	(+)
4.56	3.84	(+)
5.35	5.26	(+)
3.89	3.73	(+)
2.25	1.84	(+)
4.24	4.14	(+)

Procedure 12.1.5 The Signs Test for a Mean Difference

Step I. Gather Data and Verify Assumptions

All we need for this test is the original sample data, the sample size, and the level of significance. This procedure requires that the population of all differences $x - y$ is symmetric, and in such case the test may be used to compare two population means. If this cannot be assumed, then the test should refer to the median of all differences.

Step II. Identify Claims

For the comparison of two means, the null hypothesis is $H_0 : \mu_1 = \mu_2$ (or $\mu_d = 0$, where $\mu_d = \mu_1 - \mu_2$). The possible alternative hypotheses are $H_1 : \mu_1 > \mu_2$, $H_1 : \mu_1 < \mu_2$, and $H_1 : \mu_1 \neq \mu_2$. Alternately, null and alternate hypotheses may refer to the median difference, $\tilde{\mu}_d$.

Step III. Compute the Test Statistic and p-Value

Signs are determined first by subtracting sample y values from corresponding sample x values. Whenever the difference is negative, a negative sign (−) is applied. When the difference is positive, a positive sign (+) is applied. No sign is associated when the difference is zero.

Next, we determine the value for s.

- For a right tail test, s is the number of positive (−) signs.

- For a left tail test, s is the number of negative (+) signs.

- For a two tail test, s is the minimum of the numbers of $(+)$ and $(-)$ signs, respectively.

After finding s, we determine the test statistic and, as an option, the *p-value*. The *p-value* on a one-tail test is $P(x_s \leqslant s)$, where x_s is distributed according to the binomial distribution. The *p-value* on a two-tail test is $p\text{-}value = 2 \cdot P(x_s \leqslant s)$

- If $n \leqslant 10$, the *test statistic is simply s*. If the *p-value* is desired, we may look up $P(x_s \leqslant s)$ on Table A8. For one tail tests, this probability is our *p-value*. *Multiply $P(x_s \leqslant s)$ by two* to get the *p-value* on a two tail test.

- If $n > 10$, our test statistic is

$$T.S. : z = \frac{(s + 0.5) - 0.5 \cdot n}{0.5 \cdot \sqrt{n}}.$$

To find the optional *p-value*, look up this z-score in Table A1 to get $P(x_s \leqslant s)$. This is the *p-value* for a one-tail test. For a two tail test, multiply this by two to obtain the *p-value*.

Step IV. **Look Up the Critical Value**

- If $n \leqslant 10$, find s in Table A7 to obtain the critical value. It should be noted that the true level of significance will be slightly different from the chosen level as discussed previously.

- If $n > 10$, look up the z critical value in the bottom row of Table A2. This z critical value *should be made negative* for this test.

Step V. **Make a Decision**

If the test statistic is less than or equal to the critical value, reject the null hypothesis in favor of the alternative. Whenever the null hypothesis is rejected, the *p-value* will be less than α.

Step VI. **Make a Conclusion**

As usual, conclusions should be written in the context of the problem, without the use of symbolic notation.

Example 12.1.4 Continued: We wish to test the claim the mean lipoprotein level is lower for the patients when eating oat bran at 1% significance.

Step I. **Gather Data**

This time our sample size is $n = 14$, and our data are listed above. We assume that the population of differences of lipoprotein levels is symmetric. We have chosen the level of significance to be $\alpha = 0.01$.

Step II. **Identify Claims.** Our hypotheses are $H_0 : \mu_1 = \mu_2$, and $H_1 : \mu_1 > \mu_2$.

Step III. **Compute the Test Statistic and p-Value**

This time we are conducting a right tail test, so s is the number of $(-)$ signs. In the table above we see a total of two minus signs, so $s = 2$. Since $n > 10$, our test statistic is

$$T.S. : z = \frac{(s + 0.5) - 0.5 \cdot n}{0.5 \cdot \sqrt{n}}$$
$$= \frac{(2 + 0.5) - 0.5 \cdot 14}{0.5 \cdot \sqrt{14}}$$
$$= -2.41.$$

For the optional p-value, we look up this z-score in Table A1 and find $P(x_s \leqslant s) = P(z < -2.41) = 0.0080$. Since this is a one tail test, this value is our p-value:

$$p - value = 0.0080.$$

Step IV. **Look Up the Critical Value**

Since $n > 10$, we look up the critical value in the bottom row of Table A2 under $\alpha = 0.01$, with one tail. Here we see that the corresponding z-score is $z = 2.326$, but for this test we always make this z-score negative, so our critical value is $C.V. : z = -2.326$.

Step V. **Make a Decision**

Since the test statistic is slightly less than the critical value, we reject the null hypothesis in favor of the alternative. That is, we reject H_0 in favor of H_1: $\mu_1 > \mu_2$. This is validated by the p-value which was less than $\alpha = 0.05$.

Step VI. **Make a Conclusion**

The data support the claim that the mean lipoprotein level is higher for patients who eat corn flakes for breakfast, as compared to the mean for patients who eat oat bran cereal for breakfast.

The Signs Test is a simple and clever test for testing claims about medians (and means for symmetric populations). Its weakness lies in the fact that it only considers whether values are above or below an assumed median, but not the actual distances from the assumed median. In the following section we will use a test similar to the Signs Test which will attempt to correct this deficiency by introducing *ranks* corresponding to distances from the assumed median. Sample values will be replaced by ranks, and analysis will depend on ranks rather than the sample values themselves. The advantage of this process lies in the fact that, as with the Signs Test, our new procedure will not require that the population be distributed according to any particular shape.

12.1.3 Homework

Use the Signs Test to test the following claims.

1. The data which follow are from an article on peanut butter from *Consumer Reports* (Sept. 1990), which reported the following scores for various brands: {22, 30, 30, 36, 39, 40, 40, 41, 44, 45, 50, 50, 53, 56, 56, 56, 62, 65, 68}. Test the claim at 5% significance that these values are sampled from a population whose median is less than 55.

2. Prozac is a drug which is meant to elevate the moods of people who suffer from depression. In the following sample (provided by the *WebStat* website of the University of New England, School of Psychology), mood levels are given for patients after taking Prozac. In these values, a higher score means a less depressed psychological state: { 5, 1, 5, 7, 10, 9, 7, 11, 8 }. Test the claim at 5% significance that the median mood level is greater than 5.

3. To determine the effect that exercise has on lactic acid levels in the blood, the authors of "A Descriptive Analysis of Elite-Level Racquetball" (Research Quarterly for Exercise and Sport, 1991, pp. 109 – 114), listed the following blood lactic acid levels of 8 different players after playing three games of racquetball: { 18, 37, 40, 35, 30, 20, 33, 19 }. Test the claim at 5% significance that the median blood lactic acid level is greater than 20 after exercise equivalent to three games of racquetball.

4. Do physical characteristics of twin babies affect their birth order? Is it possible that the firstborn has an advantage in some way over the second born? The data below give first and second born head circumferences for 10 pairs of twins. The mean head circumference for the firstborn babies was 56.25 cm, while the mean for the second born babies was 56.00 cm. The firstborns have a slightly larger mean head circumference, but is this statistically significant?

First Born	54.7	53.0	57.8	56.6	53.1	57.2	57.2	57.2	57.2	58.5
Second born	54.2	52.9	56.9	55.3	54.8	57.2	57.2	55.8	56.5	59.2

Test the claim at 6% significance that the mean head circumference is greater for the firstborn babies. Assume that the population of all differences is symmetric and continuous.

5. The following values are hours worked per week for randomly selected students at Mt. San Antonio College in Walnut, California. { 15, 20, 40, 16, 20, 25, 19, 40, 40, 20, 20, 40, 0, 40, 35, 30, 35, 25, 29, 0, 0, 40, 16, 0, 40, 30, 25, 16, 35, 0 }. Test the claim at 6% significance that the median hours worked by students at this college is less than 32 hours per week.

6. The following data are from Conover WJ, *Practical Nonparametric Statistics* (3rd edition). Wiley 1999. The values represent aggressivity scores for 12 pairs of monozygotic twins.

First Born	86	71	77	68	91	72	77	91	70	71	88	87
Second Born	88	77	76	64	96	72	65	90	65	80	81	72

Test the claim at 5% significance that the mean aggressivity score is greater for the firstborn twin. Assume that the population of all differences is symmetric and continuous.

12.2 The Wilcoxon Signed Rank Test

The preceding section gave a procedure for testing claims regarding the median of a single population, or regarding the medians of two populations using matched-pair samples. The test is a nice first example of a distribution free test because it is simple to conduct and uses the probabilities of a distribution which we understand well, namely the binomial distribution.

The Signs Test has a weakness, however, in that it only considers whether values are above or below an assumed median, but does not consider how extreme the values are in their distance from the assumed median. This weakness has the effect of lowering the *power* of the Sign Test, and so we attempt to remedy this weakness here. Our remedy is the *Wilcoxon Signed Rank Test*, where we will continue to associate negative (−) signs to values which are below an assumed median, and positive (+) signs for values which are above an assumed median, but with these signs we will now also associate *ranks*. A *rank* is simply the position which a given value occupies in a sample, after the sample is sorted according to distance from the assumed median. In the one sample Signed Rank Test, we determine ranks for each sample member, and apply the negative (−) sign to the ranks corresponding to values which are below the value of the assumed median. Once these ranks are applied, we will compute the sum of the ranks corresponding to either the positive or negative signs, to arrive at the *rank sums*, $R_{(+)}$ or $R_{(-)}$, respectively. We will make a decision about our null hypothesis based on whether the rank sum is unusually large or small.

The Signed Rank Test has higher *power* than the Sign Test. For a given level of significance, the Signed Rank Test is more likely than the Sign Test to bring about a rejection of a false null hypothesis. That is good, but as always, there is a trade-off. The test is more powerful, but it has a requirement – the populations which are of interest should have symmetric distributions, with *skewness* equal to zero. This is the requirement of the Signed Rank Test, but there is an added benefit. Symmetric distributions have the property that the mean of the distribution is exactly equal to the median of the distribution. Because of this fact, claims tested by the Signed Rank Test may be directed equivalently toward the mean or median of the distribution.

The difficulty of the Signed Rank Test is that the probabilities associated with the rank sums, $R_{(+)}$ or $R_{(-)}$, are not distributed according to any distribution which we have studied so far, and the derivation of these probabilities is quite difficult. Because of this difficulty, we will rely heavily on new tables to determine critical values and *p-values* for the rank sum.

Example 12.2.1 The following data are taken from "Synovial Fluid pH, Lactate, Oxygen and Carbon Dioxide Partial Pressure in Various Joint Diseases" (Arthritis and Rheumatism, 1971, pp. 476–477): { 7.10, 7.35, 7.77, 6.95, 7.09 }. The values represent pH values of synovial fluid (a joint lubricant) taken from the knees of patients suffering from arthritis. We would like to test the claim at 5% significance that the mean pH is greater than 7.04. Thus, our null hypothesis will be $H_0 : \mu = 7.04$, and the alternative hypothesis is $H_1 : \mu > 7.04$. We will first determine the distance from each value to the assumed mean $\mu = 7.04$, by using $|x - 7.04|$, then assign ranks according to the distance from 7.04, from smallest to largest.

x	6.95	7.09	7.10	7.35	7.77
$|x-7.04|$	0.09	0.04	0.06	0.31	0.73
Ranks	3	1	2	4	5

After this, we assign the negative symbol $(-)$ to the ranks associated with sample values below 7.04, and the positive symbol $(+)$ to ranks associated with values greater than 7.04, our assumed mean.

x	6.95	7.09	7.10	7.35	7.77
$|x-7.04|$	0.09	0.04	0.06	0.31	0.73
Ranks	3	1	2	4	5
Signs	$(-)$	$(+)$	$(+)$	$(+)$	$(+)$

Once we have signs and ranks, we are able to compute the rank sums, $R_{(+)}$ or $R_{(-)}$. This test claims that the mean is greater than 7.04, and so we would expect the sum of ranks corresponding to values greater than 7.04 $R_{(+)}$, to be large, and the sum of ranks corresponding to values less than 7.04, $R_{(-)}$, to be small.

Because of the nature of the table of *p-values* for this test, we will use $R_{(-)}$ to measure significance on a right tail test, and $R_{(+)}$ to measure significance on a left tail test.

In this example, $R_{(+)} = 1 + 2 + 4 + 5 = 12$, while $R_{(-)} = 3$. With $H_1 : \mu > 7.04$ we will be conducting a right tail test, and the rank sum we will use will be $R_{(-)} = 3$. The question now becomes whether this rank sum is unusually low or not. The only way that this question can be answered is to consider all possible outcomes for the rank sums.

In summing randomly selected ranks ranging from 1 to 5, it will be important to know what the largest possible sum could be. In computing $R_{(-)}$, the largest possible sum would occur if all values are below the assumed mean, having $(-)$ signs associated in each case. In this case the sum of ranks is $1 + 2 + 3 + 4 + 5 = 15$. When n is large it will be convenient to know that the sum of all integers from 1 to n is equal to $n(n+1)/2$. To verify this in our particular case with $n = 5$, the formula becomes $5 \cdot (5+1)/2 = 15$, which is consistent with the sum computed above. The smallest sum possible would occur when no values are below the assumed mean, and in such a case we would assign 0 to the rank sum, $R_{(-)}$.

Thus, rank sums ranging from 0 to 15 are possible, and now we consider each, and how many ways they could be obtained.

Remember, in computing the rank sum, $R_{(-)}$, we only add ranks corresponding to $(-)$ signs.

x_R=Rank Sum	Signs associated to the ranks 1, 2, 3, 4, 5	Ways	$P(x_R)$
0	(+)(+)(+)(+)(+)	1	1/32
1	(-)(+)(+)(+)(+)	1	1/32
2	(+)(-)(+)(+)(+)	1	1/32
3	(-)(-)(+)(+)(+),(+)(+)(-)(+)(+)	2	2/32
4	(-)(+)(-)(+)(+),(+)(+)(+)(-)(+)	2	2/32
5	(-)(+)(+)(-)(+),(+)(-)(-)(+)(+),(+)(+)(+)(+)(-)	3	3/32
6	(-)(+)(+)(+)(-),(-)(-)(-)(+)(+),(+)(-)(+)(-)(+)	3	3/32
7	(-)(-)(+)(-)(+),(+)(-)(+)(+)(-),(+)(+)(-)(-)(+)	3	3/32
8	(-)(-)(+)(+)(-),(-)(+)(-)(-)(+),(-)(-)(+)(+)(-)	3	3/32
9	(-)(+)(-)(+)(-),(+)(-)(-)(-)(+),(+)(+)(+)(-)(-)	3	3/32
10	(-)(+)(+)(-)(-),(+)(-)(-)(+)(-),(-)(-)(-)(-)(+)	3	3/32
11	(-)(-)(-)(+)(-),(+)(-)(+)(-)(-)	2	2/32
12	(-)(-)(+)(-)(-),(+)(+)(-)(-)(-)	2	2/32
13	(-)(+)(-)(-)(-)	1	1/32
14	(+)(-)(-)(-)(-)	1	1/32
15	(-)(-)(-)(-)(-)	1	1/32
	Total	32	32/32

Notice the rising and falling of probabilities as we work our way through the possible outcomes. One might wonder if a discussion regarding the normality of the rank sums is coming!

12.2.1 Determining p-Values

The *p-value* for this test is simply

$$
\begin{aligned}
P\left(x_R \leqslant R_{(-)}\right) &= P\left(x_R \leqslant 3\right) \\
&= (1 + 1 + 1 + 2)/32 \\
&= 5/32 \\
&= 0.15625,
\end{aligned}
$$

which indicates that this rank sum is not unusually low, and thus we will not support the alternative hypothesis. We will conduct this test more formally below, after we generalize the approach to any value of n.

The construction of all possible ways to arrive at all possible rank sums is a tedious process even with n as small as 5. Clearly we need a better approach than counting by listing all possible outcomes. We begin by examining the denominators of the probabilities above. Before reducing, each probability has a denominator equal to 32 because there were 32 possible outcomes. It is not difficult to see how the 32 arises – with 5 trials and 2 possible outcomes per trial, there are $2^5 = 32$ possible outcomes. Constructing the numerators is much more difficult, and there is no direct formula for computing them. There are indirect approaches for computing these, however, and they are really quite ingenious.

One indirect approach comes from a field of math referred to as *combinatorics* which is concerned with methods of counting. We need to count the number of ways any particular rank sum could arise – this number of ways will give the numerator for each of our probabilities. In combinatorics, a special tool called a *generating function* is sometimes used for generating a sequence of numbers. We will use a generating function to develop the numerators of our probabilities.

We have seen generating functions before. In chapter four we learned that binomial probabilities from n trials can be constructed by raising the binomial $(q + p)$ to the nth power. For example, with $n = 3$, we compute

$$(q + p)^3 = q^3 + 3pq^2 + 3p^2q + p^3.$$

The entire binomial distribution is generated from the terms of this product: $P(0) = q^3$, $P(1) = 3pq^2$, $P(2) = 3p^2q$, and $P(3) = p^3$.

The generating function that generates the numbers of ways to get every possible rank sum from 0 to 15 with a sample of size $n = 5$ is $(x^1 + 1)(x^2 + 1)(x^3 + 1)(x^4 + 1)(x^5 + 1)$. After multiplying and collecting on this polynomial, we arrive at

$$\left(x^1+1\right)\left(x^2+1\right)\left(x^3+1\right)\left(x^4+1\right)\left(x^5+1\right)$$
$$= x^{15}+x^{14}+x^{13}+2x^{12}+2x^{11}+3x^{10}+3x^9+3x^8+3x^7+3x^6+3x^5+2x^4+2x^3+x^2+x+1.$$

The coefficients of these terms are $\{1,\ 1,\ 1,\ 2,\ 2,\ 3,\ 3,\ 3,\ 3,\ 3,\ 3,\ 2,\ 2,\ 1,\ 1,\ 1\}$. This sequence of numbers is exactly the sequence constructed above in counting the number of ways to get each of the 16 rank sums. These are the numerators of each probability listed, respectively.

In general, for a sample of size n, there are $n(n + 1)/2$ total possible rank sums for either $R_{(+)}$ or $R_{(-)}$. The probabilities all have denominators equal to 2^n, and the numerators are the coefficients of the product $(x^1 + 1)(x^2 + 1)\cdots(x^n + 1)$, after multiplying, combining like terms, and arranging in descending powers of x.

From these probabilities, a cumulative probability distribution is created, and the result is Table A10, giving *p-values* for each possible rank sum for all values of n up to $n = 20$.

12.2.2 Determining Critical Values

Now we turn our attention to critical values. Because the distribution of possible rank sums, x_R, is a *discrete* probability distribution, finding exact critical values according to a particular value for α becomes impossible. Still, it is possible to give critical values which *nearly* correspond to a chosen α value, as is demonstrated below.

Suppose in our example that we wish to determine a critical value which corresponds as nearly as possible to a given level of significance, α. To do this, we sum probabilities starting at $x_R = 0$ as we did for the *p-value* but only add probabilities until we come to a value c where $P(x_R \leqslant c) \approx \alpha$. The sum of probabilities from $x_R = 0$ to $x_R = c$ is called a *cumulative probability*. When we determine a value c where the cumulative probability $P(x_R \leqslant c) \approx \alpha$, we have found our critical value. Suppose we wish to use $\alpha = 0.05$ from our example above. To find our critical value, we search for a cumulative probability that is closest to $\alpha = 0.05$. The probabilities and cumulative probabilities for all possible rank sums, x_R, starting at zero, and continuing though the last possible rank sum $n(n+1)/2 = 5 \cdot (5+1)/2 = 15$, are given below.

$x_R = R_{(-)}$	Signs associated to the ranks 1, 2, 3, 4, 5	Ways	$P\left(x_R = R_{(-)}\right)$	$P\left(x_R \leqslant R_{(-)}\right)$
0	(+)(+)(+)(+)(+)	1	1/32	1/32=0.03125
1	(−)(+)(+)(+)(+)	1	1/32	2/32=0.06250
2	(+)(−)(+)(+)(+)	1	1/32	3/32=0.09375
3	(−)(−)(+)(+)(+), (+)(+)(−)(+)(+)	2	2/32	5/32=0.15625
4	(−)(+)(−)(+)(+), (+)(+)(+)(−)(+)	2	2/32	7/32=0.21875
5	(−)(+)(+)(−)(+), (+)(−)(−)(+)(+), (+)(+)(+)(+)(−)	3	3/32	10/32=0.31250
6	(−)(+)(+)(+)(−), (−)(−)(−)(+)(+), (+)(−)(+)(−)(+)	3	3/32	13/32=0.40625
7	(−)(−)(+)(−)(+), (+)(−)(+)(+)(−), (+)(+)(−)(−)(+)	3	3/32	16/32=0.50000
8	(−)(−)(+)(+)(−), (−)(+)(−)(−)(+), (−)(−)(+)(+)(−)	3	3/32	19/32=0.59375
9	(−)(+)(−)(+)(−), (+)(−)(−)(−)(+), (+)(+)(+)(−)(−)	3	3/32	22/32=0.68750
10	(−)(+)(+)(−)(−), (+)(−)(−)(+)(−), (−)(−)(−)(−)(+)	3	3/32	25/32=0.78125
11	(−)(−)(−)(+)(−), (+)(−)(+)(−)(−)	2	2/32	27/32=0.84375
12	(−)(−)(+)(−)(−), (+)(+)(−)(−)(−)	2	2/32	29/32=0.90625
13	(−)(+)(−)(−)(−)	1	1/32	30/32=0.93750
14	(+)(−)(−)(−)(−)	1	1/32	31/32=0.96875
15	(−)(−)(−)(−)(−)	1	1/32	32/32=1.00000
Totals		32	32/32	

The closest cumulative probability to our $\alpha = 0.05$ is 0.0625, and this corresponds to $x_R = 1$. This will be our critical value. Since our rank sum was $R_{(-)} = 3$, we will fail to reject the null hypothesis.

It is important to note that the cumulative probability corresponding to $x_R = 1$ was not exactly equal to our chosen $\alpha = 0.05$. Because of this fact, the true level of significance for this test is the cumulative probability, which was 0.0625.

Critical values for the Signed Rank Test for $n \leqslant 20$ are included on Table A9. One might wonder why the table stops at $n = 20$. The answer is most easily demonstrated by examining the case where $n = 20$ with probabilities associated to all possible rank sums from 0 to $n(n+1)/2 = 20 \cdot (20+1)/2 = 210$.

These probabilities are plotted as a histogram in Figure 12.1.

It seems that the rank sums approach a normal distribution with large values of n. To give evidence in support of this possibility, the *skewness* of the above distribution is zero (identical to the normal distribution), and the *kurtosis* is 2.825, quite near to the normal

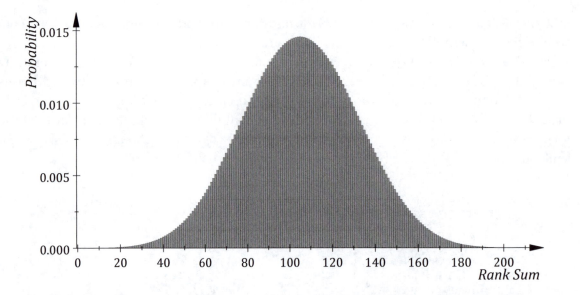

Figure 12.1: The probabilities of all possible rank sums, corresponding to $n = 20$. The distribution appears normal.

standard of 3.0. In fact, we will say that the rank sums are approximately normal as long as $n > 20$. When $n > 20$ we will use the normal distribution to estimate the *p-value* associated with a particular rank sum. That sounds easy enough, but we cannot proceed in using the normal approximation of the distribution of rank sums unless we know the mean, variance and the standard deviation of the rank sums.

12.2.3   Rank Sums' Mean, Variance, and Standard Deviation

We now derive formulas for the mean, variance and the standard deviation of the rank sums using the properties of expected value.

For a sample of size n we remember that there are n ranks of values distributed according to the distance from the assumed value of the mean. If the assumed value of the mean is correct, there is a 50% chance that any given rank lies below the mean, which is equal to the median, and a 50% chance that it lies above. Suppose we are computing $R_{(-)}$. If a rank, say i, lies below the mean, we will add it to the rank sum, but if i lies above the mean, it will not be counted, so we will add zero to the rank sum. Each outcome has a 50% chance.

Thus, for each possible rank, i, $1 \leqslant i \leqslant n$, we define a random variable W_i, equal to 0 or i, each with a 50% probability. The values, 0 or i, of the random variable are appropriate because either a particular rank, i, will be counted in the rank sum (if it is associated with a $(-)$ sign), or it will not, in which case we add a zero instead. The probability that W_i is equal to i is 50% because when our sample contains a value whose rank is i, there is a 50% chance that the value is below the mean, which is assumed to be equal to the median, and half of all values lie below the median. The probability that W_i is equal to 0 is also 50% by the rule of complements.

The expected value of W_i is

$$E(W_i) = \sum W_i \cdot P(W_i)$$
$$= i \cdot 0.50 + 0 \cdot 0.50$$
$$= 0.50 \cdot i.$$

The variance of W_i is $Var(W_i) = E(W_i^2) - E(W_i)^2$. To compute this, we first compute $E(W_i^2)$.

$$E(W_i^2) = \sum W_i^2 \cdot P(W_i)$$
$$= i^2 \cdot 0.50 + 0^2 \cdot 0.50$$
$$= 0.50 \cdot i^2$$

We have already computed $E(W_i)$, so

$$Var(W_i) = E(W_i^2) - E(W_i)^2$$
$$= 0.50 \cdot i^2 - (0.5 \cdot i)^2$$
$$= 0.50i^2 - 0.25i^2$$
$$= 0.25i^2.$$

As mentioned, we add ranks corresponding to negative $(-)$ signs and zeros corresponding to positive $(+)$ signs to compute the rank sum, $R_{(-)} = \sum W_i$, and so

$$E(R_{(-)}) = E\left(\sum W_i\right)$$
$$= \sum E(W_i)$$
$$= \sum 0.50 \cdot i.$$

We have already mentioned that the sum of values i, for $1 \leqslant i \leqslant n$, is equal to $\frac{n(n+1)}{2}$.

$$E(R_{(-)}) = \sum 0.50 \cdot i$$
$$= 0.50 \cdot \sum i$$
$$= \frac{1}{2} \frac{n(n+1)}{2} = \frac{n(n+1)}{4}$$

To compute the variance, we need the fact from algebra that the sum of all values i^2, for $1 \leqslant i \leqslant n$, is equal to $\frac{n(n+1)(2n+1)}{6}$.

Thus,

$$
\begin{aligned}
Var\left(R_{(-)}\right) &= Var\left(\sum W_i\right) \\
&= \sum Var\left(W_i\right) \\
&= \sum 0.25 \cdot i^2 \\
&= 0.25 \cdot \sum i^2 \\
&= \frac{1}{4} \cdot \frac{n\left(n+1\right)\left(2n+1\right)}{6} \\
&= \frac{n\left(n+1\right)\left(2n+1\right)}{24}.
\end{aligned}
$$

Applying a square root to the variance gives the standard deviation, and now we may now take the z-score of a rank sum, $R_{(-)}$. For $R_{(+)}$ the formula is similar.

For $R_{(-)}$ use

$$
z = \frac{R_{(-)} - n\left(n+1\right)/4}{\sqrt{n\left(n+1\right)\left(2n+1\right)/24}}.
$$

For $R_{(+)}$ use

$$
z = \frac{R_{(+)} - n\left(n+1\right)/4}{\sqrt{n\left(n+1\right)\left(2n+1\right)/24}}.
$$

It might be wondered why we did not include a continuity correction in this test statistic as we did in the z-score from the Signs Test. The reason for this is because the variance for the rank sums $R_{(-)}$ and $R_{(+)}$ is so large when $n > 20$ that the small continuity correction has almost no effect on the value of the test statistic.

With all of our theory out of the way, we may finally describe a procedure for the Wilcoxon Signed Rank Test.

Procedure 12.2.2 The Wilcoxon Signed Rank Test for a Single Population Mean

Step I. Gather Data and Verify Requirements
For this test we need the sample data, sample size, and level of significance. This test requires that the population is symmetric.

Step II. Identify Claims
The null hypothesis is $H_0 : \mu = a$. The alternative hypotheses are $H_1 : \mu > a$, $H_1 : \mu < a$, or $H_1 : \mu \neq a$.

Step III. Compute the Test Statistic and p-Value
First, for each x, compute the distances from the assumed mean $|x - a|$. Use these distances to assign ranks 1 through n, according to the distances from the mean $|x - a|$, from small to large. When two or more values *tie* because they are equal in distance from the assumed mean, assign the mean of the

ranks which would be applied to these values. Next, to each rank we associate a negative sign (−) for each corresponding x which is below the mean and a positive sign (+) to each rank corresponding to an x which is above the mean. No sign is associated to values which are equal to the assumed mean. Next we compute a rank sum R (either $R_{(−)}$ or $R_{(+)}$ depending on the type of test).

- For a right tail test, compute $R = R_{(−)} = sum\ of\ ranks$ associated with a (−) sign.

- For a left tail test, compute $R = R_{(+)} = sum\ of\ ranks$ associated with a (+) sign.

- For a two tail test, use the minimum of the two rank sums,

$$R = minimum \left\{ R_{(−)}, R_{(+)} \right\}.$$

Next, we determine the test statistic and, optionally, we may look up the corresponding *p-value*.

- If $n \leqslant 20$, the test statistic is simply R, the rank sum computed above. If desired we may look this up on Table A10 to determine the *p-value*. For one tail tests, the value found in this table is our *p-value*. Multiply by two to get the *p-value* on a two tail test.

- If $n > 20$, we use the normal approximation with R equal to the rank sum mentioned above. The test statistic is

$$T.S. : z = \frac{R - n\,(n+1)\,/4}{\sqrt{n\,(n+1)\,(2n+1)\,/24}},$$

and the optional *p-value* is found by looking up this z-score on Table A1. The value found in the table is the *p-value* for a one tail test – multiply by two for the *p-value* of a two tail test.

Step IV. **Look Up the Critical Value**

- If $n \leqslant 20$, find the critical value, x_R, corresponding to the appropriate level of significance and sample size in Table A9. It should be noted that the true level of significance will be slightly different from the chosen level as discussed previously.

- If $n > 20$, look up the z critical value in the bottom row of Table A2. The z critical value *should be made negative* for this test.

Step V. **Make a Decision**

If the test statistic is less than or equal to the critical value, reject the null hypothesis in favor of the alternative. Whenever the null hypothesis is rejected, the *p-value* will be less than α.

Step VI. **Make a Conclusion**

As usual, conclusions should be written in the context of the problem, without the use of symbolic notation.

Example 12.2.1 Continued: We now apply the formal process given above to Example 12.2.1. As a reminder, these values represent pH values of synovial fluid (a joint lubricant) taken from the knees of patients suffering from arthritis. We are testing the claim at 5% significance that the mean pH is greater than 7.04.

Step I. **Gather Data and Verify Requirements**

Our sample size is $n = 5$, and our sample is as before: { 7.10, 7.35, 7.77, 6.95, 7.09 }. We assume that the sample is taken from a symmetric population. Our level of significance is $\alpha = 0.05$.

Step II. **Identify Claims**

Our null hypothesis will be $H_0 : \mu = 7.04$, and the alternative hypothesis is $H_1 : \mu > 7.04$.

Step III. **Compute the Test Statistic and p-Value**

The ranks and signs are given below, as previously computed.

x	6.95	7.09	7.10	7.35	7.77
$\lvert x-7.04 \rvert$	0.09	0.04	0.06	0.31	0.73
Ranks	3	1	2	4	5
Signs	(−)	(+)	(+)	(+)	(+)

We are conducting a right tail test, so the rank sum of interest is $R = R_{(-)} = 3$. This time our n is less than 20, so we look up this sum in Table A10 to find the p-value. Part of this table is reproduced below. The first column is R, representing whatever rank sum is being used.

R	$n = 1$	$n = 2$	$n = 3$	$n = 4$	$n = 5$	$n = 6$
0	0.5000	0.2500	0.1250	0.0625	0.0313	0.0156
1	1.0000	0.5000	0.2500	0.1250	0.0625	0.0313
2		0.7500	0.3750	0.1875	0.0938	0.0469
3		1.0000	0.6250	0.3125	0.1563	0.0781
4			0.7500	0.4375	0.2188	0.1094
5			0.8750	0.5625	0.3125	0.1563

For a one tail test, the p-value is found directly in the table, so we have p-value = 0.1563.

Step IV. **Look Up the Critical Value**

In this example, $n \leqslant 10$, so we turn to Table A9 for the critical value.

n	One Tail Applications			
	$\alpha \approx 0.005$	$\alpha \approx 0.01$	$\alpha \approx 0.025$	$\boldsymbol{\alpha \approx 0.05}$
4	–	–	–	0
5	–	–	0	**1**
6	–	–	1	2

As before, our critical value is $x_R = 1$.

Step V. **Make a Decision**

Since our test statistic $R = R_{(-)} = 3$ is not less than or equal to our critical value $x_R = 1$, we cannot reject the null hypothesis. Here the *p-value* of 0.1563 is not less than $\alpha = 0.05$, affirming that the null hypothesis should not be rejected.

Step VI. **Make a Conclusion**

The data are not in support of the claim that the mean pH level is greater than 7.04.

Example 12.2.3 In Section 12.1, we gave the miles per gallon measurements for ten round-trip commutes in a Toyota Prius Hybrid vehicle from Lake Arrowhead to Walnut, California. The miles per gallon measurements were: {46.2, 44.9, 47.0, 46.3, 45.6, 46.0, 44.6, 45.9, 46.3, 45.6}. We used the Signs Test to test the claim at 5% significance that the median miles per gallon on this round-trip commute is equal to 45 miles per gallon against the claim that the mean is not 45 miles per gallon. In this section we would like to test the same claim, but now we will use Signed Ranks Test as we proceed.

Step I. **Gather Data and Verify Requirements.**

Our sample size is $n = 10$, and our sample is {46.2, 44.9, 47.0, 46.3, 45.6, 46.0, 44.6, 45.9, 46.3, 45.6}. We assume that this sample is taken from a symmetric population. Our level of significance is again $\alpha = 0.05$.

Step II. **Identify Claims**

Our claims are $H_0 : \mu = 45$, and $H_1 : \mu \neq 45$.

Step III. **Compute the Test Statistic and *p*-Value**

The distances from the assumed mean are given below.

x	44.6	44.9	45.6	45.6	45.9	46	46.2	46.3	46.3	47
$\lvert x-45 \rvert$	0.4	0.1	0.6	0.6	0.9	1.0	1.2	1.3	1.3	2.0

Note that there are two ties – there are two 0.6 distances, and two 1.3 distances. The ranks will be averaged for these, so that they each receive the same rank.

The first two x values are below the assumed mean of 45, so their ranks will be associated with (−) signs. All other x values are greater than 45, so their ranks will be associated with (+) signs.

x	44.6	44.9	45.6	45.6	45.9	46	46.2	46.3	46.3	47
$\|x-45\|$	0.4	0.1	0.6	0.6	0.9	1.0	1.2	1.3	1.3	2.0
Ranks	2	1	3.5	3.5	5	6	7	8.5	8.5	10
Sign	(−)	(−)	(+)	(+)	(+)	(+)	(+)	(+)	(+)	(+)

This is a two tail test, so we need to pick the minimum of the rank sums $R_{(-)}$ and $R_{(+)}$.

$$R_{(-)} = 2 + 1 = 3,$$
$$R_{(+)} = 3.5 + 3.5 + 5 + 6 + 7 + 8.5 + 8.5 + 10 = 52,$$
$$R = minimum\{R_{(-)}, R_{(+)}\} = 3.$$

For the *p-value*, we choose the smaller of these rank sums, which is $R = R_{(-)}$ = 3. With this, we again use Table A10 (since $n \leqslant 20$) to look up the *p-value*.

R	$n=6$	$n=7$	$n=8$	$n=9$	$n=10$	$n=11$
0	0.0156	0.0078	0.0039	0.0020	0.0010	0.0005
1	0.0313	0.0156	0.0078	0.0039	0.0020	0.0010
2	0.0469	0.0234	0.0117	0.0059	0.0029	0.0015
3	0.0781	0.0391	0.0195	0.0098	0.0049	0.0024
4	0.1094	0.0547	0.0273	0.0137	0.0068	0.0034
5	0.1563	0.0781	0.0391	0.0195	0.0098	0.0049

This time, the probability of interest is 0.0049, but since this is a two tail test, we multiply by two, to arrive at $p\text{-}value = 2 \cdot 0.0049 = 0.0098$.

Step IV. **Look Up the Critical Value**

In this example, $n \leqslant 10$, so we turn to Table A9 for the critical value corresponding to $n = 10$ and $\alpha = 0.05$.

	Two Tail Applications			
n	$\alpha \approx 0.01$	$\alpha \approx 0.02$	$\boldsymbol{\alpha \approx 0.05}$	$\alpha \approx 0.10$
8	0	1	4	6
9	2	3	6	8
10	3	4	**8**	11
11	5	7	11	14

This time we see a critical value of $x_R = 8$.

Step V. **Make a Decision**

Our test statistic was $R = R_{(-)} = 3$, and the critical value was $x_R = 8$. Since the test statistic is less than or equal to the critical value, we will reject the null hypothesis in favor of the alternative. This time our $p\text{-value} = 0.0098$ is less than $\alpha = 0.05$, emphasizing that the rejection comes through highly significant sample evidence.

Step VI. **Make a Conclusion**

The data do not support the claim that the mean miles per gallon for the commute mentioned above is not equal to 45.

Example 12.2.4 For our next example, we consider grade point averages for randomly selected students at Mt. San Antonio College in Walnut, California. { 2.7, 1.7, 2.0, 3.7, 3.1, 4.0, 3.4, 3.7, 2.5, 2.8, 3.0, 3.6, 3.3, 3.2, 3.1, 3.2, 2.6, 2.8, 2.0, 4.0, 2.7, 3.9, 3.1, 2.5, 3.7, 2.9, 3.5, 3.7, 2.2, 3.8 }. Test the claim at 1% significance that the mean G.P.A. of students at this college is less than 3.60

Step I. **Gather Data and Verify Requirements.** We have $n = 30$, from the above sample taken from a population assumed to be symmetric. This time we use $\alpha = 0.01$.

Step II. **Identify Claims**: Our hypotheses are $H_0 : \mu = 3.60$, and $H_1 : \mu < 3.60$.

Step III. **Compute the Test Statistic and p-Value**: This is a left tail test, so we will be computing $R_{(+)}$. The distances, ranks, signs, and ranks associated with positive $(+)$ signs are given below.

x	$|x-3.60|$	Rank	Sign	$(+)$ Ranks
2.70	0.90	22	$(-)$	
1.70	1.90	30	$(-)$	
2.00	1.60	28	$(-)$	
3.68	0.08	3	$(+)$	3
3.10	0.50	15	$(-)$	
4.00	0.40	13	$(+)$	13
3.40	0.20	8	$(-)$	
3.74	0.14	6	$(+)$	6
2.50	1.10	25	$(-)$	
2.80	0.80	20	$(-)$	
3.00	0.60	18	$(-)$	
3.56	0.04	1	$(-)$	
3.33	0.27	9	$(-)$	
3.20	0.40	11	$(-)$	
3.10	0.50	16	$(-)$	
3.20	0.40	12	$(-)$	
2.60	1.00	24	$(-)$	
2.80	0.80	21	$(-)$	
2.00	1.60	29	$(-)$	
4.00	0.40	14	$(+)$	14
2.70	0.90	23	$(-)$	
3.88	0.28	10	$(+)$	10
3.10	0.50	17	$(-)$	
2.50	1.10	26	$(-)$	
3.65	0.05	2	$(+)$	2
2.90	0.70	19	$(-)$	
3.50	0.10	5	$(-)$	
3.68	0.08	4	$(+)$	4
2.20	1.40	27	$(-)$	
3.80	0.20	7	$(+)$	7
		Total:		59

The sum of the ranks associated with positive signs is $R = R_{(+)} = 3 + 13 + 6 + 14 + 10 + 2 + 4 + 7 = 59$. This time our sample size is greater than 20, so we use the normal approximation, and compute a z test statistic:

$$T.S. : z = \frac{R - n(n+1)/4}{\sqrt{n(n+1)(2n+1)/24}}$$
$$= \frac{59 - 30(30+1)/4}{\sqrt{30(30+1)(2 \cdot 30+1)/24}}$$
$$= -3.57.$$

Looking up this value in Table A1, we see that the *p-value* is 0.0002.

Step IV. **Look Up the Critical Value**

Again, since our sample size is greater 20, we note that we are using the normal approximation to the distribution of rank sums, so we turn to the bottom row of Table A2 for critical values. Under $\alpha = 0.01$ with a one tail test, we see the z-score of 2.326. Remembering that z critical values are always made negative for this test, we will use $C.V. : z = -2.326$.

Step V. **Make a Decision**

We again reject the null hypothesis since our test statistic $T.S. : z = -3.57$ is less than the critical value $C.V. : z = -2.326$. Our *p-value* = 0.0002 is less than $\alpha = 0.01$, again indicating a highly significant departure from the null hypothesis.

Step VI. **Make a Conclusion**

The data support the claim that the mean G.P.A. for Mt. San Antonio College students is less than 3.60.

12.2.4   The Signed Rank Test for a Mean Difference

The Signed Rank test, like the Sign Test, can also be used to compare means from two populations, using matched pair samples. The process is similar to the single sample test, but is slightly modified for this application below. Also, as with the Signs Test, the test assumes that the population of differences is symmetric and continuous. This assumption is guaranteed if both distributions individually have the same shape and variation, differing only in their mean, but may be true even when the distributions don't have the same shape.

Procedure 12.2.5 The Signed Rank Test for a Mean Difference

Step I. **Gather Data and Verify Requirements**

All we need for this test are the paired samples, the sample size, and the level of significance. The population of differences is assumed to be symmetric and continuous.

Step II. **Identify Claims**

The null hypothesis is $H_0 : \mu_1 = \mu_2$ (or $\mu_d = 0$, where $\mu_d = \mu_1 - \mu_2$). Possible alternative hypotheses are $H_1 : \mu_1 > \mu_2$, $H_1 : \mu_1 < \mu_2$, and $H_1 : \mu_1 \neq \mu_2$.

Step III. **Compute the Test Statistic and p-Value**

First, for each (x, y) pair, compute the distances $x - y$. Use these distances to assign ranks 1 through n, from the smallest distance to the largest. When two or more values tie, assign the average of the ranks which would be applied to these values.

Next, to each rank we associate a negative sign ($-$) for each corresponding x, y pair where $x < y$, and a positive sign ($+$) to each rank corresponding to an x, y pair where $x > y$. No sign is associated to pairs where $x = y$. Next we compute a rank sum R (either $R_{(-)}$ or $R_{(+)}$ depending on the type of test).

- For a right tail test, compute $R = R_{(-)} = sum\ of\ ranks$ associated with a ($-$) sign.
- For a left tail test, compute $R = R_{(+)} = sum\ of\ ranks$ associated with a ($+$) sign.
- For a two tail test, use the minimum of the two rank sums,

$$R = minimum \left\{ R_{(-)}, R_{(+)} \right\}.$$

Next, we determine the test statistic and, optionally, we may look up the corresponding p-value.

- If $n \leqslant 20$, the test statistic is simply R, the rank sum computed above. If desired we may look this up on Table A10 to determine the p-value. For one tail tests, the value found in this table is our p-value. Multiply by two to get the p-value on a two tail test.
- If $n > 20$, we use the normal approximation with R equal to the rank sum mentioned above. The test statistic is

$$T.S. : z = \frac{R - n(n+1)/4}{\sqrt{n(n+1)(2n+1)/24}},$$

and the optional p-value is found by looking up this z-score on Table A1. The value found in the table is the p-value for a one tail test – multiply by two for the p-value of a two tail test.

Step IV. **Look Up the Critical Value**

- If $n \leqslant 20$, find the critical value, x_R, corresponding to the appropriate level of significance and sample size in Table A9. It should be noted that the true level of significance will be slightly different from the chosen level as discussed previously.

- If $n > 20$, look up the z critical value in the bottom row of Table A2. The z critical value *should be made negative* for this test.

Step V. **Make a Decision**

If the test statistic is less than or equal to the critical value, reject the null hypothesis in favor of the alternative. Whenever the null hypothesis is rejected, the *p-value* will be less than α.

Step VI. **Make a Conclusion**

As always, conclusions should be written in the context of the problem, without the use of symbolic notation.

We now revisit an example from the previous section, this time making our comparison using the Signed Rank Test.

Example 12.2.6 A study was conducted to investigate whether oat bran lowers serum cholesterol levels in people (Pagano & Gauvreau, 1993, p. 252). Fourteen individuals were randomly assigned a diet that included either oat bran or corn flakes. After two weeks on the initial diet, serum low-density lipoprotein levels (mg/dl) were measured. Each subject was then 'crossed-over" to the alternate diet. After two-weeks on this second diet, low-density lipoprotein levels were once again recorded. The data are recorded below.

Corn Flake Lipoprotein (x)	Oat Bran Lipoprotein (y)
4.61	3.84
6.42	5.57
5.40	5.85
4.54	4.8
3.98	3.68
3.82	2.96
5.01	4.41
4.34	3.72
3.80	3.49
4.56	3.84
5.35	5.26
3.89	3.73
2.25	1.84
4.24	4.14

As before, we test the claim that the mean lipoprotein level is lower for patients who eat oat bran for breakfast, compared to patients who eat corn flakes for breakfast. We will use $\alpha = 0.01$.

Step I. **Gather Data.** Our data are above, and our sample size is $n = 14$. For this example, we have chosen $\alpha = 0.01$.

Step II. **Identify Claims.** Our claims are H_0: $\mu_1 = \mu_2$, and H_1: $\mu_1 > \mu_2$.

Step III. **Compute the Test Statistic and p-Value**

The distances, $|x - y|$, ranks, and associated signs are given below.

Corn Flake Lipoprotein	Oat Bran Lipoprotein	$|x-y|$	Ranks	Signs
4.61	3.84	0.77	12	(+)
6.42	5.57	0.85	13	(+)
5.40	5.85	0.45	8	(−)
4.54	4.80	0.26	4	(−)
3.98	3.68	0.30	5	(+)
3.82	2.96	0.86	14	(+)
5.01	4.41	0.60	9	(+)
4.34	3.72	0.62	10	(+)
3.80	3.49	0.31	6	(+)
4.56	3.84	0.72	11	(+)
5.35	5.26	0.09	1	(+)
3.89	3.73	0.16	3	(+)
2.25	1.84	0.41	7	(+)
4.24	4.14	0.10	2	(+)

We are conducting a right tail test, so the rank sum which we need is again $R = R_{(-)} = 8 + 4 = 12$. Turning again to Table A10, we look up our p-value. This time our p-value is 0.0043, which is again highly significant.

R	$n = 12$	$n = 13$	$n = 14$	$n = 15$
11	0.0134	0.0067	0.0034	0.0017
12	0.0171	0.0085	**0.0043**	0.0021
13	0.0212	0.0107	0.0054	0.0027
14	0.0261	0.0133	0.0067	0.0034

Step IV. **Look Up the Critical Value**

With $n \leqslant 20$, we use Table A9 to find the critical value corresponding to $n = 12$, $\alpha = 0.01$, and a one tail test. In this case the critical value is $x_R = 16$.

	One Tail Applications			
n	$\alpha \approx 0.005$	$\alpha \approx 0.01$	$\alpha \approx 0.025$	$\alpha \approx 0.05$
13	10	13	17	21
14	13	16	21	26
15	16	21	25	30
16	19	24	30	36

Step V. **Make a Decision**

With a rank-sum $R = R_{(-)} = 12$ and a critical value of $x_R = 16$, we reject H_0 in favor of H_1. Our *p-value* = 0.0043 is less than $\alpha = 0.01$, indicating a highly significant departure from the null hypothesis.

Step VI. **Make a Conclusion**

The data support the claim that the mean lipoprotein level is lower for people who eat oat bran for breakfast, compared to people who eat corn flakes.

12.2.5 Historical Note

Frank Wilcoxon (1882 – 1965) was a chemist who became interested in statistics as he searched for chemical fungicides. As a chemist, he was involved with a group whose members were learning the methods taught in the ground-breaking book "Statistical Methods for Research Workers," by Ronald Fisher. Fisher was the discoverer of many widely used concepts in statistics, including analysis of variance.

Figure 12.2: Frank Wilcoxon (*Photo from A R Sampson and B Spencer, A conversation with I. Richard Savage, Statistical Science p136, 1999*).

Frank Wilcoxon, pictured in Figure 12.2, was the first to suggest the replacement of sample values by ranks as we have done in this section. The result of this clever idea was the beginning of a new type of statistics – *distribution free*, or *non-parametric* statistics.

Wilcoxon was born in Glengarriffe Castle, near Cork, Ireland. He grew up in the United States, ran away to sea, worked in the oil industry and as a tree surgeon, and attended military academy before entering college at the age of 26 to study chemistry. He was an avid bicyclist and motorcyclist.

12.2.6   Homework

Use the Wilcoxon Signed Rank to test the following claims. Assume that the samples are drawn from populations with zero *skewness*. Compare your results here to your results from the previous section and note any differences.

1. The data which follow are from an article on peanut butter from *Consumer Reports* (Sept. 1990), which reported the following scores for various brands: {22, 30, 30, 36, 39, 40, 40, 41, 44, 45, 50, 50, 53, 56, 56, 56, 62, 65, 68}. Test the claim at 5% significance that these values are sampled from a population whose mean is less than 55.

2. Prozac is a drug which is meant to elevate the moods of people who suffer from depression. In the following sample (provided by the *WebStat* website of the University of New England, School of Psychology), mood levels are given for patients after taking Prozac. In these values, a higher score means a less depressed psychological state: { 5, 1, 5, 7, 10, 9, 7, 11, 8 }. Test the claim at 5% significance that the mean mood level is greater than 5.

3. To determine the effect that exercise has on lactic acid levels in the blood, the authors of "A Descriptive Analysis of Elite-Level Racquetball" (Research Quarterly for Exercise and Sport, 1991, pp. 109 – 114), listed the following blood lactic acid levels of 8 different players after playing three games of racquetball: { 18, 37, 40, 35, 30, 20, 33, 19 }. Test the claim at 5% significance that the mean blood lactic acid level is greater than 20 after exercise equivalent to three games of racquetball.

4. Do physical characteristics of twin babies affect their birth order? Is it possible that the firstborn has an advantage in some way over the second born? The data below give head circumferences for 10 pairs of twins. The mean head circumference for the firstborn babies was 56.25 cm, while the mean for the second born babies was 56.00 cm. The firstborns have a slightly larger mean head circumference, but is this statistically significant? Test the claim at 5% significance that the mean head circumference is greater for the firstborn babies.

First Born	54.7	53.0	57.8	56.6	53.1	57.2	57.2	57.2	57.2	58.5
Second born	54.2	52.9	56.9	55.3	54.8	57.2	57.2	55.8	56.5	59.2

5. The following values are hours worked per week for randomly selected students at Mt. San Antonio College in Walnut, California. { 15, 20, 40, 16, 20, 25, 19, 40, 40, 20, 20, 40, 0, 40, 35, 30, 35, 25, 29, 0, 0, 40, 16, 0, 40, 30, 25, 16, 35, 0 }. Test the claim at 6% significance that the mean hours worked by students at this college is less than 32 hours per week.

6. The following data are from Conover WJ, *Practical Nonparametric Statistics* (3rd edition). Wiley 1999. The values represent aggressivity scores for 12 pairs of monozygotic twins.

First Born	86	71	77	68	91	72	77	91	70	71	88	87
Second Born	88	77	76	64	96	72	65	90	65	80	81	72

Test the claim at 5% significance that the mean aggressivity score is greater for the firstborn twin.

Part V

Reference Materials

Appendix A
Formulas and Tables

Formulas

Sample Mean/Standard Deviation

$$\bar{x} = \frac{\sum x}{n}, \text{ and } s = \sqrt{\frac{\sum x^2 - \frac{1}{n}\cdot(\sum x)^2}{n-1}}$$

$\bar{x}$ and μ From a Frequency Table

$$\bar{x} \approx \frac{\sum f\cdot x}{n}, \text{ where } n = \sum f,$$

$$s = \sqrt{\frac{\sum f\cdot x^2 - \frac{1}{n}\cdot(\sum f\cdot x)^2}{n-1}}$$

Percentiles

Percentile for $x = \frac{values\ less\ than\ x}{n} \cdot 100$

Transforming x to and from z

$$z = \frac{x-\mu}{\sigma}, \text{ and } x = z\,\sigma + \mu$$

Basic Probability Rules

$$P(S) = x/n$$

If A and B are not mutually exclusive,
$$P(A \text{ or } B) = P(A) + P(B) - P(A \& B)$$

If A and B are mutually exclusive,
$$P(A \text{ or } B) = P(A) + P(B)$$

$$P(S) = 1 - P(\bar{S})$$

If A and B are independent,
$$P(A \& B) = P(A) \cdot P(B)$$

If A and B are dependent,
$$P(A \& B) = P(A) \cdot P(B|A)$$

μ and σ for a Discrete Distribution

$$\mu = E(x) = \sum x \cdot P(x),$$

$$\sigma = \sqrt{\sum x^2 \cdot P(x) - \mu^2}$$

The Binomial Probability Distribution

$$P(x) = {}_nC_x\, p^x\, q^{n-x} = \frac{n!}{x!\,(n-x)!}p^x\, q^{n-x}$$

μ and σ for Binomial Distribution

$$\mu = n\,p \text{ and } \sigma = \sqrt{npq}$$

Confidence Interval for p

$$E = z_{\alpha/2}\sqrt{\frac{\hat{p}\hat{q}}{n}}, \quad \hat{p} - E < p < \hat{p} + E$$

Confidence Interval for μ (σ Known)

$$E = z_{\alpha/2}\frac{\sigma}{\sqrt{n}}, \quad \bar{x} - E < \mu < \bar{x} + E$$

Confidence Interval for μ (σ Unknown)

$$E = t_{\alpha/2}\frac{s}{\sqrt{n}}, \quad \bar{x} - E < \mu < \bar{x} + E$$

Sample Size

Estimating p ($\hat{p}$ known)	Estimating p ($\hat{p}$ unknown)	Estimating μ
$n = \frac{z_{\alpha/2}^2\,\hat{p}\hat{q}}{E^2}$	$n = \frac{z_{\alpha/2}^2(0.25)}{E^2}$	$n = \left(\frac{z_{\alpha/2}\cdot\sigma}{E}\right)^2$

Test Statistics

One Proportion	One Mean (σ known)	One Mean (σ unknown)
$z = \frac{\hat{p}-p}{\sqrt{\frac{pq}{\nu}}}$	$z = \frac{\bar{x}-\mu}{\frac{\sigma}{\sqrt{n}}}$	$t = \frac{\bar{x}-\mu}{\frac{s}{\sqrt{n}}}$

Paired Samples t test: $t = \frac{\bar{d}-\mu_d}{s_d/\sqrt{n}}$

Two Proportions Test

$$z = \frac{(\hat{p}_1-\hat{p}_2)-(p_1-p_2)}{\sqrt{\frac{\bar{p}\bar{q}}{n_1}+\frac{\bar{p}\bar{q}}{n_2}}}, \text{ where } \bar{p} = \frac{x_1+x_2}{n_1+n_2}$$

Two Independent Means t Test

$$t = \frac{(\bar{x}_1-\bar{x}_2)-(\mu_1-\mu_2)}{\sqrt{\frac{s_1^2}{n_1}+\frac{s_2^2}{n_2}}}, \quad df = \frac{(V_1+V_2)^2}{\frac{V_1^2}{n_1-1}+\frac{V_2^2}{n_2-1}}$$

$$\text{where } V_1 = \frac{s_1^2}{n_1} \text{ and } V_2 = \frac{s_2^2}{n_2}$$

Correlation and Regression

$$r = \frac{\sum xy - \frac{1}{n}(\sum x)(\sum y)}{\sqrt{\sum x^2 - \frac{1}{n}(\sum x)^2}\sqrt{\sum y^2 - \frac{1}{n}(\sum y)^2}}$$

Test for correlation:

$$\text{T.S.: } t = \frac{r}{\sqrt{\frac{1-r^2}{n-2}}}, \quad df = n - 2$$

$$a = \frac{\sum xy - \frac{1}{n}(\sum x)(\sum y)}{\sum x^2 - \frac{1}{n}(\sum x)^2}, \quad b = \bar{y} - a\cdot\bar{x}$$

Goodness of Fit

$$\chi^2 = \sum \frac{(O-E)^2}{E}$$

Multinomial Experiment: $df = k - 1$

Contingency Table:
$$df = (rows - 1)(columns - 1)$$

Analysis of Variance

$$F = \frac{n\cdot s_{\bar{x}}^2}{s_p^2}, \quad df_1 = k - 1, \, df_2 = k(n-1)$$

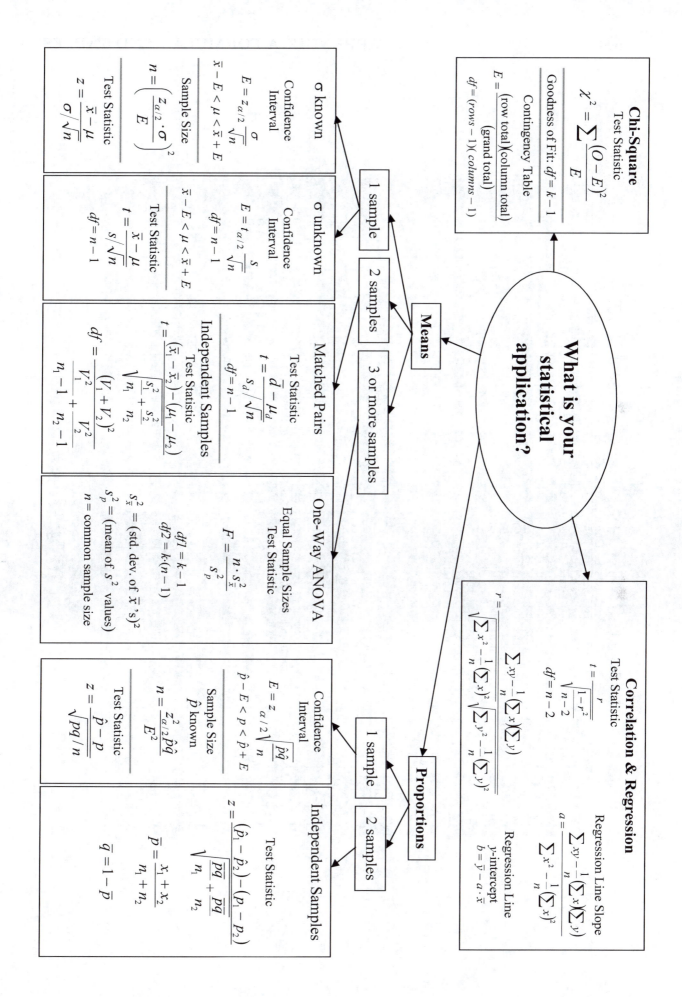

Table A1 – The Standard Normal Distribution – Cumulative Probabilities[1]

Areas to the left of z are given.

z	.00	.01	.02	.03	.04	.05	.06	.07	.08	.09
0.0	.5000	.5040	.5080	.5120	.5160	.5199	.5239	.5279	.5319	.5359
0.1	.5398	.5438	.5478	.5517	.5557	.5596	.5636	.5675	.5714	.5753
0.2	.5793	.5832	.5871	.5910	.5948	.5987	.6026	.6064	.6103	.6141
0.3	.6179	.6217	.6255	.6293	.6331	.6368	.6406	.6443	.6480	.6517
0.4	.6554	.6591	.6628	.6664	.6700	.6736	.6772	.6808	.6844	.6879
0.5	.6915	.6950	.6985	.7019	.7054	.7088	.7123	.7157	.7190	.7224
0.6	.7257	.7291	.7324	.7357	.7389	.7422	.7454	.7486	.7517	.7549
0.7	.7580	.7611	.7642	.7673	.7704	.7734	.7764	.7794	.7823	.7852
0.8	.7881	.7910	.7939	.7967	.7995	.8023	.8051	.8078	.8106	.8133
0.9	.8159	.8186	.8212	.8238	.8264	.8289	.8315	.8340	.8365	.8389
1.0	.8413	.8438	.8461	.8485	.8508	.8531	.8554	.8577	.8599	.8621
1.1	.8643	.8665	.8686	.8708	.8729	.8749	.8770	.8790	.8810	.8830
1.2	.8849	.8869	.8888	.8907	.8925	.8944	.8962	.8980	.8997	.9015
1.3	.9032	.9049	.9066	.9082	.9099	.9115	.9131	.9147	.9162	.9177
1.4	.9192	.9207	.9222	.9236	.9251	.9265	.9279	.9292	.9306	.9319
1.5	.9332	.9345	.9357	.9370	.9382	.9394	.9406	.9418	.9429	.9441
1.6	.9452	.9463	.9474	.9484	.9495	.9505	.9515	.9525	.9535	.9545
1.7	.9554	.9564	.9573	.9582	.9591	.9599	.9608	.9616	.9625	.9633
1.8	.9641	.9649	.9656	.9664	.9671	.9678	.9686	.9693	.9699	.9706
1.9	.9713	.9719	.9726	.9732	.9738	.9744	.9750	.9756	.9761	.9767
2.0	.9772	.9778	.9783	.9788	.9793	.9798	.9803	.9808	.9812	.9817
2.1	.9821	.9826	.9830	.9834	.9838	.9842	.9846	.9850	.9854	.9857
2.2	.9861	.9864	.9868	.9871	.9875	.9878	.9881	.9884	.9887	.9890
2.3	.9893	.9896	.9898	.9901	.9904	.9906	.9909	.9911	.9913	.9916
2.4	.9918	.9920	.9922	.9925	.9927	.9929	.9931	.9932	.9934	.9936
2.5	.9938	.9940	.9941	.9943	.9945	.9946	.9948	.9949	.9951	.9952
2.6	.9953	.9955	.9956	.9957	.9959	.9960	.9961	.9962	.9963	.9964
2.7	.9965	.9966	.9967	.9968	.9969	.9970	.9971	.9972	.9973	.9974
2.8	.9974	.9975	.9976	.9977	.9977	.9978	.9979	.9979	.9980	.9981
2.9	.9981	.9982	.9982	.9983	.9984	.9984	.9985	.9985	.9986	.9986
3.0	.9987	.9987	.9987	.9988	.9988	.9989	.9989	.9989	.9990	.9990
3.1	.9990	.9991	.9991	.9991	.9992	.9992	.9992	.9992	.9993	.9993
3.2	.9993	.9993	.9994	.9994	.9994	.9994	.9994	.9995	.9995	.9995
3.3	.9995	.9995	.9995	.9996	.9996	.9996	.9996	.9996	.9996	.9997
3.4	.9997	.9997	.9997	.9997	.9997	.9997	.9997	.9997	.9997	.9998
3.5	.9998	.9998	.9998	.9998	.9998	.9998	.9998	.9998	.9998	.9998
3.6	.9998	.9998	.9999	.9999	.9999	.9999	.9999	.9999	.9999	.9999
3.7	.9999	.9999	.9999	.9999	.9999	.9999	.9999	.9999	.9999	.9999
3.8	.9999	.9999	.9999	.9999	.9999	.9999	.9999	.9999	.9999	.9999
3.9	.9999	.9999	.9999	.9999	.9999	.9999	.9999	.9999	.9999	.9999
4.0	.9999	.9999	.9999	.9999	.9999	.9999	.9999	.9999	.9999	.9999

z	.00	.01	.02	.03	.04	.05	.06	.07	.08	.09
-4.0	.0001	.0001	.0001	.0001	.0001	.0001	.0001	.0001	.0001	.0001
-3.9	.0001	.0001	.0001	.0001	.0001	.0001	.0001	.0001	.0001	.0001
-3.8	.0001	.0001	.0001	.0001	.0001	.0001	.0001	.0001	.0001	.0001
-3.7	.0001	.0001	.0001	.0001	.0001	.0001	.0001	.0001	.0001	.0001
-3.6	.0002	.0002	.0001	.0001	.0001	.0001	.0001	.0001	.0001	.0001
-3.5	.0002	.0002	.0002	.0002	.0002	.0002	.0002	.0002	.0002	.0002
-3.4	.0003	.0003	.0003	.0003	.0003	.0003	.0003	.0003	.0003	.0002
-3.3	.0005	.0005	.0005	.0004	.0004	.0004	.0004	.0004	.0004	.0003
-3.2	.0007	.0007	.0006	.0006	.0006	.0006	.0006	.0005	.0005	.0005
-3.1	.0010	.0009	.0009	.0009	.0008	.0008	.0008	.0008	.0007	.0007
-3.0	.0013	.0013	.0013	.0012	.0012	.0011	.0011	.0011	.0010	.0010
-2.9	.0019	.0018	.0018	.0017	.0016	.0016	.0015	.0015	.0014	.0014
-2.8	.0026	.0025	.0024	.0023	.0023	.0022	.0021	.0021	.0020	.0019
-2.7	.0035	.0034	.0033	.0032	.0031	.0030	.0029	.0028	.0027	.0026
-2.6	.0047	.0045	.0044	.0043	.0041	.0040	.0039	.0038	.0037	.0036
-2.5	.0062	.0060	.0059	.0057	.0055	.0054	.0052	.0051	.0049	.0048
-2.4	.0082	.0080	.0078	.0075	.0073	.0071	.0069	.0068	.0066	.0064
-2.3	.0107	.0104	.0102	.0099	.0096	.0094	.0091	.0089	.0087	.0084
-2.2	.0139	.0136	.0132	.0129	.0125	.0122	.0119	.0116	.0113	.0110
-2.1	.0179	.0174	.0170	.0166	.0162	.0158	.0154	.0150	.0146	.0143
-2.0	.0228	.0222	.0217	.0212	.0207	.0202	.0197	.0192	.0188	.0183
-1.9	.0287	.0281	.0274	.0268	.0262	.0256	.0250	.0244	.0239	.0233
-1.8	.0359	.0351	.0344	.0336	.0329	.0322	.0314	.0307	.0301	.0294
-1.7	.0446	.0436	.0427	.0418	.0409	.0401	.0392	.0384	.0375	.0367
-1.6	.0548	.0537	.0526	.0516	.0505	.0495	.0485	.0475	.0465	.0455
-1.5	.0668	.0655	.0643	.0630	.0618	.0606	.0594	.0582	.0571	.0559
-1.4	.0808	.0793	.0778	.0764	.0749	.0735	.0721	.0708	.0694	.0681
-1.3	.0968	.0951	.0934	.0918	.0901	.0885	.0869	.0853	.0838	.0823
-1.2	.1151	.1131	.1112	.1093	.1075	.1056	.1038	.1020	.1003	.0985
-1.1	.1357	.1335	.1314	.1292	.1271	.1251	.1230	.1210	.1190	.1170
-1.0	.1587	.1562	.1539	.1515	.1492	.1469	.1446	.1423	.1401	.1379
-0.9	.1841	.1814	.1788	.1762	.1736	.1711	.1685	.1660	.1635	.1611
-0.8	.2119	.2090	.2061	.2033	.2005	.1977	.1949	.1922	.1894	.1867
-0.7	.2420	.2389	.2358	.2327	.2296	.2266	.2236	.2206	.2177	.2148
-0.6	.2743	.2709	.2676	.2643	.2611	.2578	.2546	.2514	.2483	.2451
-0.5	.3085	.3050	.3015	.2981	.2946	.2912	.2877	.2843	.2810	.2776
-0.4	.3446	.3409	.3372	.3336	.3300	.3264	.3228	.3192	.3156	.3121
-0.3	.3821	.3783	.3745	.3707	.3669	.3632	.3594	.3557	.3520	.3483
-0.2	.4207	.4168	.4129	.4090	.4052	.4013	.3974	.3936	.3897	.3859
-0.1	.4602	.4562	.4522	.4483	.4443	.4404	.4364	.4325	.4286	.4247
-0.0	.5000	.4960	.4920	.4880	.4840	.4801	.4761	.4721	.4681	.4641

Common Values from Interpolation

Area	Critical Value	Area	Critical Value
0.0500	-1.645	0.9500	1.645
0.0050	-2.576	0.9950	2.576

Table A2 – Critical Values for Student's t and Normal Distributions[2]

	One Tailed Applications									
	$\alpha{=}0.005$	$\alpha{=}0.01$	$\alpha{=}0.02$	$\alpha{=}0.025$	$\alpha{=}0.03$	$\alpha{=}0.04$	$\alpha{=}0.05$	$\alpha{=}0.06$	$\alpha{=}0.08$	$\alpha{=}0.10$
	Two Tailed Applications									
df	$\alpha{=}0.01$	$\alpha{=}0.02$	$\alpha{=}0.04$	$\alpha{=}0.05$	$\alpha{=}0.06$	$\alpha{=}0.08$	$\alpha{=}0.10$	$\alpha{=}0.12$	$\alpha{=}0.16$	$\alpha{=}0.20$
1	63.657	31.821	15.895	12.706	10.579	7.916	6.314	5.242	3.895	3.078
2	9.925	6.965	4.849	4.303	3.896	3.320	2.920	2.620	2.189	1.886
3	5.841	4.541	3.482	3.182	2.951	2.605	2.353	2.156	1.859	1.638
4	4.604	3.747	2.999	2.776	2.601	2.333	2.132	1.971	1.723	1.533
5	4.032	3.365	2.757	2.571	2.422	2.191	2.015	1.873	1.649	1.476
6	3.707	3.143	2.612	2.447	2.313	2.104	1.943	1.812	1.603	1.440
7	3.499	2.998	2.517	2.365	2.241	2.046	1.895	1.770	1.572	1.415
8	3.355	2.896	2.449	2.306	2.189	2.004	1.860	1.740	1.549	1.397
9	3.250	2.821	2.398	2.262	2.150	1.973	1.833	1.718	1.532	1.383
10	3.169	2.764	2.359	2.228	2.120	1.948	1.812	1.700	1.518	1.372
11	3.106	2.718	2.328	2.201	2.096	1.928	1.796	1.686	1.507	1.363
12	3.055	2.681	2.303	2.179	2.076	1.912	1.782	1.674	1.498	1.356
13	3.012	2.650	2.282	2.160	2.060	1.899	1.771	1.664	1.490	1.350
14	2.977	2.624	2.264	2.145	2.046	1.887	1.761	1.656	1.484	1.345
15	2.947	2.602	2.249	2.131	2.034	1.878	1.753	1.649	1.478	1.341
16	2.921	2.583	2.235	2.120	2.024	1.869	1.746	1.642	1.474	1.337
17	2.898	2.567	2.224	2.110	2.015	1.862	1.740	1.637	1.469	1.333
18	2.878	2.552	2.214	2.101	2.007	1.855	1.734	1.632	1.466	1.330
19	2.861	2.539	2.205	2.093	2.000	1.850	1.729	1.628	1.462	1.328
20	2.845	2.528	2.197	2.086	1.994	1.844	1.725	1.624	1.459	1.325
21	2.831	2.518	2.189	2.080	1.988	1.840	1.721	1.621	1.457	1.323
22	2.819	2.508	2.183	2.074	1.983	1.835	1.717	1.618	1.454	1.321
23	2.807	2.500	2.177	2.069	1.978	1.832	1.714	1.615	1.452	1.319
24	2.797	2.492	2.172	2.064	1.974	1.828	1.711	1.612	1.450	1.318
25	2.787	2.485	2.167	2.060	1.970	1.825	1.708	1.610	1.448	1.316
26	2.779	2.479	2.162	2.056	1.967	1.822	1.706	1.608	1.446	1.315
27	2.771	2.473	2.158	2.052	1.963	1.819	1.703	1.606	1.445	1.314
28	2.763	2.467	2.154	2.048	1.960	1.817	1.701	1.604	1.443	1.313
29	2.756	2.462	2.150	2.045	1.957	1.814	1.699	1.602	1.442	1.311
30	2.750	2.457	2.147	2.042	1.955	1.812	1.697	1.600	1.441	1.310
31	2.744	2.453	2.144	2.040	1.952	1.810	1.696	1.599	1.440	1.309
32	2.738	2.449	2.141	2.037	1.950	1.808	1.694	1.597	1.439	1.309
33	2.733	2.445	2.138	2.035	1.948	1.806	1.692	1.596	1.437	1.308
34	2.728	2.441	2.136	2.032	1.946	1.805	1.691	1.595	1.436	1.307
35	2.724	2.438	2.133	2.030	1.944	1.803	1.690	1.594	1.436	1.306
36	2.719	2.434	2.131	2.028	1.942	1.802	1.688	1.593	1.435	1.306
37	2.715	2.431	2.129	2.026	1.940	1.800	1.687	1.592	1.434	1.305
38	2.712	2.429	2.127	2.024	1.939	1.799	1.686	1.591	1.433	1.304
39	2.708	2.426	2.125	2.023	1.937	1.798	1.685	1.590	1.432	1.304
40	2.704	2.423	2.123	2.021	1.936	1.796	1.684	1.589	1.432	1.303
41	2.701	2.421	2.121	2.020	1.934	1.795	1.683	1.588	1.431	1.303
42	2.698	2.418	2.120	2.018	1.933	1.794	1.682	1.587	1.430	1.302
43	2.695	2.416	2.118	2.017	1.932	1.793	1.681	1.586	1.430	1.302
44	2.692	2.414	2.116	2.015	1.931	1.792	1.680	1.586	1.429	1.301
45	2.690	2.412	2.115	2.014	1.929	1.791	1.679	1.585	1.429	1.301
46	2.687	2.410	2.114	2.013	1.928	1.790	1.679	1.584	1.428	1.300
47	2.685	2.408	2.112	2.012	1.927	1.789	1.678	1.584	1.428	1.300
48	2.682	2.407	2.111	2.011	1.926	1.789	1.677	1.583	1.427	1.299
49	2.680	2.405	2.110	2.010	1.925	1.788	1.677	1.582	1.427	1.299
50	2.678	2.403	2.109	2.009	1.924	1.787	1.676	1.582	1.426	1.299
75	2.643	2.377	2.090	1.992	1.910	1.775	1.665	1.573	1.419	1.293
100	2.626	2.364	2.081	1.984	1.902	1.769	1.660	1.568	1.416	1.290
250	2.596	2.341	2.065	1.969	1.889	1.758	1.651	1.560	1.409	1.285
500	2.586	2.334	2.059	1.965	1.885	1.754	1.648	1.557	1.407	1.283
1000	2.581	2.330	2.056	1.962	1.883	1.752	1.646	1.556	1.406	1.282
Large df (z)	2.576	2.326	2.054	1.960	1.881	1.751	1.645	1.555	1.405	1.282

Table A3 – t-Distribution p-Values

T.S.	1	2	3	4	5	6	7	8	9	10	11	12	13	14	15	16	17	18	19	20	21	22	23	24	25	26	27	28	29	30	40	50	60	70	80	90	100	∞
	Degrees of Freedom																																					
0.0	.500	.500	.500	.500	.500	.500	.500	.500	.500	.500	.500	.500	.500	.500	.500	.500	.500	.500	.500	.500	.500	.500	.500	.500	.500	.500	.500	.500	.500	.500	.500	.500	.500	.500	.500	.500	.500	.500
0.1	.532	.535	.537	.537	.538	.538	.538	.539	.539	.539	.539	.539	.539	.539	.539	.539	.539	.539	.539	.539	.539	.539	.539	.539	.539	.539	.540	.540	.540	.540	.540	.540	.540	.540	.540	.540	.540	.540
0.2	.563	.570	.573	.574	.575	.576	.576	.577	.577	.577	.577	.578	.578	.578	.578	.578	.578	.578	.578	.578	.578	.578	.578	.578	.579	.579	.579	.579	.579	.579	.579	.579	.579	.579	.579	.579	.579	.579
0.3	.593	.604	.608	.610	.612	.613	.614	.614	.615	.615	.615	.615	.616	.616	.616	.616	.616	.616	.616	.616	.616	.617	.617	.617	.617	.617	.617	.617	.617	.617	.617	.617	.617	.618	.618	.618	.618	.618
0.4	.621	.636	.642	.645	.647	.649	.650	.650	.651	.651	.652	.652	.652	.652	.653	.653	.653	.653	.653	.653	.653	.654	.654	.654	.654	.654	.654	.654	.654	.654	.654	.655	.655	.655	.655	.655	.655	.655
0.5	.648	.667	.674	.678	.681	.683	.684	.685	.686	.686	.687	.687	.687	.688	.688	.688	.688	.688	.689	.689	.689	.689	.689	.689	.689	.689	.689	.690	.690	.690	.690	.690	.691	.691	.691	.691	.691	.692
0.6	.672	.695	.705	.710	.713	.715	.716	.717	.718	.719	.720	.720	.721	.721	.721	.722	.722	.722	.722	.722	.723	.723	.723	.723	.723	.723	.723	.723	.723	.724	.724	.724	.725	.725	.725	.725	.725	.726
0.7	.694	.722	.733	.739	.742	.745	.747	.748	.749	.750	.751	.751	.752	.752	.753	.753	.753	.754	.754	.754	.754	.754	.755	.755	.755	.755	.755	.755	.755	.755	.756	.756	.757	.757	.757	.757	.757	.758
0.8	.715	.746	.759	.766	.770	.773	.775	.777	.778	.779	.780	.780	.781	.782	.782	.782	.783	.783	.783	.783	.784	.784	.784	.784	.784	.785	.785	.785	.785	.785	.786	.786	.787	.787	.787	.787	.787	.788
0.9	.733	.768	.783	.791	.795	.799	.801	.803	.804	.805	.806	.807	.808	.808	.809	.809	.810	.810	.810	.811	.811	.811	.811	.812	.812	.812	.812	.812	.812	.812	.813	.814	.814	.814	.815	.815	.815	.816
1.0	.750	.789	.805	.813	.818	.822	.825	.827	.828	.830	.831	.832	.832	.833	.833	.834	.834	.835	.835	.835	.836	.836	.836	.836	.837	.837	.837	.837	.837	.837	.838	.839	.839	.840	.840	.840	.840	.841
1.1	.765	.807	.824	.834	.839	.843	.846	.848	.850	.851	.853	.854	.854	.855	.856	.856	.857	.857	.858	.858	.858	.858	.859	.859	.859	.859	.860	.860	.860	.860	.861	.862	.862	.862	.863	.863	.863	.864
1.2	.779	.824	.842	.852	.858	.862	.865	.868	.870	.871	.872	.873	.874	.875	.876	.876	.877	.877	.878	.878	.878	.879	.879	.879	.879	.880	.880	.880	.880	.880	.881	.882	.883	.883	.883	.883	.884	.885
1.3	.791	.838	.858	.868	.875	.879	.883	.885	.887	.889	.890	.891	.892	.893	.893	.894	.895	.895	.895	.896	.896	.897	.897	.897	.897	.898	.898	.898	.898	.898	.900	.900	.901	.901	.901	.902	.902	.903
1.4	.803	.852	.872	.883	.890	.895	.898	.901	.903	.904	.906	.907	.908	.908	.909	.910	.910	.911	.911	.912	.912	.912	.913	.913	.913	.913	.914	.914	.914	.914	.915	.916	.917	.917	.917	.918	.918	.919
1.5	.813	.864	.885	.896	.903	.908	.911	.914	.916	.918	.919	.920	.921	.922	.923	.924	.924	.925	.925	.925	.926	.926	.926	.927	.927	.927	.927	.928	.928	.928	.929	.930	.931	.931	.931	.931	.932	.933
1.6	.822	.875	.896	.908	.915	.920	.923	.926	.928	.930	.931	.932	.933	.934	.935	.935	.936	.937	.937	.937	.938	.938	.938	.939	.939	.939	.939	.940	.940	.940	.941	.942	.943	.943	.943	.943	.944	.945
1.7	.831	.884	.906	.918	.925	.930	.934	.936	.938	.940	.941	.943	.944	.944	.945	.946	.946	.947	.947	.948	.948	.948	.949	.949	.949	.950	.950	.950	.950	.950	.952	.952	.953	.953	.954	.954	.954	.955
1.8	.839	.893	.915	.927	.934	.939	.943	.945	.947	.949	.950	.952	.953	.953	.954	.955	.955	.956	.956	.957	.957	.957	.958	.958	.958	.958	.959	.959	.959	.959	.960	.961	.962	.962	.962	.962	.963	.964
1.9	.846	.901	.923	.935	.942	.947	.950	.953	.955	.957	.958	.959	.960	.961	.962	.962	.963	.963	.964	.964	.964	.965	.965	.965	.966	.966	.966	.966	.966	.967	.968	.968	.969	.969	.970	.970	.970	.971
2.0	.852	.908	.930	.942	.949	.954	.957	.960	.962	.963	.965	.966	.967	.967	.968	.969	.969	.970	.970	.970	.971	.971	.971	.972	.972	.972	.972	.972	.973	.973	.974	.975	.975	.975	.976	.976	.976	.977
2.1	.859	.915	.937	.948	.955	.960	.963	.966	.967	.969	.970	.971	.972	.973	.974	.974	.975	.975	.975	.976	.976	.976	.977	.977	.977	.977	.977	.978	.978	.978	.979	.980	.980	.980	.981	.981	.981	.982
2.2	.864	.921	.942	.954	.961	.965	.968	.971	.972	.974	.975	.976	.977	.977	.978	.979	.979	.979	.980	.980	.980	.981	.981	.981	.981	.982	.982	.982	.982	.982	.983	.984	.984	.984	.985	.985	.985	.986
2.3	.870	.926	.948	.959	.965	.969	.973	.975	.977	.978	.979	.980	.981	.981	.982	.982	.983	.983	.984	.984	.984	.984	.985	.985	.985	.985	.985	.985	.986	.986	.987	.987	.988	.988	.988	.988	.988	.989
2.4	.874	.931	.952	.963	.969	.973	.976	.978	.980	.981	.982	.983	.984	.985	.985	.986	.986	.986	.987	.987	.987	.987	.988	.988	.988	.988	.988	.988	.989	.989	.989	.990	.990	.991	.991	.991	.991	.992
2.5	.879	.935	.956	.967	.973	.977	.980	.982	.983	.984	.985	.986	.987	.987	.988	.988	.989	.989	.989	.989	.990	.990	.990	.990	.990	.991	.991	.991	.991	.991	.992	.992	.992	.993	.993	.993	.993	.994
2.6	.883	.939	.960	.970	.976	.980	.982	.984	.986	.987	.988	.988	.989	.990	.990	.990	.991	.991	.991	.991	.992	.992	.992	.992	.992	.992	.993	.993	.993	.993	.994	.994	.994	.994	.995	.995	.995	.995
2.7	.887	.943	.963	.973	.979	.982	.985	.987	.988	.989	.990	.990	.991	.991	.992	.992	.992	.993	.993	.993	.993	.994	.994	.994	.994	.994	.994	.994	.994	.994	.995	.995	.996	.996	.996	.996	.996	.997
2.8	.891	.946	.966	.976	.981	.984	.987	.988	.990	.991	.991	.992	.993	.993	.993	.994	.994	.994	.994	.995	.995	.995	.995	.995	.995	.995	.995	.995	.996	.996	.996	.996	.997	.997	.997	.997	.997	.997
2.9	.894	.949	.969	.978	.983	.986	.989	.990	.991	.992	.993	.993	.994	.994	.995	.995	.995	.995	.995	.996	.996	.996	.996	.996	.996	.996	.996	.996	.997	.997	.997	.997	.997	.998	.998	.998	.998	.998
3.0	.898	.952	.971	.980	.985	.988	.990	.992	.993	.993	.994	.995	.995	.995	.996	.996	.996	.996	.996	.997	.997	.997	.997	.997	.997	.997	.997	.997	.997	.997	.998	.998	.998	.998	.998	.998	.998	.999
3.1	.901	.955	.973	.982	.987	.989	.991	.993	.994	.994	.995	.995	.996	.996	.996	.997	.997	.997	.997	.997	.997	.997	.998	.998	.998	.998	.998	.998	.998	.998	.998	.998	.999	.999	.999	.999	.999	.999
3.2	.904	.957	.975	.984	.988	.991	.993	.994	.995	.995	.996	.996	.997	.997	.997	.997	.997	.998	.998	.998	.998	.998	.998	.998	.998	.998	.998	.998	.998	.998	.999	.999	.999	.999	.999	.999	.999	.999
3.3	.906	.960	.977	.985	.989	.992	.993	.995	.995	.996	.997	.997	.997	.997	.998	.998	.998	.998	.998	.998	.998	.998	.998	.999	.999	.999	.999	.999	.999	.999	.999	.999	.999	.999	.999	.999	.999	.999
3.4	.909	.962	.979	.986	.990	.993	.994	.995	.996	.997	.997	.997	.998	.998	.998	.998	.998	.998	.999	.999	.999	.999	.999	.999	.999	.999	.999	.999	.999	.999	.999	.999	.999	.999	.999	.999	.999	.999
3.5	.911	.964	.980	.988	.991	.994	.995	.996	.997	.997	.998	.998	.998	.998	.998	.999	.999	.999	.999	.999	.999	.999	.999	.999	.999	.999	.999	.999	.999	.999	.999	.999	.999	.999	.999	.999	.999	.999
3.6	.914	.965	.982	.989	.992	.994	.996	.997	.997	.998	.998	.998	.998	.999	.999	.999	.999	.999	.999	.999	.999	.999	.999	.999	.999	.999	.999	.999	.999	.999	.999	.999	.999	.999	.999	.999	.999	.999
3.7	.916	.967	.983	.990	.993	.995	.996	.997	.998	.998	.998	.999	.999	.999	.999	.999	.999	.999	.999	.999	.999	.999	.999	.999	.999	.999	.999	.999	.999	.999	.999	.999	.999	.999	.999	.999	.999	.999
3.8	.918	.969	.984	.990	.994	.996	.997	.997	.998	.998	.999	.999	.999	.999	.999	.999	.999	.999	.999	.999	.999	.999	.999	.999	.999	.999	.999	.999	.999	.999	.999	.999	.999	.999	.999	.999	.999	.999
3.9	.920	.970	.985	.991	.994	.996	.997	.998	.998	.999	.999	.999	.999	.999	.999	.999	.999	.999	.999	.999	.999	.999	.999	.999	.999	.999	.999	.999	.999	.999	.999	.999	.999	.999	.999	.999	.999	.999
4.0	.922	.971	.986	.992	.995	.996	.997	.998	.998	.999	.999	.999	.999	.999	.999	.999	.999	.999	.999	.999	.999	.999	.999	.999	.999	.999	.999	.999	.999	.999	.999	.999	.999	.999	.999	.999	.999	.999
4.1	.924	.973	.987	.993	.995	.997	.998	.998	.999	.999	.999	.999	.999	.999	.999	.999	.999	.999	.999	.999	.999	.999	.999	.999	.999	.999	.999	.999	.999	.999	.999	.999	.999	.999	.999	.999	.999	.999
4.2	.926	.974	.988	.993	.996	.997	.998	.999	.999	.999	.999	.999	.999	.999	.999	.999	.999	.999	.999	.999	.999	.999	.999	.999	.999	.999	.999	.999	.999	.999	.999	.999	.999	.999	.999	.999	.999	.999
4.3	.927	.975	.988	.994	.996	.998	.998	.999	.999	.999	.999	.999	.999	.999	.999	.999	.999	.999	.999	.999	.999	.999	.999	.999	.999	.999	.999	.999	.999	.999	.999	.999	.999	.999	.999	.999	.999	.999
4.4	.929	.976	.989	.994	.997	.998	.998	.999	.999	.999	.999	.999	.999	.999	.999	.999	.999	.999	.999	.999	.999	.999	.999	.999	.999	.999	.999	.999	.999	.999	.999	.999	.999	.999	.999	.999	.999	.999
4.5	.930	.977	.990	.995	.997	.998	.999	.999	.999	.999	.999	.999	.999	.999	.999	.999	.999	.999	.999	.999	.999	.999	.999	.999	.999	.999	.999	.999	.999	.999	.999	.999	.999	.999	.999	.999	.999	.999
4.6	.932	.978	.990	.995	.997	.998	.999	.999	.999	.999	.999	.999	.999	.999	.999	.999	.999	.999	.999	.999	.999	.999	.999	.999	.999	.999	.999	.999	.999	.999	.999	.999	.999	.999	.999	.999	.999	.999
4.7	.933	.979	.991	.995	.997	.998	.999	.999	.999	.999	.999	.999	.999	.999	.999	.999	.999	.999	.999	.999	.999	.999	.999	.999	.999	.999	.999	.999	.999	.999	.999	.999	.999	.999	.999	.999	.999	.999
4.8	.935	.980	.991	.996	.998	.999	.999	.999	.999	.999	.999	.999	.999	.999	.999	.999	.999	.999	.999	.999	.999	.999	.999	.999	.999	.999	.999	.999	.999	.999	.999	.999	.999	.999	.999	.999	.999	.999
4.9	.936	.980	.992	.996	.998	.999	.999	.999	.999	.999	.999	.999	.999	.999	.999	.999	.999	.999	.999	.999	.999	.999	.999	.999	.999	.999	.999	.999	.999	.999	.999	.999	.999	.999	.999	.999	.999	.999
5.0	.937	.981	.992	.996	.998	.999	.999	.999	.999	.999	.999	.999	.999	.999	.999	.999	.999	.999	.999	.999	.999	.999	.999	.999	.999	.999	.999	.999	.999	.999	.999	.999	.999	.999	.999	.999	.999	.999

Table A3 – t-Distribution p-Values

Degrees of Freedom

T.S.	1	2	3	4	5	6	7	8	9	10	11	12	13	14	15	16	17	18	19	20	21	22	23	24	25	26	27	28	29	30	40	50	60	70	80	90	100	∞
-0.0	.500	.500	.500	.500	.500	.500	.500	.500	.500	.500	.500	.500	.500	.500	.500	.500	.500	.500	.500	.500	.500	.500	.500	.500	.500	.500	.500	.500	.500	.500	.500	.500	.500	.500	.500	.500	.500	.500
-0.1	.468	.465	.463	.463	.462	.462	.462	.461	.461	.461	.461	.461	.461	.461	.461	.461	.461	.461	.461	.461	.461	.461	.461	.461	.461	.461	.461	.461	.461	.461	.460	.460	.460	.460	.460	.460	.460	.460
-0.2	.437	.430	.427	.426	.425	.424	.424	.423	.423	.423	.423	.422	.422	.422	.422	.422	.422	.422	.422	.422	.422	.422	.422	.422	.422	.422	.422	.422	.421	.421	.421	.421	.421	.421	.421	.421	.421	.421
-0.3	.407	.396	.392	.390	.388	.387	.386	.386	.386	.385	.385	.385	.385	.384	.384	.384	.384	.384	.384	.384	.384	.384	.383	.383	.383	.383	.383	.383	.383	.383	.383	.383	.383	.383	.383	.382	.382	.382
-0.4	.379	.364	.358	.355	.353	.352	.351	.350	.349	.349	.348	.348	.348	.348	.347	.347	.347	.347	.347	.347	.347	.347	.346	.346	.346	.346	.346	.346	.346	.346	.346	.345	.345	.345	.345	.345	.345	.345
-0.5	.352	.333	.326	.322	.319	.317	.316	.315	.315	.314	.314	.313	.313	.312	.312	.312	.312	.312	.311	.311	.311	.311	.311	.311	.311	.311	.311	.311	.310	.310	.310	.310	.310	.309	.309	.309	.309	.309
-0.6	.328	.305	.295	.290	.287	.285	.284	.283	.282	.281	.280	.280	.279	.279	.279	.279	.278	.278	.278	.278	.278	.277	.277	.277	.277	.277	.277	.277	.277	.277	.276	.276	.275	.275	.275	.275	.275	.274
-0.7	.306	.278	.267	.261	.258	.255	.253	.252	.251	.250	.249	.249	.248	.248	.247	.247	.247	.246	.246	.246	.246	.246	.246	.245	.245	.245	.245	.245	.245	.245	.244	.244	.243	.243	.243	.243	.243	.242
-0.8	.285	.254	.241	.234	.230	.227	.225	.223	.222	.221	.220	.220	.219	.219	.218	.218	.217	.217	.217	.217	.216	.216	.216	.216	.216	.216	.215	.215	.215	.215	.214	.214	.213	.213	.213	.213	.213	.212
-0.9	.267	.232	.217	.210	.205	.201	.199	.197	.196	.195	.194	.193	.192	.192	.191	.191	.190	.190	.190	.189	.189	.189	.189	.189	.188	.188	.188	.188	.188	.188	.187	.186	.186	.186	.185	.185	.185	.184
-1.0	.250	.211	.196	.187	.182	.178	.175	.173	.172	.170	.169	.169	.168	.167	.167	.166	.166	.165	.165	.165	.164	.164	.164	.164	.163	.163	.163	.163	.163	.163	.162	.161	.161	.160	.160	.160	.160	.159
-1.1	.235	.193	.176	.167	.161	.157	.154	.152	.150	.149	.147	.147	.146	.145	.144	.144	.143	.143	.143	.142	.142	.142	.141	.141	.141	.141	.141	.140	.140	.140	.139	.138	.138	.138	.137	.137	.137	.136
-1.2	.221	.177	.158	.148	.142	.138	.135	.132	.130	.129	.128	.127	.126	.125	.124	.124	.123	.123	.122	.122	.122	.122	.121	.121	.121	.121	.120	.120	.120	.120	.119	.118	.117	.117	.117	.117	.117	.115
-1.3	.209	.162	.142	.132	.125	.121	.117	.115	.113	.111	.110	.109	.108	.107	.107	.106	.106	.105	.105	.104	.104	.104	.103	.103	.103	.103	.102	.102	.102	.102	.101	.100	.099	.099	.099	.099	.098	.097
-1.4	.197	.148	.128	.117	.110	.106	.102	.100	.098	.096	.095	.093	.093	.092	.091	.090	.090	.089	.089	.088	.088	.088	.087	.087	.087	.087	.086	.086	.086	.086	.085	.084	.083	.083	.083	.083	.082	.081
-1.5	.187	.136	.115	.104	.097	.092	.089	.086	.084	.082	.081	.080	.079	.078	.077	.077	.076	.076	.075	.075	.074	.074	.074	.073	.073	.073	.073	.072	.072	.072	.071	.070	.069	.069	.069	.069	.068	.067
-1.6	.178	.125	.104	.092	.085	.080	.077	.074	.072	.070	.069	.068	.067	.066	.065	.065	.064	.064	.063	.063	.062	.062	.062	.061	.061	.061	.061	.060	.060	.060	.059	.058	.057	.057	.057	.057	.056	.055
-1.7	.169	.116	.094	.082	.075	.070	.067	.064	.062	.060	.059	.057	.057	.056	.055	.054	.054	.053	.053	.052	.052	.052	.051	.051	.051	.051	.050	.050	.050	.050	.048	.048	.047	.047	.047	.046	.046	.045
-1.8	.161	.107	.085	.073	.066	.061	.057	.055	.053	.051	.050	.049	.048	.047	.046	.045	.045	.044	.044	.044	.043	.043	.043	.042	.042	.042	.042	.041	.041	.041	.040	.039	.038	.038	.038	.038	.037	.036
-1.9	.154	.099	.077	.065	.058	.053	.050	.047	.045	.043	.042	.041	.040	.039	.038	.038	.037	.037	.036	.036	.036	.035	.035	.035	.035	.034	.034	.034	.034	.034	.032	.032	.031	.031	.031	.030	.030	.029
-2.0	.148	.092	.070	.058	.051	.046	.043	.040	.038	.037	.035	.034	.033	.033	.032	.031	.031	.030	.030	.030	.029	.029	.029	.029	.028	.028	.028	.028	.028	.027	.026	.026	.025	.025	.024	.024	.024	.023
-2.1	.142	.085	.063	.052	.045	.040	.037	.035	.033	.031	.030	.029	.028	.027	.027	.026	.026	.025	.025	.024	.024	.024	.023	.023	.023	.023	.023	.022	.022	.022	.021	.020	.020	.020	.019	.019	.019	.018
-2.2	.136	.079	.058	.046	.040	.035	.032	.030	.028	.026	.025	.024	.023	.023	.022	.021	.021	.021	.020	.020	.020	.019	.019	.019	.019	.018	.018	.018	.018	.018	.017	.016	.016	.016	.015	.015	.015	.014
-2.3	.131	.074	.053	.042	.035	.031	.028	.025	.024	.022	.021	.020	.019	.019	.018	.018	.017	.017	.017	.016	.016	.016	.015	.015	.015	.015	.015	.015	.014	.014	.013	.013	.013	.012	.012	.012	.012	.011
-2.4	.126	.069	.048	.037	.031	.027	.024	.022	.020	.019	.018	.017	.016	.015	.015	.015	.014	.014	.013	.013	.013	.013	.012	.012	.012	.012	.012	.012	.012	.011	.011	.010	.010	.009	.009	.009	.009	.008
-2.5	.121	.065	.044	.033	.027	.023	.021	.019	.017	.016	.015	.014	.013	.013	.012	.012	.012	.011	.011	.011	.010	.010	.010	.010	.010	.009	.009	.009	.009	.009	.008	.008	.008	.007	.007	.007	.007	.006
-2.6	.117	.061	.040	.030	.024	.020	.018	.016	.014	.013	.012	.012	.011	.011	.010	.010	.009	.009	.009	.009	.008	.008	.008	.008	.008	.008	.007	.007	.007	.007	.006	.006	.006	.006	.005	.005	.005	.005
-2.7	.113	.057	.037	.027	.021	.018	.015	.014	.012	.011	.010	.010	.009	.009	.008	.008	.008	.007	.007	.007	.007	.006	.006	.006	.006	.006	.006	.006	.006	.006	.005	.005	.004	.004	.004	.004	.004	.003
-2.8	.109	.054	.034	.024	.019	.016	.013	.012	.010	.009	.009	.008	.007	.007	.007	.006	.006	.006	.006	.005	.005	.005	.005	.005	.005	.005	.005	.005	.004	.004	.004	.004	.003	.003	.003	.003	.003	.003
-2.9	.106	.051	.031	.022	.017	.014	.012	.010	.009	.008	.007	.007	.006	.006	.005	.005	.005	.005	.005	.004	.004	.004	.004	.004	.004	.004	.004	.004	.003	.003	.003	.003	.003	.002	.002	.002	.002	.002
-3.0	.102	.048	.029	.020	.015	.012	.010	.008	.007	.007	.006	.005	.005	.005	.004	.004	.004	.004	.004	.003	.003	.003	.003	.003	.003	.003	.003	.003	.003	.003	.002	.002	.002	.002	.002	.002	.002	.001
-3.1	.099	.045	.027	.018	.013	.011	.009	.007	.006	.006	.005	.005	.004	.004	.004	.003	.003	.003	.003	.003	.003	.003	.002	.002	.002	.002	.002	.002	.002	.002	.002	.002	.001	.001	.001	.001	.001	.001
-3.2	.096	.043	.025	.017	.012	.009	.007	.006	.005	.005	.004	.004	.003	.003	.003	.003	.003	.002	.002	.002	.002	.002	.002	.002	.002	.002	.002	.002	.002	.002	.001	.001	.001	.001	.001	.001	.001	.001
-3.3	.094	.040	.023	.015	.011	.008	.007	.005	.005	.004	.003	.003	.003	.003	.002	.002	.002	.002	.002	.002	.002	.002	.002	.001	.001	.001	.001	.001	.001	.001	.001	.001	.001	.001	.001	.001	.001	.001
-3.4	.091	.038	.021	.014	.010	.007	.006	.005	.004	.003	.003	.003	.002	.002	.002	.002	.002	.002	.001	.001	.001	.001	.001	.001	.001	.001	.001	.001	.001	.001	.001	.001	.001	.001	.001	.001	.001	.001
-3.5	.089	.036	.020	.012	.009	.006	.005	.004	.003	.003	.002	.002	.002	.002	.002	.001	.001	.001	.001	.001	.001	.001	.001	.001	.001	.001	.001	.001	.001	.001	.001	.001	.001	.001	.001	.001	.001	.001
-3.6	.086	.035	.018	.011	.008	.006	.004	.003	.003	.002	.002	.002	.002	.001	.001	.001	.001	.001	.001	.001	.001	.001	.001	.001	.001	.001	.001	.001	.001	.001	.001	.001	.001	.001	.001	.001	.001	.001
-3.7	.084	.033	.017	.010	.007	.005	.004	.003	.002	.002	.002	.001	.001	.001	.001	.001	.001	.001	.001	.001	.001	.001	.001	.001	.001	.001	.001	.001	.001	.001	.001	.001	.001	.001	.001	.001	.001	.001
-3.8	.082	.031	.016	.010	.006	.004	.003	.003	.002	.002	.001	.001	.001	.001	.001	.001	.001	.001	.001	.001	.001	.001	.001	.001	.001	.001	.001	.001	.001	.001	.001	.001	.001	.001	.001	.001	.001	.001
-3.9	.080	.030	.015	.009	.006	.004	.003	.002	.002	.001	.001	.001	.001	.001	.001	.001	.001	.001	.001	.001	.001	.001	.001	.001	.001	.001	.001	.001	.001	.001	.001	.001	.001	.001	.001	.001	.001	.001
-4.0	.078	.029	.014	.008	.005	.004	.003	.002	.002	.001	.001	.001	.001	.001	.001	.001	.001	.001	.001	.001	.001	.001	.001	.001	.001	.001	.001	.001	.001	.001	.001	.001	.001	.001	.001	.001	.001	.001
-4.1	.076	.027	.013	.007	.005	.003	.002	.002	.001	.001	.001	.001	.001	.001	.001	.001	.001	.001	.001	.001	.001	.001	.001	.001	.001	.001	.001	.001	.001	.001	.001	.001	.001	.001	.001	.001	.001	.001
-4.2	.074	.026	.012	.007	.004	.003	.002	.001	.001	.001	.001	.001	.001	.001	.001	.001	.001	.001	.001	.001	.001	.001	.001	.001	.001	.001	.001	.001	.001	.001	.001	.001	.001	.001	.001	.001	.001	.001
-4.3	.073	.025	.012	.006	.004	.002	.002	.001	.001	.001	.001	.001	.001	.001	.001	.001	.001	.001	.001	.001	.001	.001	.001	.001	.001	.001	.001	.001	.001	.001	.001	.001	.001	.001	.001	.001	.001	.001
-4.4	.071	.024	.011	.006	.003	.002	.002	.001	.001	.001	.001	.001	.001	.001	.001	.001	.001	.001	.001	.001	.001	.001	.001	.001	.001	.001	.001	.001	.001	.001	.001	.001	.001	.001	.001	.001	.001	.001
-4.5	.070	.023	.010	.005	.003	.002	.001	.001	.001	.001	.001	.001	.001	.001	.001	.001	.001	.001	.001	.001	.001	.001	.001	.001	.001	.001	.001	.001	.001	.001	.001	.001	.001	.001	.001	.001	.001	.001
-4.6	.068	.022	.010	.005	.003	.002	.001	.001	.001	.001	.001	.001	.001	.001	.001	.001	.001	.001	.001	.001	.001	.001	.001	.001	.001	.001	.001	.001	.001	.001	.001	.001	.001	.001	.001	.001	.001	.001
-4.7	.067	.021	.009	.005	.003	.002	.001	.001	.001	.001	.001	.001	.001	.001	.001	.001	.001	.001	.001	.001	.001	.001	.001	.001	.001	.001	.001	.001	.001	.001	.001	.001	.001	.001	.001	.001	.001	.001
-4.8	.065	.020	.009	.004	.002	.001	.001	.001	.001	.001	.001	.001	.001	.001	.001	.001	.001	.001	.001	.001	.001	.001	.001	.001	.001	.001	.001	.001	.001	.001	.001	.001	.001	.001	.001	.001	.001	.001
-4.9	.064	.020	.008	.004	.002	.001	.001	.001	.001	.001	.001	.001	.001	.001	.001	.001	.001	.001	.001	.001	.001	.001	.001	.001	.001	.001	.001	.001	.001	.001	.001	.001	.001	.001	.001	.001	.001	.001
-5.0	.063	.019	.008	.004	.002	.001	.001	.001	.001	.001	.001	.001	.001	.001	.001	.001	.001	.001	.001	.001	.001	.001	.001	.001	.001	.001	.001	.001	.001	.001	.001	.001	.001	.001	.001	.001	.001	.001

Table A4 – Chi-Square Critical Values

	Area to the Right of Critical Value					
df	0.99	0.95	0.90	0.10	0.05	0.01
1	0.0002	0.0039	0.0158	2.7055	3.8415	6.6349
2	0.0201	0.1026	0.2107	4.6052	5.9915	9.2103
3	0.1148	0.3518	0.5844	6.2514	7.8147	11.3449
4	0.2971	0.7107	1.0636	7.7794	9.4877	13.2767
5	0.5543	1.1455	1.6103	9.2364	11.0705	15.0863
6	0.8721	1.6354	2.2041	10.6446	12.5916	16.8119
7	1.2390	2.1673	2.8331	12.0170	14.0671	18.4753
8	1.6465	2.7326	3.4895	13.3616	15.5073	20.0902
9	2.0879	3.3251	4.1682	14.6837	16.9190	21.6660
10	2.5582	3.9403	4.8652	15.9872	18.3070	23.2093
11	3.0535	4.5748	5.5778	17.2750	19.6751	24.7250
12	3.5706	5.2260	6.3038	18.5493	21.0261	26.2170
13	4.1069	5.8919	7.0415	19.8119	22.3620	27.6882
14	4.6604	6.5706	7.7895	21.0641	23.6848	29.1412
15	5.2293	7.2609	8.5468	22.3071	24.9958	30.5779
16	5.8122	7.9616	9.3122	23.5418	26.2962	31.9999
17	6.4078	8.6718	10.0852	24.7690	27.5871	33.4087
18	7.0149	9.3905	10.8649	25.9894	28.8693	34.8053
19	7.6327	10.1170	11.6509	27.2036	30.1435	36.1909
20	8.2604	10.8508	12.4426	28.4120	31.4104	37.5662
21	8.8972	11.5913	13.2396	29.6151	32.6706	38.9322
22	9.5425	12.3380	14.0415	30.8133	33.9244	40.2894
23	10.1957	13.0905	14.8480	32.0069	35.1725	41.6384
24	10.8564	13.8484	15.6587	33.1962	36.4150	42.9798
25	11.5240	14.6114	16.4734	34.3816	37.6525	44.3141
26	12.1981	15.3792	17.2919	35.5632	38.8851	45.6417
27	12.8785	16.1514	18.1139	36.7412	40.1133	46.9629
28	13.5647	16.9279	18.9392	37.9159	41.3371	48.2782
29	14.2565	17.7084	19.7677	39.0875	42.5570	49.5879
30	14.9535	18.4927	20.5992	40.2560	43.7730	50.8922
31	15.6555	19.2806	21.4336	41.4217	44.9853	52.1914
32	16.3622	20.0719	22.2706	42.5847	46.1943	53.4858
33	17.0735	20.8665	23.1102	43.7452	47.3999	54.7755
34	17.7891	21.6643	23.9523	44.9032	48.6024	56.0609
35	18.5089	22.4650	24.7967	46.0588	49.8018	57.3421
36	19.2327	23.2686	25.6433	47.2122	50.9985	58.6192
37	19.9602	24.0749	26.4921	48.3634	52.1923	59.8925
38	20.6914	24.8839	27.3430	49.5126	53.3835	61.1621
39	21.4262	25.6954	28.1958	50.6598	54.5722	62.4281
40	22.1643	26.5093	29.0505	51.8051	55.7585	63.6907
41	22.9056	27.3256	29.9071	52.9485	56.9424	64.9501
42	23.6501	28.1440	30.7654	54.0902	58.1240	66.2062
43	24.3976	28.9647	31.6255	55.2302	59.3035	67.4593
44	25.1480	29.7875	32.4871	56.3685	60.4809	68.7095
45	25.9013	30.6123	33.3504	57.5053	61.6562	69.9568
46	26.6572	31.4390	34.2152	58.6405	62.8296	71.2014
47	27.4158	32.2676	35.0814	59.7743	64.0011	72.4433
48	28.1770	33.0981	35.9491	60.9066	65.1708	73.6826
49	28.9406	33.9303	36.8182	62.0375	66.3386	74.9195
50	29.7067	34.7643	37.6886	63.1671	67.5048	76.1539
55	33.5705	38.9580	42.0596	68.7962	73.3115	82.2921
60	37.4849	43.1880	46.4589	74.3970	79.0819	88.3794
65	41.4436	47.4496	50.8829	79.9730	84.8206	94.4221
70	45.4417	51.7393	55.3289	85.5270	90.5312	100.4252
75	49.4750	56.0541	59.7946	91.0615	96.2167	106.3929
80	53.5401	60.3915	64.2778	96.5782	101.8795	112.3288
85	57.6339	64.7494	68.7772	102.0789	107.5217	118.2357
90	61.7541	69.1260	73.2911	107.5650	113.1453	124.1163
95	65.8984	73.5198	77.8184	113.0377	118.7516	129.9727
100	70.0649	77.9295	82.3581	118.4980	124.3421	135.8067
150	112.6676	122.6918	128.2751	172.5812	179.5806	193.2077
200	156.4320	168.2786	174.8353	226.0210	233.9943	249.4451
250	200.9386	214.3916	221.8059	279.0504	287.8815	304.9396
300	245.9725	260.8781	269.0679	331.7885	341.3951	359.9064

Table A5 F-Distribution Right-Tail Critical Values Corresponding to $\alpha = 0.01$

Denom. d.f.	1	2	3	4	5	6	7	8	9	10	11	12	13	14	15	16	17	18
									Numerator d.f.									
1	4052.18	4999.50	5403.35	5624.58	5763.65	5858.99	5928.36	5981.07	6022.47	6055.85	6083.32	6106.32	6125.86	6142.67	6157.28	6170.10	6181.43	6191.53
2	98.5025	99.0000	99.1662	99.2494	99.2993	99.3326	99.3564	99.3742	99.3881	99.3992	99.4083	99.4159	99.4223	99.4278	99.4325	99.4367	99.4404	99.4436
3	34.1162	30.8165	29.4567	28.7099	28.2371	27.9107	27.6717	27.4892	27.3452	27.2287	27.1326	27.0518	26.9831	26.9238	26.8722	26.8269	26.7867	26.7509
4	21.1977	18.0000	16.6944	15.9770	15.5219	15.2069	14.9758	14.7989	14.6591	14.5459	14.4523	14.3736	14.3065	14.2486	14.1982	14.1539	14.1146	14.0795
5	16.2582	13.2739	12.0600	11.3919	10.9670	10.6723	10.4555	10.2893	10.1578	10.0510	9.9626	9.8883	9.8248	9.7700	9.7222	9.6802	9.6429	9.6096
6	13.7450	10.9248	9.7795	9.1483	8.7459	8.4661	8.2600	8.1017	7.9761	7.8741	7.7896	7.7183	7.6575	7.6049	7.5590	7.5186	7.4827	7.4507
7	12.2464	9.5466	8.4513	7.8466	7.4604	7.1914	6.9928	6.8400	6.7188	6.6201	6.5382	6.4691	6.4100	6.3590	6.3143	6.2750	6.2401	6.2089
8	11.2586	8.6491	7.5910	7.0061	6.6318	6.3707	6.1776	6.0289	5.9106	5.8143	5.7343	5.6667	5.6089	5.5589	5.5151	5.4766	5.4423	5.4116
9	10.5614	8.0215	6.9919	6.4221	6.0569	5.8018	5.6129	5.4671	5.3511	5.2565	5.1779	5.1114	5.0545	5.0052	4.9621	4.9240	4.8902	4.8599
10	10.0443	7.5594	6.5523	5.9943	5.6363	5.3858	5.2001	5.0567	4.9424	4.8491	4.7715	4.7059	4.6496	4.6008	4.5581	4.5204	4.4869	4.4569
11	9.6460	7.2057	6.2167	5.6683	5.3160	5.0692	4.8861	4.7445	4.6315	4.5393	4.4624	4.3974	4.3416	4.2932	4.2509	4.2134	4.1801	4.1503
12	9.3302	6.9266	5.9525	5.4120	5.0643	4.8206	4.6395	4.4994	4.3875	4.2961	4.2198	4.1553	4.0999	4.0518	4.0096	3.9724	3.9392	3.9095
13	9.0738	6.7010	5.7394	5.2053	4.8616	4.6204	4.4410	4.3021	4.1911	4.1003	4.0245	3.9603	3.9052	3.8573	3.8154	3.7783	3.7452	3.7156
14	8.8616	6.5149	5.5639	5.0354	4.6950	4.4558	4.2779	4.1399	4.0297	3.9394	3.8640	3.8001	3.7452	3.6975	3.6557	3.6187	3.5857	3.5561
15	8.6831	6.3589	5.4170	4.8932	4.5556	4.3183	4.1415	4.0045	3.8948	3.8049	3.7299	3.6662	3.6115	3.5639	3.5222	3.4852	3.4523	3.4228
16	8.5310	6.2262	5.2922	4.7726	4.4374	4.2016	4.0259	3.8896	3.7804	3.6909	3.6162	3.5527	3.4981	3.4506	3.4089	3.3720	3.3391	3.3096
17	8.3997	6.1121	5.1850	4.6690	4.3359	4.1015	3.9267	3.7910	3.6822	3.5931	3.5185	3.4552	3.4007	3.3533	3.3117	3.2748	3.2419	3.2124
18	8.2854	6.0129	5.0919	4.5790	4.2479	4.0146	3.8406	3.7054	3.5971	3.5082	3.4338	3.3706	3.3162	3.2689	3.2273	3.1904	3.1575	3.1280
19	8.1849	5.9259	5.0103	4.5003	4.1708	3.9386	3.7653	3.6305	3.5225	3.4338	3.3596	3.2965	3.2422	3.1949	3.1533	3.1165	3.0836	3.0541
20	8.0960	5.8489	4.9382	4.4307	4.1027	3.8714	3.6987	3.5644	3.4567	3.3682	3.2941	3.2311	3.1769	3.1296	3.0880	3.0512	3.0183	2.9887
21	8.0166	5.7804	4.8740	4.3688	4.0421	3.8117	3.6396	3.5056	3.3981	3.3098	3.2359	3.1730	3.1187	3.0715	3.0300	2.9931	2.9602	2.9306
22	7.9454	5.7190	4.8166	4.3134	3.9880	3.7583	3.5867	3.4530	3.3458	3.2576	3.1837	3.1209	3.0667	3.0195	2.9779	2.9411	2.9082	2.8786
23	7.8811	5.6637	4.7649	4.2636	3.9392	3.7102	3.5390	3.4057	3.2986	3.2106	3.1368	3.0740	3.0199	2.9727	2.9311	2.8943	2.8613	2.8317
24	7.8229	5.6136	4.7181	4.2184	3.8951	3.6667	3.4959	3.3629	3.2560	3.1681	3.0944	3.0316	2.9775	2.9303	2.8887	2.8519	2.8189	2.7892
25	7.7698	5.5680	4.6755	4.1774	3.8550	3.6272	3.4568	3.3239	3.2172	3.1294	3.0558	2.9931	2.9389	2.8917	2.8502	2.8133	2.7803	2.7506
26	7.7213	5.5263	4.6366	4.1400	3.8183	3.5911	3.4210	3.2884	3.1818	3.0941	3.0205	2.9578	2.9038	2.8566	2.8150	2.7781	2.7451	2.7153
27	7.6767	5.4881	4.6009	4.1056	3.7848	3.5580	3.3882	3.2558	3.1494	3.0618	2.9882	2.9256	2.8715	2.8243	2.7827	2.7458	2.7127	2.6830
28	7.6356	5.4529	4.5681	4.0740	3.7539	3.5276	3.3581	3.2259	3.1195	3.0320	2.9585	2.8959	2.8418	2.7946	2.7530	2.7160	2.6830	2.6532
29	7.5977	5.4204	4.5378	4.0449	3.7254	3.4995	3.3303	3.1982	3.0920	3.0045	2.9311	2.8685	2.8144	2.7672	2.7256	2.6886	2.6555	2.6257
30	7.5625	5.3903	4.5097	4.0179	3.6990	3.4735	3.3045	3.1726	3.0665	2.9791	2.9057	2.8431	2.7890	2.7418	2.7002	2.6632	2.6301	2.6003
40	7.3141	5.1785	4.3126	3.8283	3.5138	3.2910	3.1238	2.9930	2.8876	2.8005	2.7274	2.6648	2.6107	2.5634	2.5216	2.4844	2.4511	2.4210
50	7.1706	5.0566	4.1993	3.7195	3.4077	3.1864	3.0202	2.8900	2.7850	2.6981	2.6250	2.5625	2.5083	2.4609	2.4190	2.3816	2.3481	2.3178
100	6.8953	4.8239	3.9837	3.5127	3.2059	2.9877	2.8233	2.6943	2.5898	2.5033	2.4302	2.3676	2.3132	2.2654	2.2230	2.1852	2.1511	2.1203
200	6.7633	4.7129	3.8810	3.4143	3.1100	2.8933	2.7298	2.6012	2.4971	2.4106	2.3375	2.2747	2.2201	2.1721	2.1294	2.0913	2.0569	2.0257
250	6.7373	4.6911	3.8609	3.3950	3.0912	2.8748	2.7114	2.5830	2.4789	2.3925	2.3193	2.2565	2.2018	2.1537	2.1110	2.0728	2.0384	2.0071
500	6.6858	4.6478	3.8210	3.3569	3.0540	2.8381	2.6751	2.5469	2.4429	2.3565	2.2833	2.2204	2.1656	2.1174	2.0746	2.0362	2.0016	1.9702
1000	6.6603	4.6264	3.8012	3.3380	3.0355	2.8200	2.6572	2.5290	2.4250	2.3386	2.2655	2.2025	2.1477	2.0994	2.0565	2.0180	1.9834	1.9519

Table A5 F-Distribution Critical Values Corresponding to $\alpha = 0.01$.

Denom. d.f.	Numerator d.f. 19	20	21	22	23	24	25	26	27	28	29	30	40	50	100	250	500	1000
1	6200.58	6208.73	6216.12	6222.84	6228.99	6234.63	6239.83	6244.62	6249.07	6253.20	6257.05	6260.65	6286.78	6302.52	6334.11	6353.14	6359.50	6362.68
2	99.4465	99.4492	99.4516	99.4537	99.4557	99.4575	99.4592	99.4607	99.4621	99.4635	99.4647	99.4658	99.4742	99.4792	99.4892	99.4952	99.4972	99.4982
3	26.7188	26.6898	26.6635	26.6396	26.6176	26.5975	26.5790	26.5618	26.5460	26.5312	26.5174	26.5045	26.4108	26.3542	26.2402	26.1713	26.1483	26.1367
4	14.0480	14.0196	13.9938	13.9703	13.9488	13.9291	13.9109	13.8940	13.8784	13.8639	13.8503	13.8377	13.7454	13.6896	13.5770	13.5088	13.4859	13.4745
5	9.5797	9.5526	9.5281	9.5058	9.4853	9.4665	9.4491	9.4331	9.4182	9.4043	9.3914	9.3793	9.2912	9.2378	9.1299	9.0644	9.0424	9.0314
6	7.4219	7.3958	7.3722	7.3506	7.3309	7.3127	7.2960	7.2805	7.2661	7.2527	7.2402	7.2285	7.1432	7.0915	6.9867	6.9229	6.9015	6.8908
7	6.1808	6.1554	6.1324	6.1113	6.0921	6.0743	6.0580	6.0428	6.0287	6.0157	6.0034	5.9920	5.9084	5.8577	5.7547	5.6918	5.6707	5.6601
8	5.3840	5.3591	5.3364	5.3157	5.2967	5.2793	5.2631	5.2482	5.2344	5.2214	5.2094	5.1981	5.1156	5.0654	4.9633	4.9009	4.8799	4.8694
9	4.8327	4.8080	4.7856	4.7651	4.7463	4.7290	4.7130	4.6982	4.6845	4.6717	4.6598	4.6486	4.5666	4.5167	4.4150	4.3527	4.3317	4.3211
10	4.4299	4.4054	4.3831	4.3628	4.3441	4.3269	4.3111	4.2963	4.2827	4.2700	4.2581	4.2469	4.1653	4.1155	4.0137	3.9513	3.9302	3.9196
11	4.1234	4.0990	4.0769	4.0566	4.0380	4.0209	4.0051	3.9904	3.9768	3.9641	3.9522	3.9411	3.8596	3.8097	3.7077	3.6450	3.6238	3.6131
12	3.8827	3.8584	3.8363	3.8161	3.7976	3.7805	3.7647	3.7500	3.7364	3.7237	3.7119	3.7008	3.6192	3.5692	3.4668	3.4037	3.3823	3.3716
13	3.6888	3.6646	3.6425	3.6224	3.6038	3.5868	3.5710	3.5563	3.5427	3.5300	3.5182	3.5070	3.4253	3.3752	3.2723	3.2087	3.1871	3.1763
14	3.5294	3.5052	3.4832	3.4630	3.4445	3.4274	3.4116	3.3969	3.3833	3.3706	3.3587	3.3476	3.2656	3.2153	3.1118	3.0477	3.0260	3.0150
15	3.3961	3.3719	3.3498	3.3297	3.3111	3.2940	3.2782	3.2635	3.2499	3.2372	3.2253	3.2141	3.1319	3.0814	2.9772	2.9126	2.8906	2.8795
16	3.2829	3.2587	3.2367	3.2165	3.1979	3.1808	3.1650	3.1503	3.1366	3.1238	3.1119	3.1007	3.0182	2.9675	2.8627	2.7974	2.7752	2.7641
17	3.1857	3.1615	3.1394	3.1192	3.1006	3.0835	3.0676	3.0529	3.0392	3.0264	3.0145	3.0032	2.9205	2.8694	2.7639	2.6981	2.6757	2.6644
18	3.1013	3.0771	3.0550	3.0348	3.0161	2.9990	2.9831	2.9683	2.9546	2.9418	2.9298	2.9185	2.8354	2.7841	2.6779	2.6115	2.5889	2.5775
19	3.0274	3.0031	2.9810	2.9607	2.9421	2.9249	2.9089	2.8941	2.8804	2.8675	2.8555	2.8442	2.7608	2.7093	2.6023	2.5353	2.5124	2.5009
20	2.9620	2.9377	2.9156	2.8953	2.8766	2.8594	2.8434	2.8286	2.8148	2.8019	2.7898	2.7785	2.6947	2.6430	2.5353	2.4677	2.4446	2.4329
21	2.9039	2.8796	2.8574	2.8370	2.8183	2.8010	2.7850	2.7702	2.7563	2.7434	2.7313	2.7200	2.6359	2.5838	2.4755	2.4073	2.3840	2.3722
22	2.8518	2.8274	2.8052	2.7849	2.7661	2.7488	2.7328	2.7179	2.7040	2.6910	2.6789	2.6675	2.5831	2.5308	2.4217	2.3530	2.3294	2.3175
23	2.8049	2.7805	2.7583	2.7378	2.7191	2.7017	2.6856	2.6707	2.6568	2.6438	2.6316	2.6202	2.5355	2.4829	2.3732	2.3038	2.2800	2.2680
24	2.7624	2.7380	2.7157	2.6953	2.6765	2.6591	2.6430	2.6280	2.6140	2.6010	2.5888	2.5773	2.4923	2.4395	2.3291	2.2591	2.2351	2.2230
25	2.7238	2.6993	2.6770	2.6565	2.6377	2.6203	2.6041	2.5891	2.5751	2.5620	2.5498	2.5383	2.4530	2.3999	2.2888	2.2183	2.1941	2.1818
26	2.6885	2.6640	2.6416	2.6211	2.6022	2.5848	2.5686	2.5536	2.5395	2.5264	2.5141	2.5026	2.4170	2.3637	2.2519	2.1809	2.1564	2.1440
27	2.6561	2.6316	2.6092	2.5887	2.5697	2.5522	2.5360	2.5209	2.5069	2.4937	2.4814	2.4699	2.3840	2.3304	2.2180	2.1464	2.1217	2.1092
28	2.6263	2.6017	2.5793	2.5587	2.5398	2.5223	2.5060	2.4909	2.4768	2.4636	2.4513	2.4397	2.3535	2.2997	2.1867	2.1145	2.0896	2.0769
29	2.5987	2.5742	2.5517	2.5311	2.5121	2.4946	2.4783	2.4631	2.4490	2.4358	2.4234	2.4118	2.3253	2.2714	2.1577	2.0850	2.0598	2.0471
30	2.5732	2.5487	2.5262	2.5055	2.4865	2.4689	2.4526	2.4374	2.4233	2.4100	2.3976	2.3860	2.2992	2.2450	2.1307	2.0575	2.0321	2.0192
40	2.3937	2.3689	2.3461	2.3252	2.3059	2.2880	2.2714	2.2559	2.2415	2.2280	2.2153	2.2034	2.1142	2.0581	1.9383	1.8602	1.8329	1.8189
50	2.2903	2.2652	2.2423	2.2211	2.2016	2.1835	2.1667	2.1510	2.1363	2.1226	2.1097	2.0976	2.0066	1.9490	1.8248	1.7425	1.7133	1.6984
100	2.0923	2.0666	2.0431	2.0214	2.0012	1.9826	1.9652	1.9489	1.9337	1.9194	1.9059	1.8933	1.7972	1.7353	1.5977	1.5013	1.4656	1.4468
200	1.9973	1.9713	1.9474	1.9252	1.9047	1.8857	1.8679	1.8512	1.8356	1.8210	1.8071	1.7941	1.6945	1.6295	1.4811	1.3711	1.3277	1.3040
250	1.9786	1.9525	1.9285	1.9063	1.8857	1.8665	1.8487	1.8319	1.8163	1.8015	1.7876	1.7744	1.6740	1.6083	1.4571	1.3432	1.2974	1.2720
500	1.9415	1.9152	1.8910	1.8686	1.8479	1.8285	1.8105	1.7936	1.7777	1.7627	1.7486	1.7353	1.6332	1.5658	1.4084	1.2846	1.2317	1.2007
1000	1.9231	1.8967	1.8724	1.8500	1.8291	1.8096	1.7915	1.7745	1.7585	1.7435	1.7293	1.7158	1.6127	1.5445	1.3835	1.2532	1.1947	1.1586

Table A5 F-Distribution Right-Tail Critical Values Corresponding to $\alpha = 0.05$

Denom. d.f.	Numerator d.f.																	
	1	2	3	4	5	6	7	8	9	10	11	12	13	14	15	16	17	18
1	161.448	199.500	215.707	224.583	230.162	233.986	236.768	238.883	240.543	241.882	242.984	243.906	244.690	245.364	245.950	246.464	246.918	247.323
2	18.5128	19.0000	19.1643	19.2468	19.2964	19.3295	19.3532	19.3710	19.3848	19.3959	19.4050	19.4125	19.4189	19.4244	19.4291	19.4333	19.4370	19.4402
3	10.1280	9.5521	9.2766	9.1172	9.0135	8.9406	8.8867	8.8452	8.8123	8.7855	8.7633	8.7446	8.7287	8.7149	8.7029	8.6923	8.6829	8.6745
4	7.7086	6.9443	6.5914	6.3882	6.2561	6.1631	6.0942	6.0410	5.9988	5.9644	5.9358	5.9117	5.8911	5.8733	5.8578	5.8441	5.8320	5.8211
5	6.6079	5.7861	5.4095	5.1922	5.0503	4.9503	4.8759	4.8183	4.7725	4.7351	4.7040	4.6777	4.6552	4.6358	4.6188	4.6038	4.5904	4.5785
6	5.9874	5.1433	4.7571	4.5337	4.3874	4.2839	4.2067	4.1468	4.0990	4.0600	4.0274	3.9999	3.9764	3.9559	3.9381	3.9223	3.9083	3.8957
7	5.5914	4.7374	4.3468	4.1203	3.9715	3.8660	3.7870	3.7257	3.6767	3.6365	3.6030	3.5747	3.5503	3.5292	3.5107	3.4944	3.4799	3.4669
8	5.3177	4.4590	4.0662	3.8379	3.6875	3.5806	3.5005	3.4381	3.3881	3.3472	3.3130	3.2839	3.2590	3.2374	3.2184	3.2016	3.1867	3.1733
9	5.1174	4.2565	3.8625	3.6331	3.4817	3.3738	3.2927	3.2296	3.1789	3.1373	3.1025	3.0729	3.0475	3.0255	3.0061	2.9890	2.9737	2.9600
10	4.9646	4.1028	3.7083	3.4780	3.3258	3.2172	3.1355	3.0717	3.0204	2.9782	2.9430	2.9130	2.8872	2.8647	2.8450	2.8276	2.8120	2.7980
11	4.8443	3.9823	3.5874	3.3567	3.2039	3.0946	3.0123	2.9480	2.8962	2.8536	2.8179	2.7876	2.7614	2.7386	2.7186	2.7009	2.6851	2.6709
12	4.7472	3.8853	3.4903	3.2592	3.1059	2.9961	2.9134	2.8486	2.7964	2.7534	2.7173	2.6866	2.6602	2.6371	2.6169	2.5989	2.5828	2.5684
13	4.6672	3.8056	3.4105	3.1791	3.0254	2.9153	2.8321	2.7669	2.7144	2.6710	2.6347	2.6037	2.5769	2.5536	2.5331	2.5149	2.4987	2.4841
14	4.6001	3.7389	3.3439	3.1122	2.9582	2.8477	2.7642	2.6987	2.6458	2.6022	2.5655	2.5342	2.5073	2.4837	2.4630	2.4446	2.4282	2.4134
15	4.5431	3.6823	3.2874	3.0556	2.9013	2.7905	2.7066	2.6408	2.5876	2.5437	2.5068	2.4753	2.4481	2.4244	2.4034	2.3849	2.3683	2.3533
16	4.4940	3.6337	3.2389	3.0069	2.8524	2.7413	2.6572	2.5911	2.5377	2.4935	2.4564	2.4247	2.3973	2.3733	2.3522	2.3335	2.3167	2.3016
17	4.4513	3.5915	3.1968	2.9647	2.8100	2.6987	2.6143	2.5480	2.4943	2.4499	2.4126	2.3807	2.3531	2.3290	2.3077	2.2888	2.2719	2.2567
18	4.4139	3.5546	3.1599	2.9277	2.7729	2.6613	2.5767	2.5102	2.4563	2.4117	2.3742	2.3421	2.3143	2.2900	2.2686	2.2496	2.2325	2.2172
19	4.3807	3.5219	3.1274	2.8951	2.7401	2.6283	2.5435	2.4768	2.4227	2.3779	2.3402	2.3080	2.2800	2.2556	2.2341	2.2149	2.1977	2.1823
20	4.3512	3.4928	3.0984	2.8661	2.7109	2.5990	2.5140	2.4471	2.3928	2.3479	2.3100	2.2776	2.2495	2.2250	2.2033	2.1840	2.1667	2.1511
21	4.3248	3.4668	3.0725	2.8401	2.6848	2.5727	2.4876	2.4205	2.3660	2.3210	2.2829	2.2504	2.2222	2.1975	2.1757	2.1563	2.1389	2.1232
22	4.3009	3.4434	3.0491	2.8167	2.6613	2.5491	2.4638	2.3965	2.3419	2.2967	2.2585	2.2258	2.1975	2.1727	2.1508	2.1313	2.1138	2.0980
23	4.2793	3.4221	3.0280	2.7955	2.6400	2.5277	2.4422	2.3748	2.3201	2.2747	2.2364	2.2036	2.1752	2.1502	2.1282	2.1086	2.0910	2.0751
24	4.2597	3.4028	3.0088	2.7763	2.6207	2.5082	2.4226	2.3551	2.3002	2.2547	2.2163	2.1834	2.1548	2.1298	2.1077	2.0880	2.0703	2.0543
25	4.2417	3.3852	2.9912	2.7587	2.6030	2.4904	2.4047	2.3371	2.2821	2.2365	2.1979	2.1649	2.1362	2.1111	2.0889	2.0691	2.0513	2.0353
26	4.2252	3.3690	2.9752	2.7426	2.5868	2.4741	2.3883	2.3205	2.2655	2.2197	2.1811	2.1479	2.1192	2.0939	2.0716	2.0518	2.0339	2.0178
27	4.2100	3.3541	2.9604	2.7278	2.5719	2.4591	2.3732	2.3053	2.2501	2.2043	2.1655	2.1323	2.1035	2.0781	2.0558	2.0358	2.0179	2.0017
28	4.1960	3.3404	2.9467	2.7141	2.5581	2.4453	2.3593	2.2913	2.2360	2.1900	2.1512	2.1179	2.0889	2.0635	2.0411	2.0210	2.0030	1.9868
29	4.1830	3.3277	2.9340	2.7014	2.5454	2.4324	2.3463	2.2783	2.2229	2.1768	2.1379	2.1045	2.0755	2.0500	2.0275	2.0073	1.9893	1.9730
30	4.1709	3.3158	2.9223	2.6896	2.5336	2.4205	2.3343	2.2662	2.2107	2.1646	2.1256	2.0921	2.0630	2.0374	2.0148	1.9946	1.9765	1.9601
40	4.0847	3.2317	2.8387	2.6060	2.4495	2.3359	2.2490	2.1802	2.1240	2.0772	2.0376	2.0035	1.9738	1.9476	1.9245	1.9037	1.8851	1.8682
50	4.0343	3.1826	2.7900	2.5572	2.4004	2.2864	2.1992	2.1299	2.0734	2.0261	1.9861	1.9515	1.9214	1.8949	1.8714	1.8503	1.8313	1.8141
100	3.9361	3.0873	2.6955	2.4626	2.3053	2.1906	2.1025	2.0323	1.9748	1.9267	1.8857	1.8503	1.8193	1.7919	1.7675	1.7456	1.7259	1.7079
200	3.8884	3.0411	2.6498	2.4168	2.2592	2.1441	2.0556	1.9849	1.9269	1.8783	1.8368	1.8008	1.7694	1.7415	1.7166	1.6943	1.6741	1.6556
250	3.8789	3.0319	2.6407	2.4078	2.2501	2.1350	2.0463	1.9756	1.9174	1.8687	1.8271	1.7910	1.7595	1.7315	1.7065	1.6841	1.6638	1.6453
500	3.8601	3.0138	2.6227	2.3898	2.2320	2.1167	2.0279	1.9569	1.8986	1.8496	1.8078	1.7715	1.7398	1.7116	1.6864	1.6638	1.6432	1.6245
1000	3.8508	3.0047	2.6138	2.3808	2.2231	2.1076	2.0187	1.9476	1.8892	1.8402	1.7982	1.7618	1.7299	1.7017	1.6764	1.6536	1.6330	1.6142

Table A5 F-Distribution Critical Values Corresponding to $\alpha = 0.05$.

Denom. d.f.	19	20	21	22	23	24	25	26	27	28	29	30	40	50	100	250	500	1000
1	247.686	248.013	248.309	248.579	248.826	249.052	249.260	249.453	249.631	249.797	249.951	250.095	251.143	251.774	253.041	253.804	254.059	254.187
2	19.4431	19.4458	19.4481	19.4503	19.4523	19.4541	19.4558	19.4573	19.4587	19.4600	19.4613	19.4624	19.4707	19.4757	19.4857	19.4917	19.4937	19.4947
3	8.6670	8.6602	8.6540	8.6484	8.6432	8.6385	8.6341	8.6301	8.6263	8.6229	8.6196	8.6166	8.5944	8.5810	8.5539	8.5375	8.5320	8.5292
4	5.8114	5.8025	5.7945	5.7872	5.7805	5.7744	5.7687	5.7635	5.7586	5.7541	5.7498	5.7459	5.7170	5.6995	5.6641	5.6425	5.6353	5.6317
5	4.5678	4.5581	4.5493	4.5413	4.5339	4.5272	4.5209	4.5151	4.5097	4.5047	4.5001	4.4957	4.4638	4.4444	4.4051	4.3811	4.3731	4.3690
6	3.8844	3.8742	3.8649	3.8564	3.8486	3.8415	3.8348	3.8287	3.8230	3.8177	3.8128	3.8082	3.7743	3.7537	3.7117	3.6861	3.6775	3.6732
7	3.4551	3.4445	3.4349	3.4260	3.4179	3.4105	3.4036	3.3972	3.3913	3.3858	3.3806	3.3758	3.3404	3.3189	3.2749	3.2479	3.2389	3.2343
8	3.1613	3.1503	3.1404	3.1313	3.1229	3.1152	3.1081	3.1015	3.0954	3.0897	3.0844	3.0794	3.0428	3.0204	2.9747	2.9466	2.9371	2.9324
9	2.9477	2.9365	2.9263	2.9169	2.9084	2.9005	2.8932	2.8864	2.8801	2.8743	2.8688	2.8637	2.8259	2.8028	2.7556	2.7264	2.7166	2.7116
10	2.7854	2.7740	2.7636	2.7541	2.7453	2.7372	2.7298	2.7229	2.7164	2.7104	2.7048	2.6996	2.6609	2.6371	2.5884	2.5583	2.5481	2.5430
11	2.6581	2.6464	2.6358	2.6261	2.6172	2.6090	2.6014	2.5943	2.5877	2.5816	2.5759	2.5705	2.5309	2.5066	2.4566	2.4256	2.4151	2.4098
12	2.5554	2.5436	2.5328	2.5229	2.5139	2.5055	2.4977	2.4905	2.4838	2.4776	2.4718	2.4663	2.4259	2.4010	2.3498	2.3179	2.3071	2.3017
13	2.4709	2.4589	2.4479	2.4379	2.4287	2.4202	2.4123	2.4050	2.3982	2.3918	2.3859	2.3803	2.3392	2.3138	2.2614	2.2287	2.2176	2.2121
14	2.4000	2.3879	2.3768	2.3667	2.3573	2.3487	2.3407	2.3333	2.3264	2.3199	2.3139	2.3082	2.2664	2.2405	2.1870	2.1536	2.1422	2.1365
15	2.3398	2.3275	2.3163	2.3060	2.2966	2.2878	2.2797	2.2722	2.2652	2.2587	2.2525	2.2468	2.2043	2.1780	2.1234	2.0893	2.0776	2.0718
16	2.2880	2.2756	2.2642	2.2538	2.2443	2.2354	2.2272	2.2196	2.2125	2.2059	2.1997	2.1938	2.1507	2.1240	2.0685	2.0336	2.0217	2.0157
17	2.2429	2.2304	2.2189	2.2084	2.1987	2.1898	2.1815	2.1738	2.1666	2.1599	2.1536	2.1477	2.1040	2.0769	2.0204	1.9849	1.9727	1.9666
18	2.2033	2.1906	2.1791	2.1685	2.1587	2.1497	2.1413	2.1335	2.1262	2.1195	2.1131	2.1071	2.0629	2.0354	1.9780	1.9418	1.9294	1.9232
19	2.1683	2.1555	2.1438	2.1331	2.1233	2.1141	2.1057	2.0978	2.0905	2.0836	2.0772	2.0712	2.0264	1.9986	1.9403	1.9035	1.8909	1.8845
20	2.1370	2.1242	2.1124	2.1016	2.0917	2.0825	2.0739	2.0660	2.0586	2.0517	2.0452	2.0391	1.9938	1.9656	1.9066	1.8691	1.8562	1.8497
21	2.1090	2.0960	2.0842	2.0733	2.0633	2.0540	2.0454	2.0374	2.0299	2.0229	2.0164	2.0102	1.9645	1.9360	1.8761	1.8381	1.8250	1.8184
22	2.0837	2.0707	2.0587	2.0478	2.0377	2.0283	2.0196	2.0116	2.0040	1.9970	1.9904	1.9842	1.9380	1.9092	1.8486	1.8099	1.7966	1.7899
23	2.0608	2.0476	2.0356	2.0246	2.0144	2.0050	1.9963	1.9881	1.9805	1.9734	1.9668	1.9605	1.9139	1.8848	1.8234	1.7843	1.7708	1.7639
24	2.0399	2.0267	2.0146	2.0035	1.9932	1.9838	1.9750	1.9668	1.9591	1.9520	1.9453	1.9390	1.8920	1.8625	1.8005	1.7608	1.7470	1.7401
25	2.0207	2.0075	1.9953	1.9842	1.9738	1.9643	1.9554	1.9472	1.9395	1.9323	1.9255	1.9192	1.8718	1.8421	1.7794	1.7391	1.7252	1.7181
26	2.0032	1.9898	1.9776	1.9664	1.9560	1.9464	1.9375	1.9292	1.9215	1.9142	1.9074	1.9010	1.8533	1.8233	1.7599	1.7191	1.7050	1.6978
27	1.9870	1.9736	1.9613	1.9500	1.9396	1.9299	1.9210	1.9126	1.9048	1.8975	1.8907	1.8842	1.8361	1.8059	1.7419	1.7006	1.6863	1.6790
28	1.9720	1.9586	1.9462	1.9349	1.9244	1.9147	1.9057	1.8973	1.8894	1.8821	1.8752	1.8687	1.8203	1.7898	1.7251	1.6834	1.6689	1.6615
29	1.9581	1.9446	1.9322	1.9208	1.9103	1.9005	1.8915	1.8830	1.8751	1.8677	1.8608	1.8543	1.8055	1.7748	1.7096	1.6674	1.6527	1.6452
30	1.9452	1.9317	1.9192	1.9077	1.8972	1.8874	1.8782	1.8698	1.8618	1.8544	1.8474	1.8409	1.7918	1.7609	1.6950	1.6524	1.6375	1.6299
40	1.8529	1.8389	1.8260	1.8141	1.8031	1.7929	1.7835	1.7746	1.7663	1.7586	1.7513	1.7444	1.6928	1.6600	1.5892	1.5425	1.5260	1.5175
50	1.7985	1.7841	1.7709	1.7588	1.7475	1.7371	1.7273	1.7183	1.7097	1.7017	1.6942	1.6872	1.6337	1.5995	1.5249	1.4748	1.4569	1.4477
100	1.6915	1.6764	1.6626	1.6497	1.6378	1.6267	1.6163	1.6067	1.5976	1.5890	1.5809	1.5733	1.5151	1.4772	1.3917	1.3308	1.3079	1.2958
200	1.6388	1.6233	1.6090	1.5958	1.5834	1.5720	1.5612	1.5511	1.5417	1.5328	1.5243	1.5164	1.4551	1.4146	1.3206	1.2495	1.2211	1.2054
250	1.6283	1.6127	1.5983	1.5850	1.5726	1.5610	1.5502	1.5400	1.5305	1.5215	1.5130	1.5049	1.4430	1.4019	1.3058	1.2318	1.2015	1.1847
500	1.6074	1.5916	1.5770	1.5635	1.5509	1.5392	1.5282	1.5178	1.5081	1.4989	1.4903	1.4821	1.4186	1.3762	1.2753	1.1940	1.1587	1.1378
1000	1.5969	1.5811	1.5664	1.5528	1.5401	1.5282	1.5171	1.5067	1.4969	1.4876	1.4789	1.4706	1.4063	1.3632	1.2596	1.1735	1.1342	1.1097

Numerator $d.f.$

Table A6 – Ryan Joiner Test[3]

n	$\alpha=.01$	$\alpha=.05$	$\alpha=.10$
5	0.832	0.880	0.903
6	0.841	0.889	0.911
7	0.852	0.898	0.919
8	0.862	0.905	0.925
9	0.872	0.912	0.930
10	0.880	0.918	0.935
11	0.888	0.923	0.939
12	0.895	0.928	0.942
13	0.901	0.932	0.945
14	0.906	0.935	0.948
15	0.911	0.938	0.951
16	0.915	0.941	0.953
17	0.919	0.944	0.955
18	0.923	0.946	0.957
19	0.926	0.948	0.958
20	0.929	0.950	0.960
21	0.932	0.952	0.961
22	0.934	0.954	0.963
23	0.937	0.955	0.964
24	0.939	0.957	0.965
25	0.941	0.958	0.966
26	0.943	0.960	0.967
27	0.944	0.961	0.968
28	0.946	0.962	0.969
29	0.948	0.963	0.970
30	0.949	0.964	0.971
31	0.950	0.965	0.971
32	0.952	0.966	0.972
33	0.953	0.967	0.973
34	0.954	0.967	0.973
35	0.955	0.968	0.974
36	0.956	0.969	0.975
37	0.957	0.970	0.975
38	0.958	0.970	0.976
39	0.959	0.971	0.976
40	0.960	0.971	0.977
41	0.960	0.972	0.977
42	0.961	0.973	0.978
43	0.962	0.973	0.978
44	0.963	0.974	0.978
45	0.963	0.974	0.979
46	0.964	0.975	0.979
47	0.965	0.975	0.980
48	0.965	0.976	0.980
49	0.966	0.976	0.980
50	0.966	0.976	0.981
60	0.971	0.980	0.983
70	0.974	0.982	0.986
80	0.977	0.984	0.987
90	0.979	0.986	0.989
100	0.980	0.987	0.990
150	0.985	0.992	0.993
200	0.988	0.994	0.996
250	0.989	0.996	0.997
400	0.997	0.996	0.997
600	0.998	0.998	0.997
1000	0.999	0.998	0.998

Table A7
Sign Test Critical Values

	One Tail Applications			
	$\alpha \approx .005$	$\alpha \approx .01$	$\alpha \approx .025$	$\alpha \approx .05$
	Two Tail Applications			
n	$\alpha \approx .01$	$\alpha \approx .02$	$\alpha \approx .05$	$\alpha \approx .10$
5	–	–	–	0
6	–	–	0	0
7	–	0	0	0
8	0	0	0	1
9	0	0	1	1
10	0	0	1	1
11	0	1	1	2
12	1	1	2	2
13	1	1	2	3
14	1	2	2	3
15	2	2	3	3
16	2	2	3	4
17	2	3	4	4
18	3	3	4	5
19	3	4	4	5
20	3	4	5	5

Table A9
Signed Rank Test Critical Values

	One Tail Applications			
	$\alpha \approx .005$	$\alpha \approx .01$	$\alpha \approx .025$	$\alpha \approx .05$
	Two Tail Applications			
n	$\alpha \approx .01$	$\alpha \approx .02$	$\alpha \approx .05$	$\alpha \approx .10$
4	–	–	–	0
5	–	–	0	1
6	–	–	1	2
7	–	0	2	4
8	0	1	4	6
9	2	3	6	8
10	3	4	8	11
11	5	7	11	14
12	7	10	14	17
13	10	13	17	21
14	13	16	21	26
15	16	21	25	30
16	19	24	30	36
17	23	28	35	41
18	28	33	40	47
19	32	38	46	54
20	37	43	52	60

Table A8: p-Values for the Sign Test

| | | | | | $x_s = s$ | | | | | | |
n	0	1	2	3	4	5	6	7	8	9	10
1	0.5000	1.0000									
2	0.2500	0.7500	1.0000								
3	0.1250	0.5000	0.8750	1.0000							
4	0.0625	0.3125	0.6875	0.9375	1.0000						
5	0.0313	0.1875	0.5000	0.8125	0.9688	1.0000					
6	0.0156	0.1094	0.3438	0.6563	0.8906	0.9844	1.0000				
7	0.0078	0.0625	0.2266	0.5000	0.7734	0.9375	0.9922	1.0000			
8	0.0039	0.0352	0.1445	0.3633	0.6367	0.8555	0.9648	0.9961	1.0000		
9	0.0020	0.0195	0.0898	0.2539	0.5000	0.7461	0.9102	0.9805	0.9981	1.0000	
10	0.0010	0.0107	0.0547	0.1719	0.3770	0.6231	0.8281	0.9453	0.9893	0.9990	1.0000

Values above are p-values for one-tail tests. Multiply by 2 for a two-tail test p-value.

Right tail test: s = number of (–) signs. Left tail test: s = number of (+) signs.

Two tail test: s = minimum number of (+) or (–) signs.

Table A10 – p-Values for the Wilcoxon Signed Rank Tests

R	$n=1$	$n=2$	$n=3$	$n=4$	$n=5$	$n=6$	$n=7$	$n=8$	$n=9$	$n=10$	$n=11$	$n=12$	$n=13$	$n=14$	$n=15$	$n=16$	$n=17$	$n=18$	$n=19$	$n=20$
0	0.5000	0.2500	0.1250	0.0625	0.0313	0.0156	0.0078	0.0039	0.0020	0.0010	0.0005	0.0002	0.0001	0.0001	0.0001	0.0001	0.0001	0.0001	0.0001	0.0001
1	1.0000	0.5000	0.2500	0.1250	0.0625	0.0313	0.0156	0.0078	0.0039	0.0020	0.0010	0.0005	0.0002	0.0001	0.0001	0.0001	0.0001	0.0001	0.0001	0.0001
2		0.7500	0.3750	0.1875	0.0938	0.0469	0.0234	0.0117	0.0059	0.0029	0.0015	0.0007	0.0004	0.0002	0.0001	0.0001	0.0001	0.0001	0.0001	0.0001
3		1.0000	0.6250	0.3125	0.1563	0.0781	0.0391	0.0195	0.0098	0.0049	0.0024	0.0012	0.0006	0.0003	0.0002	0.0001	0.0001	0.0001	0.0001	0.0001
4			0.7500	0.4375	0.2188	0.1094	0.0547	0.0273	0.0137	0.0068	0.0034	0.0017	0.0009	0.0004	0.0002	0.0001	0.0001	0.0001	0.0001	0.0001
5			0.8750	0.5625	0.3125	0.1563	0.0781	0.0391	0.0195	0.0098	0.0049	0.0024	0.0012	0.0006	0.0003	0.0002	0.0001	0.0001	0.0001	0.0001
6			1.0000	0.6875	0.4063	0.2188	0.1094	0.0547	0.0273	0.0137	0.0068	0.0034	0.0017	0.0009	0.0004	0.0002	0.0001	0.0001	0.0001	0.0001
7				0.8125	0.5000	0.2813	0.1484	0.0742	0.0371	0.0186	0.0093	0.0046	0.0023	0.0012	0.0006	0.0003	0.0001	0.0001	0.0001	0.0001
8				0.8750	0.5938	0.3438	0.1875	0.0977	0.0488	0.0244	0.0122	0.0061	0.0031	0.0015	0.0008	0.0004	0.0002	0.0001	0.0001	0.0001
9				0.9375	0.6875	0.4219	0.2344	0.1250	0.0645	0.0322	0.0161	0.0081	0.0040	0.0020	0.0010	0.0005	0.0003	0.0001	0.0001	0.0001
10				1.0000	0.7813	0.5000	0.2891	0.1563	0.0820	0.0420	0.0210	0.0105	0.0052	0.0026	0.0013	0.0007	0.0003	0.0002	0.0001	0.0001
11					0.8438	0.5781	0.3438	0.1914	0.1016	0.0527	0.0269	0.0134	0.0067	0.0034	0.0017	0.0008	0.0004	0.0002	0.0001	0.0001
12					0.9063	0.6563	0.4063	0.2305	0.1250	0.0654	0.0337	0.0171	0.0085	0.0043	0.0021	0.0011	0.0005	0.0003	0.0001	0.0001
13					0.9375	0.7188	0.4688	0.2734	0.1504	0.0801	0.0415	0.0212	0.0107	0.0054	0.0027	0.0013	0.0007	0.0003	0.0002	0.0001
14					0.9688	0.7813	0.5313	0.3203	0.1797	0.0967	0.0508	0.0261	0.0133	0.0067	0.0034	0.0017	0.0008	0.0004	0.0002	0.0001
15					1.0000	0.8438	0.5938	0.3711	0.2129	0.1162	0.0615	0.0320	0.0164	0.0083	0.0042	0.0021	0.0010	0.0005	0.0003	0.0001
16						0.8906	0.6563	0.4219	0.2480	0.1377	0.0737	0.0386	0.0199	0.0101	0.0051	0.0026	0.0013	0.0006	0.0003	0.0002
17						0.9219	0.7109	0.4727	0.2852	0.1611	0.0874	0.0461	0.0239	0.0123	0.0062	0.0031	0.0016	0.0008	0.0004	0.0002
18						0.9531	0.7656	0.5273	0.3262	0.1875	0.1030	0.0549	0.0287	0.0148	0.0075	0.0038	0.0019	0.0010	0.0005	0.0002
19						0.9688	0.8125	0.5781	0.3672	0.2158	0.1201	0.0647	0.0341	0.0176	0.0090	0.0046	0.0023	0.0012	0.0006	0.0003
20						0.9844	0.8516	0.6289	0.4102	0.2461	0.1392	0.0757	0.0402	0.0209	0.0108	0.0055	0.0028	0.0014	0.0007	0.0004
21						1.0000	0.8906	0.6797	0.4551	0.2783	0.1602	0.0881	0.0471	0.0247	0.0128	0.0065	0.0033	0.0017	0.0008	0.0004
22							0.9219	0.7266	0.5000	0.3125	0.1826	0.1018	0.0549	0.0290	0.0151	0.0078	0.0040	0.0020	0.0010	0.0005
23							0.9453	0.7695	0.5449	0.3477	0.2065	0.1167	0.0636	0.0338	0.0177	0.0091	0.0047	0.0024	0.0012	0.0006
24							0.9609	0.8086	0.5898	0.3848	0.2324	0.1331	0.0732	0.0392	0.0206	0.0107	0.0055	0.0028	0.0014	0.0007
25							0.9766	0.8438	0.6328	0.4229	0.2598	0.1506	0.0839	0.0453	0.0240	0.0125	0.0064	0.0033	0.0017	0.0008
26							0.9844	0.8750	0.6738	0.4609	0.2886	0.1697	0.0955	0.0520	0.0277	0.0145	0.0075	0.0038	0.0020	0.0010
27							0.9922	0.9023	0.7148	0.5000	0.3188	0.1902	0.1082	0.0594	0.0319	0.0168	0.0087	0.0045	0.0023	0.0012
28							1.0000	0.9258	0.7520	0.5391	0.3501	0.2119	0.1219	0.0676	0.0365	0.0193	0.0101	0.0052	0.0027	0.0014
29								0.9453	0.7871	0.5771	0.3823	0.2349	0.1367	0.0765	0.0416	0.0222	0.0116	0.0060	0.0031	0.0016
30								0.9609	0.8203	0.6152	0.4155	0.2593	0.1527	0.0863	0.0473	0.0253	0.0133	0.0069	0.0036	0.0018
31								0.9727	0.8496	0.6523	0.4492	0.2847	0.1698	0.0969	0.0535	0.0288	0.0153	0.0080	0.0041	0.0021
32								0.9805	0.8750	0.6875	0.4829	0.3110	0.1879	0.1083	0.0603	0.0327	0.0174	0.0091	0.0047	0.0024
33								0.9883	0.8984	0.7217	0.5171	0.3386	0.2072	0.1206	0.0677	0.0370	0.0198	0.0104	0.0054	0.0028
34								0.9922	0.9180	0.7539	0.5508	0.3667	0.2274	0.1338	0.0757	0.0416	0.0224	0.0118	0.0062	0.0032
35								0.9961	0.9355	0.7842	0.5845	0.3955	0.2487	0.1479	0.0844	0.0467	0.0253	0.0134	0.0070	0.0036
36								1.0000	0.9512	0.8125	0.6177	0.4250	0.2709	0.1629	0.0938	0.0523	0.0284	0.0152	0.0080	0.0042
37									0.9629	0.8389	0.6499	0.4548	0.2939	0.1788	0.1039	0.0583	0.0319	0.0171	0.0090	0.0047
38									0.9727	0.8623	0.6812	0.4849	0.3177	0.1955	0.1147	0.0649	0.0357	0.0192	0.0102	0.0053
39									0.9805	0.8838	0.7114	0.5151	0.3424	0.2131	0.1262	0.0719	0.0398	0.0216	0.0115	0.0060
40									0.9863	0.9033	0.7402	0.5452	0.3677	0.2316	0.1384	0.0795	0.0443	0.0241	0.0129	0.0068
41									0.9902	0.9199	0.7676	0.5750	0.3934	0.2508	0.1514	0.0877	0.0492	0.0269	0.0145	0.0077
42									0.9941	0.9346	0.7935	0.6045	0.4197	0.2708	0.1651	0.0964	0.0544	0.0300	0.0162	0.0086
43									0.9961	0.9473	0.8174	0.6333	0.4463	0.2915	0.1796	0.1057	0.0601	0.0333	0.0180	0.0096
44									0.9980	0.9580	0.8398	0.6614	0.4730	0.3129	0.1947	0.1156	0.0662	0.0368	0.0201	0.0107
45									1.0000	0.9678	0.8608	0.6890	0.5000	0.3349	0.2106	0.1261	0.0727	0.0407	0.0223	0.0120
46										0.9756	0.8799	0.7153	0.5270	0.3574	0.2271	0.1372	0.0797	0.0449	0.0247	0.0133
47										0.9814	0.8970	0.7407	0.5537	0.3804	0.2444	0.1489	0.0871	0.0494	0.0273	0.0148
48										0.9863	0.9126	0.7651	0.5803	0.4039	0.2622	0.1613	0.0950	0.0542	0.0301	0.0164
49										0.9902	0.9263	0.7881	0.6066	0.4276	0.2807	0.1742	0.1034	0.0594	0.0331	0.0181
50										0.9932	0.9385	0.8098	0.6323	0.4516	0.2997	0.1877	0.1123	0.0649	0.0364	0.0200
51										0.9951	0.9492	0.8303	0.6576	0.4758	0.3193	0.2019	0.1217	0.0708	0.0399	0.0220
52										0.9971	0.9585	0.8494	0.6823	0.5000	0.3394	0.2166	0.1317	0.0770	0.0437	0.0242
53										0.9980	0.9663	0.8669	0.7061	0.5242	0.3599	0.2319	0.1421	0.0837	0.0478	0.0266
54										0.9990	0.9731	0.8833	0.7291	0.5484	0.3808	0.2477	0.1530	0.0907	0.0521	0.0291
55										1.0000	0.9790	0.8982	0.7513	0.5724	0.4020	0.2641	0.1645	0.0982	0.0567	0.0319
56											0.9839	0.9119	0.7726	0.5961	0.4235	0.2809	0.1764	0.1061	0.0616	0.0348
57											0.9878	0.9243	0.7928	0.6196	0.4452	0.2983	0.1889	0.1144	0.0668	0.0379
58											0.9907	0.9353	0.8121	0.6426	0.4670	0.3161	0.2019	0.1231	0.0723	0.0413
59											0.9932	0.9451	0.8302	0.6651	0.4890	0.3343	0.2153	0.1323	0.0782	0.0448
60											0.9951	0.9539	0.8473	0.6871	0.5110	0.3529	0.2293	0.1419	0.0844	0.0487
61											0.9966	0.9614	0.8633	0.7085	0.5330	0.3718	0.2437	0.1519	0.0909	0.0527
62											0.9976	0.9680	0.8781	0.7292	0.5548	0.3910	0.2585	0.1624	0.0978	0.0570
63											0.9985	0.9739	0.8918	0.7492	0.5765	0.4104	0.2738	0.1733	0.1051	0.0615
64											0.9990	0.9788	0.9045	0.7684	0.5980	0.4301	0.2895	0.1846	0.1127	0.0664
65											0.9995	0.9829	0.9161	0.7869	0.6192	0.4500	0.3056	0.1964	0.1206	0.0715
66											1.0000	0.9866	0.9268	0.8045	0.6401	0.4699	0.3221	0.2086	0.1290	0.0768
67												0.9895	0.9364	0.8212	0.6606	0.4900	0.3389	0.2211	0.1377	0.0825
68												0.9919	0.9451	0.8371	0.6807	0.5100	0.3559	0.2341	0.1467	0.0884
69												0.9939	0.9529	0.8521	0.7003	0.5301	0.3733	0.2475	0.1562	0.0947
70												0.9954	0.9598	0.8662	0.7193	0.5500	0.3910	0.2613	0.1660	0.1012
71												0.9966	0.9659	0.8794	0.7378	0.5699	0.4088	0.2754	0.1762	0.1081
72												0.9976	0.9713	0.8917	0.7556	0.5896	0.4268	0.2899	0.1868	0.1153
73												0.9983	0.9761	0.9031	0.7729	0.6090	0.4450	0.3047	0.1977	0.1227
74												0.9988	0.9801	0.9137	0.7894	0.6282	0.4633	0.3198	0.2090	0.1305
75												0.9993	0.9836	0.9235	0.8053	0.6471	0.4816	0.3353	0.2207	0.1387
76												0.9995	0.9867	0.9324	0.8204	0.6657	0.5000	0.3509	0.2327	0.1471
77												0.9998	0.9893	0.9406	0.8349	0.6839	0.5184	0.3669	0.2450	0.1559
78												1.0000	0.9915	0.9480	0.8486	0.7017	0.5367	0.3830	0.2576	0.1650
79													0.9933	0.9547	0.8616	0.7191	0.5550	0.3994	0.2706	0.1744
80													0.9948	0.9608	0.8738	0.7359	0.5732	0.4159	0.2839	0.1841
81													0.9960	0.9662	0.8853	0.7523	0.5912	0.4325	0.2974	0.1942
82													0.9969	0.9710	0.8961	0.7681	0.6090	0.4493	0.3113	0.2045
83													0.9977	0.9753	0.9062	0.7834	0.6267	0.4661	0.3254	0.2152
84													0.9983	0.9791	0.9156	0.7981	0.6441	0.4831	0.3397	0.2262
85													0.9988	0.9824	0.9243	0.8123	0.6611	0.5000	0.3543	0.2375
86													0.9991	0.9852	0.9323	0.8258	0.6779	0.5169	0.3690	0.2490
87													0.9994	0.9877	0.9397	0.8387	0.6944	0.5339	0.3840	0.2608
88													0.9996	0.9899	0.9465	0.8511	0.7105	0.5507	0.3991	0.2729
89													0.9998	0.9917	0.9527	0.8628	0.7262	0.5675	0.4144	0.2853
90													0.9999	0.9933	0.9584	0.8739	0.7415	0.5841	0.4298	0.2979
91													1.0000	0.9946	0.9635	0.8844	0.7563	0.6006	0.4453	0.3108
92														0.9957	0.9681	0.8943	0.7707	0.6170	0.4609	0.3238
93														0.9966	0.9723	0.9036	0.7847	0.6331	0.4765	0.3371
94														0.9974	0.9760	0.9123	0.7981	0.6491	0.4922	0.3506
95														0.9980	0.9794	0.9205	0.8111	0.6647	0.5078	0.3643
96														0.9985	0.9823	0.9281	0.8236	0.6802	0.5235	0.3781

Table A10 – p-Values for the Wilcoxon Signed Rank Tests

R	n=1	n=2	n=3	n=4	n=5	n=6	n=7	n=8	n=9	n=10	n=11	n=12	n=13	n=14	n=15	n=16	n=17	n=18	n=19	n=20
97														0.9988	0.9849	0.9351	0.8355	0.6953	0.5391	0.3921
98														0.9991	0.9872	0.9417	0.8470	0.7101	0.5547	0.4062
99														0.9994	0.9892	0.9477	0.8579	0.7246	0.5702	0.4204
100														0.9996	0.9910	0.9533	0.8683	0.7387	0.5856	0.4347
101														0.9997	0.9925	0.9584	0.8783	0.7525	0.6009	0.4492
102														0.9998	0.9938	0.9630	0.8877	0.7659	0.6160	0.4636
103														0.9999	0.9949	0.9673	0.8966	0.7789	0.6310	0.4782
104														0.9999	0.9958	0.9712	0.9050	0.7914	0.6457	0.4927
105														1.0000	0.9966	0.9747	0.9129	0.8036	0.6603	0.5073
106															0.9973	0.9778	0.9203	0.8154	0.6746	0.5218
107															0.9979	0.9807	0.9273	0.8267	0.6887	0.5364
108															0.9983	0.9832	0.9338	0.8376	0.7026	0.5508
109															0.9987	0.9855	0.9399	0.8481	0.7161	0.5653
110															0.9990	0.9875	0.9456	0.8581	0.7294	0.5796
111															0.9992	0.9893	0.9508	0.8677	0.7424	0.5938
112															0.9994	0.9909	0.9557	0.8769	0.7550	0.6079
113															0.9996	0.9922	0.9602	0.8856	0.7673	0.6219
114															0.9997	0.9935	0.9643	0.8939	0.7793	0.6357
115															0.9998	0.9945	0.9681	0.9018	0.7910	0.6494
116															0.9998	0.9954	0.9716	0.9093	0.8023	0.6629
117															0.9999	0.9962	0.9747	0.9163	0.8132	0.6762
118															0.9999	0.9969	0.9776	0.9230	0.8238	0.6892
119															0.9999	0.9974	0.9802	0.9292	0.8340	0.7021
120															1.0000	0.9979	0.9826	0.9351	0.8438	0.7147
121																0.9983	0.9847	0.9406	0.8533	0.7271
122																0.9987	0.9867	0.9458	0.8623	0.7392
123																0.9989	0.9884	0.9506	0.8710	0.7510
124																0.9992	0.9899	0.9551	0.8794	0.7625
125																0.9993	0.9913	0.9593	0.8873	0.7738
126																0.9995	0.9925	0.9632	0.8949	0.7848
127																0.9996	0.9936	0.9667	0.9022	0.7955
128																0.9997	0.9945	0.9700	0.9091	0.8058
129																0.9998	0.9953	0.9731	0.9156	0.8159
130																0.9998	0.9960	0.9759	0.9218	0.8256
131																0.9999	0.9967	0.9784	0.9277	0.8350
132																0.9999	0.9972	0.9808	0.9332	0.8441
133																0.9999	0.9977	0.9829	0.9384	0.8529
134																0.9999	0.9981	0.9848	0.9433	0.8613
135																0.9999	0.9984	0.9866	0.9479	0.8695
136																1.0000	0.9987	0.9882	0.9522	0.8773
137																	0.9990	0.9896	0.9563	0.8847
138																	0.9992	0.9909	0.9601	0.8919
139																	0.9993	0.9920	0.9636	0.8988
140																	0.9995	0.9931	0.9669	0.9053
141																	0.9996	0.9940	0.9699	0.9116
142																	0.9997	0.9948	0.9727	0.9175
143																	0.9997	0.9955	0.9753	0.9232
144																	0.9998	0.9962	0.9777	0.9285
145																	0.9999	0.9967	0.9799	0.9336
146																	0.9999	0.9972	0.9820	0.9385
147																	0.9999	0.9976	0.9838	0.9430
148																	0.9999	0.9980	0.9855	0.9473
149																	0.9999	0.9983	0.9871	0.9513
150																	0.9999	0.9986	0.9885	0.9552
151																	0.9999	0.9988	0.9898	0.9587
152																	0.9999	0.9990	0.9910	0.9621
153																	1.0000	0.9992	0.9920	0.9652
154																		0.9994	0.9930	0.9681
155																		0.9995	0.9938	0.9709
156																		0.9996	0.9946	0.9734
157																		0.9997	0.9953	0.9758
158																		0.9997	0.9959	0.9780
159																		0.9998	0.9964	0.9800
160																		0.9998	0.9969	0.9819
161																		0.9999	0.9973	0.9836
162																		0.9999	0.9977	0.9852
163																		0.9999	0.9980	0.9867
164																		0.9999	0.9983	0.9880
165																		0.9999	0.9986	0.9893
166																		0.9999	0.9988	0.9904
167																		0.9999	0.9990	0.9914
168																		0.9999	0.9992	0.9923
169																		0.9999	0.9993	0.9932
170																		0.9999	0.9994	0.9940
171																		1.0000	0.9995	0.9947
172																			0.9996	0.9953
173																			0.9997	0.9958
174																			0.9997	0.9964
175																			0.9998	0.9968
176																			0.9998	0.9972
177																			0.9999	0.9976
178																			0.9999	0.9979
179																			0.9999	0.9982
180																			0.9999	0.9984
181																			0.9999	0.9986
182																			0.9999	0.9988
183																			0.9999	0.9990
184																			0.9999	0.9992
185																			0.9999	0.9993
186																			0.9999	0.9994
187																			0.9999	0.9995
188																			0.9999	0.9996
189																			0.9999	0.9996
190																			1.0000	0.9997
191 – 193																				0.9998
194 – 198																				0.9999
> 198																				0.9999

Appendix B

Solutions to Selected Exercises

B.1   Chapter 1 Solutions

Section 1.1 Solutions

1. This is a subset of all students – a sample.
3. This is a sample of all goldfish.
5. This is a statistic, computed from a sample.
7. This is a parameter, computed from a population.
9. For a truly large population, a census is typically impossible.
11. These data are categorical.
13. These data are quantitative.
15. The bar chart is the appropriate display.

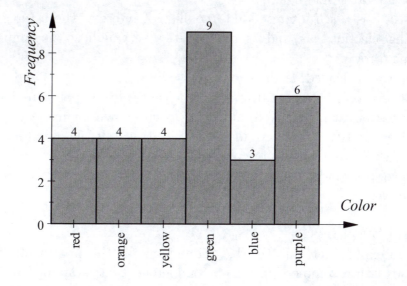

Favorite primary and secondary colors of randomly selected children.

17. The dotplot is the appropriate display.

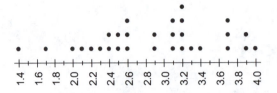

Grade point averages of random community college students.

19. Letter grades are qualitative.
21. Calendar years are quantitative.
23. The number of honors students is a discrete measurement.
25. Change in your pocket is discrete, as fractions of a cent are impossible.
27. Calendar years are interval measurements, as the zero year does not mean "no time."
29. Weights are ratio measurements, as a zero weight really means *nothing*.
31. Numbers of people who commute together are ratio measurements, as zero again means *none*.
33. Temperatures in Celsius are interval measurements, as zero does not mean *no heat*.
35. We could use the normal average where we add the values and divide by the sample size (this is the *arithmetic mean* of the sample).
37. To summarize zip codes we could give the percentage of sample members who reside in each zip code. We could also give the most frequently occurring zip code.

Section 1.2 Solutions

1. The study is a self-interest study/conflict of interest, the sample is self-selecting (yielding selection bias) and the refusal rate is very high, so the sample is not random.
3. Sampling error
5. The question is loaded, leading to response bias.
9. Correlation does not imply cause. It is true that cold temperature does correlate with higher risk of catching a cold, but it has been known for many years that colds are caused by a virus. People are more likely to transmit the virus during cold weather because they are indoors more and thus closer together.
11. It is never appropriate to exaggerate the results of statistical research, regardless of the motives.

Section 1.4 Solutions

1. This is an experiment, and the T.V. commercial is the treatment. The explanatory variables is the commercial or non-commercial that is applied to the two groups. The response variable is how people view the product after seeing (or not seeing) the commercial.
3. This is an observational study.
5. Retrospective studies use old data, cross-sectional studies use data from a single moment in time, and prospective studies gather data into the future.

7. Double blinding is important so that the people interacting with patients cannot betray information about who is receiving placebo, which would destroy the effect of the control group.

9. This is a cluster sample, as not all colleges are being selected. Since the colleges are randomly selected, any college student has some chance of being selected, but it is not clear whether the students selected will be representative of students from colleges that are not included. The sampling frame may only include all students from the involved colleges.

11. This is a stratified sample, as students are selected from each college. The sample has a good chance of representing the population of college students, so the population itself is the sampling frame.

13. This is a stratified sample that has a very good chance of representing the entire student population. The population itself is the sampling frame.

15. This is a systematic sample. It has a good chance of representing the entire population of students, so the population itself is the sampling frame.

17. This could not result in a random sample, as people from smaller states have a greater chance of being selected than people from larger states.

19. It is not clear that these groups were created by random assignment, so cause and effect conclusions cannot be made. The study is observational, not experimental.

21. Because the cars are assigned to experimental blocks randomly, it can be concluded that the higher octane gasoline is that cause of the higher mean mileage.

23. The placebo effect is always one confounding variable. In addition, it is possible that people who make it a point to drink grape juice every day are also taking care of their health in many other ways. Any such ways would be a confounding variable.

25. It is very likely that people who drink diet drinks consume more calories from other sources than those who do not drink diet drinks. After all, those who drink diet drinks are trying to reduce calories for a reason! It is also possible that diet drinks stimulate appetite for calories from other sources.

27. In non-double-blind experiments, people who interact with patients can possibly reveal whether a given treatment is a true experimental treatment or a placebo, intentionally or not. If this information is revealed to patients, the experiment is invalidated, and will be a great waste of time, effort, and money.

B.2   Chapter 2 Solutions

Section 2.1 Solutions

1. *Range* $= 12.7 - 3.9 = 8.8$. *Class width* $= 8.8/9 = 0.98$ which we round up to 1. The class boundaries are: 3.85, 4.85, 5.85, 6.85, 7.85, 8.85, 9.85, 10.85, 11.85, 12.85.

3.9	– 4.8
4.9	– 5.8
5.9	– 6.8
6.9	– 7.8
7.9	– 8.8
8.9	– 9.8
9.9	– 10.8
10.9	– 11.8
11.9	– 12.8

3.   *Range* $= 3305 - 1024 = 2281$. Class width $= 2281/6 = 380.2$ which we round up to 381. The class boundaries are: 1023.5, 1404.5, 1785.5, 2166.5, 2547.5, 2928.5, 3309.5.

1024 – 1404
1405 – 1785
1786 – 2166
2167 – 2547
2548 – 2928
2929 – 3309

5.   The frequency table and histogram are below.

Min:	2.7
Max:	27
Range:	24.3
Classes:	5
Width:	4.86
Rounded:	4.9

Class	Frequency
2.7 – 7.5	13
7.6 – 12.4	16
12.5 – 17.3	8
17.4 – 22.2	2
22.3 – 27.1	1

The class boundaries are: 2.65, 7.55, 12.45, 17.35, 22.25, 27.15.

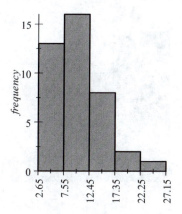

The distribution is skewed to the right.

7. The frequency table and histogram are below.

Min:	0.05
Max:	6
Range:	5.95
Classes:	8
Width:	0.74375
Rounded:	0.75

Class	Freq.
0.05 − 0.79	18
0.80 − 1.54	21
1.55 − 2.29	13
2.30 − 3.04	8
3.05 − 3.79	4
3.80 − 4.54	3
4.55 − 5.29	1
5.30 − 6.04	2

The class boundaries are: 0.045, 0.795, 1.545, 2.295, 3.045, 3.795, 4.545, 5.295, 6.045.

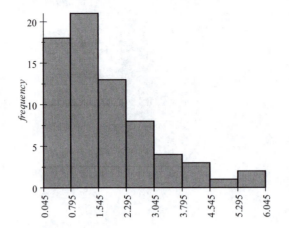

The distribution is skewed to the right.

9.   The frequency table and histogram are below.

Min:	170
Max:	21855
Range:	21685
Classes:	5
Width:	4337
Rounded	4338

Class	Freq.
170 – 4507	11
4508 – 8845	1
8846 – 13183	0
13184 – 17521	0
17522 – 21859	3

The class boundaries are: 169.5, 4507.5, 8845.5, 13183.5, 17521.5, 21859.5.

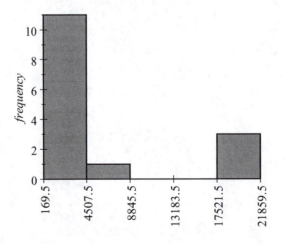

The graph is skewed to the right.

11. The frequency table and histogram are below.

Min:	0
Max:	40
Range:	40
Classes:	5
Width:	8
Rounded:	9

Class	Freq.
0 – 8	5
9 – 17	4
18 – 26	8
27 – 35	6
36 – 44	7

The class boundaries are: $-0.5, 8.5, 17.5, 26.5, 35.5, 44.5$.

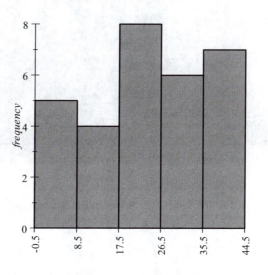

The distribution is multimodal, and not quite uniform.

13. The frequency table and histogram are below.

Min:	0.488
Max:	1.280
Range:	0.792
Classes:	10.000
Width:	0.079
Rounded:	0.080

Class	Freq.
0.488 – 0.567	12
0.568 – 0.647	59
0.648 – 0.727	172
0.728 – 0.807	348
0.808 – 0.887	425
0.888 – 0.967	342
0.968 – 1.047	116
1.048 – 1.127	28
1.128 – 1.207	17
1.208 – 1.287	13

The class boundaries are: 0.4875, 0.5675, 0.6475, 0.7275, 0.8075, 0.8875, 0.9675, 1.0475, 1.1275, 1.2075, 1.2875.

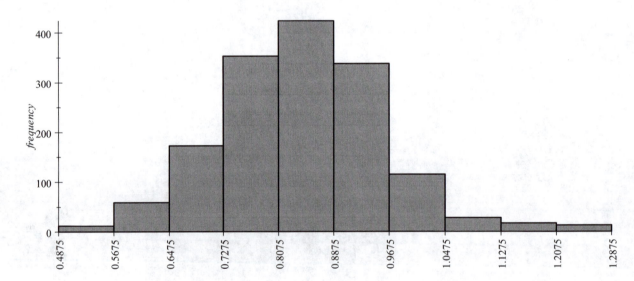

The distribution is nearly symmetric, but skewed slightly to the right, and bell shaped.

Section 2.2 Solutions

1. The stacked dotplots for these samples is given in Figure B.1.

   Step I. **Hypothesis:** One hypothesis might be that female food servers make a higher average percentage in tips than their male counterparts.

   Step II. **Summary:** In comparing the centers of each distribution of values, it appears that the average for male tips is lower than the average for females. In addition, the range of observed values is smaller for the men.

   Step III. **Shift:** The distribution of values for the female food servers seems to be shifted to the right of the males.

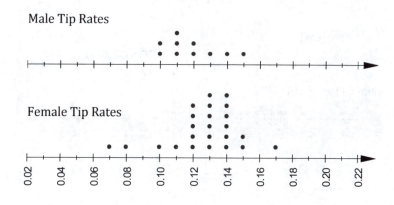

Figure B.1: The stacked dotplots for Problem 1.

Step IV. **Overlap:** The possibility of a shift in the distributions of male and female tip rates is overshadowed by the fact that the range of values for male food servers is completely contained within the larger range of values observed with the females. This means that the females have the highest and lowest average tip percentages.

Step V. **Spread:** The spread for the female sample is larger than what is apparent in the male sample. Since the spread contributes to error in estimates, we are cautious in our inferences. The spread is smaller for the males, indicating more uniformity amongst the males in this sample.

Step VI. **Sample Size:** While the female values have a larger spread, the males have a substantially smaller sample size. Unfortunately, smaller sample size also means larger error in estimates.

Step VII. **Meaning in Context:**

Step VIII. **Extreme Cases:** There is a single female that has a high tip rate, and is somewhat separated from the other females at about 17% on average. There are two females with rather low tip rates, at about 7% and 8% respectively. These low values are more extreme in terms of their distance to the center of the data set, and may be influential in lowering the overall average. The sample of males has a skewing to the right, but the values on the right are not as separated from the group as the lower values in the female sample. Still, these higher tip rates for males (no higher than 15%) will have some influence over the overall average.

Step IX. **Inferences:** The population of female food servers may earn more in tips when compared to male food servers. There are several issues that cloud this inference, such as the smallness of the given samples.

3. The stacked dotplots for these samples is given in Figure B.2.

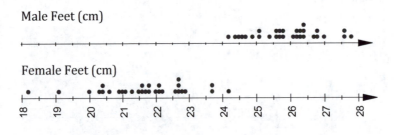

Figure B.2: The stacked dotplots for Problem 3.

Step I. **Hypothesis:** Here the most obvious hypothesis would state that the average length of male feet is larger than the average length of female feet.

Step II. **Summary:** The average of the male sample appears to be about 26 cm, while the average for the female sample is around 22 cm. These values seem to support the hypothesis.

Step III. **Shift:** The entire collection of male foot lengths is shifted to the right of the collection of female foot lengths.

Step IV. **Overlap:** Each of the male foot lengths is greater than every female foot length – there is no overlap in these samples. The closest comparison is between the largest female foot length of 24.2 cm, which nearly equal to the smallest male foot length of 24.3 cm.

Step V. **Spread:** The range of values is larger for the females than for the males. The female range is heavily influenced by two low points which might be considered outliers. Without these points, the spread of female values would be quite similar to the spread of male values.

Step VI. **Sample Size:** Each sample has 30 values. These are rather small, and weaken the inferences that we will suggest.

Step VII. **Meaning in Context:** The context of this application is simple, and seems quite obvious – the female foot lengths in the sample are all smaller than the male foot lengths. This makes sense in the context of the feet we see every day.

Step VIII. **Extreme Cases:** The only extreme cases are two female foot lengths of 18.2 cm and 19.1 cm. These would certainly influence the average foot length for females, but they do not change the fact that all female feet are shorter than all male feet in these samples.

Step IX. **Inferences:** The inference for this example is obvious – the average foot length of females is shorter than the average for males. This inference is likely to be true, due to the fact that our sample distributions do not overlap at all, and in spite of the fact that the samples are rather small.

5. The stacked dotplots for these samples is given in Figure B.3.

Convenient Classroom

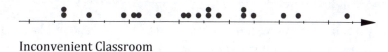

Inconvenient Classroom

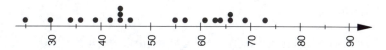

Figure B.3: The stacked dotplots for Problem 5.

Step I. **Hypothesis:** Here, it seems reasonable to hypothesize that students who attend a conveniently located classroom have a better chance of success.

Step II. **Summary:** It appears that the center of the convenient success rates is around 60%. The center of the inconvenient rates seems to be around 50%.

Step III. **Shift:** The values in the convenient success rates seem to be shifted to the right in comparison to the inconvenient success rates.

Step IV. **Overlap:** The distributions of these samples are quite overlapped, but several of the convenient success rates are greater than all inconvenient success rates, and two of the inconvenient rates are less than all of the convenient rates.

Step V. **Spread:** The spread is a bit larger for the convenient success rates, but in each case, there is a great deal of spread. Success rates range from 0% to 100%, and it appears that these samples occupy most of that range.

Step VI. **Sample Size:** Each sample has 19 values. The samples are quite small, and this smallness will decrease the reliability of our inference.

Step VII. **Meaning in Context:** Each value represents a success rate for a given class of students. The success rate is simply the proportion of enrolled students who received a C grade or better. Many of the courses in inconvenient locations did quite well, but the best success rates occurred for courses which were held in convenient locations.

Step VIII. **Extreme Cases:** There is an unusually high success rate (at 89%) for a convenient classroom. This is somewhat influential in influencing our guess of the average value of this sample. Without this outlier, we might not be so certain in our inferences.

Step IX. **Inferences:** Our inference is simply that the success rates are higher, on average, for classes which are held in convenient locations. The inference is weakened by small sample sizes and an extreme value in the convenient classroom sample.

7. The stacked dotplots for these samples is given in Figure B.4.

Step I. **Hypothesis:** One hypothesis which we might make is that the mean gain ratio is higher for schizophrenic people than for people without this disorder.

Step II. **Summary:** The centers of these samples appear to be quite similar. For the control group, the center appears to be located at around 0.80. The sample from the people with schizophrenia has a center that is slightly higher, possibly around 0.84.

Step III. **Shift:** If there is any shifting among the distributions of the samples, it may be that the control group is shifted slightly to the left of the schizophrenic group.

Step IV. **Overlap:** There is a great deal of overlap between these samples, with most values occupying the same interval on their respective axes.

Step V. **Spread:** It seems that the distribution shape is somewhat wider for the control group. This means that the eye gain ratios may demonstrate more variability among the control group. It would be a mistake, however, to miss the many large values on the right of the schizophrenic distribution which contribute to the wideness of the spread of these values.

Step VI. **Sample Size:** These samples are quite large – the schizophrenic sample size is 1532, while the control group sample size is 1374. These will contribute greatly to the strength of our inferences.

Step VII. **Meaning in Context:** Each of these observations represents the ratio of angular speed of the subject's eye to the angular speed of an object that the eye is following. When the gain ratio is less than 1, the eye is not keeping up with the object. In the cases where the ratio is larger than 1, the eye is actually turning faster than the object is moving.

Step VIII. **Extreme Cases:** The most notable extreme cases among the control group come from people whose eyes actually move quite slowly in comparison to the object they are watching. There are cases of control group subjects whose eyes move faster, but in the case of the schizophrenic people, there were many patients whose ratio was more than $1.1 = 110\%$. These people's eyes actually are moving 10% faster, or more, than the object being followed by their eyes. Of course, these extreme cases should be quite influential over the average of this sample.

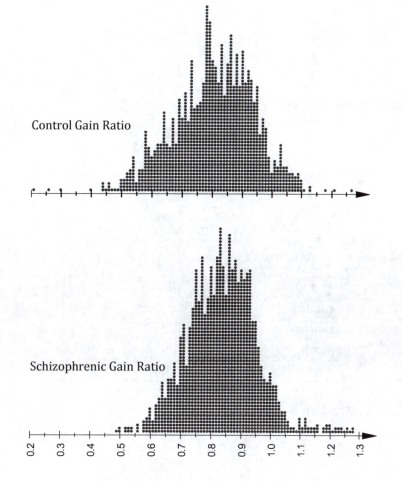

Control Gain Ratio

Schizophrenic Gain Ratio

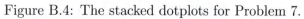

Figure B.4: The stacked dotplots for Problem 7.

Step IX. **Inferences:** Our inference will be that the average gain ratio is higher for schizophrenic people. This inference is weakened somewhat by the large overlap between these samples, but is strengthened quite a bit by the largeness of each sample.

Section 2.3 Solutions

1. The relevant statistics are below.

n	5
mode	5
median	5
midrange	8.5
mean	7.6

3. The relevant statistics are below.

n	17
mode	No Mode
median	84.73
midrange	79.81
mean	82.806

5. The relevant statistics are below.

n	40
modes	4.5, 5.1, 5.5, 7.9, 9.6, 10.8, 13.1
median	9.2
midrange	14.85
mean	10.03

7. The relevant statistics are below.

n	70
modes	0.8 and 0.9
median	1.375
midrange	3.025
mean	1.702

9. The relevant statistics are below.

n	15
mode	No Mode
median	2124
midrange	11012.5
mean	5654.8

11. The relevant statistics are below.

n	30
mode	40
median	25
midrange	20
mean	23.7

13. For this sample of size $n = 1532$, the sample size is so large we use the histogram from Section 2.1, Problem 13. The class with the highest bar is 0.808 to 0.887. We will use the midpoint of this class for the mode, which is 0.8475. The median is 0.838, the midrange is 0.884, and the mean is 0.8396.

15. The relevant statistics are below.

Class	$f = Freq.$	$x = Midpoint$	$f \cdot x$
15 – 19	4	17	68
20 – 24	8	22	176
25 – 29	16	27	432
30 – 34	14	32	448
35 – 39	8	37	296
40 – 44	6	42	252
45 – 49	1	47	47
—	57	—	1719

Here, we use $\bar{x} \approx \frac{\sum f \cdot x}{n} = \frac{\sum f \cdot x}{\sum f} = \frac{1719}{57} = 30.16 \approx 30.2$.

Section 2.4 Solutions

1. The relevant statistics are below.

Range:	13
Std. Dev.	5.3
Variance:	27.8
Lower Quartile:	3.5
Upper Quartile:	13
IQR	9.5

The usual range is from $\bar{x} - 2s$ to $\bar{x} + 2s$. Using the sample mean, $\bar{x} = 7.6$, found previously, the usual range becomes the following.

$$7.6 - 2 \cdot 5.3 \text{ to } 7.6 + 2 \cdot 5.3$$

$$-3.0 \text{ to } 18.2$$

3. The relevant statistics are below.

Range:	35.5
Std. Dev.	11.091
Variance:	123.006
Lower Quartile:	73.905
Upper Quartile:	91.835
IQR	17.93

The usual range is from $\bar{x} - 2s$ to $\bar{x} + 2s$. Using the sample mean, $\bar{x} = 82.806$, found previously, the usual range becomes the following.

$$82.806 - 2 \cdot 11.091 \text{ to } 82.806 + 2 \cdot 11.091$$
$$60.624 \text{ to } 104.99$$

5. The relevant statistics are below.

Range:	24.3
Std. Dev.	4.99
Variance:	24.87
Lower Quartile:	6.05
Upper Quartile:	12.9
IQR	6.85

The usual range is from $\bar{x} - 2s$ to $\bar{x} + 2s$. Using the sample mean, $\bar{x} = 10.03$, found previously, the usual range becomes the following.

$$10.03 - 2 \cdot 4.99 \text{ to } 10.03 + 2 \cdot 4.99$$
$$0.05 \text{ to } 20.01$$

7. The relevant statistics are below.

Range:	5.95
Std. Dev.	1.314
Variance:	1.726
Lower Quartile:	0.79
Upper Quartile:	2.35
IQR	1.56

The usual range is from $\bar{x} - 2s$ to $\bar{x} + 2s$. Using the sample mean, $\bar{x} = 1.702$, found previously, the usual range becomes the following.

$$1.702 - 2 \cdot 1.314 \text{ to } 1.702 + 2 \cdot 1.314$$
$$-0.926 \text{ to } 4.330$$

9. The relevant statistics are below.

Range:	21685
Std. Dev.	7530.7
Variance:	56710698.7
Lower Quartile:	890
Upper Quartile:	7826
IQR	6936

The usual range is from $\bar{x} - 2s$ to $\bar{x} + 2s$. Using the sample mean, $\bar{x} = 5654.8$, found previously, the usual range becomes the following.

$$5654.8 - 2 \cdot 7530.7 \text{ to } 5654.8 + 2 \cdot 7530.7$$
$$-9406.6 \text{ to } 20716.0$$

11. The relevant statistics are below.

Range:	40
Std. Dev.	13.8
Variance:	189.5
Lower Quartile:	16
Upper Quartile:	35
IQR	19

The usual range is from $\bar{x} - 2s$ to $\bar{x} + 2s$. Using the sample mean, $\bar{x} = 23.7$, found previously, the usual range becomes the following.

$$23.7 - 2 \cdot 13.8 \text{ to } 23.7 + 2 \cdot 13.8$$
$$-3.9 \text{ to } 51.3$$

13. The relevant statistics are below.

Range:	0.792
Std. Dev.	0.1177
Variance:	0.0139
Lower Quartile:	0.760
Upper Quartile:	0.913
IQR	0.153

The mean and standard deviation are $\bar{x} = 0.8396$, $s = 0.1177$.
The usual range lower limit is $\bar{x} - 2s = 0.8396 - 2 \cdot 0.1177 = 0.6042$.
The usual range upper limit is $\bar{x} + 2s = 0.8396 + 2 \cdot 0.1177 = 1.0750$.
Out of the sample of size 1532, exactly 1458 of these are within the usual range.

This represents 95.17% of the sample, which is very close to the prediction made by the empirical rule.

15. The relevant summary statistics are below.

Class	$f = Freq$	$x = Midpt$	$f \cdot x$	$f \cdot x^2$
15 – 19	4	17	68	1156
20 – 24	8	22	176	3872
25 – 29	16	27	432	11664
30 – 34	14	32	448	14336
35 – 39	8	37	296	10952
40 – 44	6	42	252	10584
45 – 49	1	47	47	2209
$\sum$	57	—	1719	54773

$$s \approx \sqrt{\frac{\sum f \cdot x^2 - \frac{1}{n}\left(\sum f \cdot x\right)^2}{n-1}} = \sqrt{\frac{54773 - \frac{1}{57}(1719)^2}{57-1}} = \sqrt{\frac{2931.5789}{56}} = \sqrt{52.3496} = 7.24$$

The variance is $s^2 = 52.3496 \approx 52.35$.

17. The relevant summary statistics are computed below.

x	$|x - \bar{x}|$	x^2
22	24.47368	484
30	16.47368	900
30	16.47368	900
36	10.47368	1296
39	7.473684	1521
40	6.473684	1600
40	6.473684	1600
41	5.473684	1681
44	2.473684	1936
45	1.473684	2025
50	3.526316	2500
50	3.526316	2500
53	6.526316	2809
56	9.526316	3136
56	9.526316	3136
56	9.526316	3136
62	15.52632	3844
65	18.52632	4225
68	21.52632	4624
883	195.4737	43853

The sample mean is next.

$$\bar{x} = \frac{\sum x}{n} = \frac{883}{19} = 46.47368.$$

Mean Absolute Deviation $= \frac{\sum |x - \bar{x}|}{n} = \frac{195.4737}{19} = 10.28809$

19. $s \approx Range/4 = (max - min)/4 = (68 - 22)/4 = 46/4 = 11.5$

21. With $n = 19$, there are $n - 1 = 19 - 1 = 18$ degrees of freedom.

23. The empirical rule predicts that 95% of values are within two standard deviations from the mean, which is the usual range. We observed that 100% of values were within this range, so it is possible that this lack of extreme or unusual values indicates a distribution which is *platykurtic*, with no outliers. Still, we are not far from what is predicted by the empirical rule – 5% of the sample is $0.05 \cdot 19 = 0.95$, so we expected only around 1 value to be unusual, when we actually observed none. The observed and expected numbers of unusual values are really quite close, thus, normality is still a possibility.

Section 2.5 Solutions

1. *Percentile for* $11 = \frac{3}{5} \cdot 100 = 60^{th}$. *z-score* $= \frac{x - \bar{x}}{s} = \frac{11 - 7.6}{5.27} = 0.65$. This is not unusual.

3. *Percentile for* $96.18 = \frac{15}{17} \cdot 100 = 88^{th}$. *z-score* $= \frac{x - \bar{x}}{s} = \frac{96.18 - 82.81}{11.09} = 1.21$. This is not unusual.

5. *Percentile for* $21 = \frac{39}{40} \cdot 100 = 98^{th}$. *z-score* $= \frac{x - \bar{x}}{s} = \frac{21 - 10.03}{4.99} = 2.20$. This *is* unusual.

7. *Percentile for* $4.61 = \frac{67}{70} \cdot 100 = 96^{th}$. *z-score* $= \frac{x - \bar{x}}{s} = \frac{4.61 - 1.702}{1.314} = 2.21$. This *is* unusual.

9. *Percentile for* $520 = \frac{1}{15} \cdot 100 = 7^{th}$. *z-score* $= \frac{x - \bar{x}}{s} = \frac{520 - 5654.8}{7530.65} = -0.68$. This is not unusual.

11. *Percentile for* $40 = \frac{23}{30} \cdot 100 = 77^{th}$. *z-score* $= \frac{x - \bar{x}}{s} = \frac{40 - 23.7}{13.8} = 1.18$. This is not unusual.

13. *Percentile for* $0.520 = \frac{3}{1533} \cdot 100 = 0^{th}$. *z-score* $= \frac{x - \bar{x}}{s} = \frac{0.520 - 0.8396}{0.1177} = -2.71$. This *is* unusual.

Section 2.6 Solutions

1. *five-point summary* $= \{ 2600, 12700, 24000, 52500, 85000 \}$. No outliers exist in this sample.

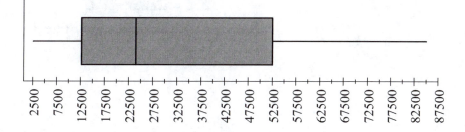

3. *five-point summary* = { 320, 458, 500.5, 545, 750 }. Outliers: 320, 680, 750.

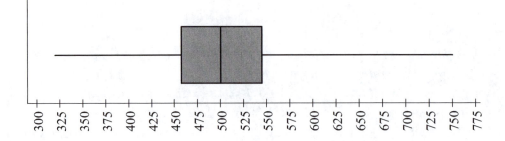

The enhanced boxplot is given below.

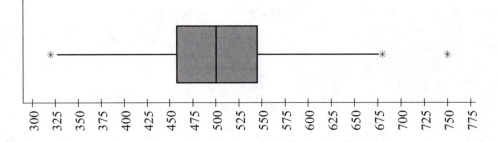

5. *five-point summary* = { 1.7, 2.7, 3.1, 3.65, 4 }. There are no outliers in this sample.

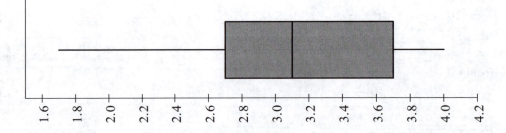

7. *five-point summary* = { 25.33, 38.3, 40.905, 45.82, 49.3 }. The only outlier is 25.33.

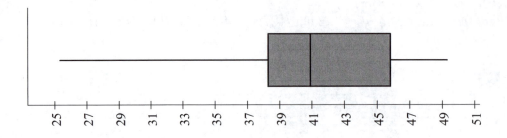

The enhanced boxplot is given below.

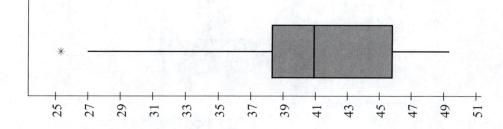

9. *five-point summary* = { 466, 1044, 2070, 5090, 10570 }. The data set has no outliers.

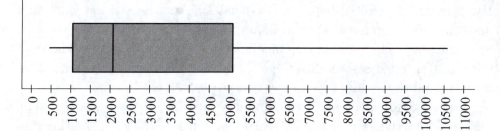

11. *five-point summary* = { 3, 4, 8, 12, 55 }. Outliers: 27, 31, 34, 55.

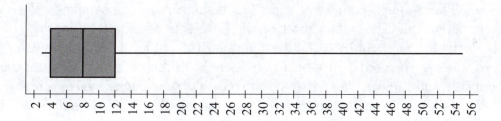

The enhanced boxplot is given below.

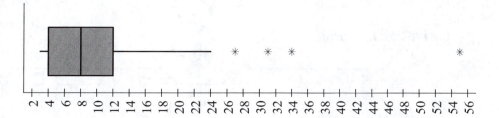

13.   Smoker Five Point Summary: {2.0, 2.6, 3.0, 3.5, 4.0}
       Non-Smoker Five Point Summary: {2.3, 3.0, 3.5, 3.8, 4.0}

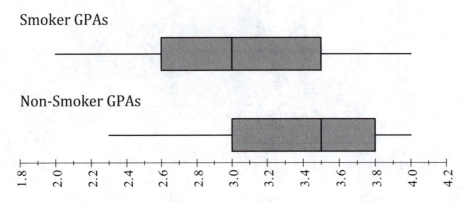

Hypothesis: A possible hypothesis is that smokers have a lower median GPA than non-smokers. Summary: From the sample data, the smokers do have a lower median, at about 3.0. The non-smoker median GPA is at 3.5. Shift: The smokers' GPAs middle 50% range is shifted to the left of the non-smokers. Overlap: The samples do overlap, and the middle 50% ranges overlap as well. Still, important comparisons can be made, including the fact that 50% of non-smokers (those above the median) have higher GPAs than the lower 75% of non-smokers (those below the third quartile). Spread: The smokers' GPAs have a wider spread than the non-smokers, making them a more varied group. There were smokers with a 4.0 GPA, but there were also smokers with a 2.0 GPA. None of the non-smokers were below a 2.3 GPA. Sample Size: There were 79 smokers, and 81 non-smokers in the samples. These are not very large, but larger than most of the data sets considered thus far. We expect that these larger samples would strengthen the value of our inferences. Meaning in Context: In context, it seems that there are differences between our samples of smokers and non-smokers. It may be true that non-smokers have a higher median GPA, but this is not to say that non-smokers individually do not score very high GPAs. Extreme Cases: There were no extreme cases in the samples. Inferences: We have some evidence that may support the claim that the median GPA is lower for smokers than for non-smokers.

15.   Corn Flake Five Point Summary: {2.3, 3.9, 4.4, 5.0, 6.4}
       Oat Bran Five Point Summary: {1.8, 3.7, 3.8, 4.8, 5.9}

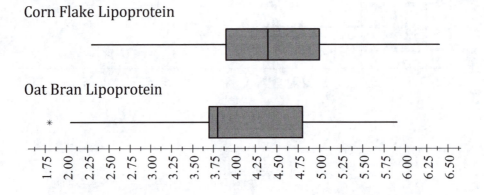

Hypothesis: We hypothesize that the lipoprotein levels are lower for people who eat oat bran for breakfast. Summary: The median lipoprotein level is lower for the oat bran eaters. Shift: There is a slight shift to the left for the distribution of oat bran eaters, when compared to the corn flake eaters. This shift, however, is very slight. Overlap: There is a great amount of overlap between the ranges of values in these samples. Additionally, the middle 50% intervals of the samples (represented by the boxes in the graph) are overlapping quite a bit. This overlap weakens our hypotheses regarding the difference between the medians. Spread: These samples are spread over similar ranges, and over similar interquartile ranges. Sample Size: Both samples are of size 14. These are very small, and diminish the quality of our inferences. Meaning in Context: The point of this study is to show the health benefit of eating oat bran as opposed to corn flakes. Does oat bran lower lipoprotein levels in comparison to other cereals? If it does, this would be valuable information for consumers. Extreme Cases: There is one low outlier for the oat bran eaters. This value will only slightly effect the value of the median, so it is not of much concern. Inferences: Because of the amount of overlap between these samples and the small sample sizes, the evidence is quite weakly in support of the claim that the median lipoprotein level is lower for oat bran eaters. More data might give better information, but the values given here give us very little certainty.

17. Affluent Five Point Summary: {0.20, 1.30, 1.80, 2.80, 4.70}
    Non-Affluent Five Point Summary: {0.30, 1.00, 1.60, 2.80, 4.40}

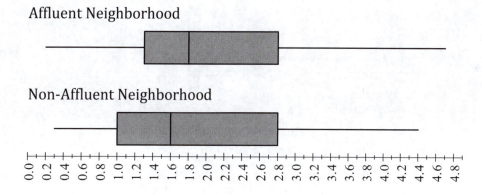

Hypothesis: We hypothesize that the median grocery price is higher in affluent neighborhoods. Summary: The median price for the affluent store is $1.80, while the median price for the non-affluent store is $1.60. Shift: The middle 50% of the affluent prices is shifted to the right. The overall range for the affluent store completely contains the range of values for the non-affluent neighborhood. This tells us that the affluent store had the highest and lowest prices. Overlap: There is a great deal of overlap between these samples. Spread: Both samples are spread over a similar range of values, but the affluent price spread is slightly larger. Sample Size: Both samples have 31 observations. These are small samples, and this smallness weakens our inferences. Meaning in Context: It is clear that different stores sell items for different prices. The question of whether grocery stores sell items at higher prices in more affluent neighborhoods may be of interest to people who shop there.

This study could be used to help customers make better decisions about where to shop. Extreme Cases: There are no outliers in either sample, so there is little concern about their influence. Inferences: While it is true that the median price is higher for the store in the affluent neighborhood, the samples exhibit a great deal of overlap, and were quite small. These facts remind us of the low level of certainty that we have in this inference.

B.3   Chapter 3 Solutions

Section 3.1 Solutions

1.

	Global Warming?			
Recycle?	Yes	Maybe	No	Totals
Yes	16	6	1	23
No	6	3	4	13
Totals	22	9	5	36

3. $P(G) = \frac{22}{36}$

5. $P(R \text{ or } G) = \frac{29}{36}$

7. $P(A) = 865/1349 = 0.641 = 64.1\%$

9. $P(A \text{ \& } B) = 0$

11. $P(A \text{ \& } D) = 479/1349 = 0.355 = 35.5\%$

13. $P(Correct\ Guess) = 1/5 = 0.200 = 20.0\%$

15. $P(On\ Time) = 278/300 = 0.927 = 92.7\%$. If 278 arrivals were on time, then $300 - 278 = 22$ arrivals were late. $P(Late) = 22/300 = 0.0733 = 7.33\%$.

17. $P(Y) = 35/60 = 0.5833$

19. $P(F) = 30/60 = 0.5000$

21. $P(Y \text{ \& } F) = 15/60 = 0.2500$

23. $P(Y \text{ \& } M) = 20/60 = 0.3333$

25. $P(N \text{ \& } F) = 15/60 = 0.2500$

27. $P(N \text{ \& } M) = 10/60 = 0.1667$

Section 3.2 Solutions

1. $P(A) = 5850/7425 = 0.788 = 78.8\%$

3. $P(C) = 1620/7425 = 0.218 = 21.8\%$

5. $P(A \text{ \& } B) = 0$

7. $P(A \text{ \& } C) = 1305/7425 = 0.176 = 17.6\%$

9. $P(C|A) = 1305/5850 = 0.223 = 22.3\%$.

11. $P(B \text{ \& } D) = 1260/7425 = 0.170 = 17.0\%$

13. $P(D|B) = 1260/1575 = 0.800 = 80.0\%$

15. $P(\bar{A}) = 1 - P(A) = 1 - 5850/7425 = 0.212 = 21.2\%$.

16. $P(\bar{D}) = 1 - P(D) = 1 - 0.782 = 0.218 = 21.8\%$.

17. $P(A \text{ \& } C) = 1305/7425 = 17.6\%$. $P(A) \cdot P(C) = (5850/7425) \cdot (1620/7425) = 0.788 \cdot 0.218 = 0.172 = 17.2\%$. $P(A \text{ \& } C) \neq P(A) \cdot P(C)$. The answers are slightly different, so the events do not appear to be independent.

19. $P(M \text{ \& } L) = 15/61 = 0.2459$

23. $P(L \text{ \& } F) = 0.0000$

27. $P(\bar{E}) = 1 - P(E) = 1 - 12/61 = 49/61 = 0.8033$

29. $P(A|F) = 7/11 = 0.6364$

31. $P(A|D) = 1/5 = 0.2000$

35. $P(A \text{ \& } D) = 1/37 = 0.0270$

37. $P\left(\bar{E}\right) = 1 - P\left(E\right) = 1 - 21/37 = 16/37 = 0.4324$

39. $P\left(A \text{ or } B\right) = P\left(A\right) + P\left(B\right) = (4/52) + (4/52) = 8/52 = 15.4\%$.

41. $P\left(A \text{ or } C\right) = P\left(A\right) + P\left(C\right) - P\left(A \& C\right) = (4/52) + (13/52) - (1/52) = 16/52 = 0.308 = 30.8\%$

43. $P\left(A \& B\right) = P\left(A\right) \cdot P\left(B\right) = 4/52 \cdot 4/52 = 0.00592 = 0.592\%$.

45. The sample space for Problems 45 through 49 is: $\{GGG, BGG, GBG, GGB, GBB, BGB, BBG, BBB\}$; a total of 8 possible outcomes. $P\left(no\ girls\right) = 1/8$, since this can happen in only one way.

47. $P\left(two\ girls\right) = 3/8$, since this can happen in three ways.

49. P (three men are the same) $= (4/4) \cdot (1/4) \cdot (1/4) = \frac{1}{16}$

51. $P\left(ring\right) = 0.95$, $P(fail) = 0.05$. $P(fail \& fail \& fail) = 0.05 \cdot 0.05 \cdot 0.05 = 0.000125$.

53. $P\left(female \& female \& female\right) = 0.498 \cdot 0.498 \cdot 0.498 = 0.124 = 12.4\%$.

55. The probability of guessing a number correctly is $1/50$. To guess correctly three times, we multiply: $P\left(correct \& correct \& correct\right) = (1/50) \cdot (1/50) \cdot (1/50) = 0.000008$.

B.4 Chapter 4 Solutions

Section 4.1 Solutions

1. $\sum P(x) = 1$, and $0 \le P(x) \le 1$ for all x, so the probabilities represent a probability distribution.

3. The variance is $\sigma^2 = Var(x) = \sum x^2 \cdot P(x) - \mu^2 = 0.801 - 0.561^2 = 0.4863$,
   The standard deviation is $\sigma = \sqrt{\sigma^2} = \sqrt{0.4863} = 0.6973$.

5. The *p-value* corresponding to $x = 0$ is *p-value* $= P(x \le 0) = P(0) = 0.547$. Thus, $x = 0$ is not significantly low.

7. The mean winning is $\mu = E(x) = \sum x \cdot P(x) = -3.20$. This means that there is an average loss of $3.20 overall.

9. The totals needed are below.

x	$P(x)$	$xP(x)$
15000	0.6	9000
-5000	0.4	-2000
Totals:	1.0	7000

   The expected return on the investment is $\mu = E(x) = \sum xP(x) = \$7,000$.

11. The first investment has a higher average return, so this is what she should pick.

13. The probability distribution is below.

x	$P(x)$
0	0.125
1	0.375
2	0.375
3	0.125

15. The variance is $\sigma^2 = Var(x) = \sum x^2 \cdot P(x) - \mu^2 = 3 - 1.5^2 = 0.75$, the standard deviation is $\sigma = \sqrt{\sigma^2} = \sqrt{0.75} = 0.8660$.

Section 4.2 Solutions

1. $\mu_x = 14.25$, $\sigma_x^2 = 2.1875$.

3. Using properties of expected value we compute μ_{x-y} first.

$$
\begin{aligned}
\mu_{x-y} &= E(x-y) \\
&= E(x) - E(y) \\
&= \mu_x - \mu_y \\
&= 14.25 - 5 \\
&= 9.25
\end{aligned}
$$

Using properties of variance, we compute σ^2_{x-y} next.

$$
\begin{aligned}
\sigma^2_{x-y} &= Var\,(x-y) \\
&= Var\,(x) + Var\,(y) \\
&= \sigma^2_x + \sigma^2_y \\
&= 2.1875 + 6.5 \\
&= 8.6875
\end{aligned}
$$

5. First we compute the mean.

$$
\begin{aligned}
\mu_{3x-2y} &= E\,(3x - 2y) \\
&= E\,(3x) - E\,(2y) \\
&= 3E\,(x) - 2E\,(y) \\
&= 3\mu_x - 2\mu_y \\
&= 3 \cdot 7 - 2 \cdot 4 \\
&= 13
\end{aligned}
$$

The variance is next.

$$
\begin{aligned}
\sigma^2_{3x-2y} &= Var\,(3x - 2y) \\
&= Var\,(3x) + Var\,(2y) \\
&= 3^2 Var\,(x) + 2^2 Var\,(y) \\
&= 9 \cdot 2^2 + 4 \cdot 3^2 \\
&= 72
\end{aligned}
$$

The standard deviation is the square root of variance.

$$
\begin{aligned}
\sigma_{3x-2y} &= \sqrt{\sigma^2_{3x-2y}} \\
&= \sqrt{72} \\
&= 6\sqrt{2} \\
&= 8.485
\end{aligned}
$$

7. First we compute the mean.

$$
\begin{aligned}
\mu_{\frac{x+y}{2}} &= E\left(\frac{1}{2}x + \frac{1}{2}y\right) \\
&= E\left(\frac{1}{2}x\right) + E\left(\frac{1}{2}y\right) \\
&= \frac{1}{2}E(x) + \frac{1}{2}E(y) \\
&= \frac{1}{2}\mu_x + \frac{1}{2}\mu_y \\
&= \frac{1}{2}\cdot 7 + \frac{1}{2}\cdot 4 \\
&= \frac{11}{2} \\
&= 5.5
\end{aligned}
$$

The variance is next.

$$
\begin{aligned}
\sigma^2_{\frac{x+y}{2}} &= Var\left(\frac{1}{2}x + \frac{1}{2}y\right) \\
&= Var\left(\frac{1}{2}x\right) + Var\left(\frac{1}{2}y\right) \\
&= \left(\frac{1}{2}\right)^2 Var(x) + \left(\frac{1}{2}\right)^2 Var(y) \\
&= \frac{1}{4}\cdot 2^2 + \frac{1}{4}\cdot 3^2 \\
&= \frac{13}{4} \\
&= 3.25
\end{aligned}
$$

The standard deviation is the square root of variance.

$$
\begin{aligned}
\sigma_{\frac{x+y}{2}} &= \sqrt{\sigma^2_{\frac{x+y}{2}}} \\
&= \sqrt{\frac{13}{4}} \\
&= \frac{\sqrt{13}}{2} \\
&= 1.8028
\end{aligned}
$$

9. $E(x^2) = E(x)^2 + Var(x) = \mu_x^2 + \sigma_x^2 = 7^2 + 2^2 = 53.$

11. The mean of $\bar{x}$ values is $\mu_{\bar{x}}$, computed as follows.

$$\begin{aligned}
\mu_{\bar{x}} &= E\left(\bar{x}\right) \\
&= E\left(\frac{x + x + x}{3}\right) \\
&= E\left(\frac{1}{3}x + \frac{1}{3}x + \frac{1}{3}x\right) \\
&= \frac{1}{3}E\left(x\right) + \frac{1}{3}E\left(x\right) + \frac{1}{3}E\left(x\right) \\
&= \left(\frac{1}{3} + \frac{1}{3} + \frac{1}{3}\right)E\left(x\right) \\
&= \mu_x \\
&= 7.
\end{aligned}$$

The standard deviation of $\bar{x}$ values begins with computing the variance.

$$\begin{aligned}
Var\left(\bar{x}\right) &= Var\left(\frac{x + x + x}{3}\right) \\
&= Var\left(\frac{1}{3}x + \frac{1}{3}x + \frac{1}{3}x\right) \\
&= \left(\frac{1}{3}\right)^2 Var\left(x\right) + \left(\frac{1}{3}\right)^2 Var\left(x\right) + \left(\frac{1}{3}\right)^2 Var\left(x\right) \\
&= \left(\frac{1}{9} + \frac{1}{9} + \frac{1}{9}\right)Var\left(x\right) \\
&= \frac{1}{3}\sigma_x^2 \\
&= \frac{1}{3}2^2 \\
&= \frac{4}{3}.
\end{aligned}$$

The standard deviation is the square root of variance, $\sigma_{\bar{x}} = \sqrt{\sigma_{\bar{x}}^2} = \sqrt{\frac{4}{3}} = \frac{2}{\sqrt{3}}$.

13. The mean of $\bar{x}$ values is $\mu_{\bar{x}}$, computed as follows.

$$\begin{aligned}
\mu_{\bar{x}} &= E\left(\bar{x}\right) \\
&= E\left(\frac{x + x + x + x + x}{5}\right) \\
&= E\left(\frac{1}{5}x + \frac{1}{5}x + \frac{1}{5}x + \frac{1}{5}x + \frac{1}{5}x\right) \\
&= \frac{1}{5}E\left(x\right) + \frac{1}{5}E\left(x\right) + \frac{1}{5}E\left(x\right) + \frac{1}{5}E\left(x\right) + \frac{1}{5}E\left(x\right) \\
&= \left(\frac{1}{5} + \frac{1}{5} + \frac{1}{5} + \frac{1}{5} + \frac{1}{5}\right)E\left(x\right) \\
&= E\left(x\right) \\
&= \mu.
\end{aligned}$$

The standard deviation of $\overline{x}$ values begins with computing the variance.

$$
\begin{aligned}
Var\left(\overline{x}\right) &= Var\left(\frac{x+x+x+x+x}{5}\right) \\
&= Var\left(\frac{1}{5}x+\frac{1}{5}x+\frac{1}{5}x+\frac{1}{5}x+\frac{1}{5}x\right) \\
&= \left(\frac{1}{5}\right)^2 Var\left(x\right)+\left(\frac{1}{5}\right)^2 Var\left(x\right)+\left(\frac{1}{5}\right)^2 Var\left(x\right)+\left(\frac{1}{5}\right)^2 Var\left(x\right)+\left(\frac{1}{5}\right)^2 Var\left(x\right) \\
&= \left(\frac{1}{25}+\frac{1}{25}+\frac{1}{25}+\frac{1}{25}+\frac{1}{25}\right) Var\left(x\right) \\
&= \frac{5}{25}\sigma^2 \\
&= \frac{\sigma^2}{5}
\end{aligned}
$$

The standard deviation is the square root of variance, $\sigma_{\overline{x}} = \sqrt{Var\left(\overline{x}\right)} = \sqrt{\frac{\sigma^2}{5}} = \frac{\sigma}{\sqrt{5}}$.

15. Based on the results of the previous problems, the mean of all sample means from samples of size n appears to be $\mu_{\overline{x}} = \mu$, while the standard deviation of all sample means from samples of size n appears to be $\sigma_{\overline{x}} = \frac{\sigma}{\sqrt{n}}$

17. The mean of the given values is $\mu = 4.0$, and the standard deviation is $\sigma = 1.5811$. The samples of size two are given first.

$$
\left\{
\begin{array}{ccccc}
\{2,2\} & \{2,3\} & \{2,4\} & \{2,5\} & \{2,6\} \\
\{3,2\} & \{3,3\} & \{3,4\} & \{3,5\} & \{3,6\} \\
\{4,2\} & \{4,3\} & \{4,4\} & \{4,5\} & \{4,6\} \\
\{5,2\} & \{5,3\} & \{5,4\} & \{5,5\} & \{5,6\} \\
\{6,2\} & \{6,3\} & \{6,4\} & \{6,5\} & \{6,6\}
\end{array}
\right\}
$$

The corresponding means follow next.

$$
\left\{
\begin{array}{ccccc}
2.0 & 2.5 & 3.0 & 3.5 & 4.0 \\
2.5 & 3.0 & 3.5 & 4.0 & 4.5 \\
3.0 & 3.5 & 4.0 & 4.5 & 5.0 \\
3.5 & 4.0 & 4.5 & 5.0 & 5.5 \\
4.0 & 4.5 & 5.0 & 5.5 & 6.0
\end{array}
\right\}
$$

Next, the dotplot is constructed.

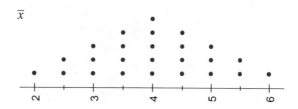

The dotplot shows a distribution that is high in the middle and low on the ends. Based on previous ideas, the mean of sample means is $\mu_{\overline{x}} = \mu = 4.0$, and the standard deviation of sample means is $\sigma_{\overline{x}} = \frac{\sigma}{\sqrt{n}} = \frac{1.5811}{\sqrt{2}} = 1.1180$.

Section 4.3 Solutions

1. This population is skewed right (positively), and platykurtic.

x	$P(x)$	$x \cdot P(x)$	$x^2 \cdot P(x)$	$(x-\mu)^3 \cdot P(x)$	$(x-\mu)^4 \cdot P(x)$
0	0.413	0.000	0.000	-0.174932	0.131374
1	0.436	0.436	0.436	0.006731	0.001676
2	0.138	0.276	0.552	0.268885	0.335837
3	0.013	0.039	0.117	0.147881	0.332584
$\sum$	1.000	0.751	1.105	0.248565	0.801471

μ	0.7510
σ	0.7355
μ_3	0.2486
μ_4	0.8015
Skewness	0.6247
Kurtosis	2.7384

3. This population is nearly symmetric, and platykurtic.

x	$P(x)$	$x \cdot P(x)$	$x^2 \cdot P(x)$	$(x-\mu)^3 \cdot P(x)$	$(x-\mu)^4 \cdot P(x)$
0	0.0325	0	0	-0.495722	1.229391
1	0.16	0.16	0.16	-0.518687	0.767656
2	0.315	0.63	1.26	-0.034836	0.016722
3	0.31	0.93	2.79	0.043588	0.022666
4	0.1525	0.61	2.44	0.535551	0.814037
5	0.0300	0.15	0.75	0.480090	1.209827
$\sum$	1.0000	2.4800	7.4000	0.009984	4.060300

μ	2.4800
σ	1.1179
μ_3	0.0010
μ_4	4.0603
Skewness	0.0071
Kurtosis	2.6003

5. The sample moments, skewness and kurtosis are given below. The sample is leptokurtic and positively skewed (to the right).

Moments	
m_2	2.1983
m_3	1.1811
m_4	9.8159
skewness	0.3624
kurtosis	2.0311

7. The sample moments, skewness and kurtosis are given below. The sample is platykurtic and negatively skewed (to the left).

Moments	
m_2	3.65
m_3	-2.325
m_4	26.6375
skewness	-0.3334
kurtosis	1.9994

Section 4.4 Solutions

1. $_{10}C_4 = \frac{10!}{4!\,(10-4)!} = 210$

3. $\mu = np = 10 \cdot 0.30 = 3.0$, $\sigma = \sqrt{npq} = \sqrt{10 \cdot 0.30 \cdot 0.70} = 1.449$

5. $_{20}C_{13} = \frac{20!}{13!\,(20-13)!} = 77,520$

7. $\mu = np = 20 \cdot 0.45 = 9.0$, $\sigma = \sqrt{npq} = \sqrt{20 \cdot 0.45 \cdot 0.55} = 2.225$

9. $n = 10$, $x = 1$, success is $S =$ "speaks English", $p = P(S) = 0.05$.

$$
\begin{aligned}
P\,(x = 1) &= \ _nC_x p^x q^{n-x} \\
&= \ _{10}C_1 \cdot 0.05^1 \cdot 0.95^{10-1} \\
&= \frac{10!}{1!\,(10-1)!} \cdot 0.05^1 \cdot 0.95^9 \\
&= 10 \cdot 0.05^1 \cdot 0.95^9 \\
&= 0.3151.
\end{aligned}
$$

11. $\mu = np = 27 \cdot 0.88 = 23.76$, $\sigma = \sqrt{npq} = \sqrt{27 \cdot 0.88 \cdot 0.12} = 1.6885$

13. $n = 12$, $x = 8$, success is $S =$ "approval", $p = P(S) = 0.60$.

$$
\begin{aligned}
P\,(x = 8) &= \ _nC_x p^x q^{n-x} = \ _{12}C_8 \cdot 0.60^8 \cdot 0.40^{12-8} \\
&= \frac{12!}{8!\,(12-8)!} \cdot 0.60^8 \cdot 0.40^4 \\
&= 495 \cdot 0.60^8 \cdot 0.40^4 \\
&= 0.2128.
\end{aligned}
$$

15. Using the rule of complements, $P\left(at\ least\ one\right) = 1 - P(none) = 1 - P(x = 0)$. $n = 5$, $x = 0$, success is $S =$ "rain", $p = P(S) = 0.20$.

$$
\begin{aligned}
P\left(x = 0\right) &= {}_nC_x p^x q^{n-x} \\
&= {}_5C_0 \cdot 0.20^0 \cdot 0.80^{5-0} \\
&= \frac{5!}{0!\,(5-0)!} \cdot 0.20^0 \cdot 0.80^5 \\
&= 1 \cdot 0.20^0 \cdot 0.80^5 \\
&= 0.3277.
\end{aligned}
$$

Returning to the original question, $P\left(at\ least\ one\right) = 1 - P(none) = 1 - P(x = 0) = 1 - 0.3277 = 0.6723$.

Section 4.5 Solutions

1. The mean is $\mu = (total\ yeast\ cells)/(number\ of\ containers) = 1872/400 = 4.68$
3. The probabilities are given below.

$x = $ Number of yeast cells	$P\left(x\right)$
0	0.0092790
1	0.0434258
2	0.1016163
3	0.1585215
4	0.1854701
5	0.1736000
6	0.1354080
7	0.0905299
8	0.0529600
9	0.0275392
10	0.0128884
11	0.0054834
12	0.0021385
13	0.0007699
14	0.0002574
15	0.0000803
16	0.0000235
Totals:	1

5. The mean is $\mu = (total\ kicks) / (number\ of\ time\ units) = 196/280 = 0.700$.

7. The probabilities are given below.

$x = Number$ of Deaths	$P(x)$
0	0.4965853
1	0.34760971
2	0.1216634
3	0.02838813
4	0.00496792
5 and over	0.00078554
Totals:	1

9. The mean is $\mu = (total\ occurrences\ of\ \text{``may''}) / (number\ of\ blocks) = 172/262 = 0.656489$

11. The probabilities are given below.

$x = Occurrences$ of "may"	$P(x)$
0	0.5186692
1	0.3405006
2	0.1117675
3	0.0244580
4	0.0040141
5	0.0005270
6 or more	0.0000636
Totals:	1

B.5   Chapter 5 Solutions

Section 5.1 Solutions

1. The density function is $y = 1/3$.

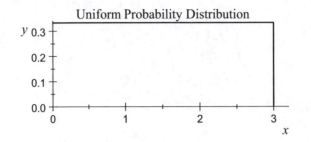

3. $P(x > 1.3) = (3 - 1.3) \cdot (1/3) = 0.567$
5. $P(0.8 < x < 2.7) = (2.7 - 0.8) \cdot (1/3) = 0.633$

Section 5.2 Solutions

1. $P(z < 0.00) = 0.5000$
3. $P(z > 2.49) = 1 - 0.9936 = 0.0064$
5. $P(z < 4.06) = 0.9999$
7. $P(z < 2.65) = 0.9960$
9. $P(z < -1.27) = 0.1020$
11. $P(1.53 < z < 2.09) = 0.9817 - 0.9730 = 0.0087$
13. The 81st percentile corresponds to a cumulative area of 0.81. The closest area to this under the positive z-scores is 0.8106, which corresponds to a z-score of $z = 0.88$.
15. Since the area to the left of $z = 0$ is 0.50, $z = 0$ the 50th percentile z-score.
17. With a right tail equal to 0.05, the area to the left is 0.95. The corresponding z-score is $z = 1.645$.
19. With a right tail equal to 0.04, the area to the left is 0.96. The corresponding z-score is $z = 1.75$.
21. Table A1 does not allow z-scores as high as $z = 5.93$, but the footnote at the bottom of the table directs us to use 0.9999 for the area corresponding to any z-score greater than 4.09. Thus, we estimate $P(z < 5.93) \approx 0.9999$.
23. Table A1 does not allow z-scores as low as $z = -11.91$, but the footnote at the bottom of the table directs us to use 0.0001 for the area corresponding to any z-score less than –11.91. Thus, we estimate $P(z < -11.91) \approx 0.0001$.

Section 5.3 Solutions

1. $z = \frac{x-\mu}{\sigma} = \frac{119-100}{15} = 1.27$; $P(x < 119) = P(z < 1.27) = 0.8980$.
3. $z = \frac{x-\mu}{\sigma} = \frac{133-100}{15} = 2.20$; $P(x > 133) = P(z > 2.20) = 1 - P(z < 2.20) = 1 - 0.9861 = 0.0139$.
5. With $x = 82$; $z = \frac{x-\mu}{\sigma} = \frac{82-100}{15} = -1.20$. With $x = 91$; $z = \frac{x-\mu}{\sigma} = \frac{91-100}{15} = -0.60$. $P(82 < x < 91) = P(-1.20 < z < -0.60) = P(z < -0.60) - P(z < -1.20) = 0.2743 - 0.1151 = 0.1592$.

7. The 96th percentile z-score is $z = 1.75$. The 96th percentile I.Q. score is: $x = z \cdot \sigma + \mu = 1.75 \cdot 15 + 100 = 126.25$.

9. The 99th percentile z-score is $z = 2.33$. The 99th percentile I.Q. score is: $x = z \cdot \sigma + \mu = 2.33 \cdot 15 + 100 = 134.95$.

11. $P(x > 73.5) = P\left(z > \frac{73.5 - 69}{3}\right) = P(z > 1.50) = 1 - 0.9332 = 0.0668$. This is not unusually tall.

13. $P(x < 60.2) = P\left(z < \frac{60.2 - 69}{3}\right) = P(z < -2.93) = 0.0017$. This is very unusual.

15. To find the z-score that separates the lower 2.5% of the population, we scan Table A1 for a region of area equal to 0.0250. The z-score that corresponds to this is $z = -1.96$, and by symmetry the z-score that separates the upper 2.5% of the population is $z = 1.96$. Converting these to heights we get the following: $x = z\sigma + \mu = -1.96 \cdot 3 + 69 = 63.12$ and $x = z\sigma + \mu = 1.96 \cdot 3 + 69 = 74.88$.

17. The z-score which most closely corresponds to an area of 0.85 is $z = 1.04$. The corresponding height is $x = z\sigma + \mu = 1.04 \cdot 3 + 69 = 72.12$.

19. $P(x > 74.5) = P\left(z > \frac{74.5 - 69}{3}\right) = P(z > 1.83) = 1 - 0.9664 = 0.0336$

B.6   Chapter 6 Solutions

Section 6.1 Solutions

1. $z_{\alpha/2} = 1.960$
3. $z_{\alpha/2} = 2.05$
5. $z_{\alpha/2} = 2.576$
7. $E = z_{\alpha/2}\sigma_{\widehat{x}} = 1.960 \cdot 2 = 3.92$.

$$\widehat{x} - E \; < \; \mu < \widehat{x} + E$$
$$10 - 3.92 \; < \; \mu < 10 + 3.92$$
$$6.08 \; < \; \mu < 13.92$$

95% of such intervals will actually contain the unknown parameter, μ.

9. $E = z_{\alpha/2}\sigma_{\widehat{x}} = 2.05 \cdot 2 = 4.1$.

$$\widehat{x} - E \; < \; \mu < \widehat{x} + E$$
$$10 - 4.1 \; < \; \mu < 10 + 4.1$$
$$5.9 \; < \; \mu < 14.1$$

96% of such intervals will actually contain the unknown parameter, μ.

11. $E = z_{\alpha/2}\sigma_{\widehat{x}} = 2.576 \cdot 2 = 5.152$.

$$\widehat{x} - E \; < \; \mu < \widehat{x} + E$$
$$10 - 5.152 \; < \; \mu < 10 + 5.152$$
$$4.848 \; < \; \mu < 15.152$$

99% of such intervals will actually contain the unknown parameter, μ.

13.   The margins of error are smaller with lower levels of confidence. The margins of error are larger with larger levels of confidence.

15.   A good estimate is an unbiased estimate, so the sample median would be a better estimate of a population median, as this estimate is unbiased.

17.   The primary problem here is that the sampling method is unlikely to represent its population well – it is a self-selected convenience sample with selection bias.

Section 6.2 Solutions

1. $n = 100$, $p = 0.65$, $q = 1 - p = 1 - 0.65 = 0.35$
Check: $np = 100 \cdot 0.65 = 65.0 \geq 10$, and $nq = 100 \cdot 0.35 = 35.0 \geq 10$.
The mean is: $\mu = np = 100 \cdot 0.65 = 65.0$.
The standard deviation is: $\sigma = \sqrt{npq} = \sqrt{100 \cdot 0.65 \cdot 0.35} = \sqrt{22.75} = 4.77$
With the continuity correction, the interval $x \geq 80$ becomes $x > 79.5$.
The z-score of 79.5 is $z = \frac{x-\mu}{\sigma} = \frac{79.5-65.0}{4.77} = 3.04$
$P(x \geq 80) = P(x > 79.5) = P(z > 3.04) = 1 - 0.9988 = 0.0012$.

3. $n = 1000$, $p = 0.60$, $q = 1 - p = 1 - 0.60 = 0.40$
Check: $np = 1000 \cdot 0.60 = 600.0 \geq 10$, and $nq = 1000 \cdot 0.40 = 400.0 \geq 10$.
The mean is: $\mu = np = 100 \cdot 0.60 = 600.0$.

The standard deviation is: $\sigma = \sqrt{npq} = \sqrt{1000 \cdot 0.60 \cdot 0.40} = \sqrt{240} = 15.492$
The relevant interval is *"more than 65%,"* and since 65% of 1000 is 650, this translates to $x \geq 651$. With the continuity correction, the interval $x \geq 651$ becomes $x > 650.5$. The z-score of 650.5 is $z = \frac{x-\mu}{\sigma} = \frac{650.5-600.0}{15.492} = 3.26$
$P(x \geq 651) = P(x > 650.5) = P(z > 3.26) = 1 - 0.9994 = 0.0006$.

5. $n = 265$, $p = 0.88$, $q = 1 - p = 1 - 0.88 = 0.12$
   Check: $np = 265 \cdot 0.88 = 233.2 \geq 10$, and $nq = 265 \cdot 0.12 = 31.8 \geq 10$.
   The mean is: $\mu = np = 265 \cdot 0.88 = 233.2$.
   The standard deviation is: $\sigma = \sqrt{npq} = \sqrt{265 \cdot 0.88 \cdot 0.12} = \sqrt{27.984} = 5.290$
   "Not enough seats" means that 251 or more people come to the flight, so this translates to $x \geq 251$. With the continuity correction, the interval $x \geq 251$ becomes $x > 250.5$.
   The z-score of 250.5 is $z = \frac{x-\mu}{\sigma} = \frac{250.5-233.2}{5.290} = 3.27$
   $P(x \geq 251) = P(x > 250.5) = P(z > 3.27) = 1 - 0.9995 = 0.0005$.

7. $z = \frac{\hat{p}-p}{\sqrt{\frac{pq}{n}}} = \frac{0.62-0.65}{\sqrt{\frac{0.65 \cdot 0.62}{500}}} = -1.057$. This is not unusually small, being only 1.057 standard deviations below the population proportion.

9. $z = \frac{\hat{p}-p}{\sqrt{\frac{pq}{n}}} = \frac{0.62-0.60}{\sqrt{\frac{0.60 \cdot 0.40}{4000}}} = 2.582$. This is unusually high, being 2.582 standard deviations above the population proportion.

11. $np = 500 \cdot 0.65 = 325 \geq 10$. $nq = 500 \cdot 0.35 = 175 \geq 10$. *p-value* $= P(z < -1.057) \approx P(z < -1.06) = 0.1446$. As before, this is not unusual.

13. $np = 4000 \cdot 0.60 = 2400 \geq 10$. $nq = 4000 \cdot 0.40 = 1600 \geq 10$. *p-value* $= P(z > 2.582) \approx P(z > 2.58) = 1 - 0.9951 = 0.0049$. This *is* unusual.

15. $\{S_1, S_1\}, \{S_1, S_2\}, \{S_1, S_3\}, \{S_1, F\}$
    $\{S_2, S_1\}, \{S_2, S_2\}, \{S_2, S_3\}, \{S_2, F\}$
    $\{S_3, S_1\}, \{S_3, S_2\}, \{S_3, S_3\}, \{S_3, F\}$
    $\{F, S_1\}, \{F, S_2\}, \{F, S_3\}, \{F, F\}$

17. $\mu_{\hat{p}} = 0.75$ and $\sigma_{\hat{p}} = 0.306186$

Section 6.3 Solutions

1. $\alpha = 1 - 0.96 = 0.04$, $1 - \alpha/2 = 0.98$. The closest value in the table is 0.9798, corresponding to $z_{\alpha/2} = 2.05$.

3. $\alpha = 1 - 0.92 = 0.08$, $1 - \alpha/2 = 0.96$. The closest value in the table is 0.9599, corresponding to $z_{\alpha/2} = 1.75$.

5. $0.476 - 0.031 < p < 0.476 + 0.031$ leads to $0.445 < p < 0.507$.

7. $\alpha = 1 - 0.98 = 0.02$, $1 - \alpha/2 = 0.99$, $\hat{p} = x/n = 382/563 = 0.679$, $\hat{q} = 1 - \hat{p} = 0.321$, so $z_{\alpha/2} = 2.33$. So $E = z_{\alpha/2}\sqrt{\frac{\hat{p} \cdot \hat{q}}{n}} = 2.33\sqrt{\frac{0.679 \cdot 0.321}{563}} = 0.0458$

9. $\alpha = 0.05$, $\hat{p} = 0.217$, $\hat{q} = 1 - \hat{p} = 0.783$, $z_{\alpha/2} = 1.96$, $E = z_{\alpha/2}\sqrt{\frac{\hat{p} \cdot \hat{q}}{n}} = 0.0266$. Confidence interval: $0.190 < p < 0.244$.

11. $n = \frac{z_{\alpha/2}^2 \cdot \hat{p} \cdot \hat{q}}{E^2} = \frac{2.05^2 \cdot 0.65 \cdot 0.35}{0.02^2} = 2390.17$ which we round up to 2391

13. There are clearly more than 10 successes and failures. $\alpha = 0.02$, $\hat{p} = 0.494$, $\hat{q} = 1 - \hat{p} = 0.506$, $z_{\alpha/2} = 2.33$, $E = z_{\alpha/2}\sqrt{\frac{\hat{p} \cdot \hat{q}}{n}} = 0.0390$. Confidence interval: $0.455 < p < 0.533$.

15.   There are clearly more than 10 successes and failures. $\alpha = 0.05$, $\hat{p} = 0.751$, $\hat{q} = 1 - \hat{p} = 0.249$, $z_{\alpha/2} = 1.96$, $E = z_{\alpha/2}\sqrt{\frac{\hat{p}\cdot\hat{q}}{n}} = 0.00946$. Confidence interval: $0.742 < p < 0.760$.

17. $n = \frac{z_{\alpha/2}^2 \cdot \hat{p}\cdot\hat{q}}{E^2} = \frac{2.33^2 \cdot 0.494 \cdot 0.506}{0.01^2} = 13570.3$ which we round up to 13,571.

19. $n = \frac{z_{\alpha/2}^2 \cdot \hat{p}\cdot\hat{q}}{E^2} = \frac{1.96^2 \cdot 0.751 \cdot 0.249}{0.02^2} = 1795.9$ which we round up to 1,796.

Section 6.4 Solutions

1. $\{1, 1\}, \{1, 2\}, \{1, 3\}, \{1, 4\}$
        $\{2, 1\}, \{2, 2\}, \{2, 3\}, \{2, 4\}$
        $\{3, 1\}, \{3, 2\}, \{3, 3\}, \{3, 4\}$
        $\{4, 1\}, \{4, 2\}, \{4, 3\}, \{4, 4\}$

3. $\mu_{\bar{x}} = 2.5$, $\sigma_{\bar{x}} = 0.7906$.

5.   First we compute the z-score of $\bar{x} = 6$: $z = \frac{\bar{x}-\mu}{\sigma/\sqrt{n}} = \frac{6-5.709}{0.197/\sqrt{5}} = 3.30$. $P(\bar{x} > 6) = P(z > 3.30) = 1 - 0.9995 = 0.0005$.

Section 6.5 Solutions

1.   97% confidence yields $\alpha = 0.03$, $1 - \alpha/2 = 0.9850$, thus $z_{\alpha/2} = 2.17$.

3. $\alpha = 0.05$, $z_{\alpha/2} = 1.96$, $E = z_{\alpha/2}\cdot\frac{\sigma}{\sqrt{n}} = 0.882$, confidence interval: $12.658 < \mu < 14.422$.

5. $\alpha = 0.06$, $z_{\alpha/2} = 1.88$, $E = z_{\alpha/2}\cdot\frac{\sigma}{\sqrt{n}} = 0.0296$, confidence interval: $3.170 < \mu < 3.230$.

7. $n = \left(\frac{z_{\alpha/2}\cdot\sigma}{E}\right)^2 = \left(\frac{2.33\cdot 0.218}{0.09}\right)^2 = 31.9$, which we round up to 32.

9. $\alpha = 0.06$, $z_{\alpha/2} = 1.88$, $E = z_{\alpha/2}\cdot\frac{\sigma}{\sqrt{n}} = 168.301$, confidence interval: $243.939 < \mu < 580.541$.

11. $n = \left(\frac{z_{\alpha/2}\cdot\sigma}{E}\right)^2 = \left(\frac{2.576\cdot 0.07901}{0.01}\right)^2 = 414.2$, which we round up to 415.

13. $n = \left(\frac{z_{\alpha/2}\cdot\sigma}{E}\right)^2 = \left(\frac{1.960\cdot 283.06}{60}\right)^2 = 85.5$, which we round up to 86.

15. $\alpha = 0.05$, $z_{\alpha/2} = 1.96$, $E = z_{\alpha/2}\cdot\frac{\sigma}{\sqrt{n}} = 0.4958$, confidence interval: $3.1292 < \mu < 4.1208$.

17. $\alpha = 0.03$, $z_{\alpha/2} = 2.17$, $E = z_{\alpha/2}\cdot\frac{\sigma}{\sqrt{n}} = 0.5627$, confidence interval: $4.4673 < \mu < 5.5927$.

Section 6.6 Solutions

1. $\alpha = 0.06$, $df = n - 1 = 37$, $t_{\alpha/2} = 1.940$.

3. $\alpha = 0.04$, $df = n - 1 = 10$, $t_{\alpha/2} = 2.359$, $E = t_{\alpha/2}\frac{s}{\sqrt{n}} = 3.101$.

5. $\alpha = 0.01$, $df = n - 1 = 26$, $t_{\alpha/2} = 2.779$, $E = t_{\alpha/2}\frac{s}{\sqrt{n}} = 24.177$. Confidence interval: $93.653 < \mu < 142.007$.

7. $\alpha = 0.02$, $df = n - 1 = 100$, $t_{\alpha/2} = 2.364$, $E = t_{\alpha/2}\frac{s}{\sqrt{n}} = 91.585$. Confidence interval: $1308.165 < \mu < 1491.335$.

9. $n = 19$, $\bar{x} = 46.47$, $s = 12.51$, $\alpha = 0.05$, $df = n - 1 = 18$, $t_{\alpha/2} = 2.101$, $E = t_{\alpha/2}\frac{s}{\sqrt{n}} = 6.030$. Confidence interval: $40.440 < \mu < 52.500$.

11. $\bar{x} = 173.59$, $s = 6.05$, $\alpha = 0.04$, $t_{\alpha/2} = 2.150$, $E = t_{\alpha/2}\cdot\frac{s}{\sqrt{n}} = 2.150\cdot\frac{6.05}{\sqrt{30}} = 2.38$,

$$173.59 - 2.38 < \mu < 173.59 + 2.38,$$

$$171.21 < \mu < 175.97.$$

13. $\bar{x} = 37.7419$, $s = 21.1723$, $\alpha = 0.04$, $t_{\alpha/2} = 2.147$, $E = z_{\alpha/2}\cdot\frac{\sigma}{\sqrt{n}} = 8.1643$, $29.58 < \mu < 45.91$.

B.7   Chapter 7 Solutions

Section 7.1 Solutions

1. $H_1 : p > 2/3$, $H_0 : p = 2/3$, this is a right tail test.

3. $H_1 : \mu < 40,000$, $H_0 : \mu = 40,000$, this is a left tail test.

5. $T.S. : t = \frac{\bar{x} - \mu}{s/\sqrt{n}}$

7. $C.V. : z = 2.326$

9. $C.V. : z = 1.960$

11. $C.V. : t = -1.840$

13. $P(z < 2.67) = 0.9962$. For a right-tail test, $p\text{-}value = P(z > 2.67) = 1 - 0.9962 = 0.0038$.

15. For a 2-tail test, $p\text{-}value = 2 \cdot (1 - P(z < |2.67|)) = 2 \cdot (1 - 0.9962) = 2 \cdot 0.0038 = 0.0076$.

17. For a left-tail test $p\text{-}value = P(z < -1.84) = 0.0329$.

19. Here the p-value is less than $\alpha = 0.05$, so we would reject the null hypothesis ($H_0 : p = \frac{2}{3}$) in support of the alternative ($H_1 : p > \frac{2}{3}$). The data support the claim that more than two-thirds of dentists choose Zest. With $0.01 < p\text{-value} < 0.05$, the data provide significant evidence for this conclusion.

21. Here the p-value is greater than α, so we would fail to reject the null hypothesis ($H_0 : \mu = 40,000$). The data do not support the claim that the average is less than 40,000 miles. With $0.01 < p\text{-value} < 0.05$, the data provide the data provide significant evidence for the alternative hypothesis, but not significant enough with $\alpha = 0.01$.

23. $H_0 : \mu = 25$, $H_1 : \mu \neq 25$. The data do not support H_1 so we have failed to reject H_0. In reality, $H_1 : \mu \neq 25$ is actually true, but H_0 says that $\mu = 25$ (so this is false) and we have not rejected this. This is a Type II error.

25. This study consists only of *hunch.com* readers, and will exhibit selection bias if used to represent all people. It should not be used to make inferences about this population.

27. The sampling frame would be the hunch.com readers who respond to questions of this sort. There is no reason to thing that this sample wouldn't represent such a sampling frame.

29. Rejecting the truth that the proportion of alcohol dependency among people who consume energy drinks is 16%, the same rate as for adults.

31. Failing to reject a false claim that the rate of alcohol dependency is the same for energy drink consumers, when it is actually higher.

33. The Type II Error is worse because it results in the neglecting of a group that is at risk.

35. We reject the hypothesis that the failure proportion for the new chutes is 0.02, when this is actually true.

37. We fail to reject the claim that the failure proportion for the new chutes is 0.02, when it is actually less than this.

39. The Type I Error is worse because people may start using a parachute that is not safe.

Section 7.2 Solutions

1. $\hat{p} = x/n = 185/450 = 0.411$, $n = 450$.

    $\alpha = 0.05$

    $H_0 : p = 0.38$; $H_1 : p > 0.38$

    $np = 450 \cdot 0.38 = 171 \geq 10$

    $nq = 450 \cdot 0.62 = 279 \geq 10$

    $T.S. : z = \frac{\hat{p}-p}{\sqrt{\frac{pq}{n}}} = \frac{0.411-0.38}{\sqrt{\frac{0.38 \cdot 0.62}{450}}} = 1.360$

    $C.V. : z = 1.645$

    $p\text{-}value = 0.0870$ (moderately insignificant)

    Fail to reject the null hypothesis. The original claim stated that $p > 0.38$; this is the alternative hypothesis, and the data do not support this. *The data are insufficient to support the claim that the population proportion is greater than 38%.*

3. $\hat{p} = x/n = 0.742$, $n = 120$.

    $\alpha = 0.02$

    $H_0 : p = 0.82$; $H_1 : p < 0.82$

    $np = 120 \cdot 0.82 = 98.4 \geq 10$

    $nq = 120 \cdot 0.18 = 21.6 \geq 10$

    $T.S. : z = \frac{\hat{p}-p}{\sqrt{\frac{pq}{n}}} = \frac{0.742-0.82}{\sqrt{\frac{0.82 \cdot 0.18}{120}}} = -2.224$

    $C.V. : z = -2.054$

    $p\text{-}value = 0.0132$ (significant)

    Reject the null hypothesis. The original claim stated that $p < 0.82$; this is the alternative Hypothesis, and we are supporting this. *The data support the claim that the population proportion is less than 82%.*

5. While the p-value *is* less than α, indicating a rejection of the null hypothesis, we cannot use these results to generalize about the population of all Mac users because the sample is not random. This study cannot be used to make inferences regarding any group outside of the sampling frame, which is the population of hunch.com readers who are inclined to answer such questions.

7. $\hat{p} = x/n = 6022/8023 = 0.75059$, $n = 8023$.

    $\alpha = 0.01$

    $H_0 : p = 0.75$; $H_1 : p \neq 0.75$

    $np = 8023 \cdot 0.75 = 6017.25 \geq 10$

    $nq = 8023 \cdot 0.25 = 2005.75 \geq 10$

    $T.S. : z = \frac{\hat{p}-p}{\sqrt{\frac{pq}{n}}} = \frac{0.75059-0.75}{\sqrt{\frac{0.75 \cdot 0.25}{8023}}} = 0.122$

    $C.V. : z = \pm 2.576$

    $p\text{-}value = 0.9025$

    Fail to reject the null hypothesis. The original claim stated that $p = 0.75$; this is the null hypothesis, and since we are not rejecting the null hypothesis, we are supporting the original claim, but the support is weak since we cannot prove the null hypothesis to be true. *The data are insufficient to reject the claim that the proportion of peas that are yellow is equal to 75%.*

9. $\hat{p} = x/n = 445/450 = 0.989$, $n = 450$.

    $\alpha = 0.01$

    $H_0 : p = 0.97$; $H_1 : p > 0.97$

$np = 450 \cdot 0.97 = 436.5 \geq 10$

$nq = 450 \cdot 0.03 = 13.5 \geq 10$

$T.S. : z = \frac{\hat{p}-p}{\sqrt{\frac{pq}{n}}} = \frac{0.989-0.97}{\sqrt{\frac{0.97 \cdot 0.03}{450}}} = 2.363$

$C.V. : z = 2.326$; $p\text{-}value = 0.0091$ (highly significant)

Reject the null hypothesis. The original claim stated that $p > 0.97$; this is the alternative hypothesis, and since we are rejecting the null hypothesis, we are supporting this original claim. *The data support the claim that the literacy rate in Japan is greater than 97%. The sample proportion was highly significantly greater than 97%, as indicated by the p-value of 0.0091.*

11. $\hat{p} = x/n = 0.569$, $n = 1444$.

   $\alpha = 0.02$

   $H_0 : p = 0.542$; $H_1 : p > 0.542$

   $np = 1444 \cdot 0.542 = 782.648 \geq 10$

   $nq = 1444 \cdot 0.458 = 661.352 \geq 10$

   $T.S. : z = \frac{\hat{p}-p}{\sqrt{\frac{pq}{n}}} = \frac{0.569-0.542}{\sqrt{\frac{0.542 \cdot 0.458}{1444}}} = 2.059$

   $C.V. : z = 2.054$; $p\text{-}value = 0.0197$

   Reject the null hypothesis. The original claim stated that $p > 0.542$; this is the alternative hypothesis, and since we are rejecting the null hypothesis, we are supporting this original claim. *The data support the claim that the success rate in Pre-Algebra is greater than 54.2%. The sample proportion was significantly greater than 54.2% as indicated by the p-value of 0.0197.*

13. $\hat{p} = x/n = 67/110 = 0.609$, $n = 110$.

   $\alpha = 0.05$

   $H_0 : p = 0.50$; $H_1 : p > 0.50$

   $np = 110 \cdot 0.50 = 55 \geq 10$

   $nq = 110 \cdot 0.50 = 55 \geq 10$

   $T.S. : z = \frac{\hat{p}-p}{\sqrt{\frac{pq}{n}}} = \frac{0.609-0.50}{\sqrt{\frac{0.50 \cdot 0.50}{110}}} = 2.286$

   $C.V. : z = 1.645$

   $p - value = 1 - 0.9890 = 0.0110$ (significant)

   Reject the null hypothesis. The original claim stated that $p > 0.38$; this is the alternative hypothesis, and the data do support this.

   The data support the claim that the proportion or children of clinically depressed mothers who develop psychological illnesses is greater than 50%.

Section 7.3 Solutions

1. $\bar{x} = 185.2$, $\sigma = 45.7$, $n = 450 > 30$,

   $\alpha = 0.06$

   $H_0 : \mu = 180$, $H_1 : \mu > 180$

   $T.S. : z = \frac{\bar{x}-\mu}{\sigma/\sqrt{n}} = \frac{185.2-180}{45.7/\sqrt{450}} = 2.414$

   $C.V. : z = 1.555$

   $p\text{-}value = 0.0080$ (highly significant)

   Reject the null hypothesis. The original claim stated that $\mu > 180$; this is the alternative hypothesis, and since we are rejecting the null hypothesis, we are supporting original claim. *The data support the claim that the population mean is greater than*

180. The sample mean was highly significantly greater than 180, as indicated by the p-value of 0.0079.

3. $\bar{x} = 1.61$, $\sigma = 0.21$, $n = 34 > 30$,
    $\alpha = 0.05$
    $H_0 : \mu = 1.59$, $H_1 : \mu < 1.59$
    $T.S. : z = \frac{\bar{x}-\mu}{\sigma/\sqrt{n}} = \frac{1.61-1.59}{0.21/\sqrt{34}} = 0.555$
    $C.V. : z = -1.645$
    $p\text{-}value = 0.7123$ (insignificant)
    Fail to reject the null hypothesis. The original claim stated that $\mu < 1.59$; but since we are failing to reject the null hypothesis, we are not supporting original claim. *The data are not sufficient to support the claim that the population mean is less than 1.59.*

5. There is a problem with this study. The students sampled were from a single college, and it is not reasonable to assume that the results from this study could apply to the population of all female college students. The sampling frame is the population of female students from Mt. San Antonio College, so that is the population to which the inference will be applied. The sample data are: $\bar{x} = 21.8$, $\sigma = 1.2$, and $n = 29$. The sample size is not greater than 30. Because the histogram provided appears roughly normal, we will consider this sufficent evidence for us to assume that the population is normal.
    $\alpha = 0.01$
    $H_0 : \mu = 22.5$; $H_1 : \mu < 22.5$
    $T.S. : z = \frac{\bar{x}-\mu}{\sigma/\sqrt{n}} = \frac{21.8-22.5}{1.2/\sqrt{29}} = -3.14$.
    $C.V. : z = -2.326$
    $p\text{-}value = 0.0008$ (highly significant)
    Reject the null hypothesis. The data support the claim that the mean foot length of female students at Mt. San Antonio College is less than 22.5 in.

7. $\bar{x} = 299.8524$, $\sigma = 0.07901$, $n = 100 > 30$, $\alpha = 0.01$, $H_0 : \mu = 299.7244$, $H_1 : \mu \neq 299.7244$
    $T.S. : z = \frac{\bar{x}-\mu}{\sigma/\sqrt{n}} = \frac{299.8524-299.7244}{0.07901/\sqrt{100}} = 16.2005$
    $C.V. : z = \pm 2.576$
    $p\text{-}value = 0.0002$ (the true $p\text{-}value$ is actually much smaller than this).
    Reject the null hypothesis. The original claim stated that $\mu = 299.7244$; and since this is the null hypothesis, we are rejecting the original claim. *The data do not support the claim that the population mean is equal to 299.7244. The sample mean was highly significantly different from this, as indicated by the p-value of 0.0002.*

    Given that the null hypothesis is actually true (to within a very small margin of error), we have rejected a true null hypothesis, so we have made a Type I error. It is possible that this error is due to sampling error, but this is unlikely because the test statistic was extraordinarily large, and this is very unlikely to happen due to chance sampling error. The only other possibility is *non-sampling error*, and in this case we must assume that the technique used to measure the speed of light tended to give values that were too large, so the sampling data probably contain *measurement bias*.

9. $\bar{x} = 1.702$, $\sigma = 1.314$, $n = 70 > 30$.

   $\alpha = 0.02$

   $H_0 : \mu = 2$; $H_1 : \mu < 2$

   $T.S. : z = \frac{\bar{x}-\mu}{\sigma/\sqrt{n}} = \frac{1.702-2}{1.314/\sqrt{70}} = -1.8974$

   $C.V. : z = -2.054$

   $p\text{-}value = 0.0289$ (significant)

   Fail to reject the null hypothesis. The original claim stated that $\mu < 2$; but since we are failing to reject the null hypothesis, we are not supporting original claim. *The data are not sufficient to support the claim that the mean rainfall for seeded clouds in this region of Tasmania is less than 2 inches. Still, the sample mean was significantly less than 2 inches as indicated by the p-value of 0.0289. This was not significant enough at 2% significance.*

Section 7.4 Solutions

1. $\bar{x} = 75.3$, $s = 45.7$, $n = 33 > 30$, $\alpha = 0.05$

   $H_0 : \mu = 67.9$, $H_1 : \mu > 67.9$

   $T.S. : t = \frac{\bar{x}-\mu}{s/\sqrt{n}} = \frac{75.3-67.9}{45.7/\sqrt{33}} = 0.930$, $C.V. : t = 1.694$

   $p\text{-}value = 0.188$ (using $df = 30$)

   Fail to reject the null hypothesis. The original claim stated that $\mu > 67.9$; this is the alternative hypothesis, and since we are failing to reject the null hypothesis, we are not supporting original claim. *The data are insufficient to support the claim that the population mean is greater than 67.9. The sample mean was insignificantly greater than 67.9, as indicated by the p-value of 0.188.*

3. $\bar{x} = 1.49$, $s = 0.21$, $n = 18$, and the population is assumed to be normal.

   $\alpha = 0.06$, $H_0 : \mu = 1.59$, $H_1 : \mu < 1.59$

   $T.S. : t = \frac{\bar{x}-\mu}{s/\sqrt{n}} = \frac{1.49-1.59}{0.21/\sqrt{18}} = -2.020$, $C.V. : t = -1.637$

   $p\text{-}value = 0.031$

   Reject the null hypothesis. The original claim stated that $\mu < 1.59$; and since we are rejecting the null hypothesis, we are supporting original claim. *The data support the claim that the population mean is less than 1.59. The sample mean was significantly less than 1.59, as indicated by the p-value of 0.031.*

5. While the p-value is quite small, we cannot conclude that the data support the claim because we have no evidence that the requirements for the test are met. The sample size is too small, and we do not know if the population from which we are sampling is normal.

7. $\bar{x} = 23.70$, $s = 13.76$, $n = 30$, and the population is assumed to be normal.

   $\alpha = 0.04$; $H_0 : \mu = 28$; $H_1 : \mu < 28$

   $T.S. : t = \frac{\bar{x}-\mu}{s/\sqrt{n}} = \frac{23.70-28}{13.76/\sqrt{30}} = -1.712$; $C.V. : t = -1.814$

   $p\text{-}value = 0.050$

   Fail to reject the null hypothesis. The original claim stated that $\mu < 28$; and since we are not rejecting the null hypothesis, we are not supporting original claim. *The data do not give sufficient evidence to support the claim that the population mean is less than 28. Still, the sample mean was significantly less than 28, as indicated by the p-value of 0.050, but not significant enough at 0.04 significance.*

9. $\bar{x} = 46.47$, $s = 12.51$, $n = 19$, and the population is assumed to be normal.

    $\alpha = 0.03$, $H_0 : \mu = 52$, $H_1 : \mu < 52$

    $T.S. : t = \frac{\bar{x} - \mu}{s/\sqrt{n}} = \frac{46.47 - 52}{12.51/\sqrt{19}} = -1.927$; $C.V. : t = -2.007$, $p\text{-value} = 0.037$

    Fail to reject the null hypothesis. The original claim stated that $\mu < 52$; and since we are not rejecting the null hypothesis, we are not supporting original claim. *The data do not give sufficient evidence to support the claim that the population mean is less than 52. Still, the sample mean was significantly less than 52, as indicated by the p-value of 0.037, but not significant enough at 0.03 significance.*

11. $\bar{x} = 85.43$, $s = 26.53$, $n = 35 > 30$, $\alpha = 0.03$

    $H_0 : \mu = 75$, $H_1 : \mu > 75$

    $T.S. : t = \frac{\bar{x} - \mu}{s/\sqrt{n}} = \frac{85.43 - 75}{26.53/\sqrt{35}} = 2.326$

    $C.V. : t = 1.946$

    $p\text{-value} = 0.014$ (Using 30 degrees of freedom)

    Reject the null hypothesis. The original claim stated that $\mu > 75$; this is the alternative hypothesis, and since we are rejecting the null hypothesis, we are supporting original claim. *The data support the claim that the population mean is greater than 75. The sample mean was significantly greater than 75, as indicated by the p-value of 0.014.*

Section 7.5 Solutions

1. The differences, $d = x - y$, are computed first.

x	y	$d = x - y$
75	79	−4
62	67	−5
81	82	−1
58	61	−3
90	89	1
73	76	−3
69	75	−6

    $\bar{d} = -3.00$, $s_d = 2.38$, $n = 7$, and we have assumed that the population of differences is normal.

    $\alpha = 0.01$; $H_0 : \mu_1 = \mu_2$ (or $\mu_d = 0$); $H_1 : \mu_1 < \mu_2$ (or $\mu_d < 0$)

    $T.S. : t = \frac{\bar{d} - \mu_d}{s_d/\sqrt{n}} = \frac{-3.00 - 0}{2.38/\sqrt{7}} = -3.334$; $C.V. : t = -3.413$; $p\text{-value} = 0.008$.

    Conclusion: reject the null hypothesis. Since the original claim is the alternative hypothesis, $\mu_1 < \mu_2$, we are supporting the original claim. *The data support the claim that algebra review is effective in raising students' average test score.*

3. The differences, $d = x - y$, are computed first.

x	y	$d = x - y$
0.0	−1.1	1.1
−1.1	0.5	−1.6
−1.6	0.5	−2.1
−0.3	0.0	−0.3
−1.1	−0.5	−0.6
−0.9	1.3	−2.2
−0.5	−1.4	0.9
0.7	0.0	0.7
−1.2	−0.8	−0.4

$\bar{d} = -0.5$, $s_d = 1.26$, $n = 9$, and we have assumed that the population is normal. $\alpha = 0.05$. $H_0 : \mu_1 = \mu_2$ (or $\mu_d = 0$); $H_1 : \mu_1 < \mu_2$ (or $\mu_d < 0$)

$T.S. : t = \frac{\bar{d} - \mu_d}{s_d/\sqrt{n}} = \frac{-0.5 - 0}{1.26/\sqrt{9}} = -1.191$; $C.V. : t = -1.860$; $p\text{-}value = 0.132$

Conclusion: fail to reject the null hypothesis. Since the original claim is the alternative hypothesis, $\mu_1 < \mu_2$, we are not supporting the original claim. *The data do not support the claim that the more weight is lost, on average, with the mCPP appetite suppressant than with a placebo.*

5. The differences, $d = x - y$, are computed first.

x	y	$d = x - y$
44.75	34.65	10.1
44.5	35.54	8.96
45.84	35.95	9.89
47.79	38.05	9.74
47.93	38	9.93
51.81	40.5	11.31

$\bar{d} = 9.99$, $s_d = 0.76$, $n = 6$, and we have assumed that the population is normal. $\alpha = 0.02$. $H_0: \mu_1 - \mu_2 = 9$, (or $\mu_d = 9$); $H_1: \mu_1 - \mu_2 > 9$ (or $\mu_d > 9$)

$T.S. : t = \frac{\bar{d} - \mu_d}{s_d/\sqrt{n}} = \frac{9.99 - 9}{0.76/\sqrt{6}} = 3.183$. $C.V.: t = 2.757$; $p\text{-value} = 0.012$

Conclusion: reject the null hypothesis. Since the original claim is the alternative hypothesis, $\mu_d > 9$, we are supporting the original claim. *The data support the claim that the mean body dissatisfaction level (or difference) is greater than 9.*

7. $\bar{d} = 0.894$, $s_d = 6.890$, $n = 31 > 30$. $\alpha = 0.01$.

$H_0 : \mu_1 = \mu_2$ (or $\mu_d = 0$); $H_1 : \mu_1 \neq \mu_2$ (or $\mu_d \neq 0$)

$T.S. : t = \frac{\bar{d} - \mu_d}{s_d/\sqrt{n}} = \frac{0.894 - 0}{6.890/\sqrt{31}} = 0.722$; $C.V. : t = \pm 2.750$; $p\text{-}value = 2 \cdot (1 - 0.755) = 0.490$

Conclusion: Fail to reject the null hypothesis. Since the original claim is the null hypothesis, $\mu_1 = \mu_2$, we are supporting the original claim. *The data do not give reason to reject the claim that the mean arm span is equal to the mean height.*

9. $\bar{d} = 0.102$, $s_d = 0.201$, $n = 31 > 30$.

$\alpha = 0.01$. $H_0 : \mu_1 = \mu_2$ (or $\mu_d = 0$); $H_1 : \mu_1 > \mu_2$ (or $\mu_d > 0$)

$T.S. : t = \frac{\bar{d} - \mu_d}{s_d/\sqrt{n}} = \frac{0.102 - 0}{0.201/\sqrt{31}} = 2.825$; $C.V. : t = 2.457$; $p\text{-}value = 1 - 0.996 = 0.004$.

Conclusion: Fail to reject the null hypothesis. Since the original claim is the null hypothesis, $\mu_1 = \mu_2$, we are supporting the original claim. *The data support the claim that grocery prices are more expensive, on average, in affluent neighborhoods.*

B.8   Chapter 8 Solutions

Section 8.1 Solutions

1. The test statistic is computed as follows. $\bar{p} = \frac{450+550}{940+875} = 0.551$, $\bar{q} = 1 - \bar{p} = 1 - 0.551 = 0.449$. $\hat{p}_1 = \frac{450}{940} = 0.479$. $\hat{p}_2 = \frac{550}{875} = 0.629$.

$$z = \frac{(\hat{p}_1 - \hat{p}_2) - (p_1 - p_2)}{\sqrt{\frac{\bar{p}\bar{q}}{n_1} + \frac{\bar{p}\bar{q}}{n_2}}}$$

$$= \frac{(0.479 - 0.629) - 0}{\sqrt{\frac{0.551 \cdot 0.449}{940} + \frac{0.551 \cdot 0.449}{875}}}$$

$$= -6.420$$

Each sample has at least 10 successes and failures, so there is some chance that the requirement for approximate normality is met. The difference observed is -6.420 standard deviations from the hypothetical difference of 0. This difference is quite significant.

3. The test statistic is computed as follows. $\bar{p} = \frac{175+100}{525+320} = 0.325$, $\bar{q} = 1 - \bar{p} = 1 - 0.325 = 0.675$. $\hat{p}_1 = \frac{175}{525} = 0.333$. $\hat{p}_2 = \frac{100}{320} = 0.313$.

$$z = \frac{(\hat{p}_1 - \hat{p}_2) - (p_1 - p_2)}{\sqrt{\frac{\bar{p}\bar{q}}{n_1} + \frac{\bar{p}\bar{q}}{n_2}}}$$

$$= \frac{(0.333 - 0.313) - 0.01}{\sqrt{\frac{0.325 \cdot 0.675}{525} + \frac{0.325 \cdot 0.675}{320}}}$$

$$= 0.301$$

Each sample has at least 10 successes and failures, so there is some chance that the requirement for approximate normality is met. The difference observed is 0.301 standard deviations from the hypothetical difference of 0.01. This difference is not at all significant.

5. The test statistic is computed as follows.

$$t = \frac{(\bar{x}_1 - \bar{x}_2) - (\mu_1 - \mu_2)}{\sqrt{\frac{s_1^2}{n_1} + \frac{s_2^2}{n_2}}}$$

$$= \frac{(12 - 13) - 0}{\sqrt{\frac{3^2}{75} + \frac{4^2}{95}}}$$

$$= -1.862$$

Each sample size is larger than 30, so the requirement for approximate normality is met. The difference observed is -1.862 standard deviations from the hypothetical difference of 0. This difference seems only moderately significant.

7. The test statistic is computed as follows.

$$t = \frac{(\bar{x}_1 - \bar{x}_2) - (\mu_1 - \mu_2)}{\sqrt{\frac{s_1^2}{n_1} + \frac{s_2^2}{n_2}}}$$

$$= \frac{(65-57) - 4}{\sqrt{\frac{12^2}{110} + \frac{15^2}{182}}}$$

$$= 2.507$$

Each sample size is larger than 30, so the requirement for approximate normality is met. The difference observed is 2.507 standard deviations from the hypothetical difference of 4. This difference seems significant.

Section 8.2 Solutions

1. $x_1 = 45$, $n_1 = 69$, $\hat{p}_1 = x_1/n_1 = 0.652$, $x_2 = 57$, $n_2 = 73$, $\hat{p}_2 = x_2/n_2 = 0.781$,
   $\bar{p} = \frac{x_1+x_2}{n_1+n_2} = 0.718$, $\bar{q} = 1 - \bar{p} = 0.282$, $\alpha = 0.05$.
   Each sample has more than 10 successes and 10 failures.
   $H_0 : p_1 = p_2$, $H_1 : p_1 < p_2$.
   $T.S. : z = \frac{(\hat{p}_1-\hat{p}_2)-(p_1-p_2)}{\sqrt{\frac{\bar{p}\bar{q}}{n_1} + \frac{\bar{p}\bar{q}}{n_2}}} = \frac{(0.652-0.781)-0}{\sqrt{\frac{0.718\cdot0.282}{69} + \frac{0.718\cdot0.282}{73}}} = -1.707.$
   $C.V. : z = -1.645$, $p\text{-}value = 0.0436$
   We reject the null hypothesis in support of the alternative. *The data support the claim that the proportion from population 1 is less than the proportion from population 2.*

3. $x_1 = 24$, $n_1 = 71$, $\hat{p}_1 = x_1/n_1 = 0.338$, $x_2 = 36$, $n_2 = 144$, $\hat{p}_2 = x_2/n_2 = 0.250$,
   $\bar{p} = \frac{x_1+x_2}{n_1+n_2} = 0.279$, $\bar{q} = 1 - \bar{p} = 0.721$, $\alpha = 0.08$.
   Each sample has more than 10 successes and 10 failures.
   $H_0 : p_1 = p_2$, $H_1 : p_1 \neq p_2$.
   $T.S. : z = \frac{(\hat{p}_1-\hat{p}_2)-(p_1-p_2)}{\sqrt{\frac{\bar{p}\bar{q}}{n_1} + \frac{\bar{p}\bar{q}}{n_2}}} = \frac{(0.338-0.250)-0}{\sqrt{\frac{0.279\cdot0.721}{71} + \frac{0.279\cdot0.721}{144}}} = 1.353.$
   $C.V. : z = \pm1.75$, $p\text{-}value = 0.1770$
   We fail to reject the null hypothesis. *There is not sufficient evidence to support the claim that the proportion from population 1 is not equal to the proportion from population 2.*

5. $x_1 = 35$, $n_1 = 115$, $\hat{p}_1 = x_1/n_1 = 35/115 = 0.304$, $x_2 = 93$, $n_2 = 303$, $\hat{p}_2 = x_2/n_2 = 0.307$,
   $\bar{p} = \frac{x_1+x_2}{n_1+n_2} = 115/418 = 0.306$, $\bar{q} = 1 - \bar{p} = 0.694$, $\alpha = 0.05$.
   Each sample has more than 10 successes and 10 failures.
   $H_0: p_1 = p_2$, $H_1: p_1 < p_2$.
   $T.S. : z = \frac{(\hat{p}_1-\hat{p}_2)-(p_1-p_2)}{\sqrt{\frac{\bar{p}\bar{q}}{n_1} + \frac{\bar{p}\bar{q}}{n_2}}} = \frac{(0.304-0.307)-0}{\sqrt{\frac{0.306\cdot0.694}{115} + \frac{0.306\cdot0.694}{303}}} = -0.0594.$
   $C.V. : z = -1.645$, $p\text{-}value = 1 - 0.5327 = 0.4761.$
   We fail to reject the null hypothesis. *There is not sufficient evidence to support the claim that the proportion from population 1 is less than the proportion from population 2. The difference between the sample proportions was not significant as indicated by the p-value of 0.4681.*

7. It is nearly impossible to have an all female group and an all male group which are created by random assignment. Because of this, studies which compare females and

males are observational, so causal conclusions are not appropriate.

9. This study does not use random samples (convenience samples are used with selection bias), so it does not permit us to make inferences about the populations of birthers or non-birthers.

11. $x_1 = 575$, $n_1 = 1007$, $\hat{p}_1 = x_1/n_1 = 0.571$, $x_2 = 91$, $n_2 = 158$, $\hat{p}_2 = x_2/n_2 = 0.575$, $\bar{p} = \frac{x_1+x_2}{n_1+n_2} = 0.572$, $\bar{q} = 1 - \bar{p} = 0.428$, $\alpha = 0.06$.

Each sample has more than 10 successes and 10 failures.

H_0: $p_1 = p_2$, H_1: $p_1 \neq p_2$.

$T.S. : z = \frac{(\hat{p}_1-\hat{p}_2)-(p_1-p_2)}{\sqrt{\frac{\bar{p}\bar{q}}{n_1}+\frac{\bar{p}\bar{q}}{n_2}}} = \frac{(0.571-0.575)-0}{\sqrt{\frac{0.572\cdot0.428}{1007}+\frac{0.572\cdot0.428}{158}}} = -0.094$.

$C.V. : z = \pm1.881$, $p\text{-value} = 2 \cdot (1 - 0.5359) = 2 \cdot 0.4641 = 0.9282$

We fail to reject the null hypothesis. The study is observational. *The data do not give sufficient evidence to support rejection of the claim that the proportions of students successfully completing Math 71 and 71B are the same. The difference between the sample proportions was insignificant, as shown by the p-value of 0.9282.*

13. $x_1 = 566$, $n_1 = 1086$, $\hat{p}_1 = x_1/n_1 = 0.521$, $x_2 = 67$, $n_2 = 112$, $\hat{p}_2 = x_2/n_2 = 0.598$, $\bar{p} = \frac{x_1+x_2}{n_1+n_2} = 0.528$, $\bar{q} = 1 - \bar{p} = 0.472$, $\alpha = 0.04$.

Each sample has more than 10 successes and 10 failures.

H_0: $p_1 = p_2$, H_1: $p_1 < p_2$.

$T.S. : z = \frac{(\hat{p}_1-\hat{p}_2)-(p_1-p_2)}{\sqrt{\frac{\bar{p}\bar{q}}{n_1}+\frac{\bar{p}\bar{q}}{n_2}}} = \frac{(0.521-0.598)-0}{\sqrt{\frac{0.528\cdot0.472}{1086}+\frac{0.528\cdot0.472}{112}}} = -1.554$.

$C.V. : z = -1.751$, $p\text{-value} = 0.0606$

We fail to reject the null hypothesis. The study is observational. *The data do not support the claim that the proportion of students successfully completing Math 51 is lower than the proportion of students who are successfully completing Math 51B. The difference between the two proportions was only moderately significant, with a p-value of 0.0606.*

15. $x_1 = 17$, $n_1 = 75$, $\hat{p}_1 = x_1/n_1 = 17/75 = 0.227$, $x_2 = 10$, $n_2 = 116$, $\hat{p}_2 = x_2/n_2 = 10/116 = 0.086$

$\bar{p} = \frac{x_1+x_2}{n_1+n_2} = \frac{17+10}{75+116} = 0.141$, $\bar{q} = 1 - \bar{p} = 1 - 0.141 = 0.859$, $\alpha = 0.03$.

Each sample has at least 10 successes and 10 failures.

$H_0 : p_1 = p_2$, $H_1 : p_1 > p_2$.

$T.S. : z = \frac{(\hat{p}_1-\hat{p}_2)-(p_1-p_2)}{\sqrt{\frac{\bar{p}\bar{q}}{n_1}+\frac{\bar{p}\bar{q}}{n_2}}} = \frac{(0.227-0.086)-0}{\sqrt{\frac{0.141\cdot0.859}{75}+\frac{0.141\cdot0.859}{116}}} = 2.734$.

$C.V. : z = 1.881$, $p\text{-value} = 1 - 0.9968 = 0.0032$.

We reject the null hypothesis. The study is observational. *The data support the claim that the proportion of teenage girls with depression from affluent families is greater than that proportion for teenage girls in non-affluent families. The difference between the sample proportions was highly significant as indicated by the p-value of 0.0032.*

17. $x_1 = 260$, $n_1 = 1000$, $\hat{p}_1 = x_1/n_1 = 260/1000 = 0.260$, $x_2 = 280$, $n_2 = 924$, $\hat{p}_2 = x_2/n_2 = 0.303$,

$\bar{p} = \frac{x_1+x_2}{n_1+n_2} = 0.281$, $\bar{q} = 1 - \bar{p} = 0.719$, $\alpha = 0.05$.

Each sample has more than 10 successes and 10 failures.

$H_0 : p_1 = p_2$, $H_1 : p_1 < p_2$.

$T.S. : z = \frac{(\hat{p}_1-\hat{p}_2)-(p_1-p_2)}{\sqrt{\frac{\bar{p}\bar{q}}{n_1}+\frac{\bar{p}\bar{q}}{n_2}}} = \frac{(0.260-0.303)-0}{\sqrt{\frac{0.281\cdot0.719}{1000}+\frac{0.281\cdot0.719}{924}}} = -2.097$.

$C.V. : z = -1.645$, $p\text{-}value = 0.0179$.

We reject the null hypothesis. The study is observational. *The data support the claim that the proportion of people who drink $\frac{1}{2}$ can of diet soda per day who are obese is less than that proportion for people who drink 1 can of diet soda per day. The difference between the sample proportions was significant as indicated by the p-value of 0.0179.*

19. $x_1 = 29$, $n_1 = 40$, $\hat{p}_1 = x_1/n_1 = 29/40 = 0.725$, $x_2 = 16$, $n_2 = 50$, $\hat{p}_2 = x_2/n_2 = 16/50 = 0.320$

$\bar{p} = \frac{x_1+x_2}{n_1+n_2} = \frac{29+16}{40+50} = 0.500$, $\bar{q} = 1 - \bar{p} = 1 - 0.500 = 0.500$, $\alpha = 0.02$.

Each sample has more than 10 successes and 10 failures.

$H_0 : p_1 = p_2$, $H_1 : p_1 > p_2$.

$T.S. : z = \frac{(\hat{p}_1-\hat{p}_2)-(p_1-p_2)}{\sqrt{\frac{\bar{p}\bar{q}}{n_1}+\frac{\bar{p}\bar{q}}{n_2}}} = \frac{(0.725-0.320)-0}{\sqrt{\frac{0.500\cdot0.500}{40}+\frac{0.500\cdot0.500}{50}}} = 3.82$.

$C.V. : z = 2.054$, $p\text{-}value \approx 0.0001$.

We reject the null hypothesis. The study is experimental. *The data support the claim that Fluoxetine with psychiatric treatment is effective in increasing the proportion of depressed teens who experience meaningful improvement.*

Section 8.3 Solutions

1. $\bar{x}_1 = 3.1$, $s_1 = 1.8$, $n_1 = 31 > 30$, $\bar{x}_2 = 2.4$, $s_2 = 0.9$, $n_2 = 37 > 30$, $\alpha = 0.05$.

$H_0 : \mu_1 = \mu_2$; $H_1 : \mu_1 < \mu_2$.

$T.S. : t = \frac{(\bar{x}_1-\bar{x}_2)-(\mu_1-\mu_2)}{\sqrt{\frac{s_1^2}{n_1}+\frac{s_2^2}{n_2}}} = \frac{(3.1-2.4)-(0)}{\sqrt{\frac{1.8^2}{31}+\frac{0.9^2}{37}}} = 1.969$,

$V_1 = \frac{s_1^2}{n_1} = \frac{1.8^2}{31} = 0.1045$ and $V_2 = \frac{s_2^2}{n_2} = \frac{0.9^2}{37} = 0.0219$,

$df = \frac{(V_1+V_2)^2}{\frac{V_1^2}{n_1-1}+\frac{V_2^2}{n_2-1}} = 42.3$ which we round down to 42.

$C.V. : t = -1.682$, $p\text{-}value = 0.974$ (Using 40 degrees of freedom).

We fail to reject the null hypothesis, which means we cannot support the alternative, which was the original claim. This problem was strange, claiming that the mean for population 1 is smaller, when in fact, the sample mean from this population was larger. This is why the p-value was so large. *The data do not give sufficient evidence to support the claim that the mean from the first population is less than that of the second population. This is emphasized by the large p-value of 0.974.*

3. $\bar{x}_1 = 0.096$, $s_1 = 0.0019$, $n_1 = 28$, $\bar{x}_2 = 0.095$, $s_2 = 0.0014$, $n_2 = 25$, $\alpha = 0.02$.

The data appear to come from normal populations.

$H_0 : \mu_1 = \mu_2$; $H_1 : \mu_1 \neq \mu_2$.

$T.S. : t = \frac{(\bar{x}_1-\bar{x}_2)-(\mu_1-\mu_2)}{\sqrt{\frac{s_1^2}{n_1}+\frac{s_2^2}{n_2}}} = \frac{(0.096-0.095)-(0)}{\sqrt{\frac{0.0019^2}{28}+\frac{0.0014^2}{25}}} = 2.196$,

$V_1 = \frac{s_1^2}{n_1} = \frac{0.0019^2}{28} = 0.000000129$ and $V_2 = \frac{s_2^2}{n_2} = \frac{0.0014^2}{25} = 0.0000000784$,

$df = \frac{(V_1+V_2)^2}{\frac{V_1^2}{n_1-1}+\frac{V_2^2}{n_2-1}} = 49.3$ which we round down to 49.

$C.V. : t = \pm 2.405$.

$p\text{-}value = 2 \cdot 0.016 = 0.032$ (Using 50 degrees of freedom).

Once again, we fail to reject the null hypothesis, which is contrary to the original claim. *There is insufficient evidence to support the claim that the means of these two populations are different. Still, the difference between these sample means was significant, as indicated by the p-value of 0.0324.*

5. $\bar{x}_1 = 0.294$, $s_1 = 0.012$, $n_1 = 35 > 30$, $\bar{x}_2 = 0.301$, $s_2 = 0.018$, $n_2 = 43 > 30$, $\alpha = 0.02$.

$H_0 : \mu_1 = \mu_2$; $H_1 : \mu_1 \neq \mu_2$.

$T.S. : t = \frac{(\bar{x}_1 - \bar{x}_2) - (\mu_1 - \mu_2)}{\sqrt{\frac{s_1^2}{n_1} + \frac{s_2^2}{n_2}}} = \frac{(0.294 - 0.301) - (0)}{\sqrt{\frac{0.012^2}{35} + \frac{0.018^2}{43}}} = -2.051,$

$V_1 = \frac{s_1^2}{n_1} = \frac{0.012^2}{35} = 0.0000041142857$ and $V_2 = \frac{s_2^2}{n_2} = \frac{0.018^2}{43} = 0.0000075348837,$

$df = \frac{(V_1 + V_2)^2}{\frac{V_1^2}{n_1 - 1} + \frac{V_2^2}{n_2 - 1}} = 73.37$ which we round down to 73.

$C.V. : t = \pm 2.377$ (using 75 degrees of freedom)

$p\text{-}value = 2 \cdot (1 - 0.980) = 2 \cdot 0.020 = 0.040$ (Using 70 degrees of freedom).

We fail to reject the null hypothesis, which means we cannot support the alternative, the original claim. *There is insufficient evidence to allow support of the claim that the mean from the first population is different from that of the second population. The difference between the sample means was significant, as indicated by the p-value of 0.040, but this was not significant enough when testing at a 2% level of significance.*

7. Students should be randomly assigned to one of the two sections. To minimize confounding variables, it would be best if the same teacher was used for each section, and students should take the same final exam.

9. $\bar{x}_1 = 3547.67$, $s_1 = 3229.90$, $n_1 = 15$, $\bar{x}_2 = 5654.80$, $s_2 = 7530.65$, $n_2 = 15$, $\alpha = 0.01$.

Data are assumed to come from normal populations.

$H_0 : \mu_1 = \mu_2$; $H_1 : \mu_1 \neq \mu_2$.

$T.S. : t = \frac{(\bar{x}_1 - \bar{x}_2) - (\mu_1 - \mu_2)}{\sqrt{\frac{s_1^2}{n_1} + \frac{s_2^2}{n_2}}} = \frac{(3547.67 - 5654.8) - (0)}{\sqrt{\frac{3229.9^2}{15} + \frac{7530.65^2}{15}}} = -0.996,$

$V_1 = \frac{s_1^2}{n_1} = \frac{3229.9^2}{15} = 695483.60067$ and $V_2 = \frac{s_2^2}{n_2} = \frac{7530.65^2}{15} = 3780713.62817,$

$df = \frac{(V_1 + V_2)^2}{\frac{V_1^2}{n_1 - 1} + \frac{V_2^2}{n_2 - 1}} = 18.98$ which we round down to 18.

$C.V. : t = \pm 2.878$.

$p\text{-}value = 2 \cdot (1 - 0.835) = 2 \cdot 0.165 = 0.330$ (Here we rounded the T.S. to –1.0).

Once again, we fail to reject the null hypothesis, which is the claim we were asked to test. *There is insufficient evidence to warrant rejection the claim that the over-counted votes is equal to the mean under-counted votes. The difference between these sample means was insignificant, as indicated by the p-value of 0.330.*

11. $\bar{x}_1 = 2.82$, $s_1 = 1.47$, $n_1 = 11$, $\bar{x}_2 = 0.90$, $s_2 = 0.74$, $n_2 = 10$, $\alpha = 0.05$.

We have assumed the populations are normal.

$H_0 : \mu_1 = \mu_2$; $H_1 : \mu_1 > \mu_2$. This is a right-tail test.

$T.S. : t = \frac{(\bar{x}_1 - \bar{x}_2) - (\mu_1 - \mu_2)}{\sqrt{\frac{s_1^2}{n_1} + \frac{s_2^2}{n_2}}} = \frac{(2.82 - 0.90) - (0)}{\sqrt{\frac{1.47^2}{11} + \frac{0.74^2}{10}}} = 3.831$

$V_1 = \frac{s_1^2}{n_1} = \frac{1.47^2}{11} = 0.1964$ and $V_2 = \frac{s_2^2}{n_2} = \frac{0.74^2}{10} = 0.0548$, $df = \frac{(V_1 + V_2)^2}{\frac{V_1^2}{n_1 - 1} + \frac{V_2^2}{n_2 - 1}} = 15.05$

which we round down to 15.

$C.V. : t = 1.753$

$p\text{-}value = 0.001$.

This time we clearly reject the null hypothesis in favor of the alternative, which was our original claim. *The sample data support the claim that the mean remission is longer for patients taking the experimental drug than for patients receiving the placebo. The difference between these averages was highly significant, as indicated by the small p-value of 0.001.*

13. $\bar{x}_1 = 2.519$, $s_1 = 1.359$, $n_1 = 32 > 30$, $\bar{x}_2 = 3.625$, $s_2 = 1.431$, $n_2 = 32 > 30$, $\alpha = 0.04$.

$H_0 : \mu_1 = \mu_2$; $H_1 : \mu_1 < \mu_2$.

$T.S. : t = \frac{(\bar{x}_1 - \bar{x}_2) - (\mu_1 - \mu_2)}{\sqrt{\frac{s_1^2}{n_1} + \frac{s_2^2}{n_2}}} = \frac{(2.519 - 3.625) - (0)}{\sqrt{\frac{1.359^2}{32} + \frac{1.431^2}{32}}} = -3.170$,

$V_1 = \frac{s_1^2}{n_1} = \frac{1.359^2}{32} = 0.05771503$ and $V_2 = \frac{s_2^2}{n_2} = \frac{1.431^2}{32} = 0.06399253$,

$df = \frac{(V_1 + V_2)^2}{\frac{V_1^2}{n_1 - 1} + \frac{V_2^2}{n_2 - 1}} = 61.84$ which we round down to 61.

$C.V. : t = -1.787$ (using 50 degrees of freedom)

p-value $= 0.001$ (Using 60 degrees of freedom).

We reject the null hypothesis, which means we support the alternative, the original claim. *The data support the claim that statistics students study, on average, less than calculus students. This is emphasized by the p-value of 0.001 indicating a highly significant difference between the sample means.*

15. $\bar{x}_1 = 9.2$, $s_1 = 3.3$, $n_1 = 53 > 30$, $\bar{x}_2 = 7.9$, $s_2 = 3.0$, $n_2 = 51 > 30$, $\alpha = 0.02$.

$H_0 : \mu_1 = \mu_2$; $H_1 : \mu_1 > \mu_2$.

$T.S. : t = \frac{(9.2 - 7.9) - 0}{\sqrt{\frac{3.3^2}{53} + \frac{3.0^2}{51}}} = 2.58$.

$V_1 = \frac{s_1^2}{n_1} = \frac{3.3^2}{53} = 0.205$ and $V_2 = \frac{s_2^2}{n_2} = \frac{3.0^2}{51} = 0.176$. $df = \frac{(0.205 + 0.176)^2}{\frac{0.205^2}{52} + \frac{0.176^2}{50}} = 101.7$ which

we round down to 101.

$C.V. : t = 2.081$. (using 100 degrees of freedom)

p-value $= 1 - 0.995 = 0.005$. (using 100 degrees of freedom)

We reject the null hypothesis, which means we support the alternative, the original claim. *The data support the claim that quality preschool care increases the young adult reading achievement level. This is emphasized by the p-value of 0.007 indicating a highly significant difference between the sample means.*

17. $\bar{x}_1 = 25.93$, $s_1 = 0.95$, $n_1 = 30$, $\bar{x}_2 = 21.67$, $s_2 = 1.33$, $n_2 = 30$, $\alpha = 0.01$.

Both populations are assumed to be normal.

$H_0 : \mu_1 = \mu_2$; $H_1 : \mu_1 \neq \mu_2$. This is a two-tail test.

$T.S. : t = \frac{(\bar{x}_1 - \bar{x}_2) - (\mu_1 - \mu_2)}{\sqrt{\frac{s_1^2}{n_1} + \frac{s_2^2}{n_2}}} = \frac{(25.93 - 21.67) - 0}{\sqrt{\frac{0.95^2}{30} + \frac{1.33^2}{30}}} = 14.276$,

$V_1 = \frac{s_1^2}{n_1} = \frac{0.95^2}{30} = 0.03008$ and $V_2 = \frac{s_2^2}{n_2} = \frac{1.33^2}{30} = 0.05896$,

$df = \frac{(V_1 + V_2)^2}{\frac{V_1^2}{n_1 - 1} + \frac{V_2^2}{n_2 - 1}} = \frac{(0.03008 + 0.05896)^2}{\frac{0.03008^2}{30 - 1} + \frac{0.05896^2}{30 - 1}} = 52.48$ which we round down to 52.

$C.V. : t = \pm 2.678$

p-value $= 2 \cdot (1 - 0.999) = 2 \cdot 0.001 = 0.002$ (Using 50 degrees of freedom).

We reject the null hypothesis, which means we support the alternative, the original claim. *The data support the claim that the mean foot length for male college students is different from that of female college students. The difference between the sample means was highly significant, as indicated by the p-value of 0.002.*

Section 8.4 Solutions

1. $\bar{d} = 0.36286$, $s_d = .40596$, $n = 14$, $t_{\alpha/2} = 2.160$, $E = t_{\alpha/2}\frac{s_d}{\sqrt{n}} = 0.2344$, $0.1285 < \mu_d < 0.5973$.

3. $\hat{p}_1 = x_1/n_1 = 632/752 = 0.8404$, $n_1 = 752$, $\hat{p}_2 = x_2/n_2 = 681/783 = 0.8697$, and $n_2 = 783$.

   Each sample has more than 10 successes and 10 failures.
   The maximum error is:

$$E = z_{\alpha/2} \cdot \sqrt{\frac{\hat{p}_1\hat{q}_1}{n_1} + \frac{\hat{p}_2\hat{q}_2}{n_2}}$$
$$= 2.054 \cdot \sqrt{\frac{0.8404 \cdot 0.1596}{752} + \frac{0.8697 \cdot 0.1303}{783}}$$
$$= 0.03692,$$

and the corresponding confidence interval is computed below.

$$\hat{p}_1 - \hat{p}_2 - E < p_1 - p_2 < \hat{p}_1 - \hat{p}_2 + E$$

$$0.8404 - 0.8697 - 0.03692 < p_1 - p_2 < 0.8404 - 0.8697 + 0.03692$$

$$-0.0662 < p_1 - p_2 < 0.00762$$

5. $\bar{x}_1 = 1.93$, $s_1 = 1.34$, and $n_1 = 3647$. Also, $\bar{x}_2 = 1.75$, $s_2 = 1.35$, and $n_2 = 3638$. The requirements for this procedure are met since both sample sizes are greater than 30. First, we must compute the degrees of freedom for our critical value. We begin by computing $V_1 = \frac{s_1^2}{n_1} = \frac{1.34^2}{3647} = 0.0004923$, and $V_2 = \frac{s_2^2}{n_2} = \frac{1.35^2}{3638} = 0.0005010$. This leads us to the degrees of freedom:

$$df = \frac{(V_1 + V_2)^2}{\frac{V_1^2}{n_1-1} + \frac{V_2^2}{n_2-1}}$$
$$= \frac{(0.0005010 + 0.0004948)^2}{\frac{0.0005010^2}{3647-1} + \frac{0.0004948^2}{3638-1}}$$
$$= 7282.8,$$

which we must *truncate (round down)* to $df = 7282$. Unfortunately, the table does not include this value for the degrees of freedom, and it is so large that we find ourselves using the bottom row marked *Large*. Next we compute the maximum error,

$$E = t_{\alpha/2} \cdot \sqrt{\frac{s_1^2}{n_1} + \frac{s_2^2}{n_2}}$$
$$= 1.960 \cdot \sqrt{\frac{1.34^2}{3647} + \frac{1.45^2}{3638}}$$
$$= 0.06412.$$

Finally, we compute our confidence interval.

$$\bar{x}_1 - \bar{x}_2 - E < \mu_1 - \mu_2 < \bar{x}_1 - \bar{x}_2 + E$$

$$1.93 - 1.75 - 0.06412 < \mu_1 - \mu_2 < 1.93 - 1.75 + 0.06412$$

$$0.1159 < \mu_1 - \mu_2 < 0.2441.$$

B.9 Chapter 9 Solutions

Section 9.1 Solutions

1. We answer the questions as follows.

   a. The scatterplot is given first.

   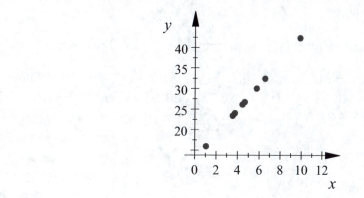

   b. The form of the relationship is linear.

   c. The direction of the relationship is positive

   d. The relationship appears perfectly strong.

   e. $\bar{x} = 5$, $s_x = 2.5679$

   f. $\bar{y} = 27.5$, $s_y = 7.7038$

   g. $r = 1$. The relationship is perfectly strong.

   h. The slope of the line of best fit is $b = \frac{r \cdot s_y}{s_x} = \frac{1 \cdot 7.7038}{2.5679} = 3$. The y-intercept is $a = \bar{y} - b\bar{x} = 27.5 - 3 \cdot 5 = 12.5$. The equation of the line of best fit is $\widehat{y} = 12.5 + 3x$.

3. We answer the questions as follows.

   a. The scatterplot is given first.

   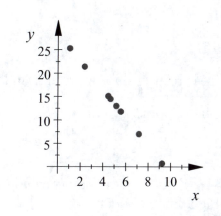

b. The form of the relationship is linear.

c. The relationship is negative.

d. The relationship appears perfectly strong.

e. $\bar{x} = 5$, $s_x = 2.5646$

f. $\bar{y} = 13.5$, $s_y = 7.6938$

g. $r = -1$. The linear relationship is perfectly strong and negative.

h. The slope of the line of best fit is $b = \frac{r \cdot s_y}{s_x} = \frac{-1 \cdot 7.6938}{2.5646} = -3$. The y–intercept is $a = \bar{y} - b\bar{x} = 13.5 - (-3)\,5 = 28.5$. The equation of the line of best fit is $\hat{y} = a + bx = 28.5 - 3x$.

5. We answer the questions as follows.

a. The scatterplot is given below.

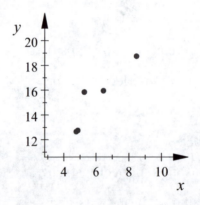

b. The form is linear.

c. The points exhibit a positive relationship.

d. The relationship seems strong.

e. $\bar{x} = 6$, $s_x = 1.5524$.

f. $\bar{y} = 15.14$, $s_y = 2.5540$.

g. $r = 0.9175$. The linear relationship between x and y is strong.

h. The slope of the line of best fit is $b = \frac{r \cdot s_y}{s_x} = \frac{0.9175 \cdot 2.5540}{1.5524} = 1.5095$. The y-intercept is $a = \bar{y} - b\bar{x} = 15.14 - 1.5095 \cdot 6 = 6.083$. The line of best fit is $\hat{y} = a + bx = 6.08 + 1.51x$.

i. $r^2 = 0.9175^2 = 0.842$. From this we may conclude that 84.2% of variation in y is explained by x using a line of best fit.

j. The predicted value of y corresponding to $x = 5$ is $\widehat{y} = 6.08 + 1.51 \cdot 5 = 13.63$.

7. We answer the questions as follows.

a. The scatterplot is given below.

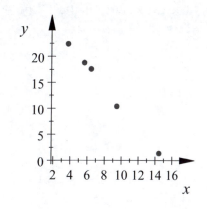

b. The form of the relationship is linear.

c. The points exhibit a negative relationship.

d. The relationship seems very strong.

e. $\overline{x} = 8$, $s_x = 4.1737$.

f. $\overline{y} = 14.02$, $s_y = 8.3398$.

g. $r = -0.9985$. The relationship between x and y is quite strong.

h. The slope of the line of best fit is $b = \frac{r \cdot s_y}{s_x} = \frac{-0.9985 \cdot 8.3398}{4.1737} = -1.9952$. The y-intercept is $a = \overline{y} - b\overline{x} = 14.02 - (-1.995)\,8 = 29.98$. The line of best fit is $\widehat{y} = a + bx = 29.98 - 2.00x$.

i. $r^2 = (-0.9985)^2 = 0.997$. From this we may conclude that 99.7% of variation in y is explained by x using a line of best fit.

j. The predicted value of y corresponding to $x = 11$ is $\widehat{y} = 29.98 - 2.00 \cdot 11 = 7.98$.

9.   The value of $r = 0.5712$ is positive, but not very close to 1. The relationship is moderately strong. We therefore give the regression line's equation: $\hat{y} = -1.0241 + 0.0971x$. The line is plotted on the scatterplot below. With $x = 55$, we predict a score of $\hat{y} = -1.0241 + 0.0971 \cdot 55 \approx 4.3$. The coefficient of determination is $r^2 = 0.5712^2 = 0.326$. This tells us that 32.6% of variability in consumer panel score is explained by the lightness measure using a line of best fit.

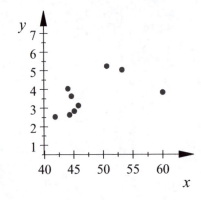

11.   The correlation coefficient is $r = 0.9721$, which indicates a strong positive relationship. Because of this, we give the equation of the regression line $\hat{y} = -25.618 + 41.129x$. With $x = 1.90$ we give our best estimate for y to be: $\hat{y} = -25.618 + 41.129 \cdot 1.90 = 52.53$. So with a mean GPA of 1.90, the success rate is predicted to be 52.5%. The scatterplot and regression line are plotted as follows. The coefficient of determination is $r^2 = 0.9721^2 = 0.945$. This tells us that 94.5% of variation in success rates in developmental math is explained by mean GPA using a line of best fit.

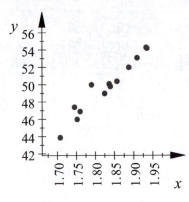

13. Here, $r = 0.7048$, and this is indicates a moderately strong positive relationship. Because of this, we give the equation of the regression line $\hat{y} = 2.5 + 0.84x$. With $x = 22$ we give our best estimate for y to be: $\hat{y} = 2.5 + 0.84 \cdot 22 = 20.98$. The scatterplot and regression line are plotted as follows. The coefficient of determination is $r^2 = 0.7048^2 = 0.4967$. This tells us that 49.7% of variability in tumor mass is explained by the CAT scan measurement using a line of best fit.

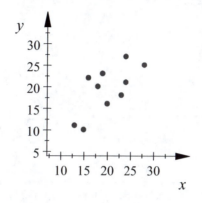

15. For this exercise, $r = 0.7076$. This is moderately strong and positive. Because of this, we give the equation of the regression line $\hat{y} = 35.449 + 0.462x$. With $x = 95$ we give our best estimate for y to be: $\hat{y} = 35.449 + 0.462 \cdot 95 = 79.34$. The scatterplot and regression line are plotted below. The coefficient of determination is $r^2 = 0.7076^2 = 0.5007$. From this we can say that about 50.1% of variability in overall percentage is explained by homework scores using a line of best fit.

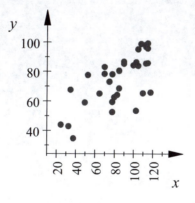

17.   The correlation coefficient is $r = 0.6228$. This is moderately strong. Because of this, we give the equation of the regression line $\hat{y} = 2.314 + 0.130x$. With $x = 159$ we give our best estimate for y to be: $\hat{y} = 2.314 + 0.130 \cdot 159 = 22.984$. The scatterplot and regression line are plotted below. The coefficient of determination is $r^2 = 0.6228^2 = 0.3879$. This tells us that 38.8% of variability in foot length is explained by height using a line of best fit.

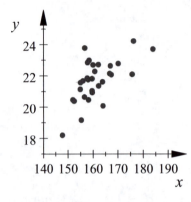

19.   For this example, $r = 0.8980$, indicating a strong positive trend, and thus the regression line, $\hat{y} = -67.164 + 0.0345x$. The scatterplot of these (x, y) pairs is as follows, with the regression line. The coefficient of determination is $r^2 = 0.8980^2 = 0.8064$. This tells us that 80.6% of variability in GPA is explained by year using a line of best fit.

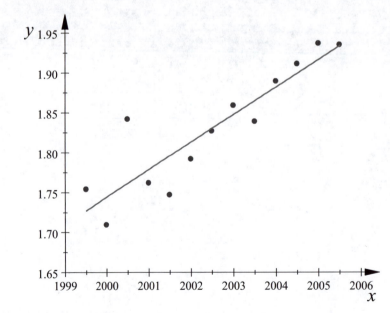

Section 9.2 Solutions

1. The mean of y values is $\overline{y} = 15.14$. The sample slope is $b = 1.51$, and the sample y-intercept is $a = 6.08$. The correlation coefficient is $r = 0.917$. With these, the relevant totals are computed as follows.

x	y	$\widehat{y}$	$(y - \widehat{y})^2$	$(y - \overline{y})^2$
5.3	15.8	14.083	2.948	0.436
4.9	12.7	13.479	0.607	5.954
8.5	18.7	18.915	0.046	12.674
6.5	15.9	15.895	0.000	0.578
4.8	12.6	13.328	0.530	6.452
$\sum$	–	–	4.131	26.094

From these we have $SSE = \sum (y - \widehat{y})^2 = 4.131$, and $SST = \sum (y - \overline{y})^2 = 26.094$. The correlation of determination is $r^2 = 1 - \frac{SSE}{SST} = 1 - \frac{4.131}{26.094} = 0.842$. If we compare this to the square of r above we have $r^2 = 0.917^2 = 0.841$. The difference between these values is due to rounding error. Finally, the standard deviation of residuals is $s_e = \sqrt{\frac{SSE}{n-2}} = \sqrt{\frac{4.131}{5-2}} = 1.173$.

3. The mean of y values is $\overline{y} = 14.02$. The sample slope is $b = -2$, and the sample y-intercept is $a = 29.98$. The correlation coefficient is $r = -0.998$. With these, the relevant totals are computed as follows.

x	y	$\widehat{y}$	$(y - \widehat{y})^2$	$(y - \overline{y})^2$
14.5	1.3	0.98	0.102	161.798
3.8	22.3	22.38	0.006	68.558
9.5	10.3	10.98	0.462	13.838
6.5	17.5	16.98	0.27	12.110
5.7	18.7	18.58	0.014	21.902
$\sum$	–	–	0.854	278.206

From these we have $SSE = \sum (y - \widehat{y})^2 = 0.854$, and $SST = \sum (y - \overline{y})^2 = 278.206$. The correlation of determination is $r^2 = 1 - \frac{SSE}{SST} = 1 - \frac{0.854}{278.206} = 0.997$. If we compare this to the square of r above we have $r^2 = (-0.998)^2 = 0.996$. The small difference between the r^2 values is due to rounding error. Finally, the standard deviation of residuals is $s_e = \sqrt{\frac{SSE}{n-2}} = \sqrt{\frac{0.854}{5-2}} = 0.533$.

5. Using technology, we compute $r^2 = 0.326$, and $s_e = 0.867$.

7. For this problem, $a = -25.6247$, $b = 41.1329$, $\sum y = 645.8$, $\sum y^2 = 32201.24$, and $\sum xy = 1185.015$. From these we may compute SSE.

$$SSE = \sum y^2 - a \sum y - b \sum xy$$
$$= 32201.24 - (-25.6247) \cdot 645.8 - 41.1329 \cdot 1185.015$$
$$= 6.568$$

The standard deviation of residuals is $s_e = \sqrt{\frac{SSE}{n-2}} = \sqrt{\frac{6.568}{13-2}} = 0.773$.

9. For this problem, $a = 2.5$, $b = 0.84$, $\sum y = 193$, $\sum y^2 = 4009$, and $\sum xy = 4028$. From these we may compute SSE.

$$SSE = \sum y^2 - a \sum y - b \sum xy$$
$$= 4009 - 2.5 \cdot 193 - 0.84 \cdot 4028$$
$$= 142.98$$

The standard deviation of residuals is $s_e = \sqrt{\frac{SSE}{n-2}} = \sqrt{\frac{142.98}{10-2}} = 4.228$.

Section 9.3 Solutions

1. This study is experimental. $H_0 : \rho = 0$, $H_1 : \rho \neq 0$; $n = 9$; $r = 0.941807$, $\alpha = 0.02$

$T.S. : t = \frac{r}{\sqrt{\frac{1-r^2}{n-2}}} = \frac{941807}{\sqrt{\frac{1-(941807)^2}{5-2}}} = 4.853$

$df = 5 - 2 = 3$.

$C.V. : t = \pm 4.541$

$p\text{-}value = 2 \cdot (1 - 0.992) = 2 \cdot 0.008 = 0.016$

Reject the null hypothesis. The data support the claim that the population of (x, y) pairs exhibits linear correlation. Since this is an experiment, we may infer that the treatment is the *cause* of the responses observed among the varying values of y.

3. This study is observational.

$H_0 : \rho = 0$, $H_1 : \rho \neq 0$; $n = 9$; $r = 0.5712$, $\alpha = 0.08$

$T.S. : t = \frac{r}{\sqrt{\frac{1-r^2}{n-2}}} = \frac{0.5712}{\sqrt{\frac{1-0.5712^2}{9-2}}} = 1.841$

$df = 9 - 2 = 7$.

$C.V. : t = \pm 2.046$

$p\text{-}value = 2 \cdot (1 - 0.943) = 0.114$

Fail to reject the null hypothesis. The data do not support the claim that the population of (x, y) pairs exhibits linear correlation.

5. This study is observational.

$H_0 : \rho = 0$, $H_1 : \rho \neq 0$; $n = 13$; $r = 0.9721$, $\alpha = 0.05$

$T.S. : t = \frac{r}{\sqrt{\frac{1-r^2}{n-2}}} = \frac{0.9721}{\sqrt{\frac{1-0.9721^2}{13-2}}} = 13.753$

$df = 13 - 2 = 11$.

$C.V. : t = \pm 2.201$

$p\text{-}value = 2 \cdot (1 - 0.999) = 0.002$ using our tables.

A more correct value is $p\text{-}value = 0.0000000283$ using computer software.

Reject the null hypothesis. The data support the claim that the population of (x, y) pairs exhibits linear correlation.

7. This study is observational. $H_0 : \rho = 0$, $H_1 : \rho \neq 0$; $n = 10$; $r = 0.7048$, $\alpha = 0.06$.

   $T.S. : t = \dfrac{r}{\sqrt{\frac{1-r^2}{n-2}}} = \dfrac{0.7048}{\sqrt{\frac{1-0.7048^2}{10-2}}} = 2.810$

   $df = 10 - 2 = 8$; $C.V. : t = \pm 2.189$

   $p\text{-value} = 2 \cdot (1 - .988) = 0.024$.

   Reject the null hypothesis. The data support the claim that the population of (x, y) pairs exhibits linear correlation.

9. This study is observational.

   $H_0 : \rho = 0$, $H_1 : \rho \neq 0$; $n = 35$; $r = 0.7076$, $\alpha = 0.04$

   $T.S. : t = \dfrac{r}{\sqrt{\frac{1-r^2}{n-2}}} = \dfrac{0.7076}{\sqrt{\frac{1-0.7076^2}{35-2}}} = 5.752$

   $df = 35 - 2 = 33$; $C.V. : t = \pm 2.138$

   $p\text{-value} = 2 \cdot (1 - 0.999) = 0.002$ using our tables.

   A more correct value is $p\text{-value} = 0.00000200$ using computer software.

   Reject the null hypothesis. The data support the claim that the population of (x, y) pairs exhibits linear correlation.

11. This study is observational.

    $H_0 : \rho = 0$, $H_1 : \rho \neq 0$; $n = 9$; $r = 0.6228$, $\alpha = 0.05$

    $T.S. : t = \dfrac{r}{\sqrt{\frac{1-r^2}{n-2}}} = \dfrac{0.6228}{\sqrt{\frac{1-0.6228^2}{30-2}}} = 4.212$

    $df = 30 - 2 = 28$.

    $C.V. : t = \pm 2.048$

    $p\text{-value} = 2 \cdot (1 - 0.999) = 0.002$

    Reject the null hypothesis. The data support the claim that the population of (x, y) pairs exhibits linear correlation.

13. The summary statistics are: $n = 14$, $\sum x = 418.4$, $\sum x^2 = 14013.62$, $\sum y = 104.4$, $\sum y^2 = 996.5$, and $\sum xy = 2640.86$. The sample slope and y-intercept are $b = -0.3175$ and $a = 16.9451$. The standard deviation of sample slope values is $s_b = 0.0603$.

$$
\begin{aligned}
s_b &= \sqrt{\frac{\sum y^2 - a \sum y - b \sum xy}{(n-2)\left(\sum x^2 - \frac{1}{n}\left(\sum x\right)^2\right)}} \\
&= \sqrt{\frac{996.5 - 16.9451 \cdot 104.4 - (-0.3175) \cdot 2640.86}{(14-2)\left(14013.62 - \frac{1}{14} \cdot 418.4^2\right)}} \\
&= 0.0603
\end{aligned}
$$

The null hypothesis is $H_0 : \beta = 0$, while the alternative is $H_1 : \beta < 0$.

$$
\begin{aligned}
T.S. : t &= \frac{b - \beta}{s_b} \\
&= \frac{-0.3175 - 0}{0.0603} \\
&= -5.265
\end{aligned}
$$

This right-tail test has $n - 2 = 12$ degrees of freedom, with critical value $C.V. : t = -2.681$. The test statistic is less than this critical value, so we will reject the null hypothesis. The *p-value is next.*

$$p\text{-}value = 0.002$$

This is highly significant. The sample data support the claim that the population slope parameter is less than zero.

15. The summary statistics are: $n = 10$, $\sum x = 404$, $\sum x^2 = 18448$, $\sum y = 545$, $\sum y^2 = 31993$, and $\sum xy = 23198$. The sample slope and y-intercept are $b = 0.5549$ and $a = 32.0809$. The standard deviation of sample slope values is $s_b = 0.3101$.

$$s_b = \sqrt{\frac{\sum y^2 - a \sum y - b \sum xy}{(n-2)\left(\sum x^2 - \frac{1}{n}\left(\sum x\right)^2\right)}}$$

$$= \sqrt{\frac{31993 - 32.0809 \cdot 545 - 0.5549 \cdot 23198}{(10-2)\left(18448 - \frac{1}{10} \cdot 404^2\right)}}$$

$$= 0.3101$$

The null hypothesis is $H_0 : \beta = 0.15$, while the alternative is $H_1 : \beta > 0.15$.

$$T.S. : t = \frac{b - \beta}{s_b}$$

$$= \frac{0.5549 - 0.15}{0.3101}$$

$$= 1.306$$

This right-tail test has $n - 2 = 8$ degrees of freedom, with critical value $C.V. : t = 2.004$. The test statistic is not greater than the critical value, so we will fail to reject the null hypothesis. The *p-value is next.*

$$p\text{-}value = 1 - 0.885$$

$$= 0.115$$

This is insignificant. The sample data do not support the claim that the population slope parameter is greater than 0.15.

17. For the confidence interval, $t_{\alpha/2} = 2.179$. Using $s_b = 0.0603$ and $b = -0.3175$ from Problem 13 gives the following interval.

$$b - t_{\alpha/2} \cdot s_b < \beta < b + t_{\alpha/2} \cdot s_b$$

$$-0.3175 - 2.179 \cdot 0.0603 < \beta < -0.3175 + 2.179 \cdot 0.0603$$

$$-0.4489 < \beta < -0.1861$$

19. For the confidence interval, $t_{\alpha/2} = 2.449$. Using $s_b = 0.3101$ and $b = 0.5549$ from Problem 15 gives the following interval.

$$b - t_{\alpha/2} \cdot s_b < \beta < b + t_{\alpha/2} \cdot s_b$$

$$0.5549 - 2.449 \cdot 0.3101 < \beta < 0.5549 + 2.449 \cdot 0.3101$$

$$-0.2045 < \beta < 1.3143$$

Section 9.4 Solutions

1. Nine values, normally distributed, would be separated by ten regions under the normal distribution of equal area. Since the total area is 1, each region would have area equal to $1/10 = 0.1000$. The areas accumulate as we move across the distribution, so the cumulative areas are 0.1000, 0.2000, 0.3000, 0.4000, 0.5000, 0.6000, 0.7000, 0.8000, and 0.9000. Looking up the z-scores which correspond to these gives z: { -1.28, -0.84, -0.52, -0.25, 0.00, 0.25, 0.52, 0.84, 1.28 }.

3. The normal quantile plot results are given below.

Sorted Sample Data	Corresponding z-score
2.8	-1.56472647
5.5	-1.18683143
5.5	-0.92889949
5.5	-0.72152228
5.6	-0.54139509
5.6	-0.37739194
6.8	-0.22300783
9.6	-0.07379127
10.5	0.07379127
11	0.22300783
12	0.37739194
12.8	0.54139509
13.3	0.72152228
15.5	0.92889949
19	1.18683143
19.5	1.56472647

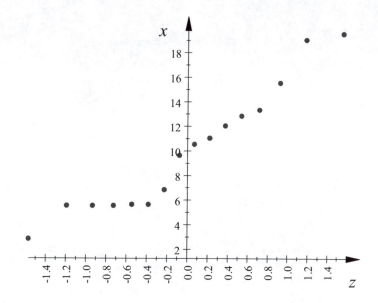

The critical value, using $\alpha = .01$ is $r = 0.9153$. The correlation coefficient is $r = 0.9665$. The sample does not give sufficient evidence to warrant rejection of the claim that it is sampled from a normal population.

5. The normal quantile plot results are given below.

Sorted Sample Data	Corresponding z-score
1.7	-1.84859629
2.0	-1.51792916
2.0	-1.30015343
2.2	-1.13097761
2.5	-0.98916863
2.5	-0.86489436
2.6	-0.75272879
2.7	-0.64932391
2.7	-0.55244258
2.8	-0.46049454
2.8	-0.37228936
2.9	-0.28689392
3.0	-0.20354423
3.1	-0.12158738
3.1	-0.04044051
3.1	0.04044051
3.2	0.12158738
3.2	0.20354423
3.3	0.28689392
3.4	0.37228936
3.5	0.46049454
3.6	0.55244258
3.7	0.64932391
3.7	0.75272879
3.7	0.86489436
3.7	0.98916863
3.8	1.13097761
3.9	1.30015343
4.0	1.51792916
4.0	1.84859629

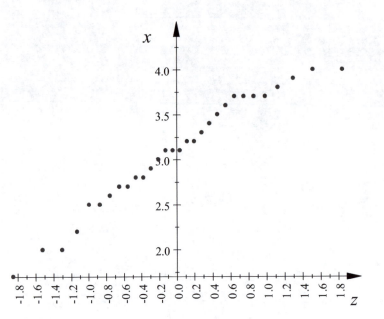

The critical value, using $\alpha = .01$ is $r = 0.9490$. The correlation coefficient is $r = 0.9860$. Sample data do not give sufficient evidence to warrant rejection of the claim that these values come from a normal population.

7. The normal quantile plot results are given below.

Sorted Sample Data	Corresponding z-score
170	-1.53412054
520	-1.15034938
790	-0.88714656
890	-0.67448975
994	-0.48877641
1039	-0.31863936
1134	-0.15731068
2124	0.00000000
2531	0.15731068
3640	0.31863936
4258	0.48877641
7826	0.67448975
17833	0.88714656
19218	1.15034938
21855	1.53412054

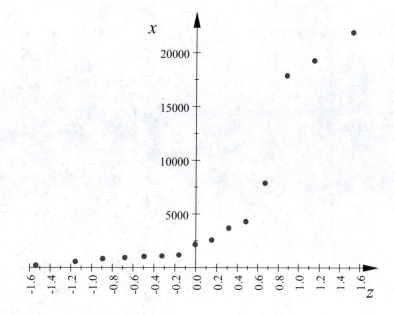

The critical value, using $\alpha = .01$ is $r = 0.9109$. The correlation coefficient is $r = 0.8430$. Sample data do not support the claim that these values come from a normal population.

9.   The critical value, using $\alpha = .01$ is $r = 0.9948$. The correlation coefficient is $r = 0.9935$.

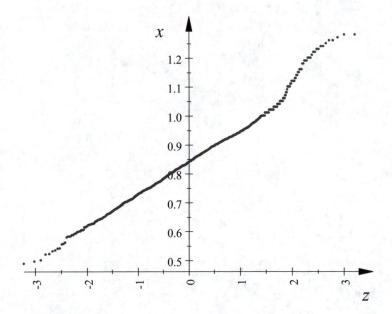

Sample data do not support the claim that these values come from a normal population.

B.10   Chapter 10 Solutions

Section 10.1 Solutions

1. H_0: The proportions of frequencies for the entire population are the same for each category.

H_1: At least one proportion is different from the others.

The sample size is $n = \sum O = 86$. Dividing this into equal proportions amongst 5 classes gives an expected frequency of $E = 86/5 = 17.2$.

O	E	$(O-E)^2/E$
13	17.2	1.025581395
25	17.2	3.537209302
11	17.2	2.234883721
18	17.2	0.037209302
19	17.2	0.188372093
Total:		7.023255813

Each expected frequency is at least 5, as required.

The test statistic is $TS : \chi^2 = \sum (O-E)^2/E = 7.0233$. There are $k = 5$ categories resulting $df = 4$. The level of significance is $\alpha = 0.05$, the critical value is $CV : \chi^2 = 9.4877$. Using computer software, *p-value* = 0.1347.

We fail to reject the null hypothesis. *The data are insufficient to allow rejection of the claim that the population proportions corresponding to the five given categories are all the same.*

3. H_0: The proportions of students signing up for Math 72 are the same for each semester.

H_1: At least one proportion is different from the others.

The sample size is $n = \sum O = 530$. Dividing this into equal proportions amongst classes gives an expected frequency of $E = 530/7 = 75.714$

O	E	$(O-E)^2/E$
41	75.714	15.91617251
75	75.714	0.006738544
159	75.714	91.61428571
67	75.714	1.00296496
112	75.714	17.38975741
37	75.714	19.79541779
39	75.714	17.80296496
Total:		163.528301

Each expected frequency is at least 5, as required.

The test statistic is $TS : \chi^2 = \sum (O - E)^2 / E = 163.5283$. There are $k = 7$ categories resulting $df = 6$. The level of significance is $\alpha = 0.05$, the critical value is $CV : \chi^2 = 12.5916$. Using computer software, $p\text{-}value = 1.05925 \cdot 10^{-32}$.

We reject the null hypothesis. *The data do not support the claim that the proportions of students signing up for Math 72 were the same for each semester offered.*

5. H_0 : The sample is drawn from a normal population, so proportions in the specified intervals match the empirical rule.

H_1 : The values are not distributed as determined by the empirical rule.

O	E	$(O - E)^2 / E$
29	38.30	2.258
198	206.82	0.376
543	520.88	0.939
564	520.88	3.570
153	206.82	14.005
45	38.30	1.172
Total:		22.321

Each expected frequency is at least 5, as required.

The test statistic is $\chi^2 = 22.321$. There are $k = 6$ categories resulting $df = 5$. $\alpha = 0.005$. The critical value is $\chi^2 = 16.7496$. Using computer software, $p\text{-}value = 0.000455$.

7. H_0: The coin is unbiased, so proportions of numbers of heads out of five should be those predicted by the binomial distribution with $n = 5$, and $p = 0.5$.

H_1: The coin is biased, so the proportions do not match those predicted by the binomial distribution.

O	E	$(O - E)^2 / E$
4	5.0	0.2000
34	38.8	0.5938
124	136.3	1.1010
289	273.8	0.8438
327	341.3	0.5992
287	273.8	0.6364
127	136.3	0.6346
51	38.8	3.8361
7	5.0	0.8000
Total:		9.2538

Each expected frequency is at least 5, as required.
The test statistic is

$$TS : \chi^2 = \frac{\sum (O - E)^2}{E}$$
$$= 9.2538.$$

There are $k = 8$ categories resulting $df = 7$. The level of significance is $\alpha = 0.05$. The critical value is $CV : \chi^2 = 15.5073$. Using computer software, $p\text{-}value = 0.3213$. We fail to reject the null hypothesis. *We are unable to reject the claim that the coin is unbiased, so proportions of numbers of heads out of 8 are not significantly different from those predicted by the binomial distribution with $n = 8$, and $p = 0.5$.*

Section 10.2 Solutions

1. H_0: Marijuana use is independent of political views.
   H_1: They are dependent.

Expected Frequencies			
	Never	*Rarely*	*Frequently*
Liberal	494.3773	151.4566	125.166
Conservative	176.9755	54.21794	44.80652
Other	193.6471	59.32543	49.02743

Each expected frequency is at least 5 as required.

$(O - E)^2 / E$			
	Never	*Rarely*	*Frequently*
Liberal	0.478302	3.064353	0.303758
Conservative	7.745764	0.960912	19.82811
Other	2.41986	3.459189	26.39392

The test statistic is $TS : \chi^2 = \sum (O - E)^2 / E = 64.6542$. There are 3 rows and 3 columns resulting in $df = (3 - 1)(3 - 1) = 4$. $\alpha = 0.01$. The critical value is $CV : \chi^2 = 13.2767$; using software, $p\text{-}value = 3.04 \cdot 10^{-13} = 0.000000000000304$. Reject the null hypothesis. The data do not support the claim that marijuana use is independent of political views. The dependency is highly significant and indicated by the extraordinarily small p-value.

3. H_0: Grades are independent of gender.
   H_1: They are dependent.

Expected Frequencies		
	female	male
A	8.35	9.65
B	51.49	59.51
C	55.67	64.33
D	19.48	22.52

Each expected frequency is at least 5 as required.

$(O-E)^2/E$		
	female	*male*
A	0.050599	0.043782
B	0.589631	0.510168
C	0.031775	0.027497
D	2.872197	2.484476

The test statistic is $TS : \chi^2 = \sum (O-E)^2/E = 6.6101$. There are 4 rows and 2 columns resulting in $df = (4-1)(2-1) = 3$. $\alpha = 0.01$. The critical value is $CV : \chi^2 = 11.3449$; using computer software, *p-value* $= 0.0854$.

Fail to reject the null hypothesis. *The data make it evident that grades are not dependent on gender.*

5. H_0: Survival is independent of gender.

   H_1: They are dependent.

Expected Frequencies		
	Survived	*Died*
Male	29.02326	22.97674
Female	18.97674	15.02326

Each expected frequency is at least 5 as required.

$(O-E)^2/E$		
	Survived	*Died*
Male	1.250019	1.578971
Female	1.911793	2.414897

The test statistic is $TS : \chi^2 = \sum (O-E)^2/E = 7.1557$. There are 2 rows and 2 columns resulting in $df = (2-1)(2-1) = 1$. $\alpha = 0.01$. The critical value is $CV : \chi^2$

$= 6.6349$; using computer software, p-value $= 0.00747$. Reject the null hypothesis. The data give evidence that survival is dependent on gender.

7. H_0: Car accidents are independent of gender.

   H_1: They are dependent.

Expected Frequencies		
	Yes	No
Female	17.5	12.5
Male	17.5	12.5

Each expected frequency is at least 5 as required.

$(O - E)^2/E$		
	Yes	No
Female	0.3571429	0.5
Male	0.3571429	0.5

The test statistic is $TS : \chi^2 = \sum (O - E)^2/E = 1.7143$. There are 2 rows and 2 columns resulting in $df = (2 - 1)(2 - 1) = 1$. $\alpha = 0.05$. The critical value is $CV : \chi^2 = 3.8415$; using software, p-value $= 0.1904$.

Fail to reject the null hypothesis. *The data support the claim that car accidents are independent of gender.*

9. H_0: Course grades are independent of units attempted.

   H_1: They are dependent.

Expected Frequencies			
	A	B	$\leq C$
0 – 12 Units	15.25	16.67	7.08
12 – 13 Units	52.80	57.70	24.50
> 13 Units	28.94	31.63	13.43

Each expected frequency is at least 5 as required.

$(O - E)^2/E$			
	A	B	$\leq C$
0 – 12 Units	2.5615	1.1247	0.5207
12 – 13 Units	1.1523	0.4868	0.2551
> 13 Units	6.8308	2.9319	1.4613

The test statistic is $TS : \chi^2 = \sum (O - E)^2 / E = 17.3251$. There are 3 rows and 3 columns resulting in $df = (3 - 1)(3 - 1) = 4$. $\alpha = 0.05$. The critical value is $CV : \chi^2 = 9.4877$; using software, $p\text{-value} = 0.00167$.

Reject the null hypothesis. *The data do not support the claim that course grades are independent of units attempted.*

B.11   Chapter 11 Solutions

Section 11.1 Solutions

1. $H_0 : \mu_1 = \mu_2 = \mu_3 = \mu_4$
    H_1: At least one mean differs from the others.

Sample Data		
$\bar{x}$	s^2	n
60.857143	60.142857	7
65.857143	64.142857	7
64.285714	88.571429	7
62.857143	28.142857	7

N	$GrandMean : \bar{\bar{x}}$
28	63.46428571
k	$SS(treatment)$
4	94.96428571
Test Stat.	$MS(treatment)$
0.5254	31.6547619
C.V.	$SS(error)$
3.0088	1446
p-value	$MS(error)$
0.66902581	60.25

$T.S. : F = \frac{n \cdot s_{\bar{x}}^2}{s_p^2} = \frac{7 \cdot 4.5221}{60.25} = 0.525390239.$
$df_1 = k - 1 = 4 - 1 = 3$
$df_2 = k \cdot (n - 1) = 4 \cdot (7 - 1) = 24$
$\alpha = 0.05$
$C.V. : F = 3.0088$
p-value $= 0.6690$ (using computer software).
We fail to reject the null hypothesis. *There is insufficient evidence to warrant rejection of the claim that these populations of bacteria colony sizes have equal means. This is emphasized by the large p-value of 0.6690, which is calculated using computer software.*

3. $H_0 : \mu_1 = \mu_2 = \mu_3 = \mu_4 = \mu_5 = \mu_6 = \mu_7$
    H_1 : At least one mean differs from the others.

Sample Data		
$\bar{x}$	s^2	n
22.29	0.8849984	10
22.91	0.9550451	10
23.1	1.2534397	10
22.81	0.595259	10
22.14	0.641526	10
23.17	0.9900056	10
21.18	0.8011103	10

N	$Grand\ Mean : \bar{\bar{x}}$
70	22.51428571
k	$SS(treatment)$
7	29.87771429
$Test\ Stat.$	$MS(treatment)$
5.694354788	4.979619048
$C.V.$	$SS(error)$
3.102766752	55.09245765
$p\text{-}value$	$MS(error)$
$8.90588 \cdot 10^{-05}$	0.874483455

$T.S. : F = \frac{n \cdot s_{\bar{x}}^2}{s_p^2} = \frac{10 \cdot 0.497962}{0.874483} = 5.6944$

$df_1 = k - 1 = 7 - 1 = 6$

$df_2 = k \cdot (n - 1) = 7 \cdot (10 - 1) = 63$

$\alpha = 0.01$

$C.V. : F = 3.1864$ (This is based on $df_2 = 50$ degrees of freedom – the correct C.V. is above).

$p\text{-}value = 0.0000891$ (using computer software).

We reject the null hypothesis. *The data do not support the claim that the lengths of eggs in these different populations have equal means. This conclusion is strongly supported by the p-value of 0.0000891, which was computed using computer software.*

5. $H_0 : \mu_1 = \mu_2 = \mu_3 = \mu_4 = \mu_5$

  H_1 : At least one mean differs from the others.

Sample Data		
$\bar{x}$	s^2	n
98.3	1.21	7
96.2	1.22333333	7
96.83	0.99555714	7
95.3	1.22333333	7
96.2	1.28666667	7

N	k
35	5
$SS(treatment)$	$MS(treatment)$
34.64044571	8.660111429
$SS(error)$	$MS(error)$
35.63334286	1.187778095
$Test\ Stat.$	$C.V.$
7.291017963	4.017876837
$p\text{-}value$	$GrandMean : \bar{\bar{x}}$
0.000316169	96.56657143

$T.S. : F = \frac{n \cdot s_{\bar{x}}^2}{s_p^2} = \frac{7 \cdot 1.237159}{1.187778} = 7.2910$

$df1 = k - 1 = 5 - 1 = 4$

$df2 = k \cdot (n - 1) = 5 \cdot (7 - 1) = 30$

$\alpha = 0.01$

$C.V. : F = 4.0179$

$p\text{-}value = 0.000316$ (using computer software).

We reject the null hypothesis. *The data do not support the claim that the temperature responses to the different treatments have equal means. This conclusion is strongly supported by the p-value of 0.000316, which was computed using computer software.*

B.12   Chapter 12 Solutions

Section 12.1 Solutions

1. $n = 19$, $\alpha = 0.05$. $H_0 : \tilde{\mu} = 55$, $H_1 : \tilde{\mu} < 55$. The number of $(+)$ signs is $s = 6$. With $n > 10$ we use the normal approximation and compute

$$T.S. : z = \frac{(s + 0.5) - 0.5 \cdot n}{0.5 \cdot \sqrt{n}} = \frac{(6 + 0.5) - 0.5 \cdot 19}{0.5 \cdot \sqrt{19}} = -1.38.$$

From this we get $p\text{-}value = P(x_s \leq s) = P(z < -1.38) = 0.0838$. $C.V. : z = -1.645$. Since our p-value is not less than $\alpha = 0.01$, we will fail to reject the null hypothesis. The data do not support the claim that the median peanut butter score is less than 55.

3. $n = 8$, $\alpha = 0.05$. $H_0 : \tilde{\mu} = 20$, $H_1 : \tilde{\mu} > 20$. The number of $(-)$ signs is $s = 2$, and since $n \leq 10$, we use this as our test statistic. We look up the $p\text{-}value$ on Table A8: $p\text{-}value = P(x_s \leq s) = 0.144531$. $C.V. : x_s = 1$. Since our test statistic is greater than our critical value, we will fail reject the null hypothesis. The data do not support the claim that the median blood lactic acid level is greater than 20 for people after exercising the equivalent of 3 racquetball games.

5. $n = 30$, $\alpha = 0.04$. $H_0 : \tilde{\mu} = 28$, $H_1 : \tilde{\mu} < 32$. This is a left tail test, so we count the number of $(+)$ signs, which is $s = 10$. With $n > 10$ we use the normal approximation and compute

$$
\begin{aligned}
T.S. : z &= \frac{(s + 0.5) - 0.5 \cdot n}{0.5 \cdot \sqrt{n}} \\
&= \frac{(10 + 0.5) - 0.5 \cdot 30}{0.5 \cdot \sqrt{30}} \\
&= -1.643.
\end{aligned}
$$

From this we get $p\text{-}value = P(x_s \leq s) = P(z < -1.64) = 0.0505$. $C.V. : z = -1.751$. Our $T.S.$ is not less than the $C.V.$, so we fail to reject the null hypothesis. The data do not support the claim that the median hours worked per week at students of this college is less than 32.

Section 12.2 Solutions

1. $n = 19$, $\alpha = 0.05$, $H_0 : \mu = 45$, and $H_1 : \mu < 55$, $T.S. : R = R_{(+)} = 33$, p-$value = 0.0054$. Critical value: $x_R = 54$. We reject the null hypothesis in favor of the alternative. The data support the claim that the mean score for peanut butter is less than 55.

x	$|x - 55|$	$Ranks$	$Sign$
22	33	19	$(-)$
30	25	17	$(-)$
30	25	18	$(-)$
36	19	16	$(-)$
39	16	15	$(-)$
40	15	13	$(-)$
40	15	14	$(-)$
41	14	12	$(-)$
44	11	10	$(-)$
45	10	8	$(-)$
50	5	5	$(-)$
50	5	6	$(-)$
53	2	4	$(-)$
56	1	1	$(+)$
56	1	2	$(+)$
56	1	3	$(+)$
62	7	7	$(+)$
65	10	9	$(+)$
68	13	11	$(+)$

3. $n = 8$, $\alpha = 0.05$, $H_0 : \mu = 45$, and $H_1 : \mu > 20$, $T.S. : R = R_{(-)} = 5$, p-$value = 0.0391$. Critical value: $x_R = 6$. We reject the null hypothesis in favor of the alternative. The data support the claim that the mean blood lactic acid level is greater than 20 after exercise equivalent to three games of racquetball.

x	18	37	40	35	30	20	33	19
$|x - 20|$	2	17	20	15	10	0	13	1
$Ranks$	3	7	8	6	4	1	5	2
$Sign$	$(-)$	$(+)$	$(+)$	$(+)$	$(+)$	$none$	$(+)$	$(-)$

5. $n = 30$, $\alpha = 0.06$, $H_0 : \mu = 45$, and $H_1 : \mu < 32$, $R = R_{(+)} = 104$,

$$
\begin{aligned}
T.S. : z &= \frac{R - n(n+1)/4}{\sqrt{n(n+1)(2n+1)/24}} \\
&= \frac{104 - 30(30+1)/4}{\sqrt{30(30+1)(2 \cdot 30 + 1)/24}} \\
&= -2.64,
\end{aligned}
$$

p-value $= 0.0041$. Critical value $z = -1.555$. We reject the null hypothesis in favor of the alternative. The data support the claim that the mean hours worked by students at this college is less than 32 hours per week.

x	$|x - 32|$	Ranks	Sign
15	17	25	$(-)$
20	12	17	$(-)$
40	8	10	$(+)$
16	16	22	$(-)$
20	12	18	$(-)$
25	7	7	$(-)$
19	13	21	$(-)$
40	8	11	$(+)$
40	8	12	$(+)$
20	12	19	$(-)$
20	12	20	$(-)$
40	8	13	$(+)$
0	32	26	$(-)$
40	8	14	$(+)$
35	3	3	$(+)$
30	2	1	$(-)$
35	3	4	$(+)$
25	7	8	$(-)$
29	3	5	$(-)$
0	32	27	$(-)$
0	32	28	$(-)$
40	8	15	$(+)$
16	16	23	$(-)$
0	32	29	$(-)$
40	8	16	$(+)$
30	2	2	$(-)$
25	7	9	$(-)$
16	16	24	$(-)$
35	3	6	$(+)$
0	32	30	$(-)$

Index

p-value
    informal, 151
z-score, 97

analysis of variance, 539

addition rule, 136
alternative hypothesis, 292
analysis of variance
    one-way, 539
arithmetic mean, 65

bar chart, 3, 5
bell-shaped, 50
Bernoulli, Jacob, 181
bias, 12
    measurement bias, 13
    non-response bias, 13
    response bias, 13
    selection bias, 12
binomial experiment
    defined, 173
    normal approximation, 239
    skewness and kurtosis, 180
bivariate data, 401
blinding, 34
blocks, 34
boxplot, 102
bubble plot, 503

categorical data, 3, 6
census, 3
central limit theorem
    sample means, 266
    sample proportions, 247
Chebyshev's theorem, 91
class boundaries, 46

class limits, 46
class midpoints, 46
class width, 45
cluster sampling, 30
coefficient of determination, 415
combination, 174
complement, 136
compound event, 125
conditional probability, 129, 139
confidence interval
    difference of means, 396
    difference of proportions, 394
    generalized, 235
    mean difference-matched pairs, 392
    population mean (σ known), 273
    population mean (σ unknown), 283
    population proportion, 255
conflicts of interest, 14
confounding, 32
confounding variable, 32
contingency table, 523
continuity correction, 240
continuous quantitative data, 6
continuous random variable, 203
control group, 34
controlled experiment, 35
convenience samples, 12
correlate, 401
correlation coefficient, 416
correlation vs. causation, 14
critical value, 219, 293
cross-sectional studies, 31

data, 1
De Moivre, Abraham, 210
degrees of freedom, 79

descriptive statistics, 3
deviation
    explained, 406
    total, 406
    unexplained, 406
deviations from the mean, 77
discrete probability distribution, 148
discrete quantitative data, 6
discrete random variable, 147
distorted representations, 15
distribution, 3
distributional thinking, 3
dotplot, 3
double blind, 35

empirical rule, 89
    using z-scores, 98
error, 11
    non-sampling error, 12
    sampling error, 11
event, 124
    compound, 125
    simple, 125
expected value
    discrete distribution, 149
    finite population, 66
experiment, 124
experimental blocks, 34
experimental study, 32
explained deviation, 406
explanatory variable, 32, 401

Fisher's kurtosis
    population, 167
    sample, 104
Fisher, Ronald A., 537
five point summary, 102
frequency tables, 44

Galton, Francis, 502
goodness of fit, 507

histogram, 44
histograms, 46

independent events, 137

inference, 2
inferential statistics, 3
interquartile range, 76
interval data, 7

kurtosis
    population, 166
    sample, 104

left tail, 219
left-tail test, 292
level of confidence, 234
level of significance, 291
levels of measurement, 7
line of best fit, 401
linear correlation, 401
linear correlation coefficient, 416
loaded questions, 14
longitudinal studies, 32
lower class limits, 46

mean
    finite population, 66
    of a discrete probability distribution, 149
    sample
      listed data, 67
      sample-frequency table, 68
mean absolute deviation, 79
median, 69
midrange, 70
multinomial experiment, 509
multiplication rule, 140
mutually exclusive, 135

nominal data, 7
normal distribution, 205
    density function, 207
    skewness and kurtosis, 210
    standard normal, 209
normal quantile plot, 479
normal sample, 50
null hypothesis, 291

observational study, 31
ordinal data, 7
outlier, 60

outliers, 110

p-value
    formal, 294
    interpretation, 297
parameter, 2
Pearson's kurtosis
    population, 166
    sample, 104
Pearson's linear correlation coefficient, 416
Pearson, Karl, 507
percentile, 96
pie chart, 54
placebo, 33
placebo effect, 33
Poisson distribution
    definition, 184
    skewness and kurtosis, 188
Poisson, Siméon Denis, 188
population, 2
power, 299
probability
    addition rule, 136
    basic definition, 125
    classical approach, 126
    conditional, 129, 139
    density function, 203
    discrete distribution, 148
    estimates using relative frequencies, 127
    multiplication rule, 140
    normal distribution, 205
    rule of complements, 136
    unconditional, 129
    uniform distribution, 203
prospective studies, 32

quadratic mean, 72
qualitative data, 6
quantitative data, 3, 6
quartile
    lower quartile, 76
    upper quartile, 76

random sample, 29
random selection, 29
random variable

continuous, 203
    discrete, 147
randomized block design, 36
range, 76
range rule of thumb, 88
ratio data, 7
refusal rate, 13
regression
    SSE, 496
    assumptions, 453
regression line, 401
relative frequencies, 50
relative frequency histogram, 50
residual plot, 440
response, 32
response variable, 32, 401
retrospective studies, 31
right tail, 219
right-tail test, 292
rounding rule
    correlation statistics, 421
    probabilities, proportions, percents, 126
    sample mean, 67
    standard deviation, 84
    variance, 84
rule for significance
    using standard deviations, 91
    using z-scores, 98
rule for unusualness
    using standard deviations, 91
    using z-scores, 98
rule of complements, 136

sample size
    population mean, 275
    population proportion, 258
sample space, 125
sampling distribution, 233
    differences between proportions, 362
    sample means, 266
    sample proportions, 247
sampling frame, 31, 300
scatterplot, 402
self interest studies, 14
self-selecting samples, 12

Signed Rank Test, 573
Signs Test, 557
simple event, 125
simple random sample, 29
skewed left, 50
skewed right, 50
skewness
    from a sample, 104
    population, 166
SSE, 407
SSR, 414
SST, 413, 444
standard deviation
    discrete distribution, 149
    finite population, 80
    sample-computational formula , 85
    sample-frequency tables, 87
    sample-listed data, 80
standard error, 233
standard score, 97
statistic, 2
statistical ethics, 17
statistics, 1
stem and leaf, 53
stratified sampling, 30
Student's t-distribution, 280
sum of squares
    error (SSE), 407
    regression (SSR), 414
    total (SST), 413, 444
systematic sampling, 31

test statistic
    χ^2 goodness of fit, 512
    analysis of variance, 542
    correlation coefficient, 454, 465, 467
    difference between two means, 379
    difference between two proportions, 368
    generalized, 237
    mean difference-matched pairs, 349
    sample mean (σ known), 320
    sample mean (σ unknown), 334
    Signed Rank Test, 581
    Signs Test, 563
total deviation, 406

treatment, 32
treatment group, 34
trimmed mean, 68
two-tail test, 292
Type I Error, 298
Type II Error, 298

unbiased estimators, 65
unconditional probability, 129
unexplained deviation, 406
uniform distribution, 203
uniform histogram, 50
upper class limits, 46

variable
    confounding, 32
    explanatory, 32
    response, 32
variance
    between the samples, 541
    discrete distribution, 149
    finite population, 80
    sample-computational formula , 85
    sample-frequency tables, 87
    sample-listed data, 80
    within the samples, 540

weighted mean, 71
Wilcoxon Signed Rank Test, 573
Wilcoxon, Frank, 591

Table A3 t-Distribution p-Values

Areas to left of T.S. are given. For right-tail test subtract 1−(*given area*). For 2-tail tests ⟨…⟩

Degrees of Freedom

Positive t-Test Statistic	1	2	3	4	5	6	7	8	9	10	11	12	13	14	15	16	17	18	19	20	21	22	23	24	25	26	27	28	29
0.0	.5000	.5000	.5000	.5000	.5000	.5000	.5000	.5000	.5000	.5000	.5000	.5000	.5000	.5000	.5000	.5000	.5000	.5000	.5000	.5000	.5000	.5000	.5000	.5000	.5000	.5000	.5000	.5000	.5000
0.1	.5317	.5353	.5367	.5374	.5379	.5382	.5384	.5386	.5387	.5388	.5389	.5390	.5391	.5391	.5392	.5392	.5392	.5393	.5393	.5393	.5394	.5394	.5394	.5394	.5394	.5394	.5395	.5395	.5395
0.2	.5628	.5700	.5729	.5744	.5753	.5760	.5764	.5768	.5770	.5773	.5774	.5776	.5777	.5778	.5779	.5780	.5781	.5781	.5782	.5782	.5783	.5783	.5784	.5784	.5785	.5785	.5785	.5785	.5786
0.3	.5928	.6038	.6081	.6104	.6119	.6129	.6136	.6141	.6145	.6148	.6151	.6153	.6155	.6157	.6159	.6160	.6161	.6162	.6163	.6164	.6164	.6165	.6166	.6166	.6167	.6167	.6168	.6168	.6169
0.4	.6211	.6361	.6420	.6452	.6472	.6485	.6495	.6502	.6508	.6512	.6516	.6519	.6522	.6524	.6526	.6528	.6529	.6531	.6532	.6532	.6533	.6534	.6535	.6536	.6537	.6537	.6538	.6539	.6540
0.5	.6476	.6667	.6743	.6783	.6809	.6826	.6838	.6847	.6855	.6861	.6865	.6869	.6873	.6876	.6878	.6881	.6883	.6884	.6886	.6887	.6889	.6890	.6891	.6892	.6893	.6894	.6894	.6895	.6896
0.6	.6720	.6953	.7046	.7096	.7127	.7148	.7163	.7174	.7183	.7191	.7197	.7202	.7206	.7210	.7213	.7215	.7218	.7220	.7222	.7224	.7225	.7227	.7228	.7229	.7230	.7231	.7231	.7233	.7234
0.7	.6944	.7218	.7328	.7387	.7424	.7449	.7467	.7481	.7492	.7501	.7508	.7514	.7519	.7523	.7527	.7530	.7533	.7536	.7538	.7540	.7542	.7544	.7545	.7547	.7548	.7549	.7550	.7551	.7552
0.8	.7148	.7462	.7589	.7657	.7701	.7729	.7750	.7766	.7778	.7797	.7804	.7815	.7819	.7823	.7826	.7832	.7834	.7837	.7839	.7841	.7842	.7844	.7845	.7847	.7848	.7849	.7850	.7858	.7863
0.9	.7333	.7684	.7828	.7905	.7953	.7986	.8010	.8028	.8042	.8054	.8063	.8071	.8078	.8083	.8088	.8093	.8097	.8100	.8103	.8106	.8108	.8111	.8113	.8115	.8116	.8118	.8120	.8121	.8122
1.0	.7500	.7905	.8045	.8130	.8184	.8220	.8247	.8267	.8283	.8296	.8306	.8316	.8322	.8329	.8334	.8339	.8343	.8347	.8350	.8354	.8357	.8359	.8361	.8364	.8366	.8367	.8369	.8371	.8373
1.1	.7651	.8070	.8242	.8335	.8393	.8433	.8461	.8483	.8501	.8514	.8526	.8535	.8544	.8551	.8557	.8562	.8567	.8571	.8575	.8578	.8581	.8584	.8586	.8589	.8591	.8593	.8597	.8598	.8600
1.2	.7789	.8235	.8419	.8518	.8581	.8623	.8654	.8678	.8696	.8711	.8723	.8734	.8742	.8750	.8756	.8762	.8767	.8772	.8776	.8779	.8782	.8785	.8788	.8790	.8793	.8795	.8797	.8799	.8802
1.3	.7913	.8384	.8578	.8683	.8748	.8793	.8826	.8851	.8870	.8886	.8899	.8910	.8919	.8927	.8934	.8940	.8945	.8950	.8954	.8958	.8962	.8965	.8968	.8970	.8973	.8975	.8977	.8979	.8981
1.4	.8026	.8518	.8720	.8829	.8898	.8945	.8979	.9005	.9025	.9042	.9055	.9066	.9075	.9084	.9091	.9097	.9103	.9107	.9112	.9116	.9119	.9122	.9126	.9131	.9133	.9136	.9138	.9141	.9154
1.5	.8128	.8638	.8847	.8960	.9030	.9079	.9114	.9140	.9161	.9177	.9191	.9203	.9212	.9221	.9228	.9235	.9240	.9250	.9255	.9258	.9261	.9267	.9269	.9272	.9274	.9276	.9278	.9280	.9293
1.6	.8222	.8746	.8960	.9076	.9148	.9196	.9232	.9259	.9280	.9297	.9310	.9322	.9332	.9340	.9348	.9354	.9360	.9365	.9370	.9374	.9377	.9381	.9384	.9387	.9389	.9392	.9394	.9396	.9398
1.7	.8307	.8844	.9062	.9178	.9251	.9300	.9335	.9362	.9383	.9400	.9414	.9426	.9435	.9444	.9451	.9458	.9463	.9468	.9473	.9477	.9481	.9484	.9487	.9490	.9492	.9495	.9497	.9499	.9501
1.8	.8386	.8938	.9152	.9269	.9343	.9390	.9426	.9452	.9473	.9490	.9503	.9515	.9523	.9533	.9539	.9546	.9552	.9557	.9563	.9569	.9572	.9575	.9578	.9580	.9583	.9585	.9587	.9589	.9590
1.9	.8458	.9011	.9232	.9349	.9421	.9469	.9504	.9530	.9551	.9567	.9580	.9591	.9601	.9609	.9616	.9622	.9627	.9632	.9636	.9640	.9644	.9647	.9650	.9652	.9655	.9657	.9659	.9661	.9663
2.0	.8524	.9082	.9303	.9419	.9490	.9538	.9572	.9597	.9617	.9633	.9646	.9657	.9666	.9674	.9680	.9686	.9691	.9696	.9700	.9704	.9707	.9710	.9713	.9715	.9718	.9720	.9722	.9724	.9725
2.1	.8585	.9147	.9367	.9482	.9551	.9598	.9631	.9655	.9675	.9690	.9702	.9712	.9721	.9728	.9735	.9740	.9745	.9750	.9753	.9757	.9760	.9763	.9766	.9768	.9770	.9772	.9774	.9776	.9777
2.2	.8642	.9207	.9424	.9537	.9605	.9649	.9681	.9705	.9723	.9738	.9750	.9759	.9767	.9774	.9781	.9786	.9790	.9794	.9798	.9801	.9804	.9807	.9809	.9812	.9814	.9816	.9817	.9819	.9820
2.3	.8695	.9259	.9475	.9585	.9651	.9694	.9725	.9748	.9765	.9779	.9790	.9799	.9807	.9813	.9819	.9824	.9828	.9832	.9835	.9838	.9841	.9843	.9846	.9848	.9850	.9851	.9853	.9854	.9856
2.4	.8743	.9308	.9521	.9628	.9692	.9734	.9763	.9784	.9801	.9813	.9824	.9832	.9840	.9846	.9851	.9855	.9859	.9863	.9866	.9869	.9871	.9874	.9876	.9877	.9879	.9881	.9882	.9884	.9885
2.5	.8789	.9352	.9561	.9666	.9728	.9767	.9795	.9815	.9831	.9843	.9852	.9860	.9867	.9873	.9877	.9882	.9885	.9888	.9891	.9893	.9896	.9898	.9900	.9902	.9903	.9905	.9906	.9908	.9908
2.6	.8831	.9392	.9598	.9700	.9760	.9797	.9823	.9842	.9856	.9868	.9877	.9884	.9890	.9895	.9900	.9903	.9907	.9910	.9912	.9914	.9916	.9918	.9920	.9921	.9923	.9924	.9925	.9926	.9927
2.7	.8871	.9429	.9631	.9730	.9786	.9822	.9847	.9865	.9878	.9888	.9897	.9903	.9909	.9914	.9918	.9921	.9924	.9927	.9929	.9931	.9933	.9935	.9936	.9937	.9939	.9940	.9941	.9942	.9943
2.8	.8908	.9463	.9661	.9756	.9810	.9844	.9867	.9884	.9896	.9906	.9914	.9920	.9925	.9929	.9933	.9936	.9939	.9941	.9943	.9945	.9946	.9948	.9949	.9950	.9951	.9952	.9953	.9954	.9955
2.9	.8943	.9494	.9687	.9779	.9831	.9863	.9885	.9901	.9912	.9921	.9928	.9933	.9938	.9941	.9945	.9948	.9950	.9952	.9954	.9955	.9957	.9958	.9960	.9961	.9961	.9963	.9963	.9964	.9965
3.0	.8976	.9523	.9712	.9800	.9850	.9880	.9900	.9915	.9925	.9933	.9940	.9945	.9949	.9952	.9955	.9958	.9960	.9962	.9963	.9965	.9966	.9967	.9968	.9969	.9970	.9971	.9971	.9972	.9973
3.1	.9007	.9549	.9734	.9819	.9866	.9894	.9913	.9927	.9936	.9944	.9949	.9954	.9958	.9961	.9963	.9966	.9967	.9969	.9971	.9972	.9973	.9974	.9975	.9976	.9976	.9977	.9978	.9978	.9979
3.2	.9036	.9573	.9753	.9835	.9880	.9907	.9922	.9937	.9946	.9953	.9958	.9962	.9965	.9968	.9970	.9972	.9974	.9975	.9976	.9978	.9978	.9980	.9980	.9981	.9981	.9982	.9983	.9983	.9984
3.3	.9063	.9596	.9771	.9850	.9893	.9918	.9934	.9946	.9954	.9961	.9965	.9968	.9971	.9974	.9976	.9977	.9979	.9980	.9981	.9982	.9983	.9984	.9984	.9985	.9985	.9986	.9986	.9987	.9987
3.4	.9089	.9617	.9788	.9864	.9904	.9928	.9943	.9953	.9961	.9966	.9970	.9974	.9976	.9978	.9980	.9982	.9983	.9984	.9985	.9986	.9987	.9987	.9988	.9988	.9989	.9989	.9989	.9990	.9990
3.5	.9114	.9636	.9803	.9876	.9914	.9936	.9950	.9960	.9966	.9971	.9975	.9978	.9980	.9982	.9984	.9985	.9986	.9987	.9988	.9989	.9989	.9990	.9990	.9991	.9991	.9992	.9992	.9992	.9993
3.6	.9138	.9654	.9816	.9886	.9922	.9943	.9956	.9965	.9971	.9975	.9979	.9981	.9984	.9985	.9986	.9988	.9989	.9990	.9990	.9991	.9992	.9992	.9993	.9993	.9993	.9994	.9994	.9994	.9995
3.7	.9160	.9670	.9829	.9896	.9930	.9950	.9962	.9970	.9975	.9979	.9982	.9985	.9986	.9988	.9989	.9990	.9991	.9992	.9992	.9993	.9993	.9994	.9994	.9994	.9995	.9995	.9995	.9996	.9996
3.8	.9181	.9686	.9840	.9904	.9937	.9955	.9966	.9974	.9979	.9983	.9985	.9987	.9989	.9990	.9991	.9992	.9993	.9993	.9994	.9994	.9995	.9995	.9995	.9996	.9996	.9996	.9997	.9997	.9997
3.9	.9201	.9701	.9850	.9912	.9943	.9960	.9971	.9977	.9982	.9985	.9988	.9990	.9991	.9992	.9993	.9994	.9994	.9995	.9995	.9996	.9996	.9996	.9996	.9997	.9997	.9997	.9997	.9998	.9998
4.0	.9220	.9714	.9860	.9919	.9950	.9964	.9974	.9980	.9984	.9987	.9990	.9991	.9992	.9993	.9994	.9995	.9995	.9996	.9996	.9996	.9997	.9997	.9997	.9997	.9998	.9998	.9998	.9998	.9998
4.1	.9239	.9727	.9869	.9926	.9953	.9968	.9977	.9983	.9986	.9989	.9991	.9993	.9994	.9994	.9995	.9996	.9996	.9997	.9997	.9997	.9998	.9998	.9998	.9998	.9998	.9998	.9998	.9999	.9999
4.2	.9256	.9739	.9877	.9932	.9958	.9972	.9980	.9985	.9988	.9991	.9993	.9994	.9995	.9996	.9996	.9997	.9997	.9997	.9998	.9998	.9998	.9998	.9998	.9999	.9999	.9999	.9999	.9999	.9999
4.3	.9273	.9750	.9884	.9937	.9961	.9975	.9982	.9987	.9990	.9992	.9994	.9995	.9996	.9996	.9997	.9997	.9998	.9998	.9998	.9998	.9999	.9999	.9999	.9999	.9999	.9999	.9999	.9999	.9999
4.4	.9289	.9760	.9891	.9942	.9965	.9977	.9984	.9989	.9991	.9993	.9995	.9996	.9997	.9997	.9998	.9998	.9998	.9998	.9999	.9999	.9999	.9999	.9999	.9999	.9999	.9999	.9999	.9999	.9999
4.5	.9304	.9770	.9898	.9946	.9968	.9979	.9986	.9990	.9993	.9994	.9996	.9996	.9997	.9998	.9998	.9998	.9999	.9999	.9999	.9999	.9999	.9999	.9999	.9999	.9999	.9999	.9999	.9999	.9999
4.6	.9319	.9779	.9903	.9950	.9971	.9982	.9988	.9991	.9994	.9995	.9996	.9997	.9998	.9998	.9999	.9999	.9999	.9999	.9999	.9999	.9999	.9999	.9999	.9999	.9999	.9999	.9999	.9999	.9999
4.7	.9333	.9788	.9909	.9953	.9973	.9983	.9989	.9992	.9994	.9996	.9997	.9998	.9998	.9999	.9999	.9999	.9999	.9999	.9999	.9999	.9999	.9999	.9999	.9999	.9999	.9999	.9999	.9999	.9999
4.8	.9346	.9796	.9914	.9957	.9976	.9985	.9990	.9993	.9995	.9996	.9997	.9998	.9999	.9999	.9999	.9999	.9999	.9999	.9999	.9999	.9999	.9999	.9999	.9999	.9999	.9999	.9999	.9999	.9999
4.9	.9359	.9804	.9919	.9960	.9978	.9986	.9991	.9994	.9996	.9997	.9998	.9998	.9999	.9999	.9999	.9999	.9999	.9999	.9999	.9999	.9999	.9999	.9999	.9999	.9999	.9999	.9999	.9999	.9999
5.0	.9372	.9811	.9923	.9963	.9980	.9988	.9992	.9995	.9996	.9997	.9998	.9998	.9999	.9999	.9999	.9999	.9999	.9999	.9999	.9999	.9999	.9999	.9999	.9999	.9999	.9999	.9999	.9999	.9999

Note: Any value which would normally round to 1 is recorded as .9999. For values of t greater than 5.0, use 0.9999 for the area.